Reliability Engineering and Services

Wiley Series in Quality & Reliability Engineering

Dr Andre Kleyner
Series Editor

The Wiley series in Quality & Reliability Engineering aims to provide a solid educational foundation for both practitioners and researchers in Q&R field and to expand the reader's knowledge base to include the latest developments in this field. The series will provide a lasting and positive contribution to the teaching and practice of engineering. The series coverage will contain, but is not exclusive to,

- statistical methods;
- physics of failure;
- reliability modeling;
- functional safety;
- six-sigma methods;
- lead-free electronics;
- warranty analysis/management; and
- risk and safety analysis

Wiley Series in Quality & Reliability Engineering

Reliability Engineering and Services
by Tongdan Jin
November 2018

Design for Safety
by Louis J Gullo and Jack Dixon
February 2018

Thermodynamic Degradation Science: Physics of Failure, Accelerated Testing, Fatigue, and Reliability Applications
by Alec Feinberg
October 2016

Next Generation HALT and HASS: Robust Design of Electronics and Systems
by Kirk A. Gray and John J. Paschkewitz
May 2016

Reliability and Risk Models: Setting Reliability Requirements, 2nd Edition
by Michael Todinov
September 2015

Applied Reliability Engineering and Risk Analysis: Probabilistic Models and Statistical Inference
by Ilia B. Frenkel, Alex Karagrigoriou, Anatoly Lisnianski, and Andre V. Kleyner
September 2013

Design for Reliability
by Dev G. Raheja (Editor) and Louis J. Gullo (Editor)
July 2012

Effective FMEAs: Achieving Safe, Reliable, and Economical Products and Processes Using Failure Modes and Effects Analysis
by Carl Carlson
April 2012

Failure Analysis: A Practical Guide for Manufacturers of Electronic Components and Systems
by Marius Bazu and Titu Bajenescu
April 2011

Reliability Technology: Principles and Practice of Failure Prevention in Electronic Systems
by Norman Pascoe
April 2011

Improving Product Reliability: Strategics and Implementation
by Mark A. Levin, Ted T. Kalal
March 2003

Test Engineering: A Concise Guide to Cost-Effective Design, Development, and Manufacture
by Patrick O'Connor
April 2001

Integrated Circuit Failure Analysis: A Guide to Preparation Techniques
by Friedrich Beck
January 1998

Measurement and Calibration Requirements for Quality Assurance to ISO 9000
by Alan S. Morris
October 1997

Electronic Component Reliability: Fundamentals, Modeling, Evaluation, and Assurance
by Finn Jensen
November 1995

Reliability Engineering and Services

Tongdan Jin
Ingram School of Engineering
Texas State University, USA

Registered Offices
John Wiley & Sons, Inc., 111 River Street, Hoboken, NJ 07030, USA
John Wiley & Sons Ltd, The Atrium, Southern Gate, Chichester, West Sussex, PO19 8SQ, UK

Editorial Office
The Atrium, Southern Gate, Chichester, West Sussex, PO19 8SQ, UK

For details of our global editorial offices, customer services, and more information about Wiley products visit us at www.wiley.com.

Wiley also publishes its books in a variety of electronic formats and by print-on-demand. Some content that appears in standard print versions of this book may not be available in other formats.

Library of Congress Cataloging-in-Publication Data

Names: Jin, Tongdan, author.
Title: Reliability engineering and services / Dr. Tongdan Jin, Professor,
    Texas State University, USA.
Description: Hoboken, NJ, USA : John Wiley & Sons, Inc., [2019] | Includes
    bibliographical references and index. |
Identifiers: LCCN 2018031844 (print) | LCCN 2018032734 (ebook) | ISBN
    9781119167037 (Adobe PDF) | ISBN 9781119167044 (ePub) | ISBN 9781119167013
    (hardcover)
Subjects: LCSH: Reliability (Engineering)
Classification: LCC TS173 (ebook) | LCC TS173 .J56 2018 (print) | DDC
    620/.00452–dc23
LC record available at https://lccn.loc.gov/2018031844

Cover Design: Wiley
Cover Images: Modern commuter train © Sailorr/Shutterstock;
Wind farm © William C. Y. Chu/Getty Images;
Airplane © 06photo/Shutterstock; Drone © Mopic/Fotolia

Set in 10/12pt WarnockPro by SPi Global, Chennai, India
Printed in Singapore by C.O.S. Printers Pte Ltd

10 9 8 7 6 5 4 3 2 1

To Youping and Ankai

Contents

Series Editor's Foreword

The Wiley Series in Quality & Reliability Engineering aims to provide a solid educational foundation for researchers and practitioners in the field of quality and reliability engineering and to expand the knowledge base by including the latest developments in these disciplines.

The importance of quality and reliability to a system can hardly be disputed. Product failures in the field inevitably lead to losses in the form of repair cost, warranty claims, customer dissatisfaction, product recalls, loss of sale, and, in extreme cases, loss of life.

With each year engineering systems are becoming more and more complex, with added functions and capabilities; however, the reliability requirements remain the same or grow even more stringent due to the proliferation of functional safety standards and rising expectations of quality and reliability on the part of the product end user. The rapid development of automotive electronic systems, eventually leading to autonomous driving, also puts additional pressure on the reliability expectations for these systems.

However, despite its obvious importance, quality and reliability education is paradoxically lacking in today's engineering curriculum. Very few engineering schools offer degree programs or even a sufficient variety of courses in quality or reliability methods. The topics of accelerated testing, reliability data analysis, renewal systems, maintenance, HALT/HASS, warranty analysis and management, reliability growth and other practical applications of reliability engineering receive little coverage in today's engineering student curriculum. Therefore, the majority of quality and reliability practitioners receive their professional training from colleagues, professional seminars, and professional publications. The book you are about to read is intended to close this educational gap and provide additional learning opportunities for a wide range of readers from graduate level students to seasoned reliability professionals.

We are confident that this book, as well as this entire book series, will continue Wiley's tradition of excellence in technical publishing and provide a lasting and positive contribution to the teaching and practice of reliability and quality engineering.

Dr. Andre Kleyner
Editor of the Wiley Series in Quality & Reliability Engineering

Preface

Reliability engineering is a multidisciplinary study that deals with the lifecycle management of a product or system, ranging from design, manufacturing, and installation to maintenance and repair services. Reliability plays a key role in ensuring human safety, cost-effectiveness, and resilient operation of infrastructures and systems. It has been widely accepted as a critical performance measure in both private and public sectors, including manufacturing, healthcare, transportation, energy, chemical and oil refinery, aviation, aerospace, and defense industries. For instance, commercial airplane engines can fly over 5000 hours before the need for overhaul and maintenance. This means that the plane can cross the Pacific Ocean nearly 500 times without failure. In road transportation, China has constructed a total of 16 000 km of high-speed rail since 2008 and the annual ridership is three billion. The service reliability reaches 0.999 999 998 given the annual fatality of five passengers on average. The F-35 is the next generation of jet fighters for the US Air Force. It is anticipated that 2000 aircraft will be deployed in the next 50 years. The design and manufacturing of these aircraft will cost $350 billion, yet the maintenance and support of the fleet is expected to be $600 billion. These examples indicate the success in deploying and operating a new product is highly dependent upon the reliability, maintenance, and repair services during its use.

This book aims to offer a holistic reliability approach to product design, testing, maintenance, spares provisioning, and resilience operations. Particularly, we present an integrated product-service system with which the design for reliability, performance-based maintenance, and spare parts logistics are synthesized to maximize the reliability while lowering the cost. Such a lifecycle approach is imperative as the industry is transitioning from a product-oriented model to a service-centric paradigm. We report the fundamental knowledge and best industry practices in reliability modeling, maintenance planning, spare parts logistics, and resilience planning across a variety of engineering domains. To that end, the book is classified into four topics: (1) design for reliability; (2) maintenance and warranty planning; (3) product and service integration; and (4) engineering resilience modeling. Each topic is further illustrated below.

Chapters 1 to 5 are dedicated to the design for reliability. They cover a wide array of reliability modeling and design methods, including non-parametric models, parametric models, reliability block diagrams, min-cut, and min-path network theory, importance measures, multistate systems, reliability and redundancy allocation, multicriteria optimization, fault-tree analysis, failure mode effects and criticality analysis, latent failures, corrective action and effectiveness, multiphase reliability growth planning, power law model, and accelerated life testing.

Chapters 6 to 8 focus on maintenance and warranty planning that deals with the decision making on replacement and repair of field units. Technical subjects include renewal theory, superimposed renewal, corrective maintenance, preventive maintenance, condition-based maintenance, performance-based maintenance, health diagnostics and prognostics management, repairable system theory, no-fault-found issue, free-replacement warranty, pro-rata warranty, extended warranty services, and two-dimensional warranty policy.

Chapters 9 to 11 model and design integrated product-service offering systems. First, basic inventory models are reviewed, including economic order quantity, continuous and period review policy with deterministic and stochastic lead time, respectively. Then the analyses are directed to repairable inventory systems that face stationary (or Poisson) demand or non-stationary demand processes. Multiresolution and adaptive inventory replenishment policy are applied to cope with the time-varying demand rate. Both single-echelon and multi-echelon inventory models are analyzed. Finally, an integrated production-service system that jointly optimizes reliability, maintenance, spares inventory, and repair capacity are elaborated in the context of multiobjective, performance-based contracting.

Chapter 12 introduces the basic concepts and modeling methods in resilience engineering. Unlike reliability issues, events considered in resilience management possess two unique features: high impact with low occurrence probability and catastrophic events with cascading failure. We present several resilience performance measures derived from the resilience curve and further discuss the difference between reliability and resilience. The chapter concludes by emphasizing that prevention, survivability, and recoverability are the three main aspects in resilience management.

This book represents a collection of the recent advancements in reliability theory and applications, and is a suitable reference for senior and graduate students, researcher scientists, reliability practitioners, and corporate managers. The case studies at the end of the chapters assist readers in finding reliability solutions that bridge the theory and applications. In addition, the book also benefits the readers in the following aspects: (1) guide engineers to design reliable products at a low cost; (2) assist the manufacturing industry in transitioning from a product-oriented culture to a service-centric organization; (3) support the implementation of a data-driven reliability management system through real-time or Internet-based failure reporting, analysis, and corrective actions system; (4) achieve zero downtime equipment operation through condition-based maintenance and adaptive spare parts inventory policy; and (5) realize low-carbon and sustainable equipment operations by repairing and reusing failed parts.

In summary, reliability engineering is evolving rapidly as automation and artificial intelligence are becoming the backbone of Industry 4.0. New products and services will constantly be developed and adopted in the next 10 to 20 years, including autonomous driving, home robotics, delivery drones, unmanned aerial vehicles, electric cars, augmented virtual reality, smart grids, Internet of Things, cloud and mobile computing, and supersonic transportation, just to name a few. The introduction and deployment of these new technologies require the innovation in reliability design, modeling tools, maintenance strategy, and repair services in order to meet the changing requirements. Therefore, emerging technologies, such as big data analytics, machine learning, neural networks, renewable energy, additive and smart manufacturing, intelligent supply chain, and sustainable operations will lead the initiatives in new product introduction, manufacturing, and after-sales support.

Tongdan Jin
San Marcos, TX 78666, USA

Acknowledgement

This book received a wide range of support and assistance during its development stage. First, I would like to thank Ms Ella Mitchell, assistant editor in Electrical Engineering at Wiley. Without her early outreach and encouragement, I would not have been able to lay out the preliminary proposal and start this writing journey.

I also want to thank the early assistance from Ms Shivana Raj, Ms Deepika Miriam, and Ms Sharon Jeba Paul, who served as the editorial contacts during the formation of the first four chapters of the book. My great appreciation is given to Mr Louis Vasanth Manoharan who provided the assistance, communications, and editorial guidelines when the remaining eight chapters were finally completed.

I am also indebted to Ms Michelle Dunckley and the design team for their creation of the nice book cover. Special appreciation is given to Ms Patricia Bateson for her professional and quality editing of the entire manuscript. My thanks are also to production editor Mr. Sathishwaran Pathbanabhan for the final quality check of the editing.

Meanwhile, I would like to thank Dr Shubin Si and Dr Hongyan Dui for the discussion and formation of integrated importance measures in Chapter 2. My appreciation is extended to Dr Zhiqiang Cai who invited me to offer reliability engineering workshops at Northwestern Polytechnical University, Xian, where the materials were used by both Masters and PhD students. Very sincerely, I want to thank the Ingram School of Engineering at Texas State University where I have been teaching reliability engineering and supply chain courses since 2010. The feedback gathered from senior engineering students allowed me to improve and enhance the book content.

I am also very grateful to all the anonymous reviewers who provided constructive suggestions during the early development stage of the book, allowing me to improve and enrich the contents of the book.

My deep appreciations are given to Professor David W. Coit, Professor Elsayed Elsayed, and Professor Hoang Pham at Rutgers University. They taught, supervised, and guided my entry into the reliability engineering world when I was pursuing my graduate study. Since then I have been enjoying this dynamic and fast growing field both in my previous industry appointment and current academic position.

Last, but not least, my thanks are extended to my family members for their support, patience, and understanding during this lengthy endeavor. Special appreciations are reserved for my wife, Youping, who spent tremendous time and effort in taking care of our kid, allowing me to focus on the writing of the book. Without her persevering support this book would not have been available at this moment.

Tongdan Jin

About the Companion Website

This book is accompanied by a companion website:

www.wiley.com/go/jin/serviceengineering

The website include:

Additional codes, charts and tables

Scan this QR code to visit the companion website

1

Basic Reliability Concepts and Models

1.1 Introduction

Reliability is a statistical approach to describing the dependability and the ability of a system or component to function under stated conditions for a specified period of time in the presence of uncertainty. In this chapter, we provide the statistical definition of reliability, and further introduce the concepts of failure rate, hazard rate, bathtub curve, and their relation with the reliability function. We also present several lifetime metrics that are commonly used in industry, such as mean time between failures, mean time to failure, and mean time to repair. For repairable systems, failure intensity rate, mean time between replacements and system availability are the primary reliability measures. The role of line replaceable unit and consumable items in the repairable system is also elaborated. Finally, we discuss the parametric models commonly used for lifetime prediction and failure analysis, which include Bernoulli, binomial, Poisson, exponential, Weibull, normal, lognormal, and gamma distributions. The chapter is concluded with the reliability inference using Bayesian theory and Markov models.

1.2 Reliability Definition and Hazard Rate

1.2.1 Managing Reliability for Product Lifecycle

Reliability engineering is an interdisciplinary field that studies, evaluates, and manages the lifetime performance of components and systems, such as automobile, wind turbines (WTs), aircraft, Internet, medical devices, power system, and radars, among many others (Blischke and Murthy 2000; Chowdhury and Koval 2009). These systems and equipment are widely used in commercial and defense sectors, ranging from manufacturing, energy, transportation, healthcare, communication, and military operations.

The lifecycle of a product typically consist of five phases: design/development, new product introduction, volume shipment, market saturation, and phase-out. Figure 1.1 depicts the inter-dependency of five phases. Reliability plays a dual role across the lifecycle of a product: reliability as engineering (RAE) and reliability as services (RASs). RAE encompasses reliability design, reliability growth planning, and warranty and maintenance. RAS concentrates on the planning and management of a repairable inventory

Reliability Engineering and Services, First Edition. Tongdan Jin.
© 2019 John Wiley & Sons Ltd. Published 2019 by John Wiley & Sons Ltd.
Companion website: www.wiley.com/go/jin/serviceengineering

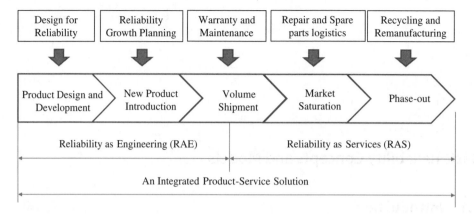

Figure 1.1 The role of reliability in the lifecycle of a product.

system, spare parts supply, and recycling and remanufacturing of end-of-life products. RAE and RAS have been studied intensively, but often separately in reliability engineering and operations management communities. The merge of RAE and RAS is driven primarily by the intense global competition, compressed product design cycle, supply chain volatility, environmental sustainability, and changing customer needs. There is a growing trend that RAE and RAS will be seamlessly integrated under the so-called product-service system, which offers a bundled reliability solution to the customers. This book aims to present an integrated framework that allows the product manufacturer to develop and market reliable products with low cost from a product's lifecycle perspective.

In many industries, reliability engineers are affiliated with a quality control group, engineering design team, supply chain logistics, and after-sales service group. Due to the complexity of a product, reliability engineers often work in a cross-functional setting in terms of defining the product reliability goal, advising corrective actions, and planning spare parts. When a new product is introduced to the market, the initial reliability could be far below the design target due to infant mortality, variable usage, latent failures, and other uncertainties. Reliability engineers must work with the hardware and software engineers, component purchasing group, manufacturing and operations department, field support and repair technicians, logistics and inventory planners, and marketing team to identify and eliminate the key root causes in a timely, yet cost-effective manner. Hence, a reliability engineer requires a wide array of skill sets ranging from engineering, physics, mathematics, statistics, and operations research to business management. Last but not the least, a reliability engineer must possess strong communication capability in order to lead initiatives for corrective actions, resolve conflicting goals among different organization units, and make valuable contributions to product design, volume production, and after-sales support.

1.2.2 Reliability Is a Probabilistic Measure

Reliability is defined as the ability of a system or component to perform its required functions under stated conditions for a specified period of time (Elsayed 2012; O'Connor

2012). It is often measured as a probability of failure or a possibility of availability. Let T be a non-negative random variable representing the lifetime of a system or component. Then the reliability function, denoted as $R(t)$, is expressed as

$$R(t) = P\{T > t\}, \quad \text{for } t = 0 \tag{1.2.1}$$

It is the probability that T exceeds an expected lifetime t which is typically specified by the manufacturer or customer. For example, in the renewable energy industry, the owner of the solar park would like to know the reliability of the photovoltaic (PV) system at the end of $t = 20$ years. Then the reliability of the solar photovoltaic system can be expressed as $R(20) = P\{T > 20\}$. As another example, as more electric vehicles (EVs) enter the market, the consumers are concerned about the reliability of the battery once the cumulative mileage reaches $100\,000$ km. In that case, $t = 100\,000$ km and the reliability of the EV battery can be expressed as $R(100\,000) = P\{T > 100\,000\}$. Depending on the actual usage profile, the lifetime T can stand for a product's calendar age, mileage, or charge–recharge cycles (e.g. EV battery). The key elements in the definition of Eq. (1.2.1) are highlighted below.

- Reliability is predicted based on "intended function" or "operation" without failure. However, if individual parts are good but the system as a whole does not achieve the intended performance, then it is still classified as a failure. For instance, a solar photovoltaic system has no power output in the night. Therefore, the reliability of energy supply is zero even if solar panels and DC–AC inverters are good.
- Reliability is restricted to operation under explicitly defined conditions. It is virtually impossible to design a system for unlimited conditions. An EV will have different operating conditions than a battery-powered golf car even if they are powered by the same type of battery. The operating condition and surrounding environment must be addressed during design and testing of a new product.
- Reliability applies to a specified period of time. This means that any system eventually will fail. Reliability engineering ensures that the system with a specified chance will operate without failure before time t.

The relationship between the time-to-failure distribution $F(t)$ and the reliability function $R(t)$ is governed by

$$F(t) = 1 - R(t) = 1 - P\{T > t\} \tag{1.2.2}$$

In statistics, $F(t)$ is also referred to as the cumulative distribution function (CDF). Let $f(t)$ be the probability density function (PDF); the relation between $R(t)$ and $f(t)$ is given as follows:

$$R(t) = \int_t^\infty f(x)dx = 1 - \int_0^t f(x)dx \tag{1.2.3}$$

Example 1.1 High transportation reliability is critical to our society because of increasing mobility of human beings. Between 2008 and 2016 China has built the world's longest high-speed rail with a total length of $25\,000$ km. The annual ridership is three billion on average. Since the inception, the cumulative death toll is 40 as of 2016 (Wikipedia 2017). Hence the annual death rate is $40/(2016 - 2008) = 5$. The reliability of the ridership is $1 - 5/(3 \times 10^9) = 0.999\,999\,998$. As another example, according to the Aviation Safety Network (ASN 2017), 2016 is the second safest year on record with 325 deaths. Given 3.5 billion passengers flying in the air in that year, the reliability of airplane ridership is $1 - 325/(3.5 \times 10^9) = 0.999\,999\,91$. This example shows that both transportation systems achieve super reliable ridership with eight "9"s for high-speed rail and seven "9"s in civil aviation.

1.2.3 Failure Rate and Hazard Rate Function

Let t be the start of an interval and Δt be the length of the interval. Given that the system is functioning at time t, the probability that the system will fail in the interval of $[t, t + \Delta t]$ is

$$P\{t < T \le t + \Delta t \mid T > t\} = \frac{P\{t < T \le t + \Delta t, \ T > t\}}{P\{T > t\}} = \frac{P\{t < T \le t + \Delta t\}}{P\{T > t\}}$$

$$= \frac{R(t) - R(t + \Delta t)}{R(t)} \tag{1.2.4}$$

The result is derived based on the Bayes theorem by realizing $P\{A, B\} = P\{A\}$, where A is the event that the system fails in the interval $[t, t + \Delta t]$ and B is the event that the system survives through t.

The failure rate, denoted as $z(t)$, is defined in a time interval $[t, t + \Delta t]$ as the probability that a failure per unit time occurs in that interval given that the system has survived up to t. That is,

$$z(t) = \frac{P\{t < T \le t + \Delta t \mid T > t\}}{(t + \Delta t) - t} = \frac{R(t) - R(t + \Delta t)}{\Delta t R(t)} \tag{1.2.5}$$

Although the failure rate $z(t)$ in Eq. (1.2.5) is often thought of as the probability that a failure occurs in a specified interval like $[t, t + \Delta t]$ given no failure before time t, it is indeed not a probability because $z(t)$ can exceed 1. For instance, given $R(t) = 0.5$, $R(t + \Delta t) = 0.4$, and $\Delta t = 0.1$, then $z(t) = 2$ failures per unit time. Hence, the failure rate represents the frequency with which a system or component fails and is expressed in failures per unit time. The actual failure rate of a product or system is closely related to the operating environment and customer usage (Cai et al. 2011).

Example 1.2 A lithium-ion battery is a rechargeable energy storage device widely used in electric transportation and utility power storage. The maximum state of charge (SOC) is commonly used to measure the lifetime of a rechargeable battery. The battery fails if the maximum SOC drops below 80% of its initial value. Assume a vehicle operates for 100 000 km and 110 000 km, the probabilities that the maximum SOC of a lithium-ion battery remains above 80% are 0.95 and 0.9, respectively. According to Eq. (1.2.5), the battery failure rate in [100 000, 110 000] can be estimated as

$$z(t) = \frac{0.95 - 0.9}{(110\,000 - 100\,000) \times 0.95} = 5.26 \times 10^{-6} \text{ failures/km}$$

The hazard function, also known as the instantaneous failure rate, is defined as the limit of the failure rate as Δt approaches zero. It is a rate per unit time similar to reading a car speedometer at a particular instant and seeing 100 km/hour. In the next instant the hazard rate may change and the testing units that have already failed have no impact because only the survivors count. By taking the limit of Δt to zero in Eq. (1.2.5), the hazard rate function $h(t)$ is obtained as follows:

$$h(t) = \lim_{\Delta t \to 0} \frac{R(t) - R(t + \Delta t)}{\Delta t R(t)} = -\frac{1}{R(t)} \lim_{\Delta t \to 0} \frac{R(t + \Delta t) - R(t)}{\Delta t}$$

$$= -\frac{1}{R(t)} \times \frac{dR(t)}{dt} = \frac{f(t)}{R(t)} \tag{1.2.6}$$

Equation (1.2.6) represents an important result as it governs the relation between $h(t)$, $f(t)$, and $R(t)$. Alternatively, from Eq. (1.2.6), the reliability function $R(t)$ can be expressed as

$$R(t) = \exp\left(-\int_0^t h(x)dx\right) \tag{1.2.7}$$

Let us denote $H(t)$ as the cumulative hazard rate function; then

$$H(t) = \int_0^t h(x)dx \tag{1.2.8}$$

By substituting Eq. (1.2.8) into (1.2.7), the reliability function $R(t)$ can also be expressed as

$$R(t) = \exp(-H(t)) \tag{1.2.9}$$

The failure rate and hazard rate are often used interchangeable in reliability literature and industry applications for modeling non-repairable systems. Both metrics are also applicable to repairable systems in which individual components are non-repairable. In addition, the values of the hazard rate and failure rate are always non-negative.

1.2.4 Bathtub Hazard Rate Curve

As the name implies, the bathtub hazard rate curve is derived from the cross-sectional shape of a bathtub. As shown in Figure 1.2, the bathtub curve consists of three different

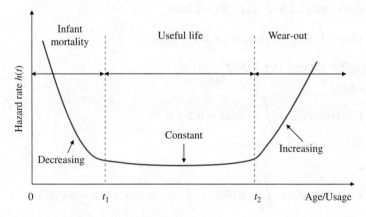

Figure 1.2 Bathtub hazard rate curve.

types of hazard rate profiles: (i) early infant mortality failures when the product is initially introduced; (ii) the constant rate during its useful period; and (iii) the increasing rate of wear-out or degradation failures as the product continues to operate at the end or beyond its design lifetime.

In military and consumer electronics industries, the infant mortality is often burned out or eliminated through the so-called environmental screening process. Namely, prior to the customer shipment, the products are tested under harsher operating conditions (e.g. temperature, humidity, vibration, and electric voltage) for a designated period of time in order to filter the weak units from the product pool. This process is adopted mainly for mission or safety critical applications as it greatly reduces the possibility of occurrence of system failures in its early life. While the bathtub curve is useful, not every product or system follows a bathtub type of hazard rate profile. For example, if units are decommissioned earlier or their usage has decreased steadily during or before the onset of the wear-out period, they will exhibit fewer failures per unit time over the chronological or calendar time (not per unit of use time) than the bathtub curve. Another case is the software products that may experience an infant mortality phase and then stabilized at a low constant hazard rate after extensive debugging and testing. This means the software product usually does not have a wear-out phase unless the hardware that runs the software application has been changed.

Example 1.3   Electronic devices usually exhibit a bathtub hazard rate profile as shown in Figure 1.2. Assume the hazard rate function is given as follows, where t is in units of months:

$$h(t) = \begin{cases} 0.1 - 0.004t, & 0 \leq t < 10 \\ 0.06, & 10 \leq t < 100 \\ 0.06 + 0.002(t - 100), & t \geq 100 \end{cases}$$

Find $H(t)$ and $R(t)$ for three phases, respectively.

Solution:
Since $h(t)$ changes in different phases, we shall derive the cumulative hazard rate and the reliability formulas separately.

For $0 \leq t < 10$, based on Eqs. (1.2.8) and (1.2.9), we have

$$H(t) = \int_0^t h(x)dx = \int_0^t (0.1 - 0.004x)dx = 0.1t - 0.002t^2$$

$$R(t) = \exp(-H(t)) = \exp(-0.1t + 0.002t^2)$$

For $10 \leq t < 100$, we have

$$H(t) = \int_0^{10} (0.1 - 0.004x)dx + \int_{10}^t 0.06dx = 0.2 + 0.06t$$

$$R(t) = \exp(-H(t)) = \exp(-0.2 - 0.06t)$$

For $t > 100$, we have

$$H(t) = \int_0^{10} (0.1 - 0.004x)dx + \int_{10}^{100} 0.06dx + \int_{100}^t [0.06 + 0.002(x - 100)]dx$$

$$= 10.2 - 0.14t + 0.001t^2$$

$$R(t) = \exp(-H(t)) = \exp(-10.2 + 0.14t - 0.001t^2)$$

Figure 1.3 Reliability and
cumulative hazard rate.

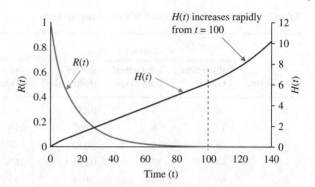

Figure 1.3 plots $H(t)$ and $R(t)$ for $0 \leq t \leq 140$. In general, $H(t)$ always monotonically increases and $R(t)$ always monastically decreases over time. Notice that $H(t)$ rapidly increases after the product enters the wear-out phase for $t > 100$.

1.2.5 Failure Intensity Rate

A repairable system, if it fails, can be repaired and restored to a good state. This is done by replacing failed components with good units. As time evolves, the frequency of failures may increase, decrease, or stay at a constant level depending on the maintenance policy, the reliability of existing components, and the new components used. Therefore, a system upon repair can be brought to one of the following conditions: as-good-as-new, as-good-as-old, and somewhere in between. The failure intensity function is a metric typically used to measure the occurrence of failures per unit time for a repairable system. A distinction shall be made between the hazard rate and the failure intensity rate. The former is used to characterize the time to the first failure of a component, while the latter deals with reoccurring failures of the same system. Hence the failure intensity rate is also referred to as the rate of occurrence of failures (ROCOFs). Let $M(t)$ be accumulative failures (or repairs) that occurred in a repairable system during $[0, t]$. The ROCOF, denoted as $m(t)$, can be estimated as

$$m(t) = \frac{M(t)}{t}, \quad \text{for } t > 0 \tag{1.2.10}$$

The unit of ROCOF is failures per unit time. For example, if a system failed three times in 300 days, then ROCOF = $3/300$ = 0.01 failure/day. Note that failures in a repairable system may happen on different component types or on the same component types, but different items.

Example 1.4   Two different repairable systems A and B are chosen to perform a lifetime test for a period of 260 days. Ten failures are observed from each system and their failure times are listed in Table 1.1. Note that failure interarrival times are the time elapse between two consecutive failures. Based on Eq. (1.2.10), the ROCOF of each system is computed and the results are listed in the table as well.

The ROCOF of systems A and B are plotted in Figure 1.4 Both systems experience the same number of failures in 260 days, but the ROCOF of System A decreases while it increases for System B. This means the repair effect on System A drives the growth of the reliability, while the repair effect on System B does not prevent it from degradation.

Table 1.1 Failure arrival times and ROCOF of two systems.

Failure No.	System A			System B		
	Failure time (days)	Interarrival time (days)	ROCOF (failure/day)	Failure time (days)	Interarrival time (days)	ROCOF (failure/day)
1	4	4	0.250	82	82	0.012
2	12	8	0.167	116	34	0.017
3	26	14	0.115	142	26	0.021
4	44	18	0.091	165	22	0.024
5	68	24	0.074	184	19	0.027
6	96	28	0.063	201	18	0.030
7	130	34	0.054	218	16	0.032
8	168	38	0.048	233	15	0.034
9	212	44	0.042	247	14	0.036
10	260	48	0.038	260	13	0.038

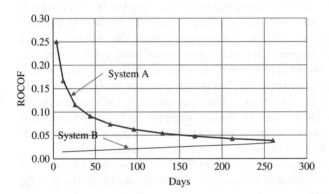

Figure 1.4 ROCOF of systems A and B.

The situation of System A is usually observed during the new product design and proto-type phase because of corrective actions and redesign. The situation of system B happens when the product enters the wear-out phase and the repair actions simply bring the system back to as-good-as-old state.

Note that $M(t)$ in Eq. (1.2.10) is a step function that jumps up each time a failure occurs and remains at the new level until the next failure. Every system will have its own observed $M(t)$ function over time. If a number of $M(t)$ curves are observed from n similar systems and the curves are averaged, we would have an estimate of $M(t)$, denoted as $\overline{M}(t)$. That is,

$$\overline{M}(t) = \int_0^t m(x)dx \qquad (1.2.11)$$

where $m(t)$ is the ROCOF function for the group of n systems. ROCOF sometimes is also called the repair rate, which is not to be confused with the length of time for performing a repair task, which will be discussed in next section.

1.3 Mean Lifetime and Mean Residual Life

1.3.1 Mean-Time-to-Failure

The mean-time-to-failure (MTTF) is a quantitative metric commonly used to assess the reliability of non-repairable systems or products. It measures the expected lifetime of a component or system before it fails. For instance, a solar photovoltaic panel is considered as a non-repairable system with a typical MTTF between 20 and 30 years. The MTTF of a tire for commercial vehicles varies between 30 000 miles and 60 000 miles (1 mile = 1.6 km). Since these items are non-repairable upon failure, they are either discarded or recycled for the environmental protection purpose. In industry, non-repairable products are also called consumable items.

Let n be the number of non-repairable systems operating in the field. The observed time-to-failure of an individual system is designated as $t_1, t_2, \ldots, t_n$. Then its MTTF can be estimated by

$$MTTF = \frac{1}{n} \sum_{i=1}^{n} t_i \tag{1.3.1}$$

If the sample size n is large enough, the time-to-failure distribution for T can be inferred statistically. Then MTTF is equivalent to the expected value of T, namely

$$MTTF = E[T] = \int_0^\infty tf(t)dt \tag{1.3.2}$$

where $f(t)$ is the PDF of the system life. MTTF can also be expressed as the integration of $R(t)$ over $[0, +\infty)$ by performing the integration by part in Eq. (1.3.2). This results in

$$MTTF = \int_0^\infty tf(t)dt = \int_0^\infty R(t)dt \tag{1.3.3}$$

Example 1.5 A WT is a complex machine comprised of multiple mechanical and electrical subsystems, including the main bearing, blades, gearbox, and generator, among others. The main bearing is a key subsystem that assists the conversion of wind kinetic energy into mechanical energy. The reliability of the main bearing degrades over time due to wear-out of its rolling balls, and cracks of the inner and outer ring races resulted from ball rotations and vibrations. The field data shows that the hazard rate function of the main bearing can be modeled as $h(t) = 0.002t$ failures/year. Estimate: (1) cumulative hazard function, (2) reliability function, (3) PDF, and (4) MTTF.

Solution:
According to Eq. (1.2.8), the cumulative hazard rate function of the main bearing is

$$H(t) = \int_0^t 0.002x \, dx = 0.001t^2 \tag{1.3.4}$$

Its reliability function $R(t)$ can be obtained by substituting $H(t)$ into Eq. (1.2.9). That is,

$$R(t) = \exp\left(-\int_0^t 0.002x \, dx\right) = \exp(-0.001t^2) \tag{1.3.5}$$

The PDF is obtained by taking the derivative with respect to t in Eq. (1.3.5) as follows:

$$f(t) = \frac{dF(t)}{dt} = \frac{d(1 - R(t))}{dt} = -\frac{dR(t)}{dt} = 0.002t \, \exp(-0.001t^2) \qquad (1.3.6)$$

Finally, the MTTF of the main bearing can be obtained from Eq. (1.3.3) as follows:

$$MTTF = \int_0^\infty e^{-0.001t^2} \, dt \cong 28.025 \text{ years} \qquad (1.3.7)$$

Unfortunately, there is no closed-form solution to Eq. (1.3.7). The result can, however, be obtained via numerical integration. Figures 1.5–1.8, respectively, depict $h(t)$, $H(t)$, $R(t)$, and $f(t)$ of the main bearing of the WT.

1.3.2 Mean-Time-Between-Failures

For a repairable system, the average time between two consecutive failures is characterized by the mean-time-between-failures (MTBFs). Both MTBF and MTTF measure the average uptime of a system, but MTBF is used for repairable systems as opposed to the

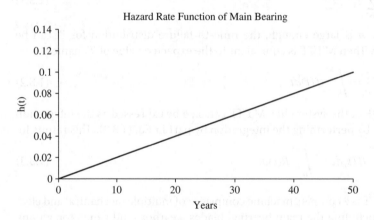

Figure 1.5 Hazard rate function of main bearing.

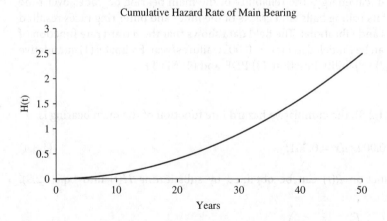

Figure 1.6 Cumulative hazard rate function of main bearing.

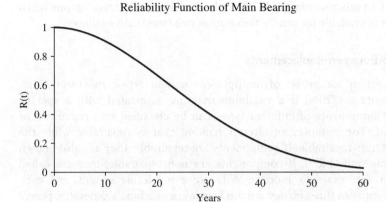

Figure 1.7 Reliability function of main bearing.

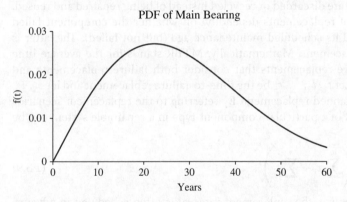

Figure 1.8 Probability density function of main bearing.

MTTF for non-repairable systems. For instance, tires are considered as a non-repairable component, and their reliability performance is characterized by MTTF. However, a car installed with four tires is a repairable system, and the reliability of a car is measured by MTBF. Let $t_1, t_2, t_3,..., t_n$ be the interarrival time between two consecutive failures. Then the MTBF of a repairable system is estimated by

$$MTBF = \frac{1}{n} \sum_{i=1}^{n} t_i \qquad (1.3.8)$$

For instance, starting from day 1, a machine failed in days 9 and 35, respectively. Assume the repair time is short and negligible; then $t_1 = 9-0 = 9$ days. As another example, after a car has run for 60 000 km, three failures have been observed. These failures correspond to one broken tire, a dead battery, and malfunction of one headlight. Then the MTBF of the car is MTBF = 60 000/3 = 20 000 km.

When estimating the MTBF, the downtime associated with waiting for repair technicians, spare parts shipping time, failure diagnostics, and administrative delay should be excluded. In other words, the MTBF only measures the average uptime when a repairable system is available for production during two consecutive failures.

1.3.3 Mean-Time-Between-Replacements

For a repairable system comprised of multiple component types, the mean-time-between-replacements (MTBRs) is a reliability measure associated with a specific component type. Components of different types can be classified into repairable or non-repairable unit. For instance, an aircraft landing gear is repairable while the tires are treated as non-repairable. If components are repairable, they are also known as a line-replaceable unit (LRU). If components are non-repairable, they are called a consumable part. For example, modern WTs are a repairable system, and each turbine typically comprises three blades, a main bearing, a gearbox, a generator, power electronics, and other control units. Components like the gearbox and generator are LRUs as they are repairable units. A failed generator after being fixed in the repair shop can be reused in other WTs. Turbine blades and bearings are treated as consumable parts. Upon failure, they are discarded or recycled instead of being repaired and reused.

There are two types of replacements depending on whether the component failed suddenly or has reached its scheduled maintenance age (but not failed). The latter is called a preventive replacement. Mathematically, MTBR stands for the average time between two consecutive replacements that consider both failure replacements and planned replacements. Let $t_1, t_2, \ldots, t_n$ be the time-to-failure replacement and let $\tau_1, \tau_2, \ldots, \tau_m$ be the time-to-planned replacement. By referring to the replacement scenarios in Figure 1.9, the MTBR of a particular component type in a repairable system can be calculated by

$$MTBR = \frac{1}{m+n} \left(\sum_{i=1}^{m} \tau_i + \sum_{i=1}^{n} t_i \right) \tag{1.3.9}$$

In preventive maintenance, the replacement interval is often scheduled in advance with a fixed length (i.e. $\tau_i = \tau$ for all i). If the hands-on replacement time is short and can be ignored, the expected value of MTBR under the constant replacement interval policy can be obtained as

$$MTBR = \tau R(\tau) + \int_0^\tau tf(t)dt = \int_0^\tau R(t)dt \tag{1.3.10}$$

Note that $\tau R(\tau)$ captures all scheduled replacement events with fixed interval τ and $\int_0^\tau tf(t)dt$ stands for the failure replacements occurring prior to τ.

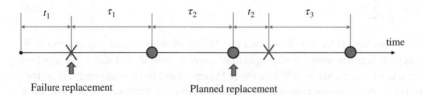

Figure 1.9 Replacement scenarios for a repairable system.

1.3.4 Mean Residual Life

In reliability engineering, the expected additional lifetime given that a component or system has survived until time t is called the mean residual life (Gupta and Bradley 2003). Let T represent the life of a component or system. The mean residual life, denoted as $L(t)$, is given as

$$L(t) = E[T - t \,|\, T \geq t] = \int_t^\infty (x - t) f_{T|T \geq t}(x) dx \qquad (1.3.11)$$

where $f_{T|T \geq t}(x)$ is the conditional PDF given that $T \geq t$. The value of $L(t)$ can be predicted or estimated based on the historical failure data, and the result is frequently used for provisioning spare parts supply or allocating repair resources in the repair shop. The conditional PDF $f_{T|T \geq t}(x)$ can be expressed as the marginal PDF $f(t)$ and reliability function $R(t)$ as follows:

$$f_{T|T \geq t}(x) = \frac{f(x)}{R(t)} \qquad (1.3.12)$$

Substituting Eq. (1.3.12) into (1.3.11), the mean residual life is obtained as

$$L(t) = \int_t^\infty (x - t) \frac{f(x)}{R(t)} dx = \frac{1}{R(t)} \int_t^\infty R(x) dx \qquad (1.3.13)$$

Example 1.6 The reliability of an electronic device can be modeled as $R(t) = \exp(-0.02\,t)$. Compute: (1) the conditional PDF $f_{T|T \geq t}(x)$ and (2) the mean residual life given $t = 10$ and 100 hours.

Solution:

(1) The marginal PDF of the lifetime T is obtained as

$$f(t) = -\frac{dR(t)}{dt} = 0.02 \exp(-0.002t) \qquad (1.3.14)$$

Now substituting Eq. (1.3.14) into (1.3.12) along with $R(t) = \exp(-0.02\,t)$, we have

$$f_{T|T \geq t}(x) = \frac{0.02 \, \exp(-0.02x)}{\exp(-0.02t)} = 0.02 \exp(-0.02(x - t)), \quad \text{for } x \geq t \quad (1.3.15)$$

Obviously the conditional PDF is still exponentially distributed with the time being shifted by t.

(2) The mean residual life of this device can be obtained by substituting $R(t) = \exp(-0.02\,t)$ into Eq. (1.3.13). That is,

$$L(t) = \frac{1}{e^{-0.02t}} \int_t^\infty e^{-0.02x} dx - t = 50 \qquad (1.3.16)$$

For $t = 10$, the mean residual life of this device is 50 hours. For $t = 100$, the mean residual life is still 50 hours. This result seems contradictory to the intuition. Indeed, this is due to the memoryless property of exponential distribution. This unique property will be discussed in Section 1.6.2.

1.4   System Downtime and Availability

1.4.1   Mean-Time-to-Repair

Mean-time-to-repair (MTTR) represents the time elapse from the moment the system is down to the moment it is resorted. MTTR encompasses the waiting time for repair technicians, the lead time of receiving spare parts, failure diagnostics time, hands-on time for replacing any faulty parts, and other downtime associated with inspections, testing, or administrative delays. A generic MTTR estimate is given below

$$MTTR = t_{ad} + t_{pt} + t_{tn} + t_{ho} + t_{ft} \tag{1.4.1}$$

where

t_{ad} = administrative delay
t_{pt} = lead time for receiving the spare part
t_{tn} = time for assembling the repair technician team
t_{ho} = hands-on time for replacing failed units
t_{ft} = failure diagnostics and testing time

For example, a plane is grounded due to the failure of an engine. Suppose the administrative delay is two days, the lead time to receive a new engine is three days, the technicians are available after two days, the hands-on time to replace the engine is two days, and failure diagnostics and final testing requires one day. Assume the delivery of a new engine and the dispatch of technicians occur concurrently; then MTTR = 2 + max{3, 2} + 2 + 1 = 8 days. Figure 1.10 graphically illustrates how to estimate the MTTR in this case. This example indicates that the actual MTTR can be shrunken if multiple activities can be executed concurrently. Hence Eq. (1.4.1) represents the upper bound estimate of MTTR.

1.4.2   System Availability

Availability is the proportion of time when a system is in a functioning condition. Reliability and maintainability jointly determine the system availability. Particularly, the former determines the length of MTBF and the latter influences the MTTR. They are related to the system availability by the following formula:

$$A = \frac{MTBF}{MTBF + MTTR} \tag{1.4.2}$$

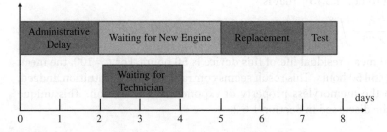

Figure 1.10 The MTTR of replacing an aircraft engine.

It is worth mentioning that two systems may have the same availability, but their MTBF and MTTR could be different. For instance, MTBF and MTTR for System A is 900 hours and 100 hours, respectively. MTBF and MTTR for System B is 450 hours and 50 hours, respectively. Obviously, the availability for both systems is 0.9, yet MTBF and MTTR of System A is twice that of System B.

1.5 Discrete Random Variable for Reliability Modeling

1.5.1 Bernoulli Distribution

In probability and statistics, a random variable can take on a set of different values, each associated with a certain probability between zero and one, in contrast to a deterministic quantity associated with unity probability. A discrete random variable can take any of a finite list of values supported by a probability mass function. If a random variable is continuous, it can take any numerical value in an interval or collection of intervals via a PDF. A CDF is the sum of the possible outcomes of a random variable, either in a discrete or continuous form. Random variables and probability theories are useful tools to model the variation of reliability because the lifetime of components and systems is influenced by various uncertainties during the design, manufacturing, and field use.

This section briefly reviews the distribution functions of three types of discrete rand variables and their statistical properties: Bernoulli distribution, binomial distribution, and Poisson distribution. Continuous random variable distributions will be discussed in Section 1.6.

If a random variable X can only take two values, either 1 or 0, with the following probability mass function (PMF)

$$P\{X = x\} = \begin{cases} p, & x = 1 \\ 1 - p, & x = 0 \end{cases} \tag{1.5.1}$$

where $0 \leq p \leq 1$, then the distribution of X is called the Bernoulli distribution. As the classical example, the outcome of flipping a coin (either head or tail) follows the Bernoulli distribution with $p = 0.5$. The probability of successfully launching a satellite using a rocket can also be modeled as a Bernoulli distribution. Typically the launch success rate p is between 0.85 and 0.98 (Guikema and Paté-Cornell 2004). The mean and the variance of X are

$$E[X] = p \tag{1.5.2}$$
$$Var(X) = p(1 - p) \tag{1.5.3}$$

In general, a Bernoulli random variable is capable of modeling the reliability of one-shot systems, such as satellite launch and missile test. Bernoulli can also be used to analyze the reliability of mission-critical systems even through the duration of the mission may last hours or days.

1.5.2 Binomial Distribution

A binomial distribution is used to describe the random outcome for situations where multiple Bernoulli tests are carried out independently at the same time. Suppose n

independent Bernoulli trials are being conducted and each trial results in a success with probability p and a failure with probability of $1 - p$. If X represents the number of successes among n trials, then X is defined as a binomial random variable with parameters $B(n, p)$. The PMF is given by

$$P\{X = k\} = \binom{n}{k} p^k (1 - p)^{n-k} = \frac{n!}{k!(n - k)!} p^k (1 - p)^{n-k}, \quad \text{for } k = 0, 1, 2, \ldots, n$$

(1.5.4)

Similarly we can obtain the mean, the second moment, and the variance of the binomial random variable X. That is,

$$E[X] = np \tag{1.5.5}$$

$$E[X^2] = np[(n - 1)p + 1] \tag{1.5.6}$$

$$Var(X) = E[X^2] - (E[X])^2 = np(1 - p) \tag{1.5.7}$$

Example 1.7   To evaluate the reliability of a ceramic capacitor, a reliability engineer randomly selected 10 identical units to perform the life test for 200 hours with 90% voltage derating. Prior to the test, data from the supplier shows that the probability for the capacitor to survive over 200 hours is $p = 0.9$ given the same voltage derating rate. Answer the following:

1) What is the probability that eight capacitors survived at 200 hours?
2) What is the probability that at least eight capacitors survived at 200 hours?
3) What are the mean and the standard deviation of failures at 200 hours?

Solution:
1) Let X be the number of survived units at $t = 200$ hours. Given $p = 0.9$ and $n = 10$, the binomial distribution is $X \sim B(10, 0.9)$. According to Eq. (1.5.4), the probability that exactly $k = 8$ units survived at the end of the test is

$$P\{X = 8\} = \frac{10!}{10!(10 - 8)!} 0.9^8 (1 - 0.9)^{10-8} = 0.1937$$

2) If at least eight capacitors survived at the end of test, it implies that k can take on any value of 8, 9, and 10. Hence the probability is estimated as

$$P\{k \geq 8\} = P\{X = 8\} + P\{X = 9\} + P\{X = 10\}$$

$$= \sum_{k=8}^{10} \binom{10}{k} 0.9^k (1 - 0.9)^{10-k} = 0.9298 \tag{1.5.8}$$

Since $P\{k \geq 8\} \gg P\{k = 8\}$, it implies that using redundant units can achieve high system reliability even if the reliability of individual components is moderate or low.
3) Let Y be the random variable representing the number of failed capacitors in the test. By realizing $Y = n - X = 10 - X$, the probability of failure is $q = 1 - p = 0.1$. This means that Y follows the binomial distribution with $Y \sim B(10, 0.1)$. Based on Eqs. (1.5.5) and (1.5.7), the mean and the standard deviation of Y can be estimated as follows:

$$E[Y] = nq = 10 \times 0.1 = 1 \tag{1.5.9}$$

$$Var(Y) = nq(1 - q) = 10 \times 0.1 \times (1 - 0.1) = 0.9 \tag{1.5.10}$$

$$\sigma_Y = \sqrt{Var(Y)} = \sqrt{0.9} = 0.949 \tag{1.5.11}$$

If we compare Eq. (1.5.7) with (1.5.10), the variance of X and Y are identical, though their expected values are different.

1.5.3 Poisson Distribution

A random variable N is regarded as a Poisson distribution with positive parameter λ when the probability mass function of X takes the following form:

$$P\{N = k\} = \frac{\lambda^k e^{-\lambda}}{k!}, \quad \text{for } k = 0, 1, 2, \dots \tag{1.5.12}$$

The Poisson distributions have a large scope of applications in statistics, engineering, science, and business (Tse 2014). Examples that may follow a Poisson include the number of phone calls received by a call center per hour, the number of decay events per second from a radioactive source, or the number of bugs remaining in a software program. Poisson distribution is also used for predicting the market size or the installed base during the new product introduction, such as WTs, semiconductor manufacturing equipment, and new airplanes (Farrel and Saloner 1986; Liao et al. 2008). The mean and variance of the Poisson distribution is given by

$$E[N] = \lambda \tag{1.5.13}$$

$$Var(N) = \lambda \tag{1.5.14}$$

It is interesting to see that the mean and variance are always identical and equal to λ for the Poisson distribution.

Example 1.8 The number of bugs embedded in a software application can be modeled as a Poisson distribution with $\lambda = 0.002$ bugs per code line. Do the following:

1) Estimate the expected number of bugs when the software program contains $m = 5000$ lines of codes.
2) If the number of bugs is required to be no more than three at 95% confidence, what is the maximum acceptable value of λ?
3) Plot the probability mass function with initial $\lambda = 0.002$ and the required λ value.

Solution:

(1) Since the average number of bugs in a line is 0.002, the expected number of bugs for a 5000-line application is given as

$$E[N] = \lambda m = (0.002)(5000) = 10 \text{ bugs} \tag{1.5.15}$$

(2) Let λ_d be the minimum acceptable bugs per code line. To achieve the target of no more than three bugs in the software with 90% confidence, the value of λ_d must satisfy the following requirement:

$$P\{N \leq 3\} = \sum_{k=0}^{3} P\{N = k\} = \sum_{k=0}^{3} \frac{(5000\lambda_d)^k e^{-5000\lambda_d}}{k!} \geq 0.9 \tag{1.5.16}$$

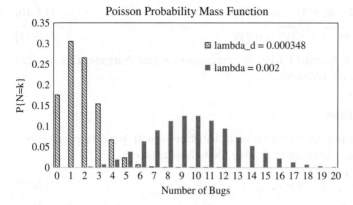

Figure 1.11 The probability mass function for software bugs.

Solving the above equation yields $\lambda_d = 0.000348$ bugs/line.

(3) The PMF corresponding to $\lambda = 0.002$ bugs/line and $\lambda_d = 0.000\ 348$ bugs/line are plotted in Figure 1.11. It is found that the PMF after the debugging is shifted to the left side. This makes sense because the probability for $k = 8, 9, 10,\ldots$ is significantly reduced upon debugging, hence fewer bugs are left in the software.

1.6 Continuous Random Variable for Reliability Modeling

1.6.1 The Uniform Distribution

A random variable is defined as uniformly distributed over the interval of $[a, b]$ when the probability of taking any value between a and b is equally likely. Let T be the random variable of the uniform distribution; the PDF is defined as

$$f(t) = \begin{cases} \frac{1}{b-a} & a < t < b \\ 0 & \text{otherwise} \end{cases} \tag{1.6.1}$$

The CDF, denoted as $F(t)$, is given by

$$F(t) = P\{T \le t\} = \begin{cases} 0 & t < a \\ \frac{t-a}{b-a} & a \le t \le b \\ 1 & t > b \end{cases} \tag{1.6.2}$$

The mean and variance of T are given by

$$E[T] = \int_a^b tf(t)dt = \int_a^b \frac{t}{b-a}dt = \frac{1}{2}(a+b) \tag{1.6.3}$$

$$E[T^2] = \int_a^b t^2 f(t)dt = \int_a^b \frac{t^2}{b-a}dt = \frac{1}{3}(a^2 + ab + b^2) \tag{1.6.4}$$

$$Var(T) = E[X^2] - (E[X])^2 = \frac{1}{12}(b-a)^2 \tag{1.6.5}$$

When $a = 0$ and $b = 1$, the PDF in Eq. (1.6.1) is called the standard uniform distribution, which is denoted as $U[0, 1]$. In Bayesian reliability inference, the standard uniform distribution is frequently used as a prior distribution for a component or system reliability estimate when the actual reliability is unknown, or is hard to deduce because of insufficient failures or testing time.

1.6.2   The Exponential Distribution

The random variable of the exponential distribution is continuous and non-negative. Hence it is an ideal random variable to model the lifetime of products and system. If T is exponentially distributed, then the PDF is given as follows:

$$f(t) = \lambda e^{-\lambda t}, \quad \text{for } t \geq 0 \tag{1.6.6}$$

where λ is the distribution parameter and $\lambda > 0$. The CDF and reliability function are defined as

$$F(t) = P\{T \leq t\} = \int_0^t \lambda e^{-\lambda x} dx = 1 - e^{-\lambda t} \tag{1.6.7}$$

$$R(t) = 1 - F(t) = e^{-\lambda t} \tag{1.6.8}$$

Both the hazard rate function and the cumulative hazard rate function are

$$h(t) = \frac{f(t)}{R(t)} = \lambda \tag{1.6.9}$$

$$H(t) = \int_0^t \lambda dx = \lambda t \tag{1.6.10}$$

For exponential lifetime distribution, the hazard rate function is a constant, or vice versa. This observation leads to an important feature of exponential random variable, that is, memoryless property. Let s be the length of time (such as hours) that the system has survived. For a random variable T possessing the memoryless property, the following equality always holds:

$$P\{T > t + s \mid T > s\} = P\{T > t\}, \quad \text{for } t > 0 \tag{1.6.11}$$

The proof of the equality is given below:

$$P\{T > t + s \mid T > s\} = \frac{P\{T > t + s, T > s\}}{P\{T > s\}}$$

$$= \frac{P\{T > t + s\}}{P\{T > s\}} = \frac{\exp(-\lambda(t + s))}{\exp(-\lambda s)}$$

$$= \exp(-\lambda t)$$

$$= P\{T > t\}$$

It states the probability that the product will survive for $s + t$ hours, given that it has survived s hours is the same as the initial probability that it survives for t hours. Finally, the mean and the variance of T are obtained and given as

$$E[T] = \int_0^\infty t f(t) dt = \int_0^\infty t \lambda e^{-\lambda t} dt = \frac{1}{\lambda} \tag{1.6.12}$$

$$E[T^2] = \int_0^\infty t^2 f(t) dt = \int_0^\infty t^2 \lambda e^{-\lambda t} dt = \frac{2}{\lambda^2} \tag{1.6.13}$$

$$Var(T) = E[T^2] - (E[T])^2 = \frac{1}{\lambda^2} \tag{1.6.14}$$

Example 1.9   Suppose that the average lifetime of a car's headlight bulb is exponentially distributed with 100000 miles. There are two identical headlights in a car. If a person makes a 5000-mile trip, what is the probability that the driver completes the trip without replacing any light bulbs? What is the probability that at least one bulb needs to be replaced because of failure?

Solution:
We define T as a random variable representing the lifetime (i.e. mileage) of the bulb. Based on Eq. (1.6.12), the parameter λ is estimated by

$$\lambda = \frac{1}{E[T]} = \frac{1}{100\,000} = 10^{-5}\text{failure/mile} \tag{1.6.15}$$

Hence the reliability function of the headlight is obtained as follows:

$$R(t) = e^{-0.00001t} \tag{1.6.16}$$

Therefore the probability that the bulb will survive at 5000 miles is

$$P\{T \geq 5,000\} = R(5000) = 0.951 \tag{1.6.17}$$

Since there are two headlight bulbs in a car, the probability that both survive up to 5000 miles is

$$P\{T_1 \geq 5000; T_2 \geq 5000\} = R^2(5000) = 0.951 \times 0.951 = 0.905 \tag{1.6.18}$$

where T_1 and T_2 are the lifetime of the two bulbs, respectively. Next, we estimate the probability that the driver needs to replace at least one bulb. This problem can be solved based on the binomial distribution. Let X be the number of failed bulbs at the end of the trip; then $X \sim B(2, 1 - 0.951)$. According to Eq. (1.5.4), we have

$$P\{X \geq 1\} = \sum_{k=1}^{2} \binom{2}{k} [1 - R(5000)]^k [R(5000)]^{2-k}$$

$$= 2 \times 0.951 \times (1 - 0.951) + (1 - 0.951)^2 = 0.0956 \tag{1.6.19}$$

1.6.3   The Weibull Distribution

The Weibull distribution perhaps is the most widely used continuous probabilistic model to analyze the time-to-failure behavior of components, systems, or equipment in a reliability community. A book dedicated to the Weibull model and its application was written by Murthy et al. (2003). Below we briefly review and Weibull distribution properties pertaining to lifetime modeling. A random variable T is said to follow a Weibull distribution if it possesses the following PDF:

$$f(t) = \frac{\beta}{\theta}\left(\frac{t}{\theta}\right)^{\beta-1} \exp\left(-\left(\frac{t}{\theta}\right)^{\beta}\right), \quad \text{for } t \geq 0 \tag{1.6.20}$$

where θ and β are the scale and shape parameters, respectively. In general, $\theta > 0$ and $0 < \beta < \infty$. The CDF is given as

$$F(t) = 1 - \exp\left(-\left(\frac{t}{\theta}\right)^{\beta}\right) \tag{1.6.21}$$

the Weibull reliability function is

$$R(t) = 1 - F(t) = \exp\left(-\left(\frac{t}{\theta}\right)^{\beta}\right) \tag{1.6.22}$$

and the hazard rate function is

$$h(t) = \frac{f(t)}{R(t)} = \frac{\beta}{\theta}\left(\frac{t}{\theta}\right)^{\beta-1} \tag{1.6.23}$$

The popularity of the Weibull distribution lies in its versatility of $h(t)$. One can model a decreasing $(0 < \beta < 1)$, constant $(\beta = 1)$, or increasing hazard rate $(\beta > 1)$ by simply changing the value of β. Figures 1.12–1.14 depict the hazard rate, PDF, and reliability function with different β. Since θ is a scale parameter, it is normalized at $\theta = 1$ in these charts.

Finally, the mean and the variance of the Weibull random variable are given as follows:

$$E[T] = \theta\Gamma(1 + \beta^{-1}) \tag{1.6.24}$$

$$Var(T) = \theta^2\Gamma(1 + 2\beta^{-2}) - (E[T])^2 \tag{1.6.25}$$

Example 1.10 The annual operating hours of a WT depends on the local wind speed. Suppose there are two identical WTs installed in different locations

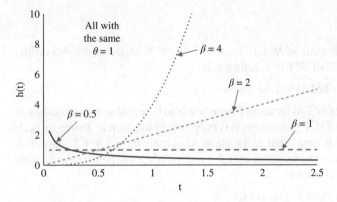

Figure 1.12 The hazard rate of the Weibull function.

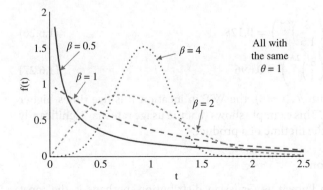

Figure 1.13 Weibull probability density function.

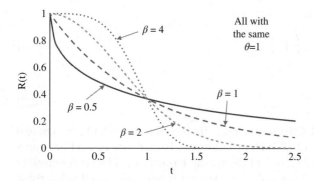

Figure 1.14 Weibull reliability function.

A and B, respectively. Location A has a high wind profile and the WT operates for 8760 hours/year. Location B has a low wind profile and the annual WT runs only 4380 hours/year. Assume a WT lifetime follows the Weibull distribution with $\theta = 1.5$ years and $\beta = 2.5$. Do the following:

1) What is the MTBF of the WT in locations A and B?
2) At the end of three years, is the WT in location B twice as reliable as the WT in location A?

Solution:

1) Let T_A and T_B be the lifetime of WT in locations A and B, respectively. According to Eq. (1.6.24), the MTBF of WT in location A is

$$E[T_A] = 1.5\Gamma(1 + 1/2.5) = 1.33 \text{ (year)}$$

To estimate the MTBF of WT in location B, one needs to take into account the equipment usage. The usage of WT in location B is only 50% in location A; hence its scale parameter, denoted θ_B, is twice that in location A, namely $\theta_B = 2\theta = 3$ years. It is commonly agreed that the shape parameter β is independent of the usage rate. Now the MTBF of WT in location B is

$$E[T_B] = 3\Gamma(1 + 1/2.5) = 2.66 \text{ (year)}$$

2) Now the reliability of both WT units at $t = 3$ years is obtained from Eq. (1.6.22) as follows:

$$R_A(t = 3) = \exp\left(-\left(\frac{3}{1.5}\right)^{2.5}\right) = 0.128 \tag{1.6.26}$$

$$R_B(t = 3) = \exp\left(-\left(\frac{3}{3}\right)^{2.5}\right) = 0.696 \tag{1.6.27}$$

By comparing $R_B(t = 3)$ with $R_A(t = 3)$, the WT in location B is five times higher than in location A in year 3. This example shows that the usage rate may significantly influence the reliability or the lifetime of a product.

1.6.4 The Normal Distribution

The normal distribution, also known as Gaussian distribution, perhaps is the most widely used continuous distribution model applied in engineering and science fields,

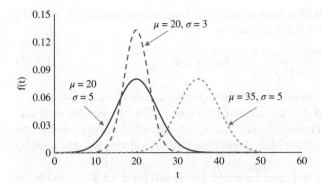

Figure 1.15 Normal PDF under different means and variances.

including a reliability analysis. A random variable T is said to be normally distributed if the PDF exhibits the following form:

$$f(t) = \frac{1}{\sqrt{2\pi}\sigma} \exp\left(-\frac{(t-\mu)^2}{2\sigma^2}\right), \quad -\infty < t < \infty \tag{1.6.28}$$

where μ and σ are the scale and the shape parameters, respectively. The normal density function has a bell-shaped curve that is symmetric around μ. Figure 1.15 plots the normal PDF for $\{\mu = 20, \sigma = 3\}$, $\{\mu = 20, \sigma = 5\}$, and $\{\mu = 35, \sigma = 5\}$ for comparison purposes.

The mean and the variance of T are equal to the parameters μ and σ^2, respectively,

$$E[T] = \int_{-\infty}^{\infty} tf(t)dt = \int_{-\infty}^{\infty} t\frac{1}{\sigma\sqrt{2\pi}} \exp\left(-\frac{(t-\mu)^2}{2\sigma^2}\right) dt = \mu \tag{1.6.29}$$

$$Var(T) = E[T^2] - (E[T])^2 = \sigma^2 \tag{1.6.30}$$

The reliability function is

$$R(t) = \int_t^{\infty} f(x)dx = 1 - \int_{-\infty}^t \frac{1}{\sqrt{2\pi}\sigma} \exp\left(-\frac{(x-\mu)^2}{2\sigma^2}\right) dx = 1 - \Phi\left(\frac{t-\mu}{\sigma}\right)$$

$$\tag{1.6.31}$$

where $\Phi(z)$ denotes the standard normal cumulative distribution with $\mu = 0$ and $\sigma = 1$. Unlike the Weibull distribution, there is no closed-form expression for Eq. (1.6.31). To estimate $R(t)$ at given t, tables are created to list the possible cumulative value for $\Phi(z)$.

1.6.5 The Lognormal Distribution

The lognormal distribution is one of the most frequently used parametric distributions in analyzing reliability data of microelectronic devices in semiconductor manufacturing industry. The reason why semiconductor life data fit the lognormal distribution well is because the lognormal distribution is formed by the multiplicative effects of random variables. This type of multiplicative interactions is often encountered in many semiconductor failure mechanisms (Oates and Lin 2009; Filippi et al. 2010). A random variable

T is said to follow a lognormal distribution when the transformed variable $X = \log(T)$ is normally distributed. The PDF for T is given by

$$f(t) = \frac{1}{t\sqrt{2\pi}\sigma} \exp\left(-\frac{(\log(t) - \mu)^2}{2\sigma^2}\right), \quad \text{for} \quad t \geq 0 \tag{1.6.32}$$

Two parameters μ and σ are required to define a lognormal distribution, where μ is called the scale parameter and σ is the shape parameter. Unlike the normal distribution where the mean and the standard deviation, respectively, equal the scale and shape parameter, the mean and the variance of the lognormal random variable are estimated by

$$E[T] = \int_0^\infty t \frac{1}{t\sqrt{2\pi}\sigma} \exp\left(-\frac{(\log(t) - \mu)^2}{2\sigma^2}\right) dt = \exp\left(\mu + \frac{\sigma^2}{2}\right) \tag{1.6.33}$$

$$E[T^2] = \int_0^\infty t^2 \frac{1}{t\sqrt{2\pi}\sigma} \exp\left(-\frac{(\log(t) - \mu)^2}{2\sigma^2}\right) dt = \exp(2\mu + 2\sigma^2) \tag{1.6.34}$$

$$Var(T) = E[T^2] - (E[T])^2 = \exp(2\mu + \sigma^2)(\exp(\sigma^2) - 1) \tag{1.6.35}$$

Figure 1.16 plots the lognormal distribution for $\sigma = 0.25$, 0.5, and 1 with common $\mu = 2$. The lognormal PDF with a smaller σ tends to resemble the bell shape of a normal distribution.

Example 1.11   An electromigration phenomenon is the transport of material caused by the movement of the ions in a conductor due to the momentum transfer between conducting electrons and diffusing metal atoms. An engineer conducted accelerated life testing on a set of samples to generate the device's life data with respect to an electromigration failure. Ten units are subject to the life test and the failure time occurred to each device is recorded and summarized in Table 1.2. Complete the following:

1) Use a probability plot to show that the time-to-failure is lognormally distributed.
2) What are the values of the scale and shape parameters?
3) What are the mean and the standard deviation of the device lifetime?

Solution:
(1) To examine whether the time-to-failure of devices are lognormally distributed, we calculate the logarithm value of these lifetime data. Let T be the device lifetime

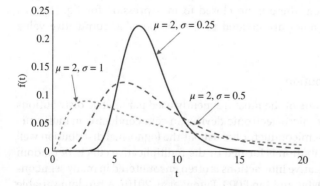

Figure 1.16 Lognormal PDF plots.

Table 1.2 Median ranks of the life data.

Sample	Lifetime T (h)	Log of lifetime X	Median rank (%)
1	57	4.04	6.7
2	152	5.02	16.3
3	272	5.61	26.0
4	512	6.24	35.6
5	862	6.76	45.2
6	1412	7.25	54.8
7	3012	8.01	64.4
8	5106	8.54	74.0
9	7002	8.85	83.7
10	9912	9.20	93.3

in hours; then $X = \ln(T)$ is the corresponding logarithmic value listed in the third column of the table. We also compute the median rank and the results are shown in the last column of the table. The formula to estimate the median rank is

$$\hat{F}(x) = \frac{i - 0.3}{n + 0.4}$$

where i is the order of failure of the ith failure data point and n is the sample size, which is 10 in our case. We then use statistical software Minitab to perform the probability plot on the data set X, and the result is shown Figure 1.17a. Given such a high P-value of 0.888, it can be concluded that X follows the normal distribution with 95% confidence. This leads to the statement that T is lognormally distributed.

(2) From Figure 1.17a, the mean and standard deviation of X are $\mu = 6.952$ and $\sigma = 1.73$. Hence the scale and shape parameters of the lifetime T are $\mu = 6.952$ and $\sigma = 1.73$ as well (but they are not the mean and standard deviation of T). The median life of the device is $\exp(6.952) = 1046$ hours. The median life is the time when 50% of the devices (i.e. five devices) failed. To verify whether the probability test in Figure 1.17a is correct, we directly input the data set T into Minitab and perform the lognormal probability plot; the resulting parameter values in Figure 1.17b are identical to Figure 1.17a.

(3) Based on Eqs. (1.6.33) and (1.6.35), the mean and the variance of the device lifetime are then obtained:

$$E[T] = \exp\left(6.953 + \frac{1.73^2}{2}\right) = 4673 \text{ hours}$$

$$Var(T) = \exp(2 \times 6.953 + 1.73^2)(\exp(1.73^2) - 1) = 4.141E + 8 \text{ or } 414{,}100{,}000$$

1.6.6 The Gamma Distribution

The gamma distribution is often used to model the number of errors in multilevel Poisson regression models, because the combination of the Poisson distribution and a gamma distribution is a negative binomial distribution. In Bayesian reliability statistics, the gamma distribution is often chosen as a conjugate prior for the distribution parameter to be estimated. For instance, the gamma distribution is the conjugate prior

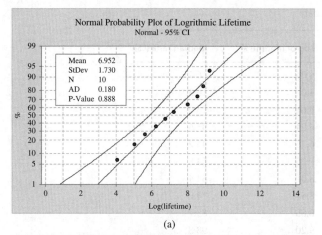

(a)

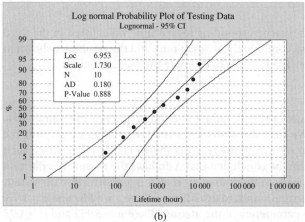

(b)

Figure 1.17 (a) Normal probability plot and (b) lognormal probability plot.

for the exponential lifetime distribution. The PDF of a random variable T with the gamma distribution is given as

$$f(t) = \frac{\lambda e^{-\lambda t}(\lambda t)^{\theta-1}}{\Gamma(\theta)}, \quad \text{for } t \geq 0 \tag{1.6.36}$$

with

$$\Gamma(\theta) = \int_0^\infty e^{-x}x^{\theta-1}dx, \quad \text{for } t > 0 \tag{1.6.37}$$

where λ and θ are the distribution parameters and both are positive values. Equation (1.6.37) is called the gamma function. If θ is a non-negative integer, then $\Gamma(\theta) = (\theta-1)!$. The mean and the variance of the Gamma random variable is (Ross 1998)

$$E[T] = \frac{\theta}{\lambda} \tag{1.6.38}$$

$$Var(T) = \frac{\theta}{\lambda^2} \tag{1.6.39}$$

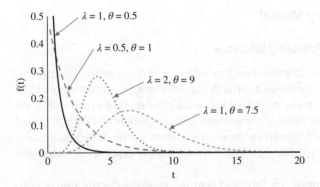

Figure 1.18 Gamma PDF plots.

Figure 1.18 plots the gamma PDF with different pairs of λ and θ. Two types of PDF curves are available depending on the values of λ and θ. For $0 < \theta \leq 1$, the PDF function declines monotonically with t. When $\theta > 1$, the PDF resembles the bell-shape type of curve. By varying the values of λ and θ, one can obtain various shapes of gamma PDF curves. Similar to the Weibull distribution, the gamma distribution is also capable of modeling a decreasing, constant, increasing failure rate corresponding to $0 < \theta < 1$, $\theta = 1$, and $\theta > 1$, respectively.

For a special case where θ are positive integers in Eq. (1.6.36), say $\theta = n = 1, 2, 3, \ldots$, the gamma distribution arises as the distribution of the amount of time one needs to wait until a total of n events has occurred. To prove that, let T_n denote the time at which the nth event occurs and note that T_n is less than or equal to t if and only if the number of events occurred by t is at least n; namely,

$$P\{T_n \leq t\} = P\{N(t) \geq n\}$$

$$= \sum_{k=n}^{\infty} P\{N(t) = k\}$$

$$= \sum_{k=n}^{\infty} \frac{e^{-\lambda t}(\lambda t)^k}{k!} \tag{1.6.40}$$

where the final identity follows, since the number of events in $[0, t]$ has a Poisson distribution with parameter λt. Taking the derivative with respect to t in Eq. (1.6.40) yields the PDF of T_n as follows:

$$f(t) = \frac{\lambda(\lambda t)^{n-1}e^{-\lambda t}}{(n-1)!} \tag{1.6.41}$$

Hence T_n is a gamma distribution with parameters (λ, n). This distribution is often referred to as the n-Erlang distribution, which is a special case of a gamma distribution with θ being positive integers. If $\theta = 1$, the gamma distribution is reduced to an exponential distribution (also see Figure 1.18). Amari and Misra (1997) derived a closed-form expression for distribution of the sum of exponential random variables, which is very useful for reliability analysis if a system is comprised of multiple components, each having a constant failure rate.

1.7 Bayesian Reliability Model

1.7.1 Concept of Bayesian Reliability Inference

Lifetime or failure models, as we discussed earlier, have one or more unknown distribution parameters. The classical statistical approach considers these parameters as fixed but unknown constants. They can be estimated using sample data taken randomly from the population. For instance, the value of parameter λ in the exponential PDF is assumed to be fixed and can be estimated. For an unknown parameter, a probabilistic statement represents the likelihood that the values calculated from a sample capture the underlying true parameter.

Unlike the classic statistical approach, Bayesian analysis treats the distribution or population parameters as random instead of being a fixed quantity. Historical data or subjective judgments can be used to determine a prior distribution for these parameters. The primary motivation to use Bayesian reliability methods is a desire to save on the test time and materials cost by leveraging the historical information of similar products or expert knowledge.

Figure 1.19 graphically shows a two-step Bayesian inference process. First, we use historical data, or subjective information, to construct a prior distribution for these unknown parameters. This model represents our initial assessment about how likely various values of the unknown parameters are. We then combine the current or new data with the prior distribution to revise this initial assessment, deriving what is called the posterior distribution model for the distribution parameters. Parameter estimates are then calculated directly from the posterior distribution. In applying the Bayes formula, conjugate prior models are a natural option to represent the parametric prior distribution. In many applications it is uncommon that historical data are available or exist to validate a chosen prior model. In that case uniform distributions are often chosen and designated as the uninformative prior probability model for the parameters to be estimated.

1.7.2 Bayes Formula

The Bayes formula is the mathematical tool that can combine prior knowledge with current data to produce a posterior distribution. Let E and F be two random events. The Bayes formula for discrete events is given as

$$P\{E \mid F\} = \frac{P\{EF\}}{P\{F\}} = \frac{P\{F \mid E\}P\{E\}}{P\{F\}} \tag{1.7.1}$$

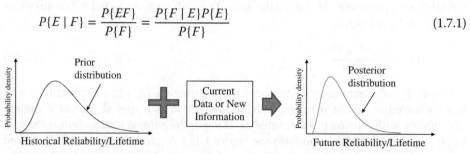

Figure 1.19 Bayesian reliability inference process.

and $P\{F\}$ in the denominator is further expanded by using the "Law of Total Probability" as

$$P\{F\} = \sum_{i=1}^{n} P\{F \mid E_i\}P\{E_i\} \tag{1.7.2}$$

with the events E_i being mutually exclusive and exhausting all possibilities with $\sum_{i=1}^{n} P\{E_i\} = 1$.

Next we present the Bayes formula for the continuous random variables. Let λ be the unknown distribution parameter to be estimated and x be the observed data. Assuming λ and x are continuous random variables, the Bayes formula in terms of PDF models takes the form as

$$f_{\lambda|x}(\lambda \mid x) = \frac{f_{x|\lambda}(x \mid \lambda)f_{\lambda}(\lambda)}{\int_0^{\infty} f_x(x \mid \lambda)f_{\lambda}(\lambda)d\lambda} \tag{1.7.3}$$

where

$f_{\lambda}(\lambda) = $ the prior distribution model for λ
$f_{\lambda|x}(\lambda|x) = $ the posterior distribution for λ given that the current data x observed
$f_{x|\lambda}(x|\lambda) = $ the likelihood function for observed data x with unknown parameter λ

When $f_{\lambda}(\lambda)$ and $f_{\lambda|x}(\lambda|x)$ both belong to the same distribution family, they are called conjugate distributions. Meanwhile $f_{\lambda}(\lambda)$ is the conjugate prior for $f_{x|\lambda}(x|\lambda)$. For example, the gamma distribution model is a conjugate prior for the hazard rate λ when failure times or repair times are sampled from an exponentially distributed population. In fact the gamma-exponential conjugate pair is used widely in Bayesian reliability inference because the posterior distribution for λ is also a gamma distribution.

Example 1.12 Suppose that the joint density function of two continuous random variables, X and Y, is given by

$$f_{X,Y}(x, y) = \frac{1}{y}e^{-x/y}e^{-y}, \quad \text{for } x > 0 \text{ and } y > 0$$

Do the following:

1) Find $f_{X|Y}(x|y)$.
2) Calculate $P\{X > 1 \mid Y = y\}$.

Solution:

(1) Using the Bayes formula in Eq. (1.7.3), we obtain the conditional density of X given $Y = y$, that is,

$$f_{X|Y}(x \mid y) = \frac{f_{X,Y}(x, y)}{f_Y(y)} = \frac{\frac{1}{y}e^{-x/y}e^{-y}}{\int_0^{\infty} \frac{1}{y}e^{-x/y}e^{-y}dx} = \frac{1}{y}e^{-x/y}$$

(2) Now the conditional CDF is

$$P\{X > 1 \mid Y = y\} = \int_1^{\infty} f_{X|Y}(x \mid y)dx = \int_1^{\infty} \frac{1}{y}e^{-x/y}dx = [-e^{-x/y}]_1^{\infty} = e^{-1/y}$$

1.8 Markov Model and Poisson Process

1.8.1 Discrete Markov Model

Let X_t be the value of the system characteristic at time t. Since X_t is not known with certainty before time t, it can be viewed as a random variable. A discrete time stochastic process is simply a sequence of random variables $X_1, X_2, \ldots$, where X_i is represents the value at different times.

$$X = \{X_1, X_2, \ldots, X_i, \ldots, X_m\} \tag{1.8.1}$$

A discrete-time stochastic process, or random sequence, is called a Markov chain if, for $i = 0, 1, 2, \ldots$, all states satisfy the following condition:

$$P\{X_{t+1} = i_{t+1} \mid X_t = i_t, X_{t-1} = i_{t-1}, \ldots, X_1 = i_1, X_0 = i_0\} = P\{X_{t+1} = i_{t+1} \mid X_t = i_t\} \tag{1.8.2}$$

This equation essentially says that the probability distribution of the state at time $t + 1$ depends only on the state at time t and does not depend on its previous states when the chain passed through on the way to state i_t at time t.

Since the Markov chain is a stationary process, the states within the Markov chain are time-invariant. Now Eq. (1.8.2) can be simplified as

$$P\{X_{t+1} = j \mid X_t = i\} = P\{X_{\tau+1} = j \mid X_\tau = i\} = p_{ij}, \quad \text{for} \quad t \neq \tau \tag{1.8.3}$$

where p_{ij} is the probability that given the system is in state i at time t, the system will be in state j at time $t + 1$. Hence p_{ij} is also called the "transition probability." Given the Markov chain has n states, all the transition probability values can be represented as an $n \times n$ square matrix:

$$\mathbf{P} = \begin{bmatrix} p_{11} & p_{12} & \cdots & p_{1n} \\ p_{21} & p_{22} & \cdots & p_{2n} \\ \vdots & \vdots & & \vdots \\ p_{n1} & p_{n2} & \cdots & p_{nn} \end{bmatrix} \tag{1.8.4}$$

Note that summation of the entries in a row must be equal to one. That is,

$$\sum_{j=1}^{n} p_{ij} = 1, \quad \text{for} \quad i = 1, 2, \ldots, n \tag{1.8.5}$$

Let $\pi = [\pi_1, \pi_2, \ldots, \pi_n]$ be the steady-state probability for state j for $j = 1, 2, \ldots, n$. For an ergodic Markov chain, the steady-state probability can be computed by (where "T" is for transpose):

$$\mathbf{P}^T \pi^T = \pi^T \tag{1.8.6}$$

where $\mathbf{P}$ is the transition matrix in Eq. (1.8.4), and "T" is for the matrix or vector transposition. The Markov model can be used to analyse the reliability of a multi-state system of which the system reliability deteriorates over time, starting from the good state initially, transitioning to a degradation state, and finally entering the failure state.

1.8.2 Birth–Death Model

The birth–death model is a continuous-time Markov chain, denoted as an $M/M/1$ queue, for which the system state at any time is a non-negative integer. Most of the Markov chains with exponential arrival and exponential service time can be modeled by the birth–death process. Figure 1.20 shows the transition diagram of the $M/M/1$ queue.

In the transition diagram, λ_j for $j = 0, 1, 2, \ldots$ is called the birth rate and μ_j for $j = 1,$ $2, 3, \ldots$ is called the death rate. They are analogous to the transition probability in the discrete Markov chain. The steady-state probability π_j for state j can be solved using the flow balance equation. The following linear equation systems can be formulated:

for $j = 0,$ $\qquad \pi_0 \lambda_0 = \pi_1 \mu_1$ $\qquad\qquad\qquad\qquad$ (1.8.7)

for $j = 1,$ $\qquad (\lambda_1 + \mu_1)\pi_1 = \lambda_0 \pi_0 + \mu_2 \pi_2$ $\qquad\qquad$ (1.8.8)

for $j = 2,$ $\qquad (\lambda_2 + \mu_2)\pi_2 = \lambda_1 \pi_1 + \mu_3 \pi_3$ $\qquad\qquad$ (1.8.9)

for $j \geq 3,$ $\qquad (\lambda_j + \mu_j)\pi_j = \lambda_{j-1}\pi_{j-1} + \mu_{j+1}\pi_{j+1}$ $\qquad$ (1.8.10)

The identity formula states that sum of all the steady-state probabilities equals unity. That is,

$$\pi_0 + \pi_1 + \pi_2 + \cdots + = 1 \qquad\qquad (1.8.11)$$

To solve this linear system, all π_j (for $j \geq 1$) can be recursively expressed as the function of π_0 based on the above flow balance equations. The final results are given below:

$$\pi_0 = \frac{1}{1 + \sum_{j=1}^{\infty} c_j} \qquad\qquad (1.8.12)$$

$$\pi_j = \frac{\lambda_0 \lambda_1 \cdots \lambda_{j-1}}{\mu_1 \mu_2 \cdots \mu_j}\pi_0, \quad \text{for } j = 1, 2, \ldots \qquad (1.8.13)$$

with

$$c_j = \frac{\lambda_0 \lambda_1 \cdots \lambda_{j-1}}{\mu_1 \mu_2 \cdots \mu_j}, \quad \text{for } j = 1, 2, \ldots \qquad (1.8.14)$$

Quite often, we are interested in the time that a customer spends in the queue system including his/her waiting time, the processing time, as well as the number of customers in waiting or under service. They are defined in Table 1.3. We are now able to link these performance measures through the powerful law called *Little's law*, which states that

$$L = \lambda W, \quad L_q = \lambda W_q, \quad L_s = \lambda W_s \qquad\qquad (1.8.15)$$

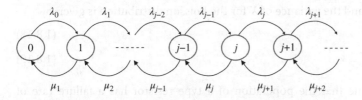

Figure 1.20 Transition diagram for the birth–death queuing system.

Table 1.3 Notation of queuing system.

Notation	Interpretation
λ	Customer arrival rate (number of persons per unit time)
L	Average number of customers in the queuing system
L_q	Average number of customers waiting
L_s	Average number of customers being processed
W	Average time a customer spends in the queuing system
W_q	Average time a customer is waiting in the queue
W_s	Average time a customer spends with the server

For any queue process, the following formulas always hold:

$$L = \sum_{j=0}^{\infty} j\pi_j, \quad L = L_q + L_s, \quad W = W_q + W_s \tag{1.8.16}$$

The birth–death queue model is perhaps the most basic queuing model. There are many different variations of queuing models, and readers are referred to the book by Winston (2004) on this topic.

1.8.3 Poisson Process

A Poisson process is a continuous-time stochastic process that counts the number of events and the time epochs at which these events occur in a given time interval, say $[0, t]$. The interarrival time between two consecutive events has an exponential distribution with parameter λ, and all these interarrival times are assumed to be mutually independent. The process is named after the Poisson distribution introduced by the French mathematician Simeon D. Poisson (Poisson 1937). In reliability engineering, the Poisson process describes the time of events of equipment failures, demand stream of spare parts, and renewal of repairable systems. It has other applications in representing the time of events in telephone calls at a call center (Willkomm et al. 2009), tracking the requests on a web server (Arlitt and Williamson 1997), and many other punctual sequences where events take place independently among each other. Let N be the number of events occurred in $[0, t]$. The PDF for a Poisson process is

$$P(N = k) = \frac{(\lambda t)^k e^{-\lambda t}}{k!}, \quad \text{for } k = 0, 1, 2, \ldots \tag{1.8.17}$$

The expected value and the variance of X for the Poisson distribution is given by

$$E[N] = \lambda t \tag{1.8.18}$$

$$Var(N) = \lambda t \tag{1.8.19}$$

Example 1.13 Assume that the population of a type resistor has a failure rate of 100 per one million operation hours (i.e. 10^{-4} failures/hour) at a given level of power derating. The component is expected to operate 40 000 hours with only two failures allowed. Using the Poisson process model calculate:

(1) The expected number of failures during the mission period of [0, 40 000].
(2) The probability that exactly 0, 1, and 2 failures occurred during the mission time.
(3) The probability that at least three failures occurred in the mission time.

Solution:

(1) The component failure rate is $\lambda = 10^{-4}$ faults/hour. Let N be the number of failures in [0, 40 000]. Based on Eq. (1.8.18), the expected failure is

$$E[N] = \lambda t = (40\,000)(1 \times 10^{-4}) = 4 \text{ failures}$$

(2) Based on Eq. (1.8.17), we can compute the probability of failure number for $k = 0$, 1, and 2 by realizing $t = 40\,000$ hours:

$$P\{N = 0\} = \frac{(\lambda t)^0 e^{-\lambda t}}{0!} = \frac{(4)^0 e^{-4}}{1} = 0.018$$

$$P\{N = 1\} = \frac{(\lambda t)^1 e^{-\lambda t}}{1!} = \frac{(4)^1 e^{-4}}{1} = 0.073$$

$$P\{N = 2\} = \frac{(\lambda t)^2 e^{-\lambda t}}{2!} = \frac{(4)^2 e^{-4}}{2} = 0.147$$

(3) The probability that at least three failures will occur during the mission time is

$$P\{X \geq 3\} = 1 - P\{X = 0\} - P\{X = 1\} - P\{X = 2\}$$
$$= 1 - 0.018 - 0.073 - 0.147 = 0.762$$

Another application of the Poisson distribution is to determine the spare parts pool that should be allocated to meet the committed service level upon request. Let D be the random demand for the spare parts and s be the base stock level. The spare parts availability, denoted as A_p, is estimated by

$$A_p = P\{D \leq s\} = \sum_{k=0}^{s} \frac{(n\lambda\tau)^k e^{-n\lambda\tau}}{k!} \tag{1.8.20}$$

where n is the number of components or parts in field operation, λ is the failure rate per unit time of individual components, s is the base stock level of spare parts, and τ is the repair turnaround time. The turnaround time is the time elapsed from when the defective item is removed from the system to when it is returned to the stockroom after repair.

Further information about statistical reliability models can be found in the books by Elsayed (2012), Blischke and Murthy (2000), Murthy et al. (2003), Nachlas (2016), and O'Connor (2012), as well as the on-line publication by NIST (2017).

Example 1.14 Assume that 600 transformers are installed in a power distribution network to supply electricity to residential consumers. The transformer's hazard rate is 10^{-5} failures/hour. The repair turnaround time is $\tau = 200$ hours per defective transformer. If the service level of the spares availability is set to be 90%, using Eq. (1.8.20) verify whether one spare transformer can meet the required service level.

Solution:

If only one spare transformer is allocated in the stock, the actual service level is

$$A_p = \sum_{k=0}^{1} \frac{[(600)(10^{-5})(200)]^k e^{-(600)(10^{-5})(200)}}{k!} = 0.66 \leq 0.9$$

Since one spare only provide 66% parts availability, one would reject the hypothesis that one back-up transformer is sufficient to meet the 90% parts availability. If we increase the spares quantity to 2, the actual A_p reaches 88%, which is still below the target level. By increasing the spare part number to 3, the actual A_p reaches 97%, which meets and exceeds the target level.

References

Amari, S.V. and Misra, R.B. (1997). Closed-form expressions for distribution of sum of exponential random variables. *IEEE Transactions on Reliability* 46 (4): 519–522.

Arlitt, M.F. and Williamson, C.L. (1997). Internet web servers: workload characterization and performance implications. *IEEE/ACM Transactions on Networking* 5 (5): 631–642.

ASN, 2017, Aviation Safety Network, available at: https://aviation-safety.net (accessed on May 30, 2017).

Blischke, W.R. and Murthy, D.N.P. (2000). *Reliability Modeling, Prediction, and Optimization*. New York, NY: Wiley.

Cai, Z., Sun, S., Si, S., and Yannou, B. (2011). Identifying product failure rate based on a conditional Bayesian network classifier. *Expert Systems with Applications* 38 (5): 5036–5043.

Chowdhury, A.A. and Koval, D.O. (2009). *Power Distribution System Reliability*. Hoboken, NJ: Wiley and IEEE Press.

Elsayed, E. (2012). *Reliability Engineering*, 2e. Hoboken, NJ: Wiley.

Farrel, J. and Saloner, G. (1986). Installed base and compatibility: innovation, product preannouncements, and predation. *The American Economic Review* 76 (5): 940–955.

Filippi, R.G., Wang, P.-C., Brendler, A., and Lloyd, J.R. (2010). Implications of a threshold failure time and void nucleation on electromigration of copper interconnects. *Journal of Applied Physics* 107 (103709): 1–7. doi: 10.1063/1.3357161.

Guikema, S.D. and Paté-Cornell, M.E. (2004). Bayesian analysis of launch vehicle success rates. *Journal of Spacecraft and Rockets* 41 (1): 93–102.

Gupta, R.C. and Bradley, D.M. (2003). Representing the mean residual life in terms of the failure rate," . *Mathematical and Computer Modelling* 37 (12/13): 1271–1280.

Jin, T. and Liao, H. (2009). Spare parts inventory control considering stochastic growth of an installed base. *Computers and Industrial Engineering* 56 (1): 452–460.

Liao, H., P. Wang, T. Jin, S. Repaka, "Spare parts management considering new sales," in *Proceedings of Annual Reliability and Maintainability Symposium*, 2008, pp. 502–507.

Murthy, D.N.P., Xie, M., and Jiang, R. (2003). *Weibull Models*. Hobken, NJ: Wiley.

Nachlas, J. (2016). *Reliability Engineering: Probability Model and Maintenance Method*, 2e. Boca Raton, FL: CRC Press.

NIST, (2017), Engineering Statistics Handbook, Chapter 8, available at: http://www.itl.nist .gov/div898/handbook/apr/section1/apr125.htm (accessed on May 20, 2017).

Oates, A.S. and Lin, M.H. (2009). Electromigration failure distributions of Cu/Low-k dual-damascene vias: impact of the critical current density and a new reliability extrapolation methodology. *IEEE Transactions on Device and Materials Reliability* 9 (2): 244–254.

O'Connor, P.D.T. (2012). *Practical Reliability Engineering*, 4e. West Sussex, UK: Wiley.

Poisson, S.D. (1937). *Probabilité des Jugements en Matière Criminelle et en Matière Civile, Précédées des Règles Générales du Calcul des Probabilitiés*, 206. Paris, France: Bachelier.

Ross, S. (1998). *A First Course in Probability*, 5e. Upper Saddle River, NJ: Prentice Hall.

Tse, K.-K. (2014). Some applications of the Poisson process. *Applied Mathematics* 5 (19): 7.

Wikepedia, (2017), "List of rail accidents in China," available at: https://en.wikipedia.org/wiki/List_of_rail_accidents_in_China (accessed on May 30, 2017).

Willkomm, D., Machiraju, S., Bolot, J., and Wolisz, A. (2009). Primary user behavior in cellular networks and implications for dynamic spectrum access. *IEEE Communications Magazine* 47 (3): 88–95.

Winston, W.L. (2004). *Operations Research: Application and Algorithm*, 4e. Belmont, CA: Brooks/Cole.

Problems

Problem 1.1 The definition of the reliability is given in Eq. (1.2.1). What are the four conditions for this probability model?

Problem 1.2 Please state the definition of hazard rate and failure intensity rate. Explain their applications in reliability analysis in the context of repairable and non-reparable systems/components. Is the bathtub curve applicable to both non-repairable and repairable systems?

Problem 1.3 Please derive the analytical expressions for $H(t)$, $R(t)$, and $f(t)$ given the following hazard rate functions: (1) $h(t) = 3$; (2) $h(t) = 3 + 2t$; and (3) $h(t) = 3t^{0.5}$.

Problem 1.4 Please plot the $h(t)$, $H(t)$, $R(t)$, and $f(t)$ in three cases in Problem 1.3, respectively.

Problem 1.5 The power law function $\lambda(t) = \alpha\beta t^{\beta-1}$ has been widely used to model the failure intensity rate for a decreasing, constant, or increasing trend. Note that α and β are positive parameters. If $0 < \beta < 1$, it represents a decreasing failure intensity rate. If $\beta = 1$, this is a constant failure intensity rate. If $\beta > 1$, it represents an increasing failure intensity rate. Please draw the power law function with $\{\alpha = 30,\ \beta = 0.5\}$, $\{\alpha = 10,\ \beta = 1\}$, $\{\alpha = 2,\ \beta = 1.5\}$, and $\{\alpha = 0.5,\ \beta = 2\}$, respectively.

Problem 1.6 If the failure intensity rate observes the power law function $\lambda(t) = \alpha\beta t^{\beta-1}$, estimate the cumulative number of failures in the intervals of $[0,\ t = 10]$ and $[t = 10,\ t = 20]$, respectively, assuming the following cases: (1) $\{\alpha = 30,\ \beta = 0.5\}$; (2) $\{\alpha = 10,\ \beta = 1\}$; and (3) $\{\alpha = 0.5,\ \beta = 2\}$.

Problem 1.7 After introducing nine identical new products to the market, the service engineer is able to collect the time-to-failure between the installation time and the failure times: 11, 21, 34, 36, 50, 60, 66, 72, 77, 111, 115, 140, 157, 209, 296, and 397 days. Do the following: (1) estimate the mean-time-to-failure; (2) estimate the standard deviation of the product lifetime; (3) estimate the cumulative distribution of the product lifetime; (4) does the lifetime follows the exponential distribution? (Hint: use the statistic software to perform a probability fit); (5) based on the fitted lifetime distribution, derive the hazard rate function; (6) derive the reliability function.

Problem 1.8 A repairable system comprised of two different types of components connected in series. The system fails if one of the components is malfunctioning. Upon failure, the faulty component is removed and replaced with a good one and the system is recovered to the operational state. The failure arrival time of two types of components over 16 months are recorded: for component type one: 9, 25, 79, 195, 280, 311, 378, 452, and 455 days and for component type two: 16, 88, 147, 196, 242, 280, 349, 400, and 443 days. Estimate: (1) the hazard rate of each component type; (2) the failure intensity rate or ROCOF of the system; (3) does the component hazard rate function possess a decreasing, constant, or increasing trend? How about the ROCOF of the system?, and (4) if the replacement of a faulty component takes 10 hours on average, estimate the steady-state system availability.

Problem 1.9 Estimate the cumulative hazard rate, reliability, PDF, and CDF given the hazard rate function: (1) $h(t) = 10$; (2) $h(t) = 5t$; (3) $h(t) = 2 + 3t$; (4) $h(t) = 4t^2$; and (5) $h(t) = 10t$ with $0 < t < 10$.

Problem 1.10 A bin of four electronic components is known to contain two defective units. The components are to be tested one at a time, in random order, until both defectives are discovered. Calculate the expected number of tests that are made.

Problem 1.11 A machine could break down for two reasons: aging of the machine or inappropriate use by the operator. To check the first possibility, it would cost c_1. The cost of repairing a breakdown machine because of aging is r_1 dollars. Similarly, there are costs of c_2 and r_2 associated with breakdown or operator errors. Let p and $1 - p$ represent, respectively, the probability of breakdown due to machine aging and inappropriate use. Should the aging check be carried out first or should the check for impropriate use be done first? In other words, determine the relationship between p, c_1, c_2, r_1, and r_2 in two different checking sequences such that the total cost is minimized.

Problem 1.12 A four-engine plane can fly reliably when only two of its engines are working. Similarly, a two-engine plane can fly if only one of them is working. Suppose that the airplane engine fails with probability $1 - p$ independently from engine to engine. For what value of p is a four-engine plane preferable to a two-engine plane?

Problem 1.13 Suppose that 5% of the microdevices produced by a semiconductor manufacturer are defective. If we purchase 100 such devices, will the number of defective devices we receive be a binomial random variable? What is the probability that at least 10 units we received are defective?

Problem 1.14 The expected number of typographical errors on a page of a book is 0.1 error/page. What is the probability that the next page you read contains (1) no error; (2) two or more typographical errors?

Problem 1.15 For a Poisson random variable N with parameter λ, show that $E[N] = \lambda$ and $Var(N) = \lambda$.

Problem 1.16 Let X be a binomial random variable with parameter (n, p). What value of p maximizes $P\{X = k\}$ where the random variable is observed to equal k with $0 \le k \le n$. If we assume n is known, then we estimate p by choosing the value for p that maximizes $P\{X = k\}$. This is known as the maximum likelihood estimation.

Problem 1.17 Let N be a Poisson random variable with parameter λ. Show that the following property is valid (Hint: use the relationship between the Poisson and binomial random variable):
$$P\{N \text{ is odd}\} = \tfrac{1}{2}(1 - e^{-2\lambda})$$

Problem 1.18 Show that
$$\sum_{k=1}^{n} \frac{\lambda^k e^{-\lambda}}{k!} = \frac{1}{n!} \int_{\lambda}^{\infty} e^{-x} x^n dx$$

Problem 1.19 Prove that the variance of the Bernoulli random variable is
$$Var(X) = p(1 - p).$$

Problem 1.20 The probability of an item failing is $p = 0.001$. What is the probability of 3 failing out of the population of $n = 170$? Solve the problem using the binomial distribution and Poisson distribution, respectively.

Problem 1.21 WTs are repairable systems. Reliability plays an important role as equipment maintenance and repair are costly. WT manufacturer A claims that MTBR of his product is exponentially distributed with 10 000 hours. Manufacturer B claims that MTBR of his product follows the Weibull distribution with $\theta = 11\,000$ hours, and $\beta = 2.5$. Assuming the repair and replacement cost of each failure are the same for two WT systems, decide which manufacturer is considered as the preferred supplier?

Problem 1.22 Based on the statement of the previous problem, the MTTR for WT of manufacturer A is five days and for B it is four days. From the equipment availability perspective, which manufacturer is more competitive in terms of sustaining the uptime of the equipment?

Problem 1.23 The main landing gear leg of a large commercial airplane usually contains six tires. Experience shows that a tire bursts on average on 1 out of 2000 landings. Assuming that a tire burst occurs independently of other tires and a safe landing is made as long as no more than two tires burst, what is the probability of an unsafe landing?

Problem 1.24 A product with an exponential lifetime distribution has a hazard rate of $h(t) = \lambda$. Let T be the random lifetime. Prove:
1) Compute $E[T] = 1/\lambda$.
2) Compute $E[T^2] = 2/\lambda^2$.
3) Estimate $E[T]$ and $E[T^2]$ given (1) $\lambda = 1$; (2) $\lambda = 0.5$; and (3) $\lambda = 3$, respectively.

Problem 1.25 A product's lifetime is exponentially distributed with parameter $\lambda = 0.001$ failures/hour. Do the following: (1) estimate the mean lifetime for this product; (2) draw the PDF; (3) draw the CDF function; (4) draw the reliability function; and (5) draw the failure rate function. (*Note*: Draw all of the curves on a separate chart.)

Problem 1.26 Suppose that the number of miles that an EV can run before its battery wears out is exponentially distributed with an average distance of 100 000 miles. If a car manufacturer decides to install this type of battery to power the EV, what percentage of the cars will fail at the end of 50 000 and 100 000 miles, respectively?

Problem 1.27 Based on the previous problem, the car manufacturer is not satisfied with the lifetime of the battery that is supplied by its subcontractor. The manufacturer wants the car to operate after running 100 000 miles with a probability of 80%. What should the minimum mean lifetime of the battery be in order to meet the new requirement?

Problem 1.28 Given the Weibull PDF, $f(t) = \alpha\beta(\alpha t)^{\beta-1} \exp(-(\alpha t)^\beta)$, where $\alpha > 0$ and $\beta > 0$ are the scale and shape parameters, respectively. Plot the PDF, CDF, reliability, and hazard rate under different values of $\beta = 0.5, 1, 1.5, 2$, and 3, respectively. In all cases, $\alpha = 2$.

Problem 1.29 (Normal) Let T be a normal random variable with $N(\mu, \sigma^2)$. Prove that $E[T] = \mu$ and $E[T^2] = \mu^2 + \sigma^2$.

Problem 1.30 (Lognormal) Let T be a lognormal random variable with scale parameter μ and shape parameter σ. Prove that
$$E[T] = \exp\left(\mu + \frac{\sigma^2}{2}\right) \quad \text{and} \quad E[T^2] = \exp(2\mu + \sigma^2)$$
(*Hint*: Using the normal moment generating function to find the mean and variance of T.)

Problem 1.31 (Gamma) Let T be a gamma random variable with parameters λ and θ. Do the following:
  (1) Prove that $E[T] = \frac{\theta}{\lambda}$ and $Var(T) = \frac{\theta}{\lambda^2}$
  (2) For $\lambda = 1$ and $\theta = 3$, find the reliability $R(T > 2)$.
  (3) For $\lambda = 1$ and $\theta = 3.5$, find the reliability $R(T > 2)$.

Problem 1.32 For exponentially distributed lifetime T, show that
$$P\{T < t + s \mid T > s\} = P\{T < t\}, \quad \text{for } t > 0 \text{ and } s > 0$$

Problem 1.33 Suppose that $p(x, y)$, the joint probability mass function of X and Y, is given by
$p(0, 0) = 0.4$, $p(0, 1) = 0.2$, $p(1, 0) = 0.1$, and $p(1, 1) = 0.3$. Calculate the conditional probability mass function of X, given that $Y = 0$ and $Y = 1$, respectively.

Problem 1.34 The joint density of X and Y is given by
$f_{X,Y}(x, y) = \frac{15}{2}x(2 - x - y)$, $0 < x < 1$, $0 < y < 1$
Find $f_{X|Y}(x \mid y)$, that is, the conditional density of X, given that $Y = y$, where $0 < y < 1$.

Problem 1.35 For the birth–death $M/M/1/\infty$ model, assume $\lambda_j = \lambda$ for all $j = 0, 1, 2$, and $\mu_j = \mu$ for all $j = 1, 2, 3, \ldots$. Define $\rho = \lambda/\mu$. Show the following results:
  (1) $\pi_0 = 1 - \rho$; (2) $L = \rho/(1 - \rho)$; (3) $L_s = \rho$.

2

Reliability Estimation with Uncertainty

2.1   Introduction

This chapter introduces the reliability block diagram and presents reliability estimation methods for a variety of system structures, including series, parallel, mixed
series-and-parallel, k-out-of-n redundancy, and networks. The focus is on modeling
and analysis of the uncertainty of component reliability estimates and its impact on
the system reliability performance. Particularly, we propose block decomposition
methods to calculate the variance of a system reliability estimate based on the moments
of a component reliability estimate. Reducing reliability estimation uncertainty is of
importance in terms of designing and operating risk-averse systems in private and
public sectors, including manufacturing, healthcare, transportation, energy, supply
chains, and defense industries, where failures often lead to costly downtime or even
human casualty. System reliability confidence intervals (CIs) are further derived under
binomial, normal, and lognormal distributions. We also extend the binary reliability
models to multistate systems. This chapter is concluded with the discussion of reliability
importance measures.

2.2   Reliability Block Diagram

A reliability block diagram (RBD) is a graphical representation to capture the functional
or operational interdependency of different components in a system. Each block represents a component or subsystem that possesses its own failure rate or reliability feature.
For instance, if blocks are connected in series, any failure along a series path causes the
whole system to fail. Parallel configuration of blocks means that the system is operational as long as one of the parallel paths is still good. Regardless of physical structure,
there exist four basic types of functional relationships between a system and its components. These are series, parallel, k-out-of-n redundancy, and network systems. Figure 2.1
shows the reliability diagrams of these systems.

The configuration of a complex system is often the mix of two or more of these basic
block diagrams. For example, reliable supply of electricity requires the generators, the
transmission/distribution lines and the transformers all to work properly. At the operational level, these components are connected and functioning in series stretching dozens
or hundreds kilometers. Figure 2.2 depicts the series connection between power generation, transmission, distribution, and consumption.

Reliability Engineering and Services, First Edition. Tongdan Jin.
© 2019 John Wiley & Sons Ltd. Published 2019 by John Wiley & Sons Ltd.
Companion website: www.wiley.com/go/jin/serviceengineering

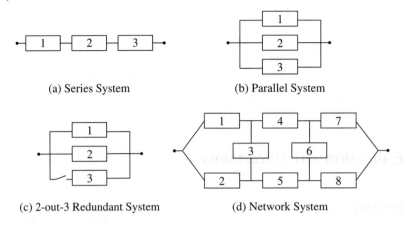

(a) Series System (b) Parallel System

(c) 2-out-3 Redundant System (d) Network System

Figure 2.1 Reliability block diagrams of basic system structures.

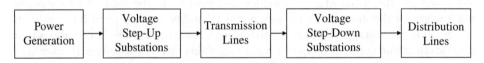

Figure 2.2 Series reliability block diagram of the electric power system.

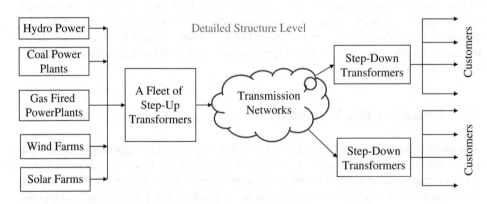

Figure 2.3 Reliability block diagram of the power system at the structure level.

Meanwhile, the generation pool consists of multiple power units and can be treated as a k-out-of-n redundant system. In a substation where many transformers work as a parallel system to escalate the voltage or reduce the voltage to facilitate the power delivery. Transmission lines are often interconnected to form a meshed network and the distribution system is created as a radial network with multiple feeders connected to one single substation. Figure 2.3 depicts the structural view of different types of reliability block diagrams across power generation, transmission, distribution, and consumption.

2.3 Series Systems

2.3.1 Reliability of Series System

A series system is comprised of multiple components interconnected in a sequential manner either physically or functionally. Typical examples of series systems include low-voltage power distribution circuits, oil pipelines, automobile assembly lines, computers, and air-conditioning systems. For a series system, failure of any component leads to the malfunction of the entire system. For example, a modern wind turbine can be treated as a series system comprised of multiple electromechanical components, including blades, main bearing/shaft, gearbox, generator, direct current (DC)–alternating current (AC) converters, and auxiliary parts. Failure of any component such as a blade or a gearbox usually brings the turbine to a down state, despite other components still remaining functional.

A series system could also be the case where the functionality of the entire system is treated as multiple components connected in series. However, components may not be physically configured in series format. For example, an automobile typically has four tires that are not formed in a linear configuration, yet they are treated as a series system in reliability analysis. This is because all tires must operate appropriately in order to ensure the normal driving function.

A reliability block diagram of a series system comprised of m components is shown in Figure 2.4. Note that $r_i(t)$ is the reliability of the ith component for $i = 1, 2, \ldots, m$. Let $R_s(t)$ be the reliability of the series system; then

$$R_s(t) = P\{T_1 > t, T_2 > t, \ldots, T_m > t\} = \prod_{i=1}^{m} P\{T_i > t\} = \prod_{i=1}^{m} r_i(t) \tag{2.3.1}$$

where T_i is the lifetime of component i and $r_i(t)$ is the associated reliability at time t. Equation (2.3.1) is derived based on the assumption that all components fail independently.

Example 2.1 Suppose at $t = 1$ year, the reliability of the individual subsystems of the electric power system in Figure 2.2 are given as follows: generation is 0.99, voltage step-up substation is 0.996, transmission line is 0.995, voltage step-down substation is 0.97, and distribution line is 0.94. Based on Eq. (2.3.1), the reliability of the entire power system is

$$R_s(t = 1) = (0.99)(0.996)(0.995)(0.97)(0.94) = 0.895 \tag{2.3.2}$$

For the series system, this example shows that the system reliability is always smaller than the reliability of individual components. The reliability of the electric distribution system in North America is required to be one outage out of 365 days for residential

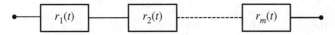

Figure 2.4 A series system comprised of m components.

service (NERC 2017); hence the minimum $R_s(t=1) = (365-1)/365 = 0.997$ is required. Obviously the system reliability in this example needs to be improved in order to meet this criterion.

2.3.2 Mean and Variance of Reliability Estimate

In reality the true component reliability $r_i(t)$ is not known and its estimate, denoted as $\hat{r}_i(t)$, is often used to infer the true value. Then the reliability estimate of the series system, denoted as $\hat{R}_s(t)$, becomes

$$\hat{R}_s(t) = \prod_{i=1}^{m} \hat{r}_i(t) \tag{2.3.3}$$

Reliability estimates are often derived from the assumption that the component or the system lifetime distribution can be represented with a parametric model, such as exponential and Weibull distributions. This assumption could be misleading if the estimation involves considerable amounts of uncertainty (Borgonov 2007). This situation usually happens when sufficient field or experimental data are not available to infer the underlying lifetime distribution. To design a risk-averse system, the uncertainty of component reliability estimates needs to be explicitly analyzed and appropriately incorporated into the system reliability estimation model.

The uncertainty of the component reliability estimate can be quantified by its moments, such as mean and variance. Figure 2.5 depicts the probability density functions of $\hat{r}_1(t)$ and $\hat{r}_2(t)$, corresponding to components 1 and 2, respectively. For a risk-neutral system design, component 1 is better than component 2 because $E[\hat{r}_1(t)] > E[\hat{r}_2(t)]$. However, if a risk-averse design is preferred, component 2 tends to be better than component 1 because its variance $Var(\hat{r}_2(t))$ is much smaller.

Different approaches are available to determine the reliability estimate and the associated variance. One common approach is to treat the number of failures as a binomial random variable in a given period of time. For instance, considering a system consisted of m different components, n_i units are allocated for reliability testing for component i for $i = 1, 2, \ldots, m$. Each unit is tested for t hours and k_i failures are observed during

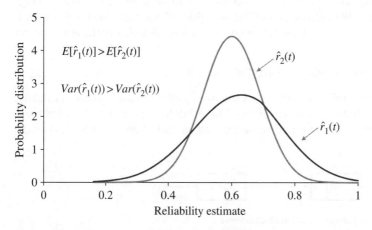

Figure 2.5 Distributions of component reliability estimates.

$[0, t]$. The status (survival or failure) of each testing unit can be treated as an independent Bernoulli trial with parameter $r_i(t)$. That is, the number of survivals X_i is a random variable following the binomial distribution $B(n_i, r_i(t))$. An unbiased estimate of $\hat{r}_i(t)$ and the associated variance can be estimated by (Coit 1997)

$$\hat{r}_i(t) = 1 - \frac{k_i}{n_i}, \quad \text{for } i = 1, 2, \dots, m \tag{2.3.4}$$

$$V\hat{a}r(\hat{r}_i(t)) = \frac{\hat{r}_i(t)(1 - \hat{r}_i(t))}{n_i - 1}, \quad \text{for } i = 1, 2, \dots, m \tag{2.3.5}$$

In statistics an estimate is said to be unbiased if the mean of the sampling distribution of that statistic is equal to the estimating parameter. To show that Eq. (2.3.5) is an unbiased estimate for $Var(\hat{r}_i(t))$, we compute the expectation of the variance estimate as follows:

$$E[V\hat{a}r(\hat{r}_i(t))] = \frac{E[\hat{r}_i(t)(1 - \hat{r}_i(t))]}{n_i - 1} = \frac{1}{n_i - 1} E\left[\frac{X_i}{n_i}\left(1 - \frac{X_i}{n_i}\right)\right]$$

$$= \frac{1}{n_i^2(n_i - 1)}(n_i E[X_i] - E[X_i^2]) \tag{2.3.6}$$

Recalling that X_i, the number of components i surviving in $[0, t]$, follows the binomial distribution with $B(n_i, r_i)$. From Eqs. (1.5.5) and (1.5.6), the first two moments of X_i are given as

$$E[X_i] = n_i r_i \tag{2.3.7}$$

$$E[X_i^2] = n_i r_i(n_i r_i + 1 - r_i) \tag{2.3.8}$$

By substituting Eqs. (2.3.7) and (2.3.8) into Eq. (2.3.6), we have

$$E[V\hat{a}r(\hat{r}_i(t))] = \frac{r_i(1 - r_i)}{n_i} \tag{2.3.9}$$

This result is equivalent to Eq. (1.5.7) divided by n_i^2. This completes the proof that Eq. (2.3.5) is the unbiased estimate of $Var(\hat{r}_i(t))$.

Given the reliability block diagram, we can further compute the mean and variance of the system reliability estimate based on the moments of component reliability estimates. Let $\hat{R}_s(t)$ be the reliability estimate of a series system in Eq. (2.3.3). The mean and the variance of $\hat{R}_s(t)$ can be computed as follows (Jin and Coit 2008):

$$E[\hat{R}_s(t)] = \prod_{i=1}^{m} E[\hat{r}_i(t)] \tag{2.3.10}$$

$$V\hat{a}r(\hat{R}_s(t)) = \prod_{i=1}^{m} \hat{r}_i^2(t) - \prod_{i=1}^{m}(\hat{r}_i^2(t) - V\hat{a}r(\hat{r}_i(t))) \tag{2.3.11}$$

Statistically, Eq. (2.3.11) measures the proportion of the unconditional variance of $\hat{R}_s(t)$ that can be attributed to the mean and the variance of $\hat{r}_i(t)$ for $i = 1, 2, \dots, m$. Both $E[\hat{R}_s(t)]$ and $V\hat{a}r(\hat{R}_s(t))$ can assist the design engineers in identifying the components that have the largest contribution to the system reliability variability. Thus remedies or proactive measures can be taken to mitigate the reliability risks by adopting more reliable or less uncertain components for system configuration.

Example 2.2 Show that Eqs. (2.3.3) and (2.3.11) are the unbiased estimate for $R_s(t)$ and $Var(\hat{R}_s(t))$, respectively.

Proof:
We first show that Eq. (2.3.3) is the unbiased estimate of series system reliability by computing its expected value as follows:

$$E[\hat{R}_s(t)] = \prod_{i=1}^{m} E[\hat{r}_i(t)] = \prod_{i=1}^{m} E\left[\frac{X_i}{n_i}\right] = \prod_{i=1}^{m} \frac{1}{n_i} E[X_i] = \prod_{i=1}^{m} r_i \tag{2.3.12}$$

The last step is obtained by realizing that $E[X_i] = n_i r_i$ where X_i follows the binomial distribution $B(n_i, r_i)$. This shows that $\hat{R}_s(t)$ is an unbiased estimate of $R_s(t)$ because $E[\hat{R}_s(t)]$ is the product of true component reliability r_i for $i = 1, 2, ..., m$.

To show that Eq. (2.3.11) is an unbiased estimate of $Var(\hat{R}_s)$, we compute the expected value of $V\hat{a}r(\hat{R}_s)$ as follows:

$$E[V\hat{a}r(\hat{R}_s(t))] = \prod_{i=1}^{m} E[\hat{r}_i^2(t)] - \prod_{i=1}^{m} E[\hat{r}_i^2(t) - V\hat{a}r(\hat{r}_i(t))]$$

$$= \prod_{i=1}^{m} \frac{1}{n_i^2} E[X_i^2] - \prod_{i=1}^{m} \left(\frac{1}{n_i^2} E[X_i^2] - E[V\hat{a}r(\hat{r}_i(t))]\right) \tag{2.3.13}$$

Substituting Eqs. (2.3.7) and (2.3.8) into Eq. (2.3.13) yields

$$E[V\hat{a}r(\hat{R}_s)] = \prod_{i=1}^{m} r_i^2 - \prod_{i=1}^{m} (r_i^2 - Var(\hat{r}_i)) = Var(\hat{R}_s) \tag{2.3.14}$$

This completes the proof that Eq. (2.3.11) is an unbiased variance estimate of $\hat{R}_s(t)$.

Example 2.3 A series system is comprised of four different components ($m = 4$). To estimate the system reliability, samples of individual components are taken and placed under a lifetime test over 100 hours, and the number of survival of each component is summarized in Table 2.1. Please estimate the mean and variance of the series system reliability estimate.

Solution:
The mean and variance of the component reliability estimate can be estimated from Eqs. (2.3.4) and (2.3.5). The results are presented in Table 2.2.

Next, we compute the mean of the system reliability estimate based on Eq. (2.3.10) as follows:

$$E[\hat{R}_s(t)] = (0.85)(0.8333)(0.92)(0.9375) = 0.611$$

Finally, the variance of the system reliability estimate is obtained by Eq. (2.3.11) as

$$V\hat{a}r(\hat{R}_s(t)) = (0.85)^2(0.8333)^2(0.92)^2(0.9375)^2$$
$$- [(0.85)^2 - 0.00671][(0.8333)^2 - 0.00479][(0.92)^2 - 0.00307]$$
$$[(0.9375)^2 - 0.00391]$$
$$= 0.00897$$

Table 2.1 Sample size and survivals at $t = 100$ hours.

Component type	$i = 1$	$i = 2$	$i = 3$	$i = 4$
Sample size n_i	20	30	25	16
Survivals k_i	17	25	23	15

Table 2.2 Mean and variance of component reliability estimates.

Component type	$i = 1$	$i = 2$	$i = 3$	$i = 4$
$E[\hat{r}_i(t)]$	0.850	0.833	0.920	0.938
$\hat{Var}(\hat{r}_i(t))$	0.00671	0.00479	0.00307	0.00391

We know that reliability of a series system is always lower than or at most equals its component reliability. However, the variance of the system reliability estimate is always larger than the variance of any component reliability estimates. For instance, $\hat{R}_s(t) = 0.611$ is smaller than $\hat{r}_2(t) = 0.833$, which is the smallest component reliability. However, $\hat{Var}(\hat{R}_s(t)) = 0.00897$ is larger than $\hat{Var}(\hat{r}_1(t)) = 0.00671$, which is the largest variance among the four components.

In general two types of uncertainties often arise in a reliability estimation; these are aleatory and epistemic uncertainties (Li et al. 2014). Aleatory uncertainty is the natural randomness in a process and is irreducible, such as operating conditions. The randomness is parameterized by the discrete and continuous probability function. A good example of aleatory uncertainty is the hourly wind speed in renewable generation and the solar irradiance on the photovoltaic (PV) system (Li and Zio 2012; Taboada et al. 2012), The epistemic uncertainty is the scientific uncertainty in the model of the process, largely due to limited data and knowledge. For instance, epistemic uncertainty in an unknown (yet a single correct) model parameter is caused by limited data or available information.

2.4 Parallel Systems

2.4.1 Reliability of Parallel Systems

A parallel system is one in which the proper operation of any one component or subsystem implies the functionality of the whole system. Typical examples of parallel systems include the multipump gas station, aircraft with dual engines, multilane highway, and distributed computing, among many others. Figure 2.6 shows the reliability block diagram of a parallel system that consists of m components, where $r_i(t)$ is the reliability of the ith component for $i = 1, 2, \ldots, m$.

A main advantage of parallel configuration is that the system can achieve fault-tolerant operation using redundant units even if the reliability of individual components is not

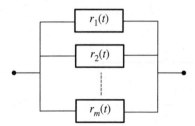

Figure 2.6 Reliability diagram for a parallel system.

high. For instance, an airplane powered with two engines is able to fly even if one of the engines fails. The reliability for a parallel system, denoted as $R_p(t)$, can be estimated using

$$R_p(t) = 1 - \prod_{i=1}^{m}(1 - r_i(t)) \tag{2.4.1}$$

For instance, a parallel system consists of three components with $r_1 = 0.5$, $r_2 = 0.6$, and $r_3 = 0.7$, respectively. The system reliability becomes $R_p = 1 - (1 - 0.5)(1 - 0.6)(1 - 0.7) = 0.94$. This example shows that high system reliability can be realized through the parallel configuration of low or medium reliability components.

2.4.2 Mean and Variance of Reliability Estimate

Let $\widehat{R}_p(t)$ be the estimate of the parallel system reliability and $\hat{r}_i(t)$ be the reliability estimate of component i for $i = 1, 2, \ldots, m$. Assuming component reliability, estimates are statistically independent. Based on Eq. (2.4.1), the mean and variance of $\widehat{R}_p(t)$ can be estimated as follows (Jin and Coit 2008):

$$E[\widehat{R}_p(t)] = 1 - \prod_{i=1}^{m}(1 - E[\hat{r}_i(t)]) \tag{2.4.2}$$

$$Var(\widehat{R}_p(t)) = \prod_{i=1}^{m}((1 - E[\hat{r}_i])^2 + Var(\hat{r}_i)) - \prod_{i=1}^{m}(1 - E[\hat{r}_i])^2 \tag{2.4.3}$$

It is worth mentioning that Eqs. (2.4.2) and (2.4.3) are unbiased estimates for $\widehat{R}_p(t)$ and $Var(\widehat{R}_p(t))$, respectively. This is left as an exercise for the readers.

Example 2.4 Suppose the four components in Table 2.1 are used to form a parallel system. Calculate the mean and variance of the reliability estimate of the parallel system.

Solution:
Note that $E[\hat{r}_i(t = 100)$ and $Var(\hat{r}_i(t = 100))$ for $i = 1, 2, 3$, and 4 are already calculated and available in Table 2.2. We directly substitute these data into Eqs. (2.4.2) and (2.4.3) to estimate the mean and variance of $\widehat{R}_p(t)$ at $t = 100$ hours.

$$E[\widehat{R}_p(t)] = 1 - (1 - 0.85)(1 - 0.8333)(1 - 0.92)(1 - 0.9375) = 0.99988$$

$$\begin{aligned}
Var(\widehat{R}_p(t)) = {}& [(1 - 0.85)^2 + 0.00671][(1 - 0.8333)^2 + 0.00479] \\
& \times [(1 - 0.92)^2 + 0.00307][(1 - 0.9375)^2 + 0.00391] \\
& - (1 - 0.85)^2(1 - 0.8333)^2(1 - 0.92)^2(1 - 0.9375)^2 \\
= {}& 5.473 \times 10^{-8}
\end{aligned}$$

Two observations are made here. First, the reliability of a parallel system is larger than any component reliability. Second, the reliability variance of a parallel system is smaller than the reliability variance of its components. For instance, $\hat{Var}(\hat{R}_p(t)) = 5.473 \times 10^{-8}$ is smaller than $\hat{Var}(\hat{r}_3(t)) = 0.00307$, which is the smallest among four components. This property is very important as it implies that parallel configuration not only increases the system reliability but also reduces the uncertainty of the system reliability.

2.5 Mixed Series and Parallel Systems

The systems discussed so far are pure series or pure parallel structures. There are many systems of which the components are configured in a mix of series and parallel modes. Generally these mixed structures can be classified into three categories: (i) series–parallel system, (ii) parallel–series system, and (iii) hybrid series and parallel system. This section presents the reliability estimation models for mixed configurations.

2.5.1 Series–Parallel System

A series–parallel system can be treated as a two-level hierarchical structure. At the system level multiple subsystems are connected in series. At the subsystem level, components are configured in a parallel structure. Figure 2.7 shows a series–parallel system that is comprised of three subsystems, and each subsystem itself consists of two or three components in parallel.

Let s be the number of subsystems and m_j be the number of components within subsystem j for $j = 1, 2, \ldots, s$. For instance, the series–parallel system in Figure 2.7 has $s = 3$, $m_1 = 2$, $m_2 = 2$, and $m_3 = 3$. The reliability estimate of a series–parallel system, denoted as $\hat{R}_{sp}(t)$, can be estimated by

$$\hat{R}_{sp}(t) = \prod_{j=1}^{s} \hat{R}_j(t) = \prod_{j=1}^{s}\left(1 - \prod_{i=1}^{m_j}(1 - \hat{r}_{ij}(t)) \right) \tag{2.5.1}$$

where $\hat{R}_j(t)$ is the reliability estimate of subsystem j and $\hat{r}_{ij}(t)$ is the reliability estimate of component i for $i = 1, 2, \ldots, m_j$ in subsystem j. Equation (2.5.1) shows that reliability estimation of a series–parallel system can be performed in two steps. In step one, we calculate the reliability of each subsystem based on parallel system reliability models

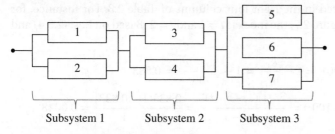

Figure 2.7 A series–parallel system.

in Eqs. (2.4.1) or (2.4.2). In step two, we multiply all the subsystem reliability models and obtain the whole system reliability using Eqs. (2.3.3) or (2.3.10). The mean of $\widehat{R}_{sp}(t)$, denoted as $E[\widehat{R}_{sp}(t)]$, can be obtained by substituting $\hat{r}_{ij}(t)$ with $E[\hat{r}_{ij}(t)]$ in Eq. (2.5.1) for all i and j. That is,

$$E[\widehat{R}_{sp}(t)] = \prod_{j=1}^{s} E[\widehat{R}_j(t)] = \prod_{j=1}^{s}\left(1 - \prod_{i=1}^{m_{ij}}(1 - E[\hat{r}_{ij}(t)])\right) \tag{2.5.2}$$

Let $Var(\widehat{R}_{sp})$ be the variance estimate of $\widehat{R}_{sp}(t)$. To obtain an unbiased estimate of $Var(\widehat{R}_{sp})$, Jin and Coit (2008) proposed a block decomposition method to propagate the variance of $\hat{r}_{ij}$ from the component level to the system level. Without providing the detailed derivation, the unbiased estimate for $Var(\widehat{R}_{sp})$ can be expressed as

$$\widehat{Var}(\widehat{R}_{sp}) = \prod_{j=1}^{s}\left(1 - \prod_{i=1}^{m_j}(1 - \hat{r}_{ij})\right)^2$$

$$- \prod_{j=1}^{s}\left(1 - 2\prod_{i=1}^{m_j}(1 - \hat{r}_{ij}) + \prod_{i=1}^{m_j}(\hat{r}_{ij}^2 - \widehat{Var}(\hat{r}_{ij}))\right)$$

$$= \prod_{j=1}^{s}\left(1 - \prod_{i=1}^{m_j}\hat{q}_{ij}\right)^2 - \prod_{j=1}^{s}\left(1 - 2\prod_{i=1}^{m_j}\hat{q}_{ij} + \prod_{i=1}^{m_j}(\hat{q}_{ij}^2 - \widehat{Var}(\hat{r}_{ij}))\right) \tag{2.5.3}$$

where $\hat{q}_{ij}$ is the unreliability estimate for component j in subsystem i, and $\hat{q}_{ij} = 1 - \hat{r}_{ij}$. Based on Goodman's (1960) results, Ramirez-Marquez and Jiang (2006) developed a recursive method to compute $Var(\widehat{R}_{sp})$, which yields the same result as the block decomposition in Equation (2.5.3). The idea is to sequentially reduce one component each time the series or parallel decomposition is applied.

Example 2.5 Let us compute $E[\widehat{R}_{sp}(t)]$ and $\widehat{Var}(\widehat{R}_{sp}(t))$ for the series–parallel system in Figure 2.7. Life testing is carried out over 200 hours on seven components. Failures of each component at $t = 100$ and 200 hours are recorded accordingly, and the results are listed in Table 2.3. Note that n_{ij} and k_{ij} are the sample size and failure amount of component j in subsystem i.

Solution:

Step 1: We estimate the mean and the variance of the component reliability estimate based on their failure data at $t = 100$ hours and 200 hours, respectively. The results are presented in the right four columns of Table 2.3. For instance, for component 2 in Figure 2.7, the index is $i = 1$ and $j = 2$. Based on Eqs. (2.3.4) and (2.3.5), we have

$$E[\hat{r}_{12}(t = 100)] = 1 - \frac{k_{12}}{m_{12}} = 1 - \frac{30 - 25}{30} = 0.833$$

$$\widehat{Var}(\hat{r}_{12}(t = 100)) = \frac{\hat{r}_{12}(t)(1 - \hat{r}_{12}(t))}{m_{12} - 1} = \frac{0.833(1 - 0.833)}{30 - 1} = 0.0048$$

Step 2: Since all subsystems belong to parallel structure, the mean and the variance of subsystem reliability estimates are obtained from Eqs. (2.4.2) and (2.4.3). For instance, to estimate the mean and variance of the reliability estimate of subsystem 2 at $t = 200$ hours, we have

$$E[\hat{R}_2(t = 200)] = 1 - (1 - 0.88)(1 - 0.875) = 0.985$$

$$\hat{Var}(\hat{R}_2(t = 200)) = [(1 - 0.88)^2 + 0.004][(1 - 0.875)^2 + 0.0073] - (1 - 0.88)^2(1 - 0.875)^2 = 2.058 \times 10^{-4}$$

Similarly, we compute the mean and variance of the reliability estimate of subsystems 1 and 3; the results are summarized in rows 2 to 4 of Table 2.4.

Step 3: Finally, we compute the mean and variance of the entire system reliability estimate based on Eqs. (2.3.10) and (2.3.11), respectively. The results are presented in the last row of Table 2.4. One can also estimate $E[\hat{R}_{sp}(t)]$ and $Var(\hat{R}_{sp}(t))$ by directly substituting the mean and variance of component reliability estimates from Table 2.3 into Eqs. (2.5.2) and (2.5.3). This will be left to the readers for a self-exercise.

Table 2.3 Component test data, reliability, and variance at $t = 100$ and 200 hours.

Component type	i	j	n_{ij}	k_{ij} ($t = 100$)	k_{ij} ($t = 200$)	$t = 100$		$t = 200$	
						$E[\hat{r}_{ij}(t)]$	$\hat{Var}(\hat{r}_{ij}(t))$	$E[\hat{r}_{ij}(t)]$	$\hat{Var}(\hat{r}_{ij}(t))$
1	1	1	20	17	16	0.850	0.0067	0.800	0.0084
2	1	2	30	25	24	0.833	0.0048	0.800	0.0055
3	2	1	25	23	22	0.920	0.0031	0.880	0.0044
4	2	2	16	15	14	0.938	0.0039	0.875	0.0073
5	3	1	20	17	16	0.850	0.0067	0.800	0.0084
6	3	2	22	20	19	0.909	0.0039	0.864	0.0056
7	3	3	18	17	16	0.944	0.0031	0.889	0.0058

Table 2.4 Mean and variance of reliability estimates for subsystems and system.

Subsystem i	$E[\hat{R}_i(t = 100)]$	$\hat{Var}(\hat{R}_i(t = 100))$	$E[\hat{R}_i(t = 200)]$	$\hat{Var}(\hat{R}_i(t = 200))$
1	0.975	3.263×10^{-4}	0.960	6.040×10^{-4}
2	0.995	4.896×10^{-5}	0.985	2.058×10^{-4}
3	0.999	1.626×10^{-6}	0.997	1.209×10^{-5}
System	0.969	3.71×10^{-4}	0.9427	7.82×10^{-4}

2.5.2 Parallel–Series System

For a parallel–series configuration, the system can be treated as an architecture with multiple subsystems configured in parallel. At the subsystem level, each subsystem is comprised of multiple components in series. Figure 2.8 shows a typical parallel–series system of which three subsystems 1, 2, and 3 are placed in parallel. Each subsystem is formed by two or three components connected in series. For instance, Subsystem 2 is formed by components 3, 4, and 5 in series.

Let s be the number of subsystems and m_j be the number of components in subsystem j for $j = 1, 2, \ldots, s$. The reliability of the parallel–series system is given as

$$R_{ps}(t) = 1 - \prod_{j=1}^{s}(1 - R_j(t)) = 1 - \prod_{j=1}^{s}\left(1 - \prod_{i=1}^{m_j} r_{ij}(t)\right) \tag{2.5.4}$$

where $R_j(t)$ is the reliability of subsystem j and $r_{ij}(t)$ is the reliability of component i in subsystem j. Let $\widehat{R}_{ps}(t)$ be the estimate of $R_{ps}(t)$. To obtain the mean of $\widehat{R}_{ps}(t)$, we substitute $r_{ij}(t)$ with its mean value $E[\widehat{r}_{ij}(t)]$ in Eq. (2.5.4); that is,

$$E[\widehat{R}_{ps}(t)] = 1 - \prod_{j=1}^{s}(1 - E[\widehat{R}_j(t)]) = 1 - \prod_{j=1}^{s}\left(1 - \prod_{i=1}^{n_j} E[\widehat{r}_{ij}(t)]\right) \tag{2.5.5}$$

Similarly, the estimate of $Var(\widehat{R}_{ps})$ is given as follows:

$$\widehat{Var}(\widehat{R}_{ps}(t)) = \prod_{j=1}^{s}\left(1 - \prod_{i=1}^{m_j} \widehat{r}_{ij}(t)\right)^2$$
$$- \prod_{j=1}^{s}\left(1 - 2\prod_{i=1}^{m_j} \widehat{r}_{ij}(t) + \prod_{i=1}^{m_j}(\widehat{r}_{ij}^2(t) - \widehat{Var}(\widehat{r}_{ij}(t)))\right) \tag{2.5.6}$$

Both Eqs. (2.5.5) and (2.5.6) are the unbiased estimates for $\widehat{R}_{ps}(t)$ and $Var(\widehat{R}_{ps})$, respectively. Again, we leave this as exercises for the readers.

2.5.3 Mixed Series–Parallel System

A mixed series–parallel system can be treated as a hierarchical structure that is embedded with series–parallel and/or parallel–parallel subsystems at different layers. Figure 2.9 shows a typical mixed series–parallel system comprised of four subsystems with a total of seven components. Subsystems 1 and 2 are pure parallel systems.

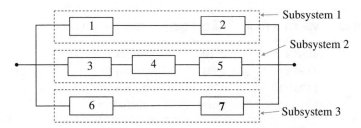

Figure 2.8 A parallel–series system.

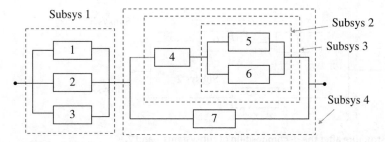

Figure 2.9 A mixed series–parallel system with seven components.

Subsystem 3 is a series–parallel structure with component 4 and subsystem 2 in series. Subsystem 4 is equivalent to a parallel–series structure where component 7 is in parallel with subsystem 3.

Given the reliability or the lifetime data at component levels, different methods such as series and parallel reduction can be applied to calculate the reliability of the mixed series–parallel system. Let $\hat{R}_{sys}$ be the reliability estimate of a mixed series–parallel system. A procedure for estimating $E[\hat{R}_{sys}]$ and $V\hat{a}r(\hat{R}_{sys})$ for any systems that can be decomposed into series, parallel, or series–parallel structures are given as follows:

Step 1: Identify the reliability blocks (i.e. subsystems) based on series and parallel decomposition rules.
Step 2: Begin with the most embedded subsystem. If it is a series configuration, compute the mean and the variance of the subsystem reliability estimate using Eqs. (2.3.10) and (2.3.11). If it is parallel, compute the mean and the variance of the subsystem reliability estimate using Eqs. (2.4.2) and (2.4.3).
Step 3: Move to the upper level subsystem and apply series and parallel reduction recursively as in Step 2 until $E[\hat{R}_{sys}]$ and $V\hat{a}r(\hat{R}_{sys})$ are obtained.

Example 2.6 The purpose of this example is to show how sequential series and parallel reduction (decomposition) can be applied to compute the mean and variance of the reliability estimate of mixed series–parallel systems in Figure 2.9. We use the component reliability data in Table 2.3 to compute $E[\hat{R}_{sys}(t)]$ and $V\hat{a}r(\hat{R}_{sys}(t))$ at $t = 100$ hours.

Solution:
Step 1: We first compute the mean and the variance of reliability estimates for subsystems 1 and 2 at $t = 100$ hours, and the resulting structure is shown in Figure 2.10. Since both are parallel subsystems, the mean and variance are obtained from Eqs. (2.4.2) and (2.4.3), respectively:

$$E[\hat{R}_1(100)] = 1 - (1 - 0.85)(1 - 0.8333)(1 - 0.92) = 0.998$$

$$\begin{aligned}
V\hat{a}r(\hat{R}_1(100)) &= [(1 - 0.85)^2 + 0.00671][(1 - 0.8333)^2 + 0.00479] \\
&\quad \times [(1 - 0.92)^2 + 0.00307] - (1 - 0.85)^2(1 - 0.8333)^2 \\
&\quad \times (1 - 0.92)^2 \\
&= 5.006 \times 10^{-6}
\end{aligned}$$

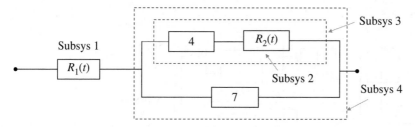

Figure 2.10 System structure after the decomposition of subsystems 1 and 2.

$$E[\widehat{R}_2(100)] = 1 - (1 - 0.85)(1 - 0.909) = 0.986$$

$$\begin{aligned}Vâr(\widehat{R}_2(100)) &= [(1 - 0.85)^2 + 0.00671][(1 - 0.909)^2 + 0.0039] \\ &\quad - (1 - 0.85)^2(1 - 0.909)^2 \\ &= 1.7 \times 10^{-5}\end{aligned}$$

Step 2: Since component 4 and subsystem 2 are in series, the system structure can be further reduced as shown in Figure 2.11. Equations (2.3.10) and (2.3.11) are used to compute the mean and the variance of the reliability of subsystem 3:

$$E[\widehat{R}_3(100)] = (0.938)(0.986) = 0.925$$

$$\begin{aligned}Vâr(\widehat{R}_3(100)) &= (0.938)^2(0.986)^2 - [(0.938)^2 - 0.0039] \\ &\quad [(0.986)^2 - 0.00017] \\ &= 0.01577\end{aligned}$$

Step 3: Since subsystem 4 comprises component 7 and subsystem 3 in parallel, the system can be further reduced and becomes a series system as shown in Figure 2.12:

$$E[\widehat{R}_4(100)] = 1 - (1 - 0.925)(1 - 0.944) = 0.993$$

$$\begin{aligned}Vâr(\widehat{R}_4(100)) &= [(1 - 0.925)^2 + 0.001577][(1 - 0.944)^2 + 0.0031] \\ &\quad - (1 - 0.925)^2(1 - 0.944)^2 \\ &= 0.0002147\end{aligned}$$

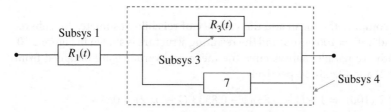

Figure 2.11 System structure after the decomposition of component 4 and subsystem 2.

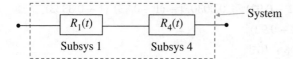

Figure 2.12 System structure after the decomposition of component 7 and subsystem 3.

Step 4: Finally, we compute the mean and the variance of the original system using Eqs. (2.3.10) and (2.3.11) by realizing that subsystems 1 and 4 are in series. The results are given below:

$$E[\hat{R}_{sys}(100)] = 1 - (1 - 0.998)(1 - 0.993) = 0.991$$

$$\hat{Var}(\hat{R}_{sys}(100)) = [(1 - 0.998)^2 + 5 \times 10^{-6}][(1 - 0.993)^2 + 2.147 \times 10^{-4}]$$
$$- (1 - 0.998)^2(1 - 0.993)^2$$
$$= 0.000219$$

2.6 Systems with *k*-out-of-*n*:G Redundancy

2.6.1 Reliability for Hot-Standby Redundant Systems

A *k*-out-of-*n*:G system, or simply called *k*-out-of-*n* system, is the one in which the proper function of any *k* components (note that $k \leq n$) or larger in the system implies the proper system function. A *k*-out-of-*n*:G system is equivalent to an $(n - k + 1)$-out-of-*n*:F system, where the system fails if $n - k + 1$ components are malfunctioning. To guarantee reliability and safety, wide-body passenger aircraft such as Boeing 747 and Airbus 380 are powered with 2-out-4 engine systems. The plane can fly reliably as long as two engines are working. In the power industry the electric grid is required to meet the criterion of $N - 1$ fault tolerance, meaning the utility company is still able to meet the demand at the loss or failure of the largest generating unit.

Three types of *k*-out-of-*n* systems are generally used in practical applications: hot standby, warm standby, and cold standby (Elsayed 2012). In hot standby mode, all *n* components are in operating mode, and the load is shared among themselves. For instance, the engine cluster in Boeing 747 belongs to the 2-out-4 hot standby redundancy as four engines simultaneously generate thrust during the flight. In warm standby, only *k* components undertake the load. The other $n - k$ components, though in online mode, do not share the load. A redundancy array of independent disks (RAID) in the data center and the dual power distribution lines (primary and secondary) are typical examples of warm standby systems. Cold standby differs from hot standby and warm standby in that all of the redundant components are in off-line mode. They need to be powered up and switched to the operating mode upon request. A good example of a cold standby system is the back-up power generator on a university campus. For a repairable system, the inventory spare parts can be treated as the cold standby unit. If the field system fails, the faulty unit can be replaced with the spare part, and the system can return to the working state in a timely manner.

Let $r(t)$ be the component reliability in the *k*-out-of-*n*:G system in a hot standby mode. Assume all components are identical. The reliability of the redundant system, denoted as $R_s(t)$, can be estimated by

$$R_s(t) = \sum_{i=k}^{n} \binom{n}{i} r^i(t)(1 - r_i(t))^{n-i} \qquad (2.6.1)$$

It is worth mentioning that series and parallel systems are special cases of *k*-out-of-*n* redundant systems. If $k = n$, Eq. (2.6.1) reduces to the series system reliability model in Eq. (2.3.3). If $k = 1$, Eq. (2.6.1) is equivalent to the parallel system reliability model in Eq. (2.4.1).

For a redundant system with warm standby components, Amari et al. (2012) derived a state-space approach to reliability characterization for a k-out-of-n warm standby system with identical components subject to exponential failure. In addition, they provide closed-form expressions for the hazard rate, probability density function, and mean residual life function. Wu et al. (2017) developed a new modeling approach for availability evaluation of a k-out-of-n:G warm standby redundancy system with repair. They showed that the performance evaluation process algebra turns out to be an efficient way to deal with the availability evaluation of systems with different groups of repairable components.

2.6.2 Application to Data Storage Systems

A redundant array of independent disks (RAID) is a data storage virtualization technology. It combines multiple disk drives to form a single logical unit for the purposes of data redundancy, parallel throughput, and information retrieval in large data centers. Figure 2.13 depicts the configuration of a RAID-5 system with four disks connected in parallel. It employs one redundant disk to achieve fault tolerance that allows up to one disk failure. For example, a data file "A" is split into multiple segments A1, A2, and A3, each stored on Disks 1, 2, and 3, respectively. Disk 4 is used to store the parity codes of file "A", denoted as PA (parity array). If one block of data fails to be retrieved from any one of Disks 1, 2, and 3, the parity codes in Disk 4 are used to reconstruct the damaged data sector by performing the reverse Boolean operation. During the reconstruction process, the bad sector of the data disk is mapped out. After the system rebuilds the damaged sector using the data from remaining good disks, the reconstructed data are saved in a good, yet unused, sector on the same disk. This self-recovery process is called media repair or data restoration. Therefore, all stored data in the RAID-5 system can be retrieved as long as three disks are functioning. Hence, a RAID-5 system can be treated as a 3-out-of-4 hot standby redundant system.

The working principle of RAID-6 is similar to RAID-5 except that it has five disk drives and two of them are used as redundant disks for data recovery. Thus RAID-6 can be treated as a 3-out-of-5 hot standby redundant system.

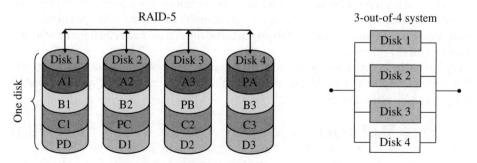

Figure 2.13 A RAID-5 system and its reliability block diagram.

Example 2.7 Assume the reliability of a single disk drive in RAID-5 and RAID-6 follows the exponential distribution with failure rate $\lambda = 10^{-6}$/hour. Compute:

1) The reliability of RAID-5 and RAID-6 after operating in one year.
2) The reliability of RAID-5 and RAID-6 after operating in 100 years.

Solution:

(1) First we compute the reliability of a single disk (i.e. component) at $t = 1$ year. To simplify the analysis, $\lambda = 10^{-6}$/hour is converted into $\lambda = 0.00876$/year:

$$r(t = 1) = \exp(-0.00876 \times 1) = 0.9913$$

RAID-5 is a 3-out-of-4 system with $k = 3$ and $n = 4$. Substituting $r(t = 1) = 0.9913$ into Eq. (2.6.1), we have

$$
\begin{aligned}
R_{RAID-5}(t = 1) &= \sum_{i=3}^{4} \binom{4}{i} 0.9913^i (1 - 0.9913)^{5-i} \\
&= \frac{4!}{3!(4-3)!} 0.9913^3 (1 - 0.9913)^1 \\
&\quad + \frac{4!}{4!(4-4)!} 0.9913^4 (1 - 0.9913)^0 \\
&= 0.99955
\end{aligned}
$$

RAID-6 is a 3-out-of-5 system with $k = 3$ and $n = 5$. Using Eq. (2.6.1), we have

$$
\begin{aligned}
R_{RAID-6}(t) &= \sum_{i=3}^{5} \binom{5}{i} 0.9913^i (1 - 0.9913)^{5-i} \\
&= 0.999\,993
\end{aligned}
$$

Obviously, adding one more redundant disk makes the reliability of RAID-6 100 times better than RAID-5 at the end of one year.

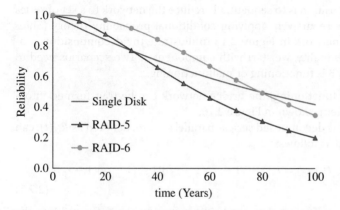

Figure 2.14 Reliability for RAID-5 and RAID-6 from years 1 to 100.

(2) Now we compute the reliability of two RAID systems over 100 years; the results are depicted in Figure 2.14. The reliability of RAID-5 and RAID-6 drops to 0.2 and 0.35 at $t = 100$ years, respectively. A common approach to preserving the long-time digital information is to regularly replace old disks with new disks as time evolves.

An interesting observation from Figure 2.14 is that the reliability of a single disk drive remains the lowest between year 1 and 30. It then exceeds RAID-5 from year 30, and further exceeds RAID-6 after year 80. This interesting phenomenon shows that for long-time data storage, placing a data file in a single disk seems to more reliable than splitting the file across multiple disks using segmentation.

2.7 Network Systems

2.7.1 Edge Decomposition

There are lots of applications where network architecture is required, such as highway transportation, power transmission, telecommunications, Internet, blood vessels of human beings, and integrated circuits. A network consists of a set of edges and nodes, and any node is connected by at least one edge. Figure 2.15 shows a five-component bridge network, while most real-world networks are far more complicated with dozens or even hundreds of thousands of components that are interconnected. Network reliability analysis deals with determining the reliability measures required for stochastic edge or node failures. Typical measures include the two-terminal, the k-terminal, and the all-terminal reliability models. Most network reliability analysis problems are as hard as the NP-complete problems (Ball 1986), namely, the computational time grows exponentially with the size of the network.

In this section, we first present the edge decomposition method as it provides the exact solution to the two-terminal network reliability estimation. Then we introduce two approximation methods: minimum path and minimum cut. The minimum path always results in the lower bound of the network reliability while the minimum cuts yield the upper bound of the network reliability estimation (Colbourn 1987). Finally, we discuss a linear-quadratic minim cut model to estimate the mean and the variance of the two-terminal network reliability estimate.

The idea of edge decomposition is to sequentially reduce the network to a set of series and/or parallel systems by recursively applying conditional probability theory (Shier 1991). We use the bridge network in Figure 2.15 to illustrate the decomposition procedure. Without loss of generality, we start with component 3. Two scenarios need to be considered: component 3 is functioning or malfunctioning.

Step 1: If component 3 is functioning, the bridge network can be transformed into a series–parallel system, shown in Figure 2.16.

The reliability of this decomposed series–parallel system, denoted as $R_{s1}(t)$, can be readily obtained as follows:

$$R_{s1}(t) = r_3 R_{s1}(t \mid r_3) = r_3[1 - (1 - r_1)(1 - r_2)][1 - (1 - r_4)(1 - r_5)]$$

$$(2.7.1)$$

Step 2: If component 3 is malfunctioning, the bridge network is transformed into the following mixed series and parallel system.

Figure 2.15 A simple bridge network with five components.

Figure 2.16 A series–parallel system conditioned on a function of component 3.

Figure 2.17 A mixed series and parallel system conditioned on the failure of component 3.

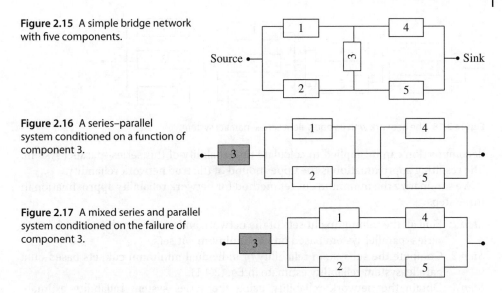

The reliability of the decomposed system in Figure 2.17, denoted as $R_{s2}(t)$, is given as follows:

$$R_{s2}(t) = q_3 R_{s2}(t \mid q_3) = (1 - r_3)[1 - (1 - r_1 r_4)(1 - r_2 r_5)] \tag{2.7.2}$$

where q_3 is the unreliability of component 3 and $q_3 = 1 - r_3$. Finally, the reliability of the original network is obtained by aggregating $R_{s1}(t)$ and $R_{s2}(t)$. That is,

$$
\begin{aligned}
R_{net}(t) &= R_{s1}(t) + R_{s2}(t) \\
&= r_1 r_4 + r_2 r_5 + r_1 r_3 r_5 + r_2 r_3 r_4 - r_1 r_3 r_4 r_5 \\
&\quad - r_2 r_3 r_4 r_5 - r_1 r_2 r_3 r_5 - r_1 r_2 r_3 r_4 - r_1 r_2 r_4 r_5 + 2 r_1 r_2 r_3 r_4 r_5
\end{aligned}
\tag{2.7.3}
$$

Though the principle of edge decomposition is straightforward, the number of substructures resulted from the decomposition often grow exponentially with the size of the network. Next we introduce minimum cut and minimum path approaches that can handle large-size networks more efficiently.

2.7.2   Minimum Cut Set

A cut vector, **y**, is a set of component status for which the corresponding system reliability becomes zero provided all these components fail. For example, in Figure 2.15 the component set $\mathbf{y} = \{1, 2, 3\}$ is a cut vector because failures of all these components render the network down. Other vectors such as $\{1, 3, 4, 5\}$ and $\{2, 3, 4, 5\}$ are also cut vectors.

A *minimum cut vector,* **x**, is a cut vector for which any vector $\mathbf{y} > \mathbf{x}$ leads to non-operation of the system. For example, component set $\mathbf{x} = \{1, 2\}$ is a minimum cut vector. If these components are not functional, the network is down. However, $\mathbf{y} = \{1, 2, 3\}$ is a cut vector, but not the minimum cut. This is because set $\{1, 2\}$ is sufficient to bring the network down, and component 3 is not needed. Other vectors such as $\{4, 5\}$, $\{1, 3, 5\}$, and $\{2, 3, 4\}$ are also minimum cuts. Given all these minimum cut sets, the bridge network can be transformed into the series–parallel system shown in Figure 2.18, where each cut set forms a parallel subsystem. Now the block

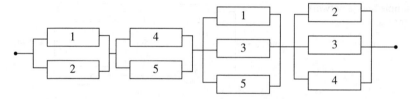

Figure 2.18 The network with its equivalent series–parallel system.

decomposition can be applied to calculate the reliability of this series–parallel system. The resulting approximation is the upper bound of the true network reliability.

We summarize the minimum cut set method for network reliability approximation in three steps:

Step 1: Find all the minimum cut sets of the network and transform the network into a series–parallel system based on the minimum cut sets.

Step 2: Compute the subsystem reliability of individual minimum cut sets based on a parallel system reliability estimate in Eq. (2.4.1).

Step 3: Obtain the network reliability using the series system reliability estimate in Eq. (2.3.3).

Example 2.8 Based on the component reliability data in Table 2.5, estimate the bridge network reliability based on minimum cut (MC) sets.

Solution:

Step 1: The bridge network in Figure 2.15 has four minimum cut sets, which are listed in Table 2.6.

Step 2: Compute the reliability of subsystems 1, 2, 3, and 4 as parallel configurations; the results are listed in the third column of Table 2.6. For example:
$$R_1 = 1 - (1 - 0.9) \times (1 - 0.8) = 0.98$$
$$R_3 = 1 - (1 - 0.9) \times (1 - 0.7) \times (1 - 0.5) = 0.985$$

Step 3: Compute the network reliability using the series system model in Eq. (2.3.3):
$$R_{net}(t) = 0.98 \times 0.8 \times 0.985 \times 0.97 = 0.749$$

Table 2.5 Component reliability.

Component	1	2	3	4	5
Reliability	0.9	0.8	0.7	0.6	0.5

Table 2.6 Reliability of minimum cut sets.

Subsystem i	Minimum cut set	Reliability R_i
1	$C_1 = \{1, 2\}$	0.98
2	$C_2 = \{4, 5\}$	0.8
3	$C_3 = \{1, 3, 5\}$	0.985
4	$C_4 = \{2, 3, 4\}$	0.97

The true reliability of the bridge network can be obtained from Eq. (2.7.3); that is,

$$R_{true}(t) = (0.9)(0.6) + (0.8)(0.5) + (0.9)(0.7)(0.6) + (0.8)(0.7)(0.5)$$
$$- (0.9)(0.7)(0.6)(0.5) - (0.8)(0.7)(0.6)(0.5)$$
$$+ 2 \times (0.9)(0.8)(0.7)(0.6)(0.5)$$
$$= 0.766$$

Note that $R_{net}(t) < R_{true}(t)$, the minimum cut results in a lower bound of the network reliability.

2.7.3 Minimum Path Set

A *path vector*, **y**, is a component status vector for which the corresponding system reliability is assured provided all these components are operational. For example, component set **y** = {1, 2, 3, 5} is a path vector. Namely, the network is functional if all these components are working. Other vectors such as {1, 3, 4, 5} and {2, 3, 4, 5} are also path vectors.

A minimum path vector, **x**, is a path vector for which any vector **x** < **y** has a corresponding zero system reliability. For example, component set **x** = {1, 4} is a minimum path vector. If these components are working, the network is operational. If we remove any component from set **x**, the system is not always functional regardless of the reliability of remaining components. Set **y** = {1, 3, 4} is a path set, but not a minimum path set. The reason is if component 3 is malfunctioning, the path formed by {1, 4} still ensures the reliable connection between the source and the terminal nodes. Other vectors such as {1, 5}, {1, 3, 5}, and {2, 3, 4} are also minimum path vectors. Based on the minimum path sets, the bridge network can be transformed into a parallel–series system, as shown in Figure 2.19.

We summarize the procedure of estimating network reliability using the minimum path sets in three steps:

Step 1: Find all the minimum path sets of the network and transform the network into a parallel–series system based on minimum path vectors.
Step 2: Compute the reliability of individual path vectors using the series system reliability model in Eq. (2.3.3).
Step 3: Obtain the network reliability using the parallel system reliability model in Eq. (2.4.1).

Example 2.9 Estimate the reliability of the bridge network using the component data in Table 2.5. Does the minimum path yield an upper bound of the network reliability?

Figure 2.19 Network with its equivalent parallel–series system.

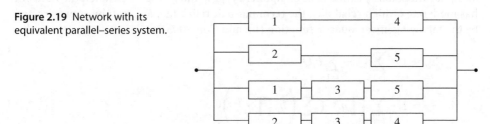

Table 2.7 Reliability of minimum path sets.

Subsystem i	Minimum path sets	Reliability R_i
1	$P_1 = \{1, 4\}$	0.54
2	$P_2 = \{2, 5\}$	0.4
3	$P_3 = \{1, 3, 5\}$	0.315
4	$P_4 = \{2, 3, 4\}$	0.336

Solution:

Step 1: The bridge network in Figure 2.15 has four minimum path sets as denoted by P_i for $i = 1, 2, 3,$ and 4.

Step 2: Compute the reliability of each subsystems using the series model. The results are shown in Table 2.7.

Step 3: Compute the network reliability based on the parallel system model in Eq. (2.4.1).

$$R_{net}(t) = 1 - (1 - 0.54) \times (1 - 0.4) \times (1 - 0.315) \times (1 - 0.336) = 0.874$$

Since $R_{net}(t)$ is larger than the true value $R_{true}(t) = 0.766$, this confirms that the minimum path set generates an upper bound of the network reliability.

2.7.4 Linear-Quadratic Approximation to Terminal-Pair Reliability

Similar to series or parallel systems, one might be also interested in modeling and computing the variance of the network reliability estimate. If the terminal-pair reliability can be explicitly expressed as a function of component reliability estimates, the variance of the network reliability estimate can be directly computed based on the mean and variance of component reliability estimates. For simple or small networks, it is not difficult to obtain the analytical network reliability estimate. For instance, Eq. (2.7.3) represents the explicit reliability estimate of the two-terminal bridge network. Deriving an explicit network reliability estimate becomes extremely difficult or time consuming as the size of the network increases. Below we present an approximation method to compute the variance of the terminal-pair reliability estimate based on the minimum cuts of component reliability estimates.

Let C_i be the minimum cut set i for $i = 1, 2, \ldots, s$ of a network system and k be the index for the component type in the network for $k = 1, 2, \ldots, n$. Let h_{ik} be the number of component type k within cut set C_i, where $h_{ik} \in \{1, 2, \ldots\}$. For example, the minimum cut sets of the bridge network is given in Table 2.6, where $s = 4$ and $h_{ik} = 1$ for all i and k. Let q_k be the unreliability estimate of component type k and $q_k = 1 - r_k$. When the network has moderate or high reliability, the terminal-pair reliability, R_{net}, can be approximated by the following linear-quartic function (Jin and Coit 2003):

$$R_{net} \approx 1 - \sum_{i=1}^{s} Q_i + \sum_{i=1}^{s} b_i Q_i^2$$

$$= 1 - \sum_{i=1}^{s} \prod_{k \in C_i} q_k^{h_{ik}} + \sum_{i=1}^{s} \left(b_i \prod_{k \in C_i} q_k^{2h_{ik}} \right) \tag{2.7.4}$$

with

$$Q_i = \prod_{k \in C_i} q_k^{h_{ik}}, \text{ unreliability of minimum cut set } C_i \qquad (2.7.5)$$

$$b_i = \sum_{j=1, j \neq i}^{s} a_{ij}, \text{ with } a_{ij} = \min\{Q_i/Q_j/Q_i\} \qquad (2.7.6)$$

Since terminal-pair reliability is approximated using the linear and quadratic terms of Q_i, this approach is referred to as the linear-quadratic reliability (LQR) model. Though the LQR model is built upon minimum cuts, due to the truncation of higher-order terms, it may not always guarantee the lower bound in the entire reliability range. However, when components or network reliability are high, which is true for most real-world applications, this situation rarely occurs.

Example 2.10 Estimate the terminal-pair reliability of the network in Figure 2.1d using minimum cut, minimum path, and the LQR model, and compare with the edge decomposition method.

Solution:
Edge decomposition. We first compute the true terminal-pair reliability using the edge decomposition. The network has eight components, and we begin with components 3 and 6. Four cases are considered depending on their reliability state.

Case 1: If components 3 and 6 are malfunctioning, the original network becomes a parallel–series system in Figure 2.20a. This is a regular parallel-series system, and its terminal-pair reliability is given as

$$R_1 = (1 - r_3)(1 - r_6)[1 - (1 - r_1 r_4 r_7)(1 - r_2 r_5 r_8)]$$

Case 2: If component 3 is malfunctioning while component 6 is working, the network becomes a mixed system in Figure 2.20b. The terminal pair reliability is

$$R_2 = (1 - r_3)r_6[1 - (1 - r_1 r_4)(1 - r_2 r_5)][1 - (1 - r_7)(1 - r_8)]$$

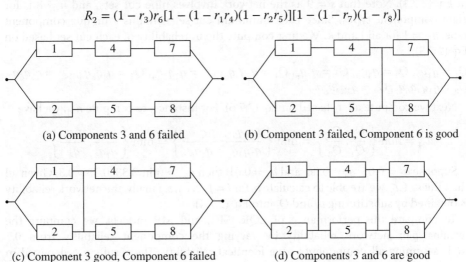

(a) Components 3 and 6 failed   (b) Component 3 failed, Component 6 is good

(c) Component 3 good, Component 6 failed   (d) Components 3 and 6 are good

Figure 2.20 Network under different status of components 3 and 6.

Case 3: If component 3 is working while component 6 is not, the network becomes a mixed system in Figure 2.20c. The terminal pair reliability is

$$R_3 = r_3(1 - r_6)[1 - (1 - r_1)(1 - r_2)][1 - (1 - r_4 r_7)(1 - r_5 r_8)]$$

Case 4: If components 3 and 6 are both working, the network becomes a series–parallel system in Figure 2.20d. The terminal pair reliability is

$$R_4 = r_3 r_6[1 - (1 - r_1)(1 - r_2)][1 - (1 - r_4)(1 - r_5)][1 - (1 - r_7)(1 - r_8)]$$

Finally, the exact terminal-pair reliability of the original network is obtained as

$$R_{true} = R_1 + R_2 + R_3 + R_4$$

Minimum paths (MPs): The minimum path sets of the original network are $P_1 = \{1, 4, 7\}$, $P_2 = \{2, 5, 8\}$, $P_3 = \{1, 3, 5, 8\}$, $P_4 = \{2, 3, 4, 7\}$, $P_5 = \{1, 4, 6, 8\}$, $P_6 = \{2, 5, 6, 7\}$, $P_7 = \{1, 3, 5, 6, 7\}$, and $P_8 = \{2, 3, 4, 6, 8\}$. The terminal-pair network is translated into a parallel–series system with its reliability as

$$R_{path} = 1 - (1 - r_1 r_4 r_7)(1 - r_2 r_5 r_8)(1 - r_1 r_3 r_5 r_8)(1 - r_2 r_3 r_4 r_7)(1 - r_1 r_4 r_6 r_8)$$
$$\times (1 - r_2 r_5 r_6 r_7)(1 - r_1 r_3 r_5 r_6 r_7)(1 - r_2 r_3 r_4 r_6 r_8)$$

Minimum cuts (MCs): The minimum cut sets of the network are $C_1 = \{1, 2\}$, $C_2 = \{4, 5\}$, $C_3 = \{7, 8\}$, $C_4 = \{1, 3, 5\}$, $C_5 = \{2, 3, 4\}$, $C_6 = \{4, 6, 8\}$, $C_7 = \{5, 6, 7\}$, $C_8 = \{1, 3, 6, 8\}$, and $C_9 = \{2, 3, 6, 7\}$. Hence, the terminal-pair network is translated into a series–parallel system with its reliability as

$$R_{cut} = (1 - q_1 q_2)(1 - q_4 q_5)(1 - q_7 q_8)(1 - q_1 q_3 q_5)(1 - q_2 q_3 q_4)$$
$$\times (1 - q_4 q_6 q_8)(1 - q_5 q_6 q_7)(1 - q_1 q_3 q_6 q_8)(1 - q_2 q_3 q_6 q_7)$$

where q_i is the unreliability of component i with $q_i = 1 - r_i$ for $i = 1, 2, \ldots, 8$.

LQR model: The above minimum cut sets are used as the input to the LQR model in Eq. (2.7.4). Note that $s = 9$ as the network involves nine cut sets, and $n = 8$ is for eight component types. Since each minimum cut set contains a unique component type, $h_{ik} = 1$ for all i and k. We first compute the unreliability of each cut set based on Eq. (2.7.5),

$Q_1 = q_1 q_2$, $Q_2 = q_4 q_5$, $Q_3 = q_7 q_8$, $Q_4 = q_1 q_3 q_5$, $Q_5 = q_2 q_3 q_4$, $Q_6 = q_4 q_6 q_8$, $Q_7 = q_5 q_6 q_7$, $Q_8 = q_1 q_3 q_6 q_8$, $Q_9 = q_2 q_3 q_6 q_7$.

Next we compute all a_{ij} based on Eq. (2.7.6). For instance, to compute a_{24}, we have

$$a_{24} = \min\left\{\frac{Q_2}{Q_4}, \frac{Q_4}{Q_2}\right\} = \min\left\{\frac{q_4 q_5}{q_1 q_3 q_5}, \frac{q_1 q_3 q_5}{q_4 q_5}\right\} = \min\left\{\frac{q_4}{q_1 q_3}, \frac{q_1 q_3}{q_4}\right\}$$

Suppose $q_1 = 0.05$, $q_3 = 0.06$, and $q_4 = 0.01$; then $a_{24} = \min\{3.33, 0.3\} = 0.3$. Given all the values of a_{ij}, we are able to calculate b_i for $i = 1, 2, \ldots, s$. Finally, the network reliability is obtained by substituting b_i and Q_i into Eq. (2.7.4).

To compare the performance of LQR, MC, and MP models, we compute the terminal-pair network reliability by varying the component reliability from 0.7 to 1, assuming all components have identical reliability. The results are depicted in Figure 2.21 along with the exact terminal-pair network reliability. As expected, the MP always overestimates the reliability, while the MC always underestimates the network

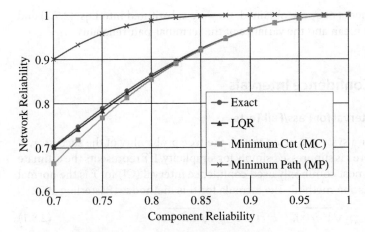

Figure 2.21 Compare with the exact network reliability.

reliability. We find that LQR in this case also underestimates the reliability, but is closer to the true reliability than MC. The approximation quality of both LQR and MC increases with the component reliability, but LQR converges faster than MC. Similar observations can be made when LQR is applied to other network structures (Jin and Coit 2003).

2.7.5 Moments of Terminal-Pair Reliability Estimate

The network reliability estimate, $\widehat{R}_{net}$, is obtained by substituting q_k with $\widehat{q}_k$ in Eq. (2.7.4). Furthermore, the mean and the variance of $\widehat{R}_{net}$ are obtained as follows:

$$E[\widehat{R}_{net}] \approx 1 - \sum_{i=1}^{s} \left(\prod_{k \in C_i} E[\widehat{q}_k^{h_{ik}}] - b_i \prod_{k \in C_i} E[\widehat{q}_k^{2h_{ik}}] \right) \tag{2.7.7}$$

$$Var(\widehat{R}_{net}) \approx \sum_{i=1}^{s} \sum_{j=1}^{s} \left[\prod_{k \in C_{ij}} E[\widehat{q}_k^{h_{ik}+h_{jk}}] - \prod_{k \in C_{ij}} E[\widehat{q}_k^{h_{ik}}] E[\widehat{q}_k^{h_{jk}}] \right]$$

$$+ \sum_{i=1}^{s} \sum_{j=1}^{s} b_i b_j \left(\prod_{k \in C_{ij}} E[\widehat{q}_k^{2h_{ik}+2h_{jk}}] - \prod_{k \in C_{ij}} E[\widehat{q}_k^{2h_{ik}}] E[\widehat{q}_k^{2h_{jk}}] \right)$$

$$- 2 \sum_{i=1}^{s} \sum_{j=1}^{s} b_i \left(\prod_{k \in C_{ij}} E[\widehat{q}_k^{h_{ik}+2h_{jk}}] - \prod_{k \in C_{ij}} E[\widehat{q}_k^{h_{ik}}] E[\widehat{q}_k^{2h_{jk}}] \right) \tag{2.7.8}$$

Both equations are obtained by realizing that reliability estimates for different component types are statistically independent. However, the advantages of the LQR model lies in the accommodation of dependency of duplicated components of the same types that appear in multiple cut sets. Take the example of the bridge network in Figure 2.15; the same unreliability estimate for component 1 appears in two MC sets of $\{1, 2\}$ and $\{1, 3, 5\}$. Iterations are required in order to determine the higher-order moments of component unreliability estimates based on their lower-order moments (Jin and Coit 2001). Once the higher-order moments of $\hat{q}_k$ are obtained, they are substituted into Eqs. (2.7.7) and (2.7.8) to compute the mean and the variance of the terminal-pair reliability.

2.8 Reliability Confidence Intervals

2.8.1 Confidence Interval for Pass/Fail Tests

When a component or a system is being tested, the true probability of the success, r, is usually not known. Here t is suppressed simply for simplicity. If $\hat{r}$ represents the estimate of r, the simplest and most commonly used confidence interval (CI) for $\hat{r}$ is the normal distribution approximation method. The formula for this method is (Easterling 1972)

$$CI = \{E[\hat{r}] - z_{1-\alpha/2}\sqrt{V\hat{a}r(\hat{r})}, \quad E[\hat{r}] + z_{1-\alpha/2}\sqrt{V\hat{a}r(\hat{r})}\} \tag{2.8.1}$$

where $z_{1-\alpha/2}$ is the $(1 - \alpha/2) \times 100$ percentile of a standard normal distribution and $E[\hat{r}]$ and $V\hat{a}r(\hat{r})$ are the sample mean and the variance of $\hat{r}$, respectively. However, Eq. (2.8.1) is not valid in extreme cases such as $\hat{r} = 0$ or 1. The latter occurs where zero failures are observed in the test. To overcome the drawback, different methods have been proposed. Among these methods, the Clopper–Pearson interval (Clopper and Pearson 1934) and the Wilson score interval (Wilson 1927) are the two most popular ones. The Wilson score interval estimates are given as

$$CI = \left\{ \frac{\hat{r} + u - vz_{1-\alpha/2}}{1 + 2u}, \quad \frac{\hat{r} + u + vz_{1-\alpha/2}}{1 + 2u} \right\} \tag{2.8.2}$$

where

$$u = \frac{1}{2n}z_{1-\alpha/2}^2 \tag{2.8.3}$$

$$v = \sqrt{\frac{\hat{r}(1 - \hat{r})}{n} + \frac{z_{1-\alpha/2}^2}{4n^2}} \tag{2.8.4}$$

Note that n is the sample size and Eq. (2.8.2) is also valid for a small number of trials and extreme values such as $\hat{r} = 0$ or 1.

Realizing that Eqs. (2.8.1) and (2.8.2) are approximation methods, Clopper and Pearson (1934) developed an exact method based on the binomial distribution. The lower and upper reliability bounds, denoted as $\hat{r}_L$ and $\hat{r}_U$, can be estimated by

$$\sum_{j=0}^{k} \binom{n}{j} \hat{r}_L^j (1 - \hat{r}_L)^{n-j} = 1 - \frac{\alpha}{2} \tag{2.8.5}$$

$$\sum_{j=k}^{n} \binom{n}{j} \hat{r}_U^j (1 - \hat{r}_U)^{n-j} = 1 - \frac{\alpha}{2} \tag{2.8.6}$$

For the binominal reliability test, k is the number of survivals among n sample units. The Clopper–Pearson bound is also known as the "exact" bound or beta-binomial bound because there are no underlying distribution assumptions on the reliability estimate $\hat{r}$ (Vollset 1993).

Example 2.11   In accelerated life testing, the components are tested under stressed conditions, such as high temperature and humidly. It is assumed that 20 new lithium-ion battery cells are charged and discharged under extreme temperature conditions. At the end of 3000 charge–discharge cycles, 4 units failed. Estimate the point reliability of the battery cells and the associated confidence interval at $(1 - \alpha) = 90\%$ confidence level.

Solution:

Since the sample size $n = 20$ and the number of survivals $k = 20 - 4 = 16$, the point reliability estimate of a battery cell is

$$\hat{r} = \frac{k}{n} = \frac{16}{20} = 0.8$$

Case 1: Use the normal approximation method to estimate CI. The variance of $\hat{r}$ is given as follows:

$$Var(\hat{r}) = \frac{\hat{r}(1 - \hat{r})}{n - 1} = \frac{0.8(1 - 0.8)}{19} = 0.008421$$

Let $\hat{r}_L$ and $\hat{r}_U$ be the lower and upper bounds of $\hat{r}$. According to Eq. (2.8.1), we have

$$\hat{r}_L = E[\hat{r}] - z_{1-0.1/2}\sqrt{V\hat{a}r(\hat{r})} = 0.8 - 1.645\sqrt{0.008421} = 0.649$$

$$\hat{r}_U = E[\hat{r}] + z_{1-0.1/2}\sqrt{V\hat{a}r(\hat{r})} = 0.8 + 1.645\sqrt{0.008421} = 0.951$$

Case 2: Use the Wilson score interval in Eq. (2.8.2) to estimate the CI of the battery reliability. We first calculate u and v as follows:

$$u = \frac{1}{2n}z_{1-\alpha/2}^2 = \frac{1}{(2)(20)}(1.645)^2 = 0.0676$$

$$v = \sqrt{\frac{0.8(1 - 0.8)}{20} + \frac{1.645^2}{4(20^2)}} = 0.0984$$

Then

$$\hat{r}_L = \frac{\hat{r} + u - vz_{1-\alpha/2}}{1 + 2u} = \frac{0.8 + 0.0676 - (0.0984)(1.645)}{1 + 2(0.0676)} = 0.622$$

$$\hat{r}_U = \frac{\hat{r} + u + vz_{1-\alpha/2}}{1 + 2u} = \frac{0.8 + 0.0676 + (0.0984)(1.645)}{1 + 2(0.0676)} = 0.907$$

Case 3: Use the binominal model to estimate the CI of the battery reliability. We have first to calculate $\hat{r}_L$ and $\hat{r}_U$ based on Eqs. (2.8.5) and (2.8.6); that is,

$$\sum_{j=0}^{16} \binom{20}{j} \hat{r}_L^j (1 - \hat{r}_L)^{20-j} = 0.95$$

$$\sum_{j=16}^{20} \binom{20}{j} \hat{r}_U^j (1 - \hat{r}_U)^{n-j} = 0.95$$

Though there is no explicit solution to both equations, we find the solution by trial-and-error, and the result is

$$\hat{r}_L = 0.656 \quad \text{and} \quad \hat{r}_U = 0.896$$

If one compares the results of three CI estimations, the normal approximation yields a wider range of CI, while the binomial method tends to generate a tiger interval. The CI from the Wilson score interval belongs between these two methods.

2.8.2 Confidence Intervals for System Reliability

For a complex system with multiple subsystems connected in series, the system reliability $R_s(t)$ is the multiplication of subsystem reliability values. Assuming that subsystems fail independently, this implies that the logarithm of the system reliability is the sum of the logarithm of individual subsystem reliability; that is,

$$\ln \widehat{R}_s(t) = \sum_{i=1}^{m} \ln \widehat{R}_i(t) \tag{2.8.7}$$

where $\widehat{R}_i(t)$ is the reliability estimate of subsystem i and m is the number of subsystems. If m is sufficiently large, the central limit theorem (CLT) can be involved. CLT states that the sum of a large number of independent random variables tends to be normally distributed regardless of the distribution of individual random variables. In other words, if m is sufficiently large, $\ln \widehat{R}_s(t)$ approximately follows the normal distribution regardless of the subsystem reliability data type or time-to-failure distribution. The lognormal based lower and upper reliability bounds, denoted as R_L and R_U, developed by Coit (1997), are restated below:

$$R_L = \widehat{R}_s(t) \exp\left(\frac{1}{2}\widehat{\sigma}^2 - z_{\alpha/2}\widehat{\sigma}\right) \tag{2.8.8}$$

$$R_U = \widehat{R}_s(t) \exp\left(\frac{1}{2}\widehat{\sigma}^2 + z_{\alpha/2}\widehat{\sigma}\right) \tag{2.8.9}$$

where

$$\widehat{\sigma}^2 = \ln\left(1 + \frac{\widehat{Var}(\widehat{R}_s(t))}{\widehat{R}_s^2(t)}\right) \tag{2.8.10}$$

where $\widehat{Var}(\widehat{R}_s(t))$ can be estimated from Eq. (2.3.11). Coit (1997) showed that this approximation is reasonably accurate for any type of complex systems that can be partitioned into at least five series or parallel subsystems. This requirement is usually met for most engineering design problems. For example, a modern wind turbine system typically consists of three blades, a gearbox, a main shaft, DC–AC converters, a generator, and some control and protection subsystems. Each of these functions can be provided by a variety of different components. For instance, a subsystem can be configured as redundant structures, but each function is required at the system level, thereby forming a series model at the system level. The lower and upper bounds of Eqs. (2.8.8) and (2.8.9) can be extended to a parallel system with multiple subsystems where the unreliability can be approximated by a lognormal distribution.

2.9 Reliability of Multistate Systems

2.9.1 Series or Parallel Systems with Three-State Components

Most of the reliability literature deals with binary systems of binary components in which the only two states are functioning and failed. The ideal of a multistate system

(MSS) reliability dates back to the early works by Barlow and Wu (1978), El-Neweihi et al. (1978), and Ross (1979) in the 1970s. These researchers treat the general case of more than two states of components and systems in many practical applications. This idea is useful because in many real-life situations components and/or systems can operate in intermediate states. For example, a diode allows the current to pass if positive voltage is applied to the P–N junction. It is also capable of blocking the current flow if negative voltage is applied to the junction. Hence, the diode possesses two operating states: it may fail open when forward current is applied or fail short if reverse voltage is applied. Another example is the highway lanes. The actual capacity of a multilane highway reduces if one lane is blocked due to an accident, yet incoming vehicles are still able to go through other lanes at a reduced capacity. In the wind power industry, depending on wind speed, a wind turbine can operate in one of four states: idle, nonlinear power generation, constant power output, and shut-down states. These are just a few examples of MSSs that are around us. Recent researchers extended the multistate concepts to solving general multistate network reliability problems using the sum of disjoint products (Zuo et al. 2007; Yeh 2007), fuzzy theory (Liu et al. 2008), and supply chain networks (Lin 2009), where the link capacity is characterized as a multilevel reliability index. Xing and Dai (2009) also generalized and extended the decision diagram method to multistate systems.

For an m-components series system where each component possesses three states, fail open, fail short, and operational, the system reliability can be estimated as (Elsayed 2012)

$$R_s(t) = \prod_{i=1}^{m}(1 - q_i^o(t)) - \prod_{i=1}^{m} q_i^s(t) \tag{2.9.1}$$

where $q_i^o(t)$ and $q_i^s(t)$ are the probabilities of fail open and fail short, respectively, at time t. It is worth mentioning that for a multistate series system, the system fails if all m components fail in short mode or any component fails in open mode.

The reliability of a parallel system is similar where components possess three states: fail open, fail short, and operational (Elsayed 2012). For a three-state parallel system, the system fails if all m components fail in open mode or any component fails in short mode:

$$R_s(t) = \prod_{i=1}^{m}(1 - q_i^s(t)) - \prod_{i=1}^{m} q_i^o(t) \tag{2.9.2}$$

2.9.2 Universal Generating Function

The universal generating function (UGF) is shown to be quite effective in reliability analysis of components or systems with more than three states. The concept of UGF was introduced by Ushakov (1986). Lisnianski et al. (1996) first applied UGF to analyze power system reliability considering multiple operating states. Levitin et al. (1998) presented an algorithm for optimizing reliability of complex multistate, series–parallel systems with propagated failures and imperfect protections. Ding et al. (2011) performed a long-term reserve planning by utilizing UGF methods for electric grid with high wind power penetration.

Let s be the number of possible states for component i and j be the index of the state of that component. Furthermore, let p_{ij} be the probability of component i in state j for

$j = 0, 1, \ldots, s - 1$. The reliability of the ith multistate component (for $i = 1, 2, \ldots, m$) in a system can be expressed by a u-function as follows:

$$u_i(z) = \sum_{j=0}^{s-1} p_{ij} z^j, \quad \text{for } i = 1, 2, \ldots, m \tag{2.9.3}$$

For example, the u-function of a binary component is $u(z) = p_1 z^0 + (1 - p_1)z^1$. For a system with m components, the u-functions of individual components represent the probability mass function (PMF) of discrete random variables $X_1, X_2, \ldots, X_m$. Given the system structure function and u-functions of components, we can obtain the u-function that represents the PMF of the system state variable X using the composition operator. That is,

$$U(z) = \underset{\varphi}{\otimes}\{u(z), u_2(z), \ldots, u_m(z)\} \tag{2.9.4}$$

The system reliability measure can be obtained as $E[X] = U'(z = 1)$.

Example 2.12   We take the example from Levitin and Lisnianski (2001) to illustrate the MSS reliability analysis using UGF. Consider two electricity generators with nominal capacity $P_{nom} = 100\,\text{MW}$ as two separate MSS. In the first generator some minor failures forces the capacity down to 60 MW and any large issues lead to the complete outage. Similarly, the power output of the second generator is reduced to 80 MW, 40 MW, or complete outage depending on the severity of the failure issues. We estimate the reliability of the aggregate power output of both generators.

Solution:
There are three possible states to characterize the first generator. These are

$$G_1^1 = 0, \quad G_2^1 = \frac{60}{100} = 0.6, \quad G_3^1 = \frac{100}{100} = 1$$

and the corresponding steady-state probabilities are

$$p_1^1 = 0.1, \quad p_2^1 = \frac{60}{100} = 0.6, \quad p_3^1 = 0.3$$

For the second generator, there are four possible states to characterize its performance:

$$G_1^2 = 0, \quad G_2^2 = \frac{40}{100} = 0.4, \quad G_3^2 = \frac{80}{100} = 0.8, \quad G_4^2 = \frac{100}{100} = 1$$

and four probability measures to characterize the four possible states of the second generator:

$$p_1^2 = 0.05, \quad p_2^2 = 0.25, \quad p_3^2 = 0.3, \quad p_4^2 = 0.4$$

Now using Eq. (2.9.3), we obtain the reliability measure for both MSS power generators

$$u_1(z) = p_1^1 z^{G_1^1} + p_2^1 z^{G_2^1} + p_3^1 z^{G_3^1} = 0.1 z^0 + 0.6 z^{0.6} + 0.3 z^1$$

$$u_2(z) = 0.05 z^0 + 0.25 z^{0.4} + 0.3 z^{0.8} + 0.4 z^1$$

If the minimum requirement for a generator's power output is $P_{min} = 50\,\text{MW}$, this implies that $G_k{}^1$ and $G_k{}^2$ will be equal to or larger than $P_{min}/P_{nom} = 50/100 = 0.5$. The stationary availability for both MSS generators is

$$A_1(P_{min} = 50) = \sum_{G_k^1 \geq 0.5} p_k^1 = 0.6 + 0.3 = 0.9$$

$$A_2(P_{min} = 50) = \sum_{G_k^2 \geq 0.5} p_k^2 = 0.3 + 0.4 = 0.7$$

Suppose a power plant consists of two MSS generators. The minimum power delivered from the power plant will not be less than $P_{min} = 100\,\text{MW}$. What is the supply reliability of this power plant? This problem can be solved by treating two MSS generators forming a parallel system. Based on Eq. (2.9.4), we have

$$
\begin{aligned}
U_{plant}(z) &= u_1(z) \times u_2(z) \\
&= (0.1z^0 + 0.6z^{0.6} + 0.3z^1) \times (0.05z^0 + 0.25z^{0.4} + 0.3z^{0.8} + 0.4z^1) \\
&= 0.005z^0 + 0.025z^{0.4} + 0.03z^{0.6} + 0.03z^{0.8} + 0.205z^1 + 0.255z^{1.4} \\
&\quad + 0.24z^{1.6} + 0.09z^{1.8} + 0.12z^2
\end{aligned}
$$

Let G_k be the state of the power plant output. To meet the minimum power supply requirement, $G_k = P_{min}/P_{nom} = 100/100 = 1$. Hence the stationary availability for the power plant with $P_{min} = 100\,\text{MW}$ and larger is

$$A_{plant}(P_{min} = 100) = \sum_{G_k \geq 1} p_k = 0.205 + 0.255 + 0.24 + 0.09 + 0.12 = 0.91$$

2.10 Reliability Importance

2.10.1 Marginal Reliability Importance

When engineers design a multicomponents system, it is useful to identify the components that are the most crucial to the system reliability. Therefore, it is necessary to define the measurements of the importance of a component to system reliability. The most fundamental measure of statistical importance of components is the Birnbaum importance (BI) measure (Barlow and Proschan, 1975; Leemis 1995). Since its inception, various importance measures haven been developed, such as the Bayesian importance measure, criticality importance measure, redundancy importance measures, and Fussell–Vesely (FV) importance measure. In general, BI and all its extensions can be categorized as marginal reliability importance (MRI) measures as opposed to joint reliability importance (JRI) measures. This section introduces several MRI measures and the concept of JRI will be elaborated in Section 2.10.2. For a comprehensive treatment of reliability importance measures and their applications, readers are referred to the book by Kuo and Zhu (2012b).

For a system with n independent components, the BI of component i, denoted as $I_B(i)$, is defined as

$$I_B(i) = \frac{\partial R(\mathbf{r})}{\partial r_i}, \quad \text{for } i = 1, 2, 3, \ldots, m \tag{2.10.1}$$

$R(\mathbf{r})$ is the system reliability expressed as a function of component reliability, and $\mathbf{r} = [r_1, r_2, \ldots, r_m]$. If the system consists of several subsystems and each subsystem contains multiple components, then the system reliability importance can also be expressed as

$$I_B(i,j) = \frac{\partial R(\mathbf{r})}{\partial R_i} \frac{\partial R_i}{\partial r_{ij}} \tag{2.10.2}$$

where R_i is the reliability of the ith subsystem and r_{ij} is the reliability of the jth component in subsystem i. This "chain-rule" property makes it possible to compute the importance of each component of a subsystem. To use this formula, the statistical independence of component reliability estimates is required.

Lambert (1975) proposed an upgrading function defined as the fractional reduction in the probability of the system failure when the component failure rate is reduced fractionally. The model assumes an exponential time-to-failure for all components and the importance measure is defined as

$$I_{UF}(i) = \frac{\lambda_i}{R(\mathbf{q})} \times \frac{\partial R(\mathbf{q})}{\partial \lambda_i}, \quad \text{for } i = 1, 2, 3, \ldots, m \tag{2.10.3}$$

where λ_i is the exponential failure rate of component I and $\mathbf{q} = (q_1, q_2, q_3, \ldots, q_m)$ with q_i being the unreliability for component i.

Gandini (1990) proposed the critical importance. This importance measure is based on the fact that it is more difficult to improve the more reliable components than the less reliable ones,

$$I_{CR}(i) = \frac{\partial R(\mathbf{q})}{\partial q_i} \times \frac{q_i}{R(\mathbf{q})} \tag{2.10.4}$$

Based on Birnbaum's definition, Griffith (1985) developed a simple sensitivity measure formula for a consecutive k-out-of-n:F system. Papastavridis (1987) also derived the same formula using a different approach. Their formula is given as follows:

$$I_B(i) = \frac{R(i-1)R'(n-i) - R(n)}{r_i} \tag{2.10.5}$$

where $R(i-1)$ is reliability of a consecutive k-out-of-$(i-1)$:F system comprised of components $1, 2, \ldots, i-1$. In addition, $R'(n-i)$ is the reliability of a consecutive k-out-of-$(i-1)$:F system comprised of components $i+1, i+2, \ldots, n$.

Example 2.13 Three components are used to construct a series system in Figure 2.1a, where the component reliability is given as $r_1(t) = e^{-0.5t}$, $r_2(t) = e^{-(0.5t)^2}$, and $r_3(t) = e^{-t^{2.5}}$. Compute the reliability importance of each component.

Solution:
The reliability of this three-component series system is given as

$$R_s(t) = r_1(t)r_2(t)r_3(t)$$

According to Eq. (2.10.1), the B-importance measure of an individual component is

$$I_{B1}(t) = \frac{dR_s(t)}{dr_1(t)} = r_2(t)r_3(t) = e^{-(0.5t)^2 - t^{2.5}}$$

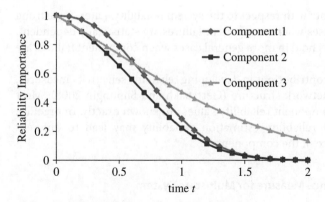

Figure 2.22 Time-dependent reliability importance.

$$I_{B2}(t) = \frac{dR_s(t)}{dr_2(t)} = r_1(t)r_3(t) = e^{-0.5-t^{2.5}}$$

$$I_{B3}(t) = \frac{dR_s(t)}{dr_3(t)} = r_1(t)r_2(t) = e^{-0.5-(0.5t)^2}$$

The measures show the reliability importance of three components changes with time. Figure 2.22 plots the BI value of three components for $t \in [0, 2]$. It is interesting to see that component 1 is the most important in $[0, 0.62]$ and its importance is overtaken by component 3 after $t = 0.62$. In fact, for a series system, there is a correlation between component reliability and the corresponding importance measure. The component with the worst reliability always has the largest importance value. Going back to the example, the reliability of component 1 is the lowest among three components in $[0, 0.62]$, while the reliability of component 3 becomes the lowest post $t = 0.62$. As Barlow and Prochan (1975) have indicated for a "series" system, the component with the lowest reliability is the most B-important, reflecting the principle of "a chain is only as strong as its weakest link."

2.10.2 Joint Reliability Importance Measure

B-importance measures and their extensions do not explain how components mutually affect the system reliability. Design engineers working within a fixed budget might need to conduct a trade-off analysis among components within a system. Using MRI alone would require calculation of MRI for all components in every trade-off being considered. The joint reliability importance (JRI) of pairs of components, introduced by Hong and Lie (1993), is more effective for making this kind of trade-off because it indicates how components mutually interact to the system reliability. The definition of JRI is given as follows:

$$I_{JRI}(i,j) = \frac{\partial^2 R(\mathbf{r})}{\partial r_i \partial r_j}, \quad \text{for} \quad i \neq j \tag{2.10.6}$$

If $I_{JRI}(i, j) \geq 0$, this implies that r_i is more important when r_j is working than when r_j has failed. If $I_{JRI}(i, j) \leq 0$, this implies that r_i is more important when r_j has failed than when r_j is working. This indicates that $I_{JRI}(i, j)$ is an appropriate metric to quantify the interactions of two components with respect to the system reliability. Later, Armstrong (1995) extended the JRI to cases where component failures are statistically dependent. He showed that similar results hold in more general cases when component failures are statistically dependent.

There are other important contributions in developing reliability sensitivity measures, such as in the context of a network structure (Gertsbakh and Shpungin 2008). Most research has assumed that component reliability values are known exactly. In practice, consideration of component reliability estimation variability may lead to different assessments of the importance of the components.

2.10.3 Integrated Importance Measure for Multistate System

The performance measure of a multistate system aims to assess all outcomes or aspects when both the system and its components may assume more than two levels of reliability states. For example, 0%, 50%, and 100% capacities of a water plant correspond to states 0, 1, and 2, respectively. For a system with n components and M system states, the system performance can be expressed as an expectation of utilities as follows:

$$U = \sum_{j=1}^{M} a_j P\{\Phi(X) = j\} = \sum_{j=1}^{M} a_j P\{\Phi(X_1, X_2, \dots, X_n) = j\} \tag{2.10.7}$$

where a_j is the system performance level when it is in state j and $\mathbf{X} = [X_1, X_2, \dots, X_n]$ with X_i being the state of component i. The probability that the system is in state j is given by $P\{\Phi(\mathbf{X}) = j\}$, where $\Phi(\mathbf{X})$ is the system structure function related to states of all components. Griffith (1980) proposed the following model to evaluate the importance of state m of component i to the overall system. That is,

$$I_m^G(i) = \sum_{j=1}^{M} (a_j - a_{j-1})[P\{\Phi(m_i, X) \geq j\} - P\{\Phi((m-1)_i, X) \geq j\}] \tag{2.10.8}$$

Equation (2.10.8) can be interpreted as the change of the system performance when component i deteriorates from current state m down to adjacent state $m - 1$.

However, it is more reasonable in practice that a component may transit from state m to any other degradation states, other than state $m - 1$. To generalize $I_m^G(i)$, Wu and Chan (2003) proposed a new importance measure that allows the arbitrary transition of component i to any degraded state as

$$I_m^{Wu}(i) = P_{im} \sum_{j=0}^{M} a_j P\{\Phi(m_i, X) = j)\} \tag{2.10.9}$$

In practice changes in the operating condition usually lead to the change of transition rates of a multistate component. For example, improvements of the repair service or the reduction of usage can affect the transition rates between a degradation state and a fully operable state. Therefore, in the operation and optimization of multistate systems, it is imperative to consider the difference of transition rates between component states. Considering the joint effect of the probability distributions, transition intensities of the target component states and the system performance, Si et al. (2012) proposed the following integrated importance measure (IIM) to study how the transition of component

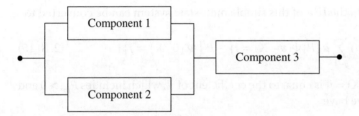

Figure 2.23 A series–parallel system with three components.

states affects the system performance:

$$I_{m,l}^{IIM}(i) = P_{im}\lambda_{ml}(i) \sum_{j=1}^{M} a_j[P\{\Phi(m_i, X) = j\} - P\{\Phi(l_i, X) = j\}], \tag{2.10.10}$$

where P_{im} is the probability that component i is in state m and $\lambda_{ml}(i)$ is its transition rate from state m to state l.

We can interpret $I_{m,l}^{IIM}(i)$ as the change of the system performance when component i transfers from state m to state l. IIM can identify which component state holds the largest responsibility of system performance loss and which state should be prioritized in order to improve the system reliability. We take a three-component system in Figure 2.23 to illustrate how to use the UGF method for IIM evaluation. Note that the state spaces of components 1 and 2 are $\{0, 1\}$ and the state space of component 3 is $\{0, 1, 2\}$. The structure function is $\Phi(X) = \Phi(X_1, X_2, X_3) = \min\{X_1 + X_2, X_3\}$. We can define the UGF of component i as follows:

$$u_1(z) = P_{10}z^0 + P_{11}z^1, \quad u_2(z) = P_{20}z^0 + P_{21}z^1, \quad u_3(z) = P_{30}z^0 + P_{31}z^1 + P_{32}z^2 \tag{2.10.11}$$

By applying the composition operators $\Phi_{1,2} = x_1 + x_2$ and $\Phi_S = \min\{\Phi_{1,2}, x_3\}$ to the three-component system, the UGF for the entire multistate system is obtained as

$$U(z) = \Phi_S(\Phi_{1,2}(u_1(z), u_2(z)), u_3(z)) \tag{2.10.12}$$

In order to find the UGF for the parallel subsystem comprised of components 1 and 2, the operator $\Phi_{1,2}$ is applied to individual UGF $u_1(z)$ and $u_2(z)$. It yields

$$\begin{aligned}\Phi_{1,2}(u_1(z), u_2(z)) &= \Phi_{1,2}(P_{10}z^0 + P_{11}z^1, \ P_{20}z^0 + P_{21}z^1)\\ &= P_{10}P_{20}z^0 + P_{11}P_{20}z^1 + P_{10}P_{21}z^1 + P_{11}P_{21}z^2 \end{aligned} \tag{2.10.13}$$

In order to find the UGF for the entire system in which component 3 is connected in series with components 1 and 2 that are configured in parallel, operator Φ_S should be applied:

$$\begin{aligned}U(z) &= \Phi_S(\Phi_{1,2}(u_1(z), u_2(z)), u_3(z))\\ &= \Phi_S(P_{10}P_{20}z^0 + P_{11}P_{20}z^1 + P_{10}P_{21}z^1 + P_{11}P_{21}z^2, P_{30}z^0 + P_{31}z^1 + P_{32}z^2)\\ &= P_{10}P_{20}P_{30}z^0 + P_{11}P_{20}P_{30}z^0 + P_{10}P_{21}P_{30}z^0 + P_{11}P_{21}P_{30}z^0\\ &\quad + P_{10}P_{20}P_{31}z^0 + P_{11}P_{20}P_{31}z^1 + P_{10}P_{21}P_{31}z^1 + P_{11}P_{21}P_{31}z^1\\ &\quad + P_{10}P_{20}P_{32}z^0 + P_{11}P_{20}P_{32}z^1 + P_{10}P_{21}P_{32}z^1 + P_{11}P_{21}P_{32}z^2\end{aligned} \tag{2.10.14}$$

Based on Eq. (2.10.10), the IIM of this simple multistate system can be converted to

$$I_{m,0}^{IIM}(i) = P_{im}\lambda_{m0}(i)\sum_{j=1}^{2} a_j[P\{\Phi(m_i, X) = j\} - P\{\Phi(0_i, X) = j\}]$$

(2.10.15)

The value of $P\{\Phi(m_i, \mathbf{X}) = j\}$ is equal to the coefficient of z^j, which includes $P_{im} > 0$ and $P_{im} = 1$. For example, we have

$$I_{2,0}^{IIM}(3) = P_{32}\lambda_{20}(3)\sum_{j=1}^{2} a_j[P\{\Phi(2_3, X) = j\} - P\{\Phi(0_3, X) = j\}]$$

$$= P_{32}\lambda_{20}(3)\{a_1[P\{\Phi(2_3, X) = 1\} - P\{\Phi(0_3, X) = 1\}]$$
$$+ a_2[P\{\Phi(2_3, X) = 2\} - P\{\Phi(0_3, X) = 2\}]\}$$
$$= P_{32}\lambda_{20}(3)[a_1(P_{11}P_{20} + P_{10}P_{21} - 0) + a_2(P_{11}P_{21} - 0)]$$
$$= a_1 P_{32}\lambda_{20}(3)(P_{11}P_{20} + P_{10}P_{21}) + a_2 P_{32}b_{20}(3)P_{11}P_{21}$$

Other important measures pertaining to multistate systems are also proposed in the literature. For instance, Ramirez-Marquez and Coit (2005) develop composite importance measures that are involved in measuring how a specific component affects multistate system reliability. Others (Levitin et al. 2003) have focused on investigating how a particular component state or set of states affects multistate system reliability. Readers can refer to the recent review paper by Kuo and Zhu (2012a) who reported the latest development in reliability importance measures of multistate systems.

2.10.4 Integrated Importance Measure for System Lifetime

System lifetime can be divided into several different stages depending on the characteristics of the failure intensity function or hazard rate curve. The importance of a component to the system reliability performance can also be different in different stages. To achieve the best system performance, the fact that the importance of a component to the system performance may change with the lifetime needs to be taken into account. It is usually not easy to explicitly measure the contribution of component reliability to the system performance in different life stages. However, the change of system performance caused by component degradation is often related to the expected number of component failures. The latter can be predicted by the renewal function. Dui et al. (2014) extended the IIM from a single time instance to different lifetime stages as follows:

$$I^{IIM}(i, t) = \frac{\int_0^t \sum_{m=l}^{M}\sum_{j=0}^{l-1}\sum_{k=1}^{M}(a_k - a_{k-1})(P\{\Phi(m_i, X(u)) \geq k\} - P\{\Phi(j_i, X(u)) \geq k\})dM_i(u)}{\sum_{i=1}^{n}\int_0^t \sum_{m=l}^{M}\sum_{j=0}^{l-1}\sum_{k=1}^{M}(a_k - a_{k-1})(P\{\Phi(m_i, X(u)) \geq k\} - P\{\Phi(j_i, X(u)) \geq k\})dM_i(u)}$$

(2.10.16)

where $M_i(t)$ is the renewal function for component i and $dM_i(t)$ is the probability that component i fails in an infinitesimal interval of $[t, t + dt]$. This generalized IIM describes which component is the most important in improving the system performance across its lifetime or at different stages. The effect of component reliability change on system performance represents a time-dependent importance measure. It is very useful for system design and maintenance planning. For example, in the design phase, this measure can be a useful tool to determine which components should be prioritized for reliability growth or maintenance support such that the best system performance across different life stages is ensured.

If we let $t \to \infty$ in the upper limit of the integrals in Eq. (2.10.16), the probability that component i causes the system performance reduction is given by

$$I^{IIM}(i) = \frac{\int_0^\infty \sum_{m=l}^{M} \sum_{j=0}^{l-1} \sum_{k=1}^{M} (a_k - a_{k-1})(P\{\Phi(m_i, X(u)) \geq k\} - P\{\Phi(j_i, X(u)) \geq k\})dM_i(u)}{\sum_{i=1}^{n} \int_0^\infty \sum_{m=l}^{M} \sum_{j=0}^{l-1} \sum_{k=1}^{M} (a_k - a_{k-1})(P\{\Phi(m_i, X(u)) \geq k\} - P\{\Phi(j_i, X(u)) \geq k\})dM_i(u)}$$

(2.10.17)

Equation (2.10.17) evaluates the importance of component i that causes the system performance reduction from the time when the system is in a perfect condition to the time when it eventually fails. During the system lifetime, $I^{IIM}(i)$ is a useful metric to determine which components should be managed in terms of lifecycle reliability.

The operation of some systems like satellites contains several missions or stages, such as the launch stage, mission stage, maintenance stage, and re-entry stage; thus the importance of a component to system performance reduction may vary in different stages. Let t_1 and t_2 represent the start and the end time of a particular stage, respectively. We can obtain the importance of component i in stage $t_1 \leq t \leq t_2$ as follows:

$$I^{IIM}(i, t_1, t_2) = \frac{\int_{t_1}^{t_2} \sum_{m=l}^{M} \sum_{j=0}^{l-1} \sum_{k=1}^{M} (a_k - a_{k-1})(P\{\Phi(m_i, X(u)) \geq k\} - P\{\Phi(j_i, X(u)) \geq k\})dM_i(u)}{\sum_{i=1}^{n} \int_{t_1}^{t_2} \sum_{m=l}^{M} \sum_{j=0}^{l-1} \sum_{k=1}^{M} (a_k - a_{k-1})(P\{\Phi(m_i, X(u)) \geq k\} - P\{\Phi(j_i, X(u)) \geq k\})dM_i(u)}$$

(2.10.18)

The importance measure in different stages is a useful guideline for managers to implement the proper actions for achieving system performance improvement at different life stages. In different life stages, the IIM of component i is only dependent on t_1 and t_2. By changing the lower or upper time limit, we can easily estimate IIM and characterize its properties in different stages.

Example 2.14 Assume the lifetime and repair time of component i are exponentially distributed. The failure rate and repair rate of component i are λ_i and μ_i, respectively. Compute the IIM for a system lifetime.

Solution:
The lifetime and repair time distributions of component i are $F_i(t) = 1 - e^{-\lambda_i t}$ and $G_i(t) = 1 - e^{-\mu_i t}$, respectively. The Laplace–Stieltjes transforms of $F_i(t)$ and $G_i(t)$ are

$$F_i^{(L)}(s) = \lambda_i/(s + \lambda_i) \quad \text{and} \quad G_i^{(L)}(s) = \mu_i/(s + \mu_i)$$

Thus, we have

$$M_i^{(L)}(s) = \frac{F_i^{(L)}(s)}{1 - F_i^{(L)}(s)G_i^{(L)}(s)} = \frac{\lambda_i(s + \mu_i)}{s(s + \lambda_i + \mu_i)}$$

$$= \frac{\lambda_i \mu_i}{\lambda_i + \mu_i} \times \frac{1}{s} + \frac{\lambda_i^2}{\lambda_i + \mu_i} \times \frac{1}{s + \lambda_i + \mu_i}$$

By performing an inverse Laplace–Stieltjes transformation, we have

$$M_i(t) = \frac{\lambda_i \mu_i t}{\lambda_i + \mu_i} t + \frac{\lambda_i^2}{(\lambda_i + \mu_i)^2}(1 - e^{-(\lambda_i + \mu_i)t})$$

and

$$dM_i(t) = \left(\frac{\lambda_i \mu_i}{\lambda_i + \mu_i} + \frac{\lambda_i^2 e^{-(\lambda_i + \mu_i)t}}{\lambda_i + \mu_i} \right) dt \tag{2.10.19}$$

Finally, substituting Eq. (2.10.19) into (2.10.17), the IIM for a system lifetime is obtained as

$$I^{IIM}(i) = \frac{\int_0^\infty \sum_{m=l}^{M} \sum_{j=0}^{l-1} \sum_{k=1}^{M} (a_k - a_{k-1})(P\{\Phi(m_i, X(t)) \geq k\} - P\{\Phi(j_i, X(t)) \geq k\}) \left(\frac{\lambda_i \mu_i}{\lambda_i + \mu_i} + \frac{\lambda_i^2}{\lambda_i + \mu_i} e^{-(\lambda_i + \mu_i)t} \right) dt}{\sum_{i=1}^{n} \int_0^\infty \sum_{m=l}^{M} \sum_{j=0}^{l-1} \sum_{k=1}^{M} (a_k - a_{k-1})(P\{\Phi(m_i, X(t)) \geq k\} - P\{\Phi(j_i, X(t)) \geq k\}) \left(\frac{\lambda_i \mu_i}{\lambda_i + \mu_i} + \frac{\lambda_i^2}{\lambda_i + \mu_i} e^{-(\lambda_i + \mu_i)t} \right) dt}$$

References

Amari, S.V., Pham, H., and Misra, R.B. (2012). Reliability characteristics of *k*-out-of-*n* warm standby systems. *IEEE Transactions on Reliability* 61 (4): 1007–1018.

Armstrong, M. (1995). Joint reliability importance of components. *IEEE Transaction on Reliability* 44 (3): 408–412.

Ball, M.O. (1986). Computational complexity of network reliability analysis: an overview. *IEEE Transaction on Reliability* 35 (3): 230–239.

Barlow, R.E. and Proschan, F. (1975). *Statistical Theory of Reliability and Life Testing*. New York: Holt, Rinehart and Winston.

Barlow, R.E. and Wu, A.S. (1978). Coherent systems with multi-state components. *Mathematical Operations Research* 3 (4): 275–281.

Borgonov, E. (2007). A new uncertainty measure. *Reliability Engineering and System Safety* 92 (6): 771–784.

Clopper, C. and Pearson, S. (1934). The use of confidence or fiducial limits illustrated in the case of the binomial. *Biometrika* 26 (4): 404–413.

Coit, D.W. (1997). System reliability confidence intervals for complex systems with estimated component reliability. *IEEE Transactions on Reliability* 46 (4): 487–493.

Colbourn, C.J. (1987). *The Combinatorics of Network Reliability*. New York, NY: Oxford University Press.

Ding, Y., Wang, P., Goel, L. et al. (2011). Long-term reserve expansion of power systems with high wind power penetration using universal generating function methods. *IEEE Transactions on Power System* 26 (2): 766–774.

Dui, H., Si, S., Cui, L. et al. (2014). Component importance for multi-state system lifetimes with renewal functions. *IEEE Transactions on Reliability* 63 (1): 105–117.

Easterling, R.G. (1972, 1972). Approximate confidence limits for system reliability. *Journal of the American Statistical Association* 67 (337): 220–222.

El-Neweihi, E., Proschan, F., and Sethuramn, J. (1978). Multi-state coherent systems. *Journal of Applied Probability* 15: 675–688.

Elsayed, E.A. (2012). *Reliability Engineering*, Chapter 2, 2e. Hoboken, NJ: Wiley.

Gandini, A. (1990). Importance and sensitivity analysis in assessing system reliability. *IEEE Transaction on Reliability* 39 (1): 61–69.

Gertsbakh, I. and Shpungin, Y. (2008). Network reliability importance measures: combinatorics and Monte Carlo based computations. *WSEAS Transactions on Computers* 7 (4): 216–227.

Goodman, L. (1960). On the exact variance of product. *Journal of American Statistical Association* 50 (292): 708–713.

Griffith, W.S. (1980). Multistate reliability models. *Journal of Applied Probability* 17 (3): 735–744.

Griffith, W.S. (1985). Consecutive k-out-of-n:F system: reliability, availability, component importance and multi-state extension. *American Journal of Mathematical and Management Science* 5 (1 and 2): 125–160.

Hong, J.S. and Lie, C.H. (1993). Joint reliability-importance of two edges in an undirected network. *IEEE Transaction on Reliability* 42 (1): 17–23.

Jin, T. and Coit, D. (2001). Variance of system reliability estimation with arbitrarily repeated components. *IEEE Transactions on Reliability* 50 (4): 409–413.

Jin, T. and Coit, D. (2003). Approximating network reliability estimates using linear and quadratic unreliability of minimal cuts. *Reliability Engineering and System Safety* 82 (1): 41–48.

Jin, T. and Coit, D. (2008). Unbiased variance estimates for system reliability estimate using block decompositions. *IEEE Transactions on Reliability* 57 (3): 458–464.

Kuo, W. and Zhu, X. (2012a). Some recent advances on importance measures in reliability. *IEEE Transactions on Reliability* 61 (2): 344–360.

Kuo, W. and Zhu, X. (2012b). *Importance Measures in Reliability, Risk, and Optimization: Principles and Applications*, 1e. Wiley.

Lambert, H. E., Fault Tree for Decision Making in System Analysis, PhD Thesis, UCRL-51829, University of California, Livermore, 1975.

Leemis, L.M. (1995). *Reliability Probabilistic Models and Statistical Methods*. Englewood Clifs, NJ: Prentice Hall, Inc.

Levitin, G. and Lisnianski, A. (2001). A new approach to solving problems of multi-state system reliability optimization. *Quality and Reliability Engineering International* 17 (2): 93–104.

Levitin, G., Lisnianski, A., Ben-Haim, H., and Elmakis, D. (1998). Redundancy optimization for series-parallel multi-state systems. *IEEE Transactions on Reliability* 47 (2): 165–172.

Levitin, G., Podofillini, L., and Zio, E. (2003). Generalised importance measures for multi-state elements based on performance level restrictions. *Reliability Engineering and System Safety* 82 (3): 287–298.

Li, Y.-F. and Zio, E. (2012). Uncertainty analysis of the adequacy assessment model of a distributed generation system. *Renewable Energy* 41: 235–244.

Li, Y.-F., Ding, Y., and Zio, E. (2014). Random fuzzy extension of the universal generating function approach for the reliability assessment of multi-state systems under aleatory and epistemic uncertainties. *IEEE Transactions on Reliability* 63 (1): 13–25.

Lin, Y.K. (2009). System reliability evaluation for a multistate supply chain network with failure nodes by using minimal paths. *IEEE Transactions on Reliability* 58 (1): 34–40.

Lisnianski, A., Levitin, G., Ben-Haim, H., and Elmakis, D. (1996). Power system structure optimization subject to reliability constraints. *Electric Power Systems Research* 39 (2): 145–152.

Liu, Y., Huang, H.-Z., and Levitin, G. (2008). Reliability and performance assessment for fuzzy multi-state elements. *Proceedings of the Institution of Mechanical Engineers, Part O: Journal of Risk and Reliability* 222 (4): 675–686.

NERC, Reliability Standards for the Bulk Electric Systems of North America, Report of North America Reliability Corporation, January 26, 2017, available at: http://www.nerc .com/pa/Stand/Reliability%20Standards%20Complete%20Set/RSCompleteSet.pdf (accessed on February 12, 2017).

Papastavridis, S. (1987). The most important component in a consecutive k-out-of-n:F system. *IEEE Transaction on Reliability* 36 (2): 266–268.

Ramirez-Marquez, J.E. and Coit, D.W. (2005). Composite importance measures for multi-state systems with multi-state components. *IEEE Transactions on Reliability* 54 (3): 517–529.

Ramirez-Marquez, J.E. and Jiang, W. (2006). An improved confidence bounds for system reliability. *IEEE Transactions on Reliability* 55 (1): 26–36.

Ross, S. (1979). Multivalued state component systems. *Annual Probability.* 7 (2): 379–383.

Shier, D. (1991). *Network Reliability and Algebraic Structures.* New York, NY: Clarendon Press.

Si, S., Dui, H., Zhao, X. et al. (2012). Integrated importance measure of component states based on loss of system performance. *IEEE Transactions on Reliability* 6 (1): 192–202.

Taboada, H., Z. Xiong, T. Jin, and J. Jimenez, (2012), "Exploring a solar photovoltaic-based energy solution for green manufacturing industry," in *Proceedings of IEEE Conference on Automation Science and Engineering*, pp. 40–45.

Ushakov, I. (1986). Universal generating function. *Soviet Journal Computer Systems Science* 24 (5): 118–129.

Vollset, S.E. (1993). Confidence intervals for a binomial proportion. *Statistics in Medicine* 12 (9): 809–824.

Wilson, E.B. (1927). Probable inference, the law of succession, and statistical inference. *Journal of the American Statistical Association* 22: 209–212.

Wu, S. and Chan, L.Y. (2003). Performance utility-analysis of multi-state systems. *IEEE Transactions on Reliability* 52 (1): 14–20.

Wu, X., Hillston, J., and Feng, C. (2017). Availability modeling of generalized k-out-of-n:G warm standby systems with PEPA. *IEEE Transactions on Systems, Man, and Cybernetics: Systems* 47 (12): 3177–3188.

Xing, L. and Dai, Y.S. (2009). A new decision-diagram-based method for efficient analysis on multistate systems. *IEEE Transactions on Dependable and Secure Computing* 6 (3): 161–174.

Yeh, W.C. (2007). An improved sum-of-disjoint-products technique for the symbolic network reliability analysis with known minimal paths. *Reliability Engineering and System Safety* 92 (2): 260–268.

Zuo, M.J., Tian, Z., and Huang, H.-Z. (2007). An efficient method for reliability evaluation of multistate networks given all minimal path vectors. *IIE Transactions* 39 (8): 811–817.

Problems

Problem 2.1   An electric vehicle manufacturer is testing a new generation of lithium-ion batteries. One hundred battery cells are tested under fast direct current charging and discharging process for 500 cycles. A battery cell fails if its maximum state-of-charge (SOC) drops below 80% of its initial capacity. The number of failed cells at 300 and 500 cycles is 10 and 25, respectively. Estimate the mean and the variance of the cell reliability.

Problem 2.2   Solar PV systems emerged as a green and sustainable energy source to mitigate climate change resulting from carbon emissions. A roof-top PV system usually consists of PV panels, a charge controller, a battery pack, and a DC–AC inverter. A PV system can also operate in islanding mode like a microgrid if the generation can fully meet the demand. Assume the lifetimes of the PV, the charger controller, the battery, and the inverter follow the exponential distribution with failure rates of 0.1, 0.01, 0.05, and 0.02 failures/year, respectively.

Questions:

(1) Draw the reliability block diagram of the roof-top PV system.

(2) Estimate the reliability of each components at $t = 1$, 10, and 20 years.

(3) Estimatestimate the reliability of the entire PV system at $t = 1$, 10, and 20 years.

Problem 2.3   A company developed a new microdevice that will be used in offshore oil drilling equipment. To mimic the harsh field operating condition, 20 units are chosen and placed in an environmental chamber to perform accelerated life testing. The following results are observed: one unit failed at $t = 60$ hours, another unit failed at $t = 80$ hours, and two additional units failed at $t = 90$ hours. The test terminated at $t = 100$ hours with a total of 16 units surviving. Answer the following:

(1) Estimate the mean and the variance of device reliability at $t = 50, 70$, and 100 hours, respectively.

(2) If 10 such devices are deployed in the field, what is the probability that exactly 8 units can survive at $t = 70$ hours?

(3) If 10 such devices are deployed in the field, what is the probability that all these units will fail at $t = 100$ hours.

Problem 2.4   A system consists of six components connected in series. Reliability tests are applied to these components over a 20-day period and the results are summarized in Table 2.8.

Table 2.8 Component failure data.

Component type	$i = 1$	$i = 2$	$i = 3$	$i = 4$	$i = 5$	$i = 6$
Sample size	20	30	25	16	25	16
Cumulative failures by day 10	0	2	1	1	1	1
Cumulative failures by day 20	1	2	2	1	2	1

Questions:
(1) Estimate the mean and the variance of the system reliability at $t = 10$ days.
(2) Estimate the mean and the variance of the system reliability at $t = 15$ days.
(3) Estimate the mean and the variance of the system reliability at $t = 20$ days.

Problem 2.5   Let $\hat{q}(t)$ be the unreliability estimate of a component and $\hat{q}(t) = 1 - \hat{r}(t)$. Show that $V\hat{ar}(\hat{q}(t)) = V\hat{ar}(\hat{r}(t))$.

Problem 2.6   A series system consists of five components. Estimate the mean and the variance of the reliability estimates of the series system at $t_1 = 1$ week and $t_2 = 2$ weeks, respectively. Note that component lifetime testing data are presented in Table 2.9.

Table 2.9 Component lifetime testing data.

i	1	2	3	4	5
Sample size	10	14	20	17	19
Failures by t_1	0	1	2	1	0
Failures by t_2	1	2	2	1	1

Problem 2.7   Show that $\hat{R}_s(t)$ in Eq. (2.3.3) is an unbiased estimate of $R_s(t)$.

Problem 2.8   Show that $V\hat{ar}(\hat{R}_s(t))$ in Eq. (2.3.11) is an unbiased estimate of $Var(\hat{R}_s(t))$.

Problem 2.9   For aircraft equipped with two identical engines, the system is reliable as long as one engine is functional. What is the minimum reliability requirement for each engine in order to attain the 0.999 999 reliability goal of the aircraft? What is the aircraft reliability if three engines are installed?

Problem 2.10 The same group of components in Problem 2.10.6 are used to construct a parallel system. Estimate the mean and the variance of the reliability estimates of a system comprised of five components in parallel at t_1 and t_2, respectively.

Problem 2.11 Show that Eqs. (2.4.2) and (2.4.3) are the unbiased estimates for $\hat{R}_p(t)$ and $Var(\hat{R}_p(t))$, respectively. Note that $\hat{R}_p(t)$ is the reliability estimate for a parallel system.

Problem 2.12 Based on the reliability data in Example 2.1, assume that a battery storage system is comprised of three battery cells connected in parallel. The battery system is able to discharge the electric energy as long as one cell is functional. Evaluate the reliability of this energy storage system in terms of mean reliability and its variance using Monte Carlo simulations.

Problem 2.13 Show that series and parallel systems are the special cases of k-out-of-n:G systems. If $k = n$, Eq. (2.6.1) is equivalent to Eq. (2.3.3) for a pure series system reliability model. If $k = 1$, it is the same as Eq. (2.4.1) for pure parallel systems.

Problem 2.14 A power grid is often designed under $N - 1$ redundancy criterion, meaning the network is able to meet the peak load even if one generator fails or is unavailable due to maintenance. Suppose the peak load of the power grid is 200 MW (megawatts) and 10 identical power generators are installed to produce the electricity for industrial, commercial, and residential users. The power capacity of each generator is 25 MW. Answer the following questions:
1) Assuming 10 generators are always available and never fail, is this power system indeed an $N - 1$ design scheme?
2) If the reliability of each generator is 0.99 and the reliability of the power grid is required to be 0.9999, is this power system an $N - 1$ design scheme?
3) Given that the reliability of each generator is 0.99 and the power system is designed under the $N - 1$ rule, what is the maximum reliability that the grid can achieve in peak hours?

Problem 2.15 Estimate the mean and the variance of the following series–parallel and parallel–series system based on component reliability data in Table 2.5 (Figure 2.24).

Problem 2.16 Please estimate the mean and the variance of the following hybrid system based on component reliability data in Table 2.5 (Figure 2.25).

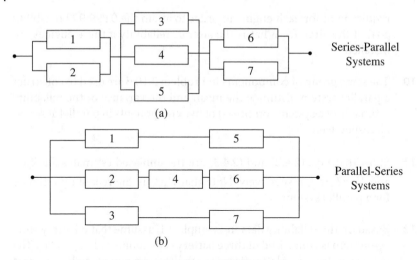

Series-Parallel
Systems

(a)

Parallel-Series
Systems

(b)

Figure 2.24 (a) Series–parallel system, (b) parallel–series system.

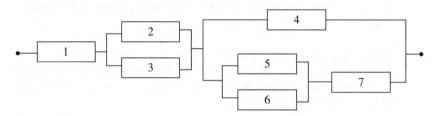

Figure 2.25 A mixed series–parallel system.

Problem 2.17 It is anticipated that by 2030 wind energy will represent 30% of the utility market in the USA. To achieve this penetration rate, large wind farms with dozens or hundreds of wind turbines are being deployed in onshore regions or offshore areas. A main challenge in operating a wind farm is the high equipment maintenance costs, which include spare parts, labor hours, and rental of cranes or boats (in offshore). Suppose an offshore wind farm consists of a fleet of 50 wind turbines and the reliability of each wind turbine is $r = 0.9$ at the end of year 1 and $r = 0.75$ at the end of year 2. The fleet availability is defined as the probability that a certain number of wind turbines must be functional at any instance of time. Compute the fleet availability if:

(1) At least 45 wind turbines must be operational during the first year.
(2) At least 45 wind turbines must be operational during the second year.
(3) If we want to ensure that at least 45 turbines are available with 0.99 confidence in year 1, what is the minimum reliability of an individual wind turbine?

Problem 2.18 Find the minimum path and minimum set of the network in Figure 2.1d.

Problem 2.19 Based on the minimum path and minimum set of the network in Figure 2.1d, compute:
1) The terminal pair reliability based on minimum paths.
2) The terminal pair reliability based on minimum cuts.
3) What is the exact terminal pair reliability of this network (using edge decomposition)?

Problem 2.20 A diode bridge is a device that changes alternating current (AC) to direct current (DC). Its reliability diagram is shown in Figure 2.15 where components 1–5 correspond to five diodes. A total of 30 devices have been made and installed at the beginning of January 2014. Failures reported from customers are collected and analyzed by the service department of the device manufacturer and the results are summarized in Table 2.10. Please estimate the mean and the variance of this AC–DC device reliability based on the linear-quadratic reliability model.

Table 2.10 Reported diode failures in each quarter in 2014.

Months	$i = 1$	$i = 2$	$i = 3$	$i = 4$	$i = 5$
1–3	0	0	0	0	1
4–6	1	0	0	0	0
7–9	1	1	1	1	0
10–12	0	1	0	0	1

Problem 2.21 Let $\hat{r}$ be the component reliability estimate. Given the mean $E[\hat{r}] = 0.9$ and the variance $Var(\hat{r}) = 0.01$, estimate the confidence interval for $\hat{r}$ based on the normal distribution approximation at $\alpha = 0.1$ and 0.05, respectively.

Problem 2.22 Given $E[\hat{r}] = 0.9$ and the sample size $n = 20$, estimate the confidence interval for $\hat{r}$ using Wilson score interval estimates at $\alpha = 0.1$ and 0.05, respectively.

Problem 2.23 Given $E[\hat{r}] = 0.9$ and the sample size $n = 20$, estimate the confidence interval of reliability estimate $\hat{r}$ using the Clopper–Pearson method at $\alpha = 0.1$ and 0.05, respectively.

Problem 2.24 Based on the system structure and the component reliability given in Problem 2.6, estimate the confidence interval of the system reliability estimate at $\alpha = 0.1$ and 0.05, respectively. The following approaches can be considered:
(1) The log-normal approximation approach.
(2) Normal approximation.
(3) How about the beta binomial method (i.e. the Clopper–Pearson method)?

Problem 2.25 Estimate the reliability of the following two systems shown in Figure 2.26 where each component possesses three states: fail open, fail short, and normal operation. The component reliability information is listed in Table 2.11.

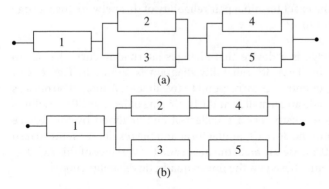

(a)

(b)

Figure 2.26 (a) A series–parallel system, (b) a mixed series–parallel system.

Table 2.11 Reliability of multistate components.

i	1	2	3	4	5
Fail open	0.1	0.1	0.1	0.1	0.1
Fail short	0.2	0.2	0.2	0.2	0.2

Problem 2.26 Three components are available to construct a multistate system and each component can operate in one of following states: 100%, 75%, 5%, or 25% of the capacity. The probabilities associated with each state are given in Table 2.12. Using the universal generating function approach to estimating the MSS reliability, estimate the system reliability if:
1) Three components are connected as a pure series system.
2) Three components are connected as a pure parallel system.

Table 2.12 Reliability of four-state components.

State (j)	$j = 1$	$j = 2$	$j = 3$	$j = 4$
Capacity (%)	100	75	50	25
Component 1	0.5	0.3	0.1	0.1
Component 2	0.4	0.2	0.2	0.2
Component 3	0.3	0.45	0.1	0.15

Problem 2.27 Compute the Birnbaun importance measures of each component within a system in Figure 2.27. The reliability of components 1 to 4 is $r_1 = 0.9$, $r_2 = 0.85$, $r_3 = 0.8$, and $r_4 = 0.75$.

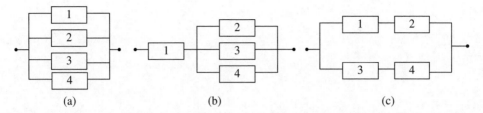

(a)   (b)   (c)

Figure 2.27 (a) Parallel system; (b) a series–parallel system; (c) parallel–series system.

Problem 2.28 Show that Eqs. (2.5.1) and (2.5.3) are, respectively, unbiased estimates of $\widehat{R}_p(t)$ and $Var(\widehat{R}_p(t))$, where $\widehat{R}_p(t)$ is the reliability estimate of a parallel system.

Problem 2.29 Show that Eqs. (2.5.5) and (2.5.6) are, respectively, unbiased estimates of $\widehat{R}_{sp}(t)$ and $Var(\widehat{R}_{sp}(t))$, where $\widehat{R}_{sp}(t)$ is the reliability estimate of a series–parallel system.

Problem 2.30 For a series system comprised of m components, the mean and variance of $\widehat{R}_i(t)$ are μ_i and σ_i^2, where $\widehat{R}_i(t)$ is the reliability estimate of component i. Show that the lower and upper bounds of the system reliability estimate, $\widehat{R}_s(t)$, are given as follows:

$$R_L = \widehat{R}_s(t) \exp\left(\tfrac{1}{2}\widehat{\sigma}^2 - z_{\alpha/2}\widehat{\sigma}\right)$$

$$R_U = \widehat{R}_s(t) \exp\left(\tfrac{1}{2}\widehat{\sigma}^2 + z_{\alpha/2}\widehat{\sigma}\right) \text{where}$$

$$\widehat{\sigma}^2 = \ln\left(1 + \tfrac{\widehat{Var}(\widehat{R}_s(t))}{\widehat{R}_s^2(t)}\right)$$

Problem 2.31 Given is a system with five components configured in parallel. To estimate the system reliability, 20 units of each are placed for reliability testing. The number of failures at $t = 50$ days and 100 days are recorded and listed in the following table. Assuming the system unreliability is lognormally distributed, estimate the following:
1) The mean and the variance of the system reliability estimate at $t = 50$ and 100 days.
2) The lower and upper bounds of the system reliability estimate at $t = 50$ and 100 days.

Component	1	2	3	4	5
n	20	20	20	20	20
k ($t = 50$ days)	1	0	2	1	2
k ($t = 100$ days)	2	1	3	2	2

3

Design and Optimization for Reliability

3.1 Introduction

Design for reliability (DFR) encompasses a set of tools that support product and process management from concept development, prototyping, new product introduction, volume production, and all the way through retirement. With affordable repair and maintenance costs, the goal of DFR is to ensure that customer expectations for reliability are fully met across a product lifecycle. In this chapter, we introduce several DFR tools that are commonly used by reliability practitioners and design engineers: reliability and redundancy allocation (RRA), failure-in-time (FIT) design, design for six-sigma (DFSS), fault-tree analysis (FTA), and failure mode, effect, and criticality analysis (FMECA). While RRA is implemented at the component level, both FIT and DFSS are often used in the detailed design of subsystems or assemblies. FTA uses a top-down approach to investigate the root cause of systems failure and to determine the ways to reduce the risk of such failures. FMECA allows the design engineers to identify and rank the potential failure modes in the early development phase, and further determine the remedies to remove or mitigate the risk of failures.

3.2 Lifecycle Reliability Optimization

3.2.1 Reliability–Design Cost

Reliability can be designed in during the product development and prototyping stage. It is generally agreed that the design and development costs increase exponentially with the reliability of the item, including hardware and software (Huang et al. 2007; Guan et al. 2010). This is mainly due to the fact that when reliability reaches a certain level, the efforts and the resources required for further reduction of the failure rate increases much faster than the reliability gain. Let $c_d(r)$ be the product design and development cost with reliability r for $0 < r \leq 1$. Mettas (2000) proposes the following exponential model to characterize the relation between the reliability and the cost:

$$c_d(r) = c_0 \exp\left((1-k)\frac{r - r_{\min}}{r_{\max} - r}\right), \quad \text{for} \quad r_{\min} \leq r < r_{\max} \tag{3.2.1}$$

The model contains four non-negative parameters, namely, $r_{\min}$, $r_{\max}$, k, and c_0. Note that $r_{\min}$ represents the minimum acceptable reliability required by the customers; $r_{\max}$

Reliability Engineering and Services, First Edition. Tongdan Jin.
© 2019 John Wiley & Sons Ltd. Published 2019 by John Wiley & Sons Ltd.
Companion website: www.wiley.com/go/jin/serviceengineering

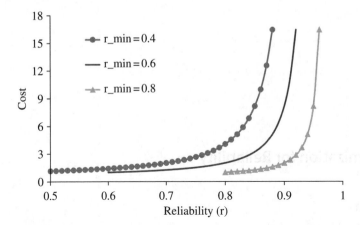

Figure 3.1 Reliability–design cost under different r_{min} with $r_{max} = 1$, $k = 0.3$ and $c_0 = 1$.

is the maximum reliability that can be achieved by the manufacturer. Obviously the condition $0 < r_{min} < r_{max} \leq 1$ always holds and k for $0 \leq k \leq 1$ is the feasibility of increasing the reliability. A large k implies that it is relatively easy, and hence less costly, to increase the reliability based on current manufacturing technology and available materials. Finally c_0 is the baseline design and development cost with $r = r_{min}$.

Figure 3.1 shows how $c_d(r)$ increases with the reliability r for different values of r_{min}. These curves are generated assuming $r_{max} = 1$, $k = 0.3$, and $c_0 = 1$. It is also found that to achieve the same reliability level, it is more costly to improve the product with lower reliability. This observation make senses because a product with a low reliability requires more resources and effort to bring it to a high level compared with a reliable product.

Depending on the nature of the product and existing technologies, reliability of certain components can be more difficult to improve compared with others. For example, in an airplane, it is relatively easy to improve the reliability of the avionics than the engine because the operation condition and the structure of the latter is much more complicated than the former. Parameter k represents the degree of difficulty to increase a product's reliability. Given $r_{min} = 0.7$, $r_{max} = 1$, and $c_0 = 1$, Figure 3.2 depicts how the reliability design cost increases with r corresponding to $k = 0.1, 0.5$, and 0.8, respectively. It shows that for a larger k, less cost is required in order to achieve the same reliability level.

Sometimes the failure rate, instead of the reliability, is more convenient to capture the relation between reliability and design cost. For instance, the failure rate is often used to measure the reliability of a software product or program. Instead of the total cost, the incremental design cost sometimes is more interesting to the manufacturer as it measures the reliability growth as a result of additional investment. Let $b(\alpha)$ be the incremental reliability–design cost for a product. Then a variant of Eq. (3.2.1) is given below

$$b(\alpha) = c_0 \left[\exp\left(\phi \frac{\alpha_{max} - \alpha}{\alpha - \alpha_{min}} \right) - 1 \right], \quad \text{for } \alpha_{min} < \alpha \leq \alpha_{max} \qquad (3.2.2)$$

where α_{max} and α_{min} are the maximum acceptable and the minimum achievable failure rate, respectively. Note that c_0 is the baseline design cost when $\alpha = \alpha_{max}$ and ϕ is a positive parameter characterizing the difficulties in reliability growth. In general, a large ϕ implies that more effort and resources are required in order to achieve the same reliability performance.

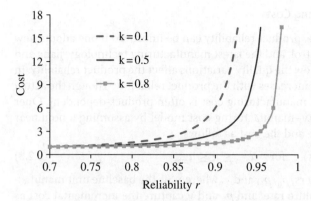

Figure 3.2 Reliability–design cost under various k with $r_{min} = 0.7$, $r_{max} = 1$ and $c_0 = 1$.

Example 3.1 Two products A and B are in the development stage. Their baseline reliability costs and the associated parameters are given in Table 3.1. Do the following: if an additional \$5000 of design budget is allocated to each product, which one will yield a higher reliability growth?

Solution:
Based on Eq. (3.2.1), we can obtain the reliability growth expression as follows:

$$r = \frac{ar_{max} + r_{min}}{1 + a}, \quad \text{with} \quad a = \frac{\ln d(r) - \ln c_0}{1 - k}$$

For Product A, we have

$$a = \frac{\ln(10\,000 + 5000) - \ln(10\,000)}{1 - 0.5} = 0.811$$

$$r = \frac{(0.811)(0.95) + 0.8}{1 + 0.811} = 0.867$$

$$r - r_{min} = 0.867 - 0.8 = 0.067$$

For Product B, we have

$$a = \frac{\ln(15\,000 + 5000) - \ln(15\,000)}{1 - 0.7} = 0.959$$

$$r = \frac{(0.959)(0.99) + 0.85}{1 + 0.959} = 0.909$$

$$r - r_{min} = 0.909 - 0.85 = 0.059$$

Therefore, the reliability growth of Product A is larger than B upon an additional \$5000 investment.

Table 3.1 Baseline design cost and reliability–cost parameters.

Product	c_0 (\$)	r_{min}	r_{max}	k
A	10 000	0.8	0.95	0.5
B	15 000	0.85	0.97	0.7

3.2.2 Reliability–Manufacturing Cost

During the manufacturing stage, product reliability can be improved if one adopts new materials, advanced quality control, and the latest manufacturing technology. Jiang and Murthy (2009) have discussed how the quality variations affect the product reliability. In general, the manufacturing cost increases with the product reliability. Though the actual relation between reliability and manufacturing cost is often product-dependent, Öner et al. (2010) propose a reliability–manufacturing cost model by assuming a nonlinear relation between the failure rate and the cost as follows:

$$c(\alpha) = p_0 + p_1(\alpha^{-v} - \alpha_{max}^{-v}), \quad \text{for} \quad \alpha_{min} \leq \alpha \leq \alpha_{max} \tag{3.2.3}$$

The model has three parameters, p_0, p_1, and v, where p_0 is the baseline unit manufacturing cost at the maximum failure rate, and p_1 and v capture the incremental cost as the failure rate α is reduced with respect to α_{max}. Particularly, v captures the degree of difficulties in reducing the failure rate given the current manufacturing technology or raw materials. This model assumes that the design phase is relatively short compared to the product useful lifetime. Hence, the impact of the learning effect is not considered. Reliability–manufacturing cost models incorporating the learning effects are available in Teng and Thompson (1996).

Figure 3.3 depicts the unit manufacturing cost versus the failure rate under $v = 0.3, 0.5$, and 0.7, respectively. Curves A, B, and C are plotted with the same values of $p_0 = \$400$, $p_1 = \$50$, and $\alpha_{max} = 0.01$. For large v, it is more costly to improve the product reliability through the manufacturing process. C and D are plotted with the same value of v, but different p_1. A larger p_1 implies that more resources are required in order to maintain the same failure rate. Therefore, a trade-off must be made between the DFR and the design for manufacturing so as to achieve the reliability target at a minimum cost. This will be discussed in the next section.

3.2.3 Minimizing Product Lifecycle Cost

A main objective in reliability optimization is to optimize the resource allocation across product design, manufacturing, and field support such that the lifecycle cost

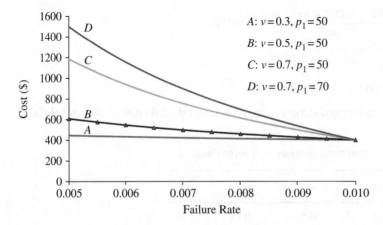

Figure 3.3 Reliability–manufacturing cost with $\alpha_{min} = 0.005$ and $\alpha_{max} = 0.01$.

is minimized. In this section, we present a DFR optimization model to facilitate the product manufacturer to achieve this objective. In our setting, we assume that the manufacturer offers free repair and replacements during the warranty period. The goal is to optimize the product reliability such that the costs associated with design, manufacturing, and warranty services are minimized. The following notation used in the model is defined:

T = length of the warranty period (year)
p_2 = annual repair or replacement cost per unit product ($/unit/year)
Q = number of products shipped to the customers

The optimization model is formulated to minimize the total costs of all installed products during the warranty period T. These costs include design, manufacturing, and warranty services.

Model 3.1

$$\text{Min}: \quad C(\alpha) = c_0 \left[\exp\left(\phi \frac{\alpha_{max} - \alpha}{\alpha - \alpha_{min}} \right) - 1 \right] + (p_0 + p_1(\alpha^{-\nu} - \alpha_{max}^{-\nu}))Q + \alpha p_2 TQ$$

(3.2.4)

Subject to:

$$\alpha_{min} < \alpha \le \alpha_{max}$$

(3.2.5)

The objective function (3.2.4) is comprised of design, manufacturing, and warranty repair costs. Particularly, the first item is the incremental design cost, while the second item stands for the manufacturing cost of Q units, also referred to as an installed base. The last item is the repair cost for the failures from Q units during T years. Constraint (3.2.5) simply defines the upper and lower bounds of the product failure rate.

Figure 3.4 graphically shows the relations between design cost, manufacturing cost, and warrant repair cost. On one hand, to decrease the failure rate, more investments are required in product design and manufacturing stages, but the warranty cost will be saved because of reduced failure returns. On the other hand, we can save the design and

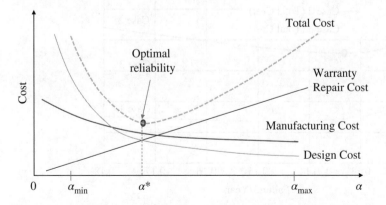

Figure 3.4 Sum of the design, manufacturing, and warranty costs.

manufacturing costs by accepting a relatively high product failure rate, but the warranty cost will escalate because of high field returns. Therefore, the manufacturer needs to determine the optimal α^* that can balance the trade-off between design, manufacturing, and warranty costs. We take the derivative with respect to α in Eq. (3.2.4) and set it to zero. It yields

$$\frac{c_0 \phi(\alpha_{min} - \alpha_{max})}{(\alpha - \alpha_{min})^2} \exp\left(\phi \frac{\alpha_{max} - \alpha}{\alpha - \alpha_{min}}\right) - v p_1 \alpha^{-v-1} + p_2 QT = 0 \qquad (3.2.6)$$

The optimality is guaranteed because $C(\alpha)$ is the convex function between $[\alpha_{min}, \alpha_{max}]$. Namely,

$$\frac{\partial^2}{\partial \alpha} C(\alpha) > 0, \quad \text{for } \alpha_{min} < \alpha \le \alpha_{max} \qquad (3.2.7)$$

Hence, the optimal value of α^* is the one that satisfies Eq. (3.2.6). Unfortunately, there is no closed-form solution to α^*. Thus one can find α^* by using the iteration search over the interval between $[\alpha_{min}, \alpha_{max}]$. Below we present a numerical example to illustrate this method.

Example 3.2 The following parameters are given for a new product design: $c_0 = \$50\,000$, $\alpha_{min} = 0.1$ failures/year, $\alpha_{max} = 0.3$ failures/year, $\phi = 0.3$, $p_0 = \$10\,000$/unit, $p_1 = \$500$, $v = 0.2$, and $p_2 = \$4000$/unit/year. Find α^* in the following cases: (1) find the optimal α^* for $T = 1$ year and $Q = 1000$ units; (2) find α^* for $T = 1$ year and $Q = 3000$ units; and (3) find α^* if $T = 2$ years and $Q = 1000$ units.

Solution:
In case 1, let the increment step $\Delta\alpha = 0.001$. We use $\alpha = \alpha_{min} + \Delta\alpha$ as our initial solution and substitute it into Eq. (3.2.4) to calculate the total cost. Then we increase $\alpha = \alpha_{min} + 2\Delta\alpha$ and calculate the new cost using Eq. (3.2.4). This process is repeated until α reaches α_{max}. Finally, we sort the cost from the lowest to the highest, with the one that results in the lowest cost chosen as the optimal α^*. In case 1, we find $\alpha^* = 0.146$ and the total cost is \$680\,605.

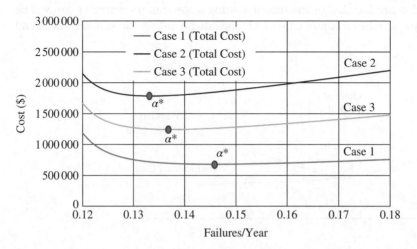

Figure 3.5 The optimal reliability of the product in three cases.

In case 2, we increase $Q = 3000$ units and keep $T = 1$ year. The iteration search results in the optimal failure rate of $\alpha^* = 0.133$ and the total cost is \$1 784 308.

In case 3, we extend T to 2 years while keeping $Q = 1000$ unit. It turns out that the optimal $\alpha^* = 0.137$ and the total cost is \$1 243 583. The results of the three cases are shown in Figure 3.5.

3.3   Reliability and Redundancy Allocation

The RRA model is a mathematical programing that guides the design engineer to achieve the following objectives: (i) minimize the product resources while attaining the reliability goal or (ii) maximize the system reliability without violating the resource constraints. We discuss the cost minimization problem in this section.

For reliability allocation models, the component reliability is treated as a decision variable. For redundancy allocation models, the decision variable is the level of redundancy and the choice of component types. Most papers focus on a single objective, while some have addressed multiobjective optimization. Various mathematical programming techniques have been used to solve reliability–redundancy optimization problems. These techniques include dynamic programming (Fyffe et al. 1968), integer programming, mixed integer programming (Ghare and Taylor 1969; Bulfin and Liu 1985), heuristics, genetic algorithms (Coit and Smith 1998; Taboada et al. 2008), and a solution space reduction procedure combined with branch-and-bound (Sung and Cho 1999), among others. Each technique is effective in solving some particular optimization problems. Readers are referred to the book by Kuo and Zuo (2003) for a comprehensive treatment on this topic. In this section, typical RRA models are reviewed in terms of problem formulation and basic solution techniques.

3.3.1   Reliability Allocation for Cost Minimization

For multicomponent system design, it is often useful to predict the reliability of the system based on individual component reliability. The question is how to meet the reliability target of the system subject to cost and operational constraints? The simplest answer to this question is to distribute the reliability goal equally among all subsystems or components. However, such allocation may not be feasible or optimal in terms of minimizing the cost, weight, and volume of a product.

Take the three-component system in Figure 3.6 as an example. If the target system reliability is $R_s = 0.95$, then we have various options to achieve the goal. For instance,

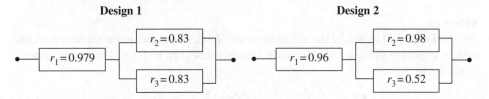

Figure 3.6 Two design options for a series–parallel system.

we may choose $r_1 = 0.979$ and $r_2 = r_3 = 0.83$, which leads to $R_s = 0.951$; or we can also assign $r_1 = 0.96$, $r_2 = 0.98$, and $r_3 = 0.52$, which yields $R_s = 0.951$. Both designs meet the system reliability target. In Design 1, components 2 and 3 have medium reliability, while in Design 2, reliability of component 2 is high, but it is low for component 3. Obviously, the total system cost is likely to be different because component costs are correlated with reliability. Reliability allocation is a method of apportioning a system's target reliability among different components. Hence, it is a useful tool for design engineers to determine the required reliability of individual components to achieve the system reliability goal with minimum cost, time, and other resource consumptions. Such requirement can be met by solving the following reliability optimization problem, denoted as Model 3.2.

Model 3.2 Minimize:

$$C(\mathbf{r}) = \sum_{i=1}^{s} c_i(r_i) \tag{3.3.1}$$

Subject to:

$$R_s(r_1, r_2, \cdots, r_s) \geq R_{\min} \tag{3.3.2}$$

$$g_j(r_1, r_2, \ldots, r_s) \leq b_j, \quad \text{for} \quad j = 1, 2, \ldots, m \tag{3.3.3}$$

$$0 < r_i \leq 1, \text{for } j = 1, 2, \ldots, m \tag{3.3.4}$$

where

s = number of component types in the system
r_i = reliability for component i, a decision variable, $i = 1, 2, 3, \ldots, s$
$c_i(r_i)$ = the cost of component i with reliability r_i
$R_{\min}$ = target system reliability
b_j = constraint limit, $j = 1, 2, 3, \ldots, m$

The decision variables are the component reliability r_i. The objective function (3.3.1) aims to minimize the total system cost, which is the sum of all s components. Constraint (3.3.2) ensures that the system achieves the target reliability. Constraint (3.3.3) stands for other constraints, such as weight or volume, that are imposed on the design. The constraint (3.3.4) is self-explanatory.

Example 3.3 We use the simple system in Figure 3.6 to demonstrate the application of Model 3.2. The reliability-cost data of components are given in Table 3.2. The target system reliability $R_{\min} = 0.9$.

Solution:
We use Eq. (3.2.1) to model the reliability-cost of individual components. Based on the system structure in Figure 3.6, the optimization model can be formulated as follows:
   Minimize:

$$C(r_1, r_2, r_3) = \sum_{i=1}^{3} c_{0,i} \exp\left((1 - k_i)\frac{r_i - r_{\min,i}}{r_{\max,i} - r_i} \right) \tag{3.3.5}$$

Table 3.2 Baseline design cost and reliability–cost parameters.

Component	c_0 ($)	r_{min}	r_{max}	k
1	10 000	0.9	0.98	0.5
2	15 000	0.85	0.95	0.7
3	8 000	0.7	0.9	0.3

Subject to:

$$r_1(r_2 + r_3 - r_2 r_3) \geq 0.9 \tag{3.3.6}$$

$$r_{min,i} \leq r_i < r_{max,i}, \quad \text{for } i = 1, 2, \text{ and } 3 \tag{3.3.7}$$

This problem can be solved using Lagrangian relaxation or by enumerations because only three design variables are involved. Since the objective is convex, we use the non-linear optimizer in Excel to directly search for the optimal component reliability. The following optimality is obtained: $r_1^* = 0.92$, $r_2^* = 0.9$, and $r_3^* = 0.77$, and the total cost is $C(\mathbf{r}^*) = \$43\,589$.

3.3.2 Reliability Allocation under Cost Constraint

Sometimes the design budget like cost is limited; then the objective becomes how to allocate the budget to individual components such that the overall system reliability is maximized. Hwang et al. (1979) proposes a reliability allocation model to maximize the reliability of complex systems subject to resources constraints. The problem is formulated as follows.

Model 3.3

$$\text{Max}: \quad f(\mathbf{r}) = R(r_1, r_2, \ldots, r_s) \tag{3.3.8}$$

Subject to:

$$g_j(\mathbf{r}) \leq b_j, \quad \text{for } j = 1, 2, \ldots, m \tag{3.3.9}$$

$$0 < r_i \leq 1, \text{ for } i = 1, 2, \ldots, s \tag{3.3.10}$$

where objective function (3.3.8) aims to maximize the system reliability and constraint (3.3.9) defines the resource constraint, such as budget and time. Sayama et al. (1973) proposed a generalized Lagrangian function to solve Model 3.3, which takes the following form:

$$L(\mathbf{r}, \lambda; \beta) = f(\mathbf{r}) - \sum_{j=1}^{m} \lambda_j(g_j(\mathbf{r}) - b_j) + \sum_{j=1}^{m} \begin{cases} \beta(g_j(\mathbf{r}) - b_j)^2, & g_j(\mathbf{r}) \leq 0 \\ \dfrac{\lambda_j \beta(g_j(\mathbf{r}) - b_j)}{\lambda_j + \beta(g_j(\mathbf{r}) - b_j)}, & g_j(\mathbf{r}) \geq 0 \end{cases} \tag{3.3.11}$$

If $\beta = 0$, Eq. (3.3.11) becomes the classical Lagrangian function. They solved two numerical examples and found the generalized Lagrangian function has a faster convergence speed than the gradient method and sequential unconstrained minimum techniques. The following example is used to illustrate how to allocate components with different reliability to maximize the reliability of a series system.

Example 3.4 A product is configured by three subsystems in series. For each subsystem, two vendors are available to supply the components. These components usually offer the same functionality, yet with different reliability and cost. The design engineer must identify one among two suppliers for each subsystem, but not both. Also each subsystem is only comprised one component. Component reliability and cost are listed in Table 3.3. Note that i is the index of the subsystem and j is the index of the component type within a subsystem. Answer the following questions: (1) What is the maximum achievable system reliability? (2) What is the maximum system reliability given the total cost budget of 6? (3) What is the maximum system reliability if the total budget is 7?

Solution:

1) If there is no cost constraint, the system achieves its maximum reliability when the reliability of individual subsystems is the highest. Hence the best solution is $r_{12} = 0.99$, $r_{21} = 0.97$, and $r_{31} = 0.98$, and the system reliability is $(0.99)(0.97)$ $(0.98) = 0.941$.
2) Given the total budget is 6, the only feasible solution is $r_{11} = 0.95$, $r_{22} = 0.96$, and $r_{32} = 0.97$. The associated cost is $c_{11} + c_{22} + c_{32} = 1 + 2 + 3 = 6$. All other potential solutions will violate the cost constraint. Hence, the best achievable system reliability is $(0.95)(0.96)(0.97) = 0.885$.
3) If the design budget is 7, $\{r_{11} = 0.95, r_{21} = 0.97, r_{33} = 0.97\}$, $\{r_{11} = 0.95, r_{22} = 0.96, r_{31} = 0.98\}$, and $\{r_{12} = 0.99, r_{22} = 0.96, r_{31} = 0.98\}$ are the feasible solutions. However, the last one is preferred as it yields a higher system reliability of 0.931.

In Example 3.4, reliability allocation is carried out by identifying the best component type among multiple choices. This design strategy ensures the maximization of the overall system reliability without violating the resource constraint. Most real-world systems are quite complex and are configured with a large amount of components, each associated with different amounts of reliability, cost, weight, and volume. In these cases, advanced solution techniques and optimization algorithms must be employed in order to solve large problems that involve dozens or hundreds of variables. For additional information on this subject, readers are referred to the book by Kuo et al. (2001).

Table 3.3 Component reliability and cost.

Subsystem i	Component type $j = 1$		Component type $j = 2$	
	Reliability (r_{ij})	Cost (c_{ij})	Reliability (r_{ij})	Cost (c_{ij})
1	0.95	1	0.99	2
2	0.97	3	0.96	2
3	0.98	4	0.97	3

3.3.3 Redundancy Allocation for Series System

A redundancy allocation problem (RAP) deals with how to build a parallel structure to effectively increase the system reliability. Hence the decision variable is the level of redundant units for a component type. This differs from the reliability allocation problem in which the component's reliability is the design variable. If multiple component types are available to a subsystem, the choice of component types is also a decision variable. For series–parallel systems, if subsystems allow multiple component type choices, Chern (1992) has proved that the redundancy allocation problem is indeed NP-hard. Dynamic programming, mixed integer programing, and heuristic algorithms have been used to solve this type of problem. Liang and Smith (2004) developed an ant colony meta-heuristic method to solve the RAP where different components can be placed in parallel. The general formulation for RAP for a series system under cost constraint is defined as follows.

Model 3.4

$$\text{Max} : R(x) = \prod_{i=1}^{s} \left(1 - \prod_{j=1}^{n_i} (1 - r_{ij})^{x_{ij}} \right) \tag{3.3.12}$$

Subject to:

$$\sum_{i=1}^{s} \sum_{j=1}^{n_i} x_{ij} c_{ij} \leq b \tag{3.3.13}$$

$$x_{ij} \in \{0, 1, 2, 3, \ldots\} \tag{3.3.14}$$

As shown in Eq. (3.3.12), the objective is to maximize the series–parallel system reliability, where x_{ij} is the decision variable, representing the number of components j in subsystem i for $i = 1, 2, \ldots, s$. Constraint (3.3.13) states that the investments of components should be confined to the available budget b, where c_{ij} is the unit cost of component j in subsystem i. Note that n_j is the number of component types for subsystem i. Constraint (3.3.14) simply ensures the non-negativity of x_{ij}. Model 3.4 is a mixed-integer nonlinear optimization (MINL) problem. Below we introduce a heuristic algorithm to solve this problem through a simple example.

Example 3.5 We use the series system in Figure 3.7 to demonstrate the application of Model 3.4. The system consists of three subsystems, and redundant units are adopted to construct each subsystem. For this problem $s = 3$ and $n_i = 2$ for $i = 1, 2,$ and 3. Reliability and cost of the candidate components for each subsystem are provided in Table 3.3. Assume the total design budget is $b = 13$ and the goal is to maximize the system reliability by determining which component(s) and how many should be configured in parallel in each subsystem.

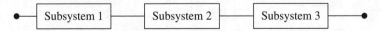

Figure 3.7 A series system consists of three subsystems.

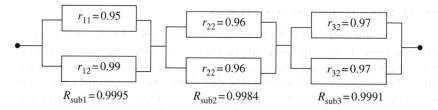

Figure 3.8 The optimal redundancy level for each subsystem.

Solution:
We use Excel solver to tackle this mixed integer programming problem and the optimal configuration of the system is depicted in Figure 3.8. It shows that subsystem 1 is formed by two components with different types, subsystem 2 consists of two identical components of type 2, and subsystem 3 is also formed by two identical components of type 2. The system reliability is 0.997 with the total available cost budget $b = 13$.

3.3.4 Redundancy Allocation for *k*-out-of-*n* Subsystems

For a series system comprised of multiple k-out-of-n subsystems, redundancy allocation aims to meet the reliability of the whole system while minimizing the total cost. Take the automated test equipment (ATE) as an example. ATE is a capital-intensive machine widely used for wafer probing and testing in the back end of semiconductor manufacturing. The downtime cost of ATE is expensive and typically in a range between \$10 000 and \$50 000/hour depending on the device under test. Hence k-out-of-n:G redundant configuration is often adopted to ensure the reliability of the ATE system. Figure 3.9 shows the reliability block diagram of an ATE system comprised of four subsystems in series: high-speed digital (HSD), analog, direct current (DC), and microwave (MW).

To test a device, a minimum number of modules (components) must be available for each subsystem. For instance, for the analog subsystem, at least k_3 modules should be good in order to generate a sufficient number of analog waveforms feeding into the device under test. The redundant modules in each subsystem usually act as a hot-standby mode, and can immediately resume the testing load if any active modules fail. Hence the reliability of a subsystem can be modeled as a k_i-out-of-n_i:G system, where k_i are the minimum required working units and n_i are the maximum components

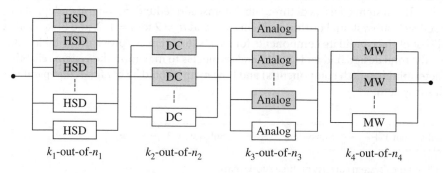

Figure 3.9 *k*-out-of-*n* subsystem redundancy allocation in ATE.

for subsystem i. The goal is to determine the redundant units for each subsystem, denoted as $x_i = n_i - k_i$, such that the ATE meets the target reliability while the cost is minimized.

Model 3.5

$$\text{Min}: \qquad g(\mathbf{x}) = \sum_{i=1}^{s} x_i c_i \qquad\qquad (3.3.15)$$

Subject to:

$$\prod_{i=1}^{s} \sum_{j=k_i}^{k_i+x_i} \binom{k_i + x_i}{k_i} r_i^j (1 - r_i)^{k_i + x_i - j} \geq R_{\min} \qquad\qquad (3.3.16)$$

$$\sum_{i=1}^{s} x_i \leq n_{\max} \qquad\qquad (3.3.17)$$

$$x_i \leq n_i^{\max}, \quad \text{for} \quad x = 1, 2, \dots, s \qquad\qquad (3.3.18)$$

$$x_i \in \{0, 1, 2, 3, \dots\} \qquad\qquad (3.3.19)$$

The objective function (3.3.15) minimizes the total cost associated with all types of redundant units in the system, where c_i is the unit cost of component type i. Constraint (3.3.16) ensures that the system reliability must be equal to or larger than $R_{\min}$, which is the target reliability. Constraint (3.3.17) and (3.3.18) define the volume or space limit of the machine, where $n_{\max}$ is the maximum number of redundant units that an ATE can accommodate. Finally, constraint (3.3.19) simply states that x_i is a non-negative integer.

Model 3.5 is a MINL problem. In general, this type of problem is difficult to solve because of its combinatorial nature coupled with the non-linearity. Below we propose a heuristic method to tackle this optimization problem. The procedure consists of six steps.

Heuristic Algorithm for Redundancy Allocation of k-out-of-n Subsystems:
Step 1: Calculate the minimum reliability required for each subsystem, i.e.
$$R_{\min}^{sub} = (R_{\min})^{1/s}.$$
Step 2: Compute the reliability of all subsystems for given k_i at $x_i = 0$ for $i = 1, 2, \dots, s$.
Step 3: If $R_i \geq R_{\min}^{sub}$ for all i, there is no need of redundancy allocation, so go to Step 5. Else go to the next step.
Step 4: Increase $x_j = x_j + 1$ and compute the reliability R_j. If $R_j > R_{\min}^{sub}$, then go to Step 5. Else let $x_j = x_j + 1$ and compute R_j, until $R_k \geq R_{\min}^{sub}$.
Step 5: Given all the feasible solution set of $\mathbf{x} = \{x_1, x_2, \dots, x_s\}$, estimate the total cost using Eq. (3.3.15).
Step 6: Choose the $\mathbf{x}$ with the lowest cost $g(\mathbf{x})$ as the optimal solution.

Below we present a numerical example to illustrate the application of the heuristic optimization algorithm.

Example 3.6 The customer who purchased the ATE machine requires the system reliability to be kept above 0.95 and the maximum $n_{\max} = 20$. The reliability of individual modules or components are given in Table 3.4.

Table 3.4 Component reliability, cost, and configuration of ATE.

Index i	Subsystem	Reliability (r_i)	Cost ($)	k	x^*	n
1	HSD	0.90	50 000	6	2	8
2	DC	0.95	40 000	4	1	5
3	Analog	0.85	70 000	3	1	4
4	MW	0.80	90 000	2	1	3

Solution:
Since the system is relatively small with only four subsystems, we implement an enumeration procedure to find the optimal x_i for each subsystem. Given $n_{max} = 20$ and $k_1 + k_1 + k_2 + k_4 = 25$, the range of x_i fall in $[0, 5]$ for all $i = 1, 2, 3$, and 4. Hence the maximum number of iterations is $(4 + 1)^4 = 625$. For a given $\mathbf{x}$, we compute the reliability and cost of the ATE system. Among all 625 possible configurations, the system that meets the reliability criterion of 0.95 with the lowest cost is the optimal design. The optimal redundancy for each subsystem is shown in the second column to the right, and the cost is $g(\mathbf{x}^*) = \$300\,000$. The corresponding subsystem reliability $R_1 = 0.99865$, $R_2 = 0.9962$, $R_3 = 0.98598$, and $R_4 = 0.972$. The system reliability is 0.9535.

Sometimes a redundant system allows both hot-standby and cold-standby subsystems. For instance, space exploration and satellite systems achieve high reliability by using both hot-standby and cold-standby redundancy for non-repairable systems (Sinaki 1994). The power grid is another good example because both hot-standby and cold-standby generating units are in place for dealing with contingent events. For exponential component failures, the system reliability at time t can be expressed as (Coit and Liu 2000)

$$R_s(t) = \left\{ \prod_{i \in H} \sum_{j=k_i}^{k_i + x_i} \binom{k_i + x_i}{k_i} (\exp(-\lambda_i t))^j (1 - \exp(-\lambda_i t))^{k_i + x_i - j} \right\}$$
$$\times \left\{ \prod_{i \in C} \left(\exp(-\lambda_i k_i t) \sum_{j=0}^{x_i} \frac{(\lambda_i k_i t)^j}{j!} \right) \right\} \tag{3.3.20}$$

where H is the set of subsystems with hot-standby components, C is the set of subsystems with cold-standby components, and λ_i is the failure rate of component type i for $i = 1, 2, \ldots, s$. Due to memoryless property, the reliability of the cold-standby redundant subsystems can be modeled as a Poisson distribution. Existing works on redundancy allocation typically focus on active or cold standby redundancies or a mix of them.

Recently Levitin et al. (2017) presented a solution methodology to determine optimal design configuration and optimal operation of heterogeneous warm-standby series–parallel systems. Jia et al. (2017) estimated the reliability of a demand-based warm-standby power generation system considering the component degradation process. Hadipour et al. (2018) also solved a redundancy allocation in series–parallel systems with warm-standby and active components in repairable subsystems. It is worth mentioning reliability modeling and optimization considering hot and warm standby components deserves further investigation in literature.

3.4   Multiobjective Reliability–Redundancy Allocation

3.4.1   Pareto Optimality

Quite often maximizing the system reliability or minimizing the cost is not sufficient as reliability estimates or system operation conditions often involve uncertainty. Therefore it is desirable to incorporate the reliability estimation uncertainties into the optimization model in order to achieve robust system design. For risk-averse system operation, it is desirable to maximize the system reliability as well as to minimize its uncertainty. This type of design requirement can be realized under a multiobjective optimization framework, where the decision-makers can choose the best comprised design from a set of non-dominant solutions. A typical multiobjective optimization problem (MOOP) can be formulated as follows.

Model 3.6

$$\text{Max: } f(\mathbf{x}) = \{f_1(\mathbf{x}), f_2(\mathbf{x}), \dots, f_n(\mathbf{x})\} \tag{3.4.1}$$

Subject to:

$$g_j(\mathbf{x}) \leq b_j, \text{ for } j = 1, 2, \dots, m \tag{3.4.2}$$

where $f_1(\mathbf{x}), f_2(\mathbf{x}), \dots, f_n(\mathbf{x})$ are the set of objective functions. They can represent the system reliability, the variance of the reliability estimate, and the cost, among others. Constraint (3.4.2) could be the target values of the subsystem failure rate, component reliability, available budget, weights, and volume limits. It is usually difficult to find one unique solution $\mathbf{x}^*$ that maximizes all the objectives in Eq. (3.4.1). In fact these objective functions are often in conflict with each other. For instance, maximizing system reliability requires more redundant components while minimizing the cost means less components should be used. Therefore, instead of identifying one optimal solution, the concept of Pareto optimality is used to define the optimal solution for a MOOP model. For a multiple-objective optimization with objective function $f_i(\mathbf{x})$ for $i = 1, 2, \dots, n$, a solution $\mathbf{x}^*$ is Pareto optimal if no $\mathbf{x} \in S$ exists, such that

$$f_i(\mathbf{x}) \leq f_i(\mathbf{x}^*) \quad \text{and} \quad f_j(\mathbf{x}) < f_j(\mathbf{x}^*) \quad \text{for at least one } i \text{ or } j \tag{3.4.3}$$

where, $i, j = 1, 2, 3, \dots, s$. A Pareto optimal solution is also called the non-dominant solution in some literature. Based on the expertise and preferences, the decision-maker can choose a best-compromised solution from the set of non-dominant solutions as the final design. Figure 3.10 shows typical Pareto solutions with the objective to maximize the system reliability $f_1(\mathbf{x})$ and minimize the total cost $f_2(\mathbf{x})$. The front line formed by the Pareto solutions is called the Pareto frontier. Any solutions above the Pareto frontier are dominated by the Pareto solutions on the frontier. For instance, solutions A and B are two non-dominant solutions because the reliability of B is larger than that of A, but its cost is larger than A. Solution C, however, is dominated by solution B. This is because the reliability of B is higher than that of C and the cost is lower than C.

Pareto optional solutions can be obtained via the weighted method. It transforms the MOOP model into a single objective optimization by assigning an appropriate non-negative weight, v_i, to each objective function. The sum of these weights does not need to be unity. However, if there are two objectives, it is useful for them to sum to

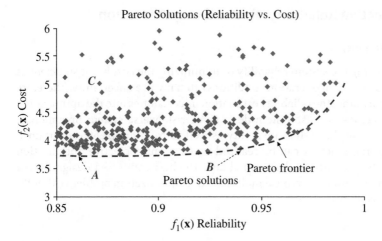

Figure 3.10 Pareto frontier and Pareto solutions.

unity. For a MOOP model with n objectives, the transformation of a single objective function is given as follows.

Model 3.7

$$\text{Max:} \quad z = \sum_{i=1}^{n} v_i f_i(\mathbf{x}) \tag{3.4.4}$$

Subject to:

$$g_j(\mathbf{x}) \leq b_j, \quad \text{for} \quad j = 1, 2, 3, \ldots, m \tag{3.4.5}$$

Solving the single objective problem obtains one non-dominant solution. Ozen (1986) stated that the optimal solution to a weighting problem is a non-dominant solution for the multi-objective optimization problem as long as weights are non-negative. By appropriately changing the weights, v_i, a set of non-dominant solutions, will be obtained.

Besides Pareto optimal solution strategy, goal programming and goal attainment are also frequently employed to tackle the MOOP model. They transform a multiobjective optimization problem into a single objective problem. These two methods are effective if the decision-maker has experience or information about the scope of objective functions and the relative weights of each objective functions. In most cases, problems are formulated when there is little knowledge of the system level implications. Then Pareto optimal solutions become an attractive alternative for multiple objective decision-making and provides the decision-maker with the best-compromised solution.

3.4.2 Maximizing Reliability and Minimizing Variance

In Chapter 2, we discussed how to estimate the system reliability based on the component reliability estimate. Since reliability estimates always involve estimation uncertainties, it is desirable to reduce the variability of the system reliability when performing reliability–redundancy allocations. Assuming that component reliability estimates are statistically independent, we present a MOOP model to maximize the reliability of a series system and to minimize its variance.

Model 3.8

$$\text{Max:} \quad \widehat{R}(\mathbf{x}) = \prod_{i=1}^{s}\left(1 - \prod_{j=1}^{n_i}\widehat{q}_{ij}^{x_{ij}}\right) \tag{3.4.6}$$

$$\text{Min:} \quad \widehat{Var}(\widehat{R}(\mathbf{x})) = \prod_{i=1}^{s}\left(1 - 2\prod_{j=1}^{n_i}\widehat{q}_{ij}^{x_{ij}} + \prod_{j=1}^{n_i}(\widehat{q}_{ij}^2 + \widehat{\sigma}_{ij}^2)^{x_{ij}}\right) - \prod_{i=1}^{s}\left(1 - \prod_{j=1}^{n_i}\widehat{q}_{ij}^{x_{ij}}\right)^2 \tag{3.4.7}$$

Subject to:

$$\sum_{i=1}^{s}\sum_{j=1}^{n_i} c_{ij}x_{ij} \leq C \tag{3.4.8}$$

$$\sum_{i=1}^{s}\sum_{j=1}^{n_i} w_{ij}x_{ij} \leq W \tag{3.4.9}$$

where

$\widehat{q}_{ij}$ = unreliability estimate of component i in subsystem j, and $\widehat{q}_{ij} = 1 - \widehat{r}_{ij}$

$\widehat{\sigma}_{ij}^2$ = variance of $\widehat{q}_{ij}$

w_{ij} = unit weight for jth type of component in subsystem i

W = total allowed weight of the system

All other notations are the same as those in Model 3.4. Note that $\widehat{R}(\mathbf{x})$ and $\widehat{Var}(\widehat{R}(\mathbf{x}))$ are the mean and variance of series–parallel system reliability estimates that are given in Chapter 2. Model 3.8 is a biobjective optimization with the goal of maximizing the system reliability and minimizing the associated variance. Note that x_{ij} is the integer decision variable representing the quantity of the jth type of component used in subsystem i. Constraints (3.4.8) and (3.4.9) represent the cost and weight criteria, respectively.

Model 3.8 can be solved effectively using a genetic algorithm (GA). Originally proposed by Holland (1975), GA is a probabilistic method that uses a set of designs rather than a single solution during the search process. It is based on the natural selection process that the fittest candidates have the greatest probability to survive in the evolution to the next generation. In the population of each generation, a feasible design is represented by a chromosome comprised of a set of genes, each corresponding to one decision variable. For instance, a design vector like $\mathbf{x} = [x_1, x_2, x_3]$ is the chromosome and x_1, x_2, and x_3 are the genes.

The GA process consists of two basic operations: crossover and mutation. *Crossover* is the exchange of two parent genes to produce new offspring genes. *Mutation* is the sudden change occurred on an individual gene. Usually the crossover is given by 50% while the mutation rate is around 10% in probability. The working principles of both operations are briefly explained below. For a detailed discussion on GA applications in reliability optimization, readers are referred to the tutorial by Konak et al. (2006).

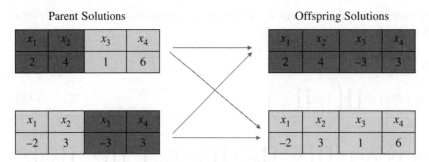

Figure 3.11 The principle of crossover operation.

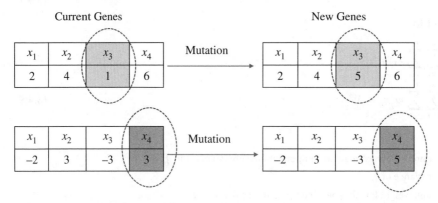

Figure 3.12 The principle of mutation.

A. Crossover Operation

Assume the design vector consists of four variables: x_1, x_2, x_3, and x_4. As shown in Figure 3.11, the parent chromosomes represent the feasible solutions of the original problem. Crossover allows the exchange of two parent genes to form new offspring chromosomes that contain the genes from previous generations.

B. Mutation Operation

Mutation is a genetic operator used to maintain the genetic diversity from one generation of chromosomes to the next. The mutation operator involves a probability event where an arbitrary gene in a chromosome is modified from its original state. A common practice of implementing the mutation operator involves generating a random variable for each gene, and decides whether or not a particular gene will be changed. Mutation allows the GA to avoid local minima by preventing the population of chromosomes from becoming too similar to each other, thus slowing or even stopping the evolution process. Figure 3.12 graphically shows the mutation operation of two chromosomes occurring on different genes.

3.4.3 Numerical Experiment

We use the numerical example in Coit et al. (2004) to demonstrate the application of Model 3.8. The system has three subsystems in series and five component choices (or types) in each subsystem. Table 3.5 lists all component data including mean reliability

Table 3.5 Component reliability, cost, and weight.

	Subsystem 1 ($i = 1$)				Subsystem 2 ($i = 2$)				Subsystem 3 ($i = 3$)			
j	r_{ij}	σ_{ij}^2	c_{ij}	w_{ij}	j	r_{ij}	σ_{ij}^2	c_{ij}	w_{ij}	j	r_{ij}	σ_{ij}^2	c_{ij}	w_{ij}
1	0.96	0.0043	8	5	1	0.98	0.0020	6	5	1	0.95	0.0059	7	6
2	0.92	0.0067	6	6	2	0.93	0.0109	3	7	2	0.89	0.0196	5	8
3	0.86	0.0120	6	8	3	0.88	0.0088	4	3	3	0.87	0.0071	4	5
4	0.77	0.0253	3	6	4	0.69	0.0238	3	5	4	0.71	0.0257	3	5
5	0.68	0.0181	3	5	5	0.6	0.0120	4	3	5	0.65	0.0190	2	4

Source: Reproduced with permission of IEEE.

r_{ij}, variance of the reliability estimate (σ_{ij}^2), unit cost and weight. Design constraints on the system level are maximum cost $C = 29$ and a maximum weight $W = 45$. For this problem, all component reliability estimates are statistically independent.

When solving Model 3.8 by the weighted objective method as described in Section 3.4.1, there are two weights, v_1 and v_2, in the objective function shown in Eq. (3.4.10) below. The values of the weights are alternated in order to obtain different non-dominant solutions of the original problem. The problem is solved repeatedly by varying the value of v_1 from 1 to 0 (or $v_2 = 1 - v_1$) using the GA algorithm. The importance of original objective functions is controlled by the relative size of two weights:

$$\text{Max}: \quad z = v_1 \widehat{R}(\mathbf{x}) - v_2 V \widehat{ar}(\widehat{R}(\mathbf{x})) \tag{3.4.10}$$

Three non-dominant solutions listed in Table 3.6 constitute the Pareto optimal solutions. For instance, if solution A is chosen, the reliability $\widehat{R}(\mathbf{x}) = 0.9844$ would be the highest, but $V\widehat{ar}(\widehat{R}(\mathbf{x})) = 0.000\,438$ is also the largest. If solution C is chosen, $\widehat{R}(\mathbf{x})$ would be the lowest, but $V\widehat{ar}(\widehat{R}(\mathbf{x}))$ is also the smallest. Therefore, A, B, and C form a set of non-dominant designs.

These three non-dominant solutions are mapped into the corresponding system configurations (see Figures 3.13–3.15). An important observation is that as v_1 decreases, it becomes more important to minimize the variance of the reliability estimate and less important to maximize the system reliability. Hence the design changes accordingly.

The topic of RRA has continued to receive attention recently. There are three new research trends that have been observed. First, the traditional RRA problem is expanded to incorporate other reliability programs, such as the reliability growth test (Heydari and

Table 3.6 Pareto optimal solutions for independent component estimates.

Solution	Subsystem 1					Subsystem 2					Subsystem 3					$\widehat{R}(\mathbf{x})$	$Var(\widehat{R}(\mathbf{x}))$
	x_{11}	x_{12}	x_{13}	x_{14}	x_{15}	x_{21}	x_{22}	x_{23}	x_{24}	x_{25}	x_{31}	x_{32}	x_{33}	x_{34}	x_{35}		
A	0	1	0	2	0	0	2	0	0	0	1	0	1	0	0	0.9844	4.38×10^{-4}
B	0	2	0	0	0	0	2	0	0	0	0	0	1	1	2	0.9842	3.80×10^{-4}
C	0	1	0	1	1	0	2	0	0	0	0	0	2	0	1	0.9834	3.53×10^{-4}

Source: Reproduced with permission of IEEE.

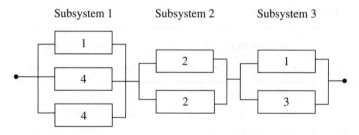

Figure 3.13 Pareto optimal solution $\widehat{R}(\mathbf{x}) = 0.9844$, $\widehat{Var}(\widehat{R}(\mathbf{x})) = 0.000438$.

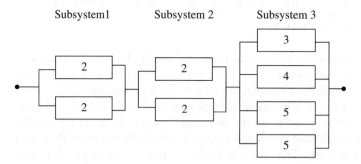

Figure 3.14 Pareto optimal solution $\widehat{R}(\mathbf{x}) = 0.9842$, $\widehat{Var}(\widehat{R}(\mathbf{x})) = 0.00038$.

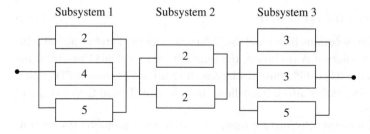

Figure 3.15 Pareto optimal solution $\widehat{R}(x) = 0.9834$, $\widehat{Var}(\widehat{R}(\mathbf{x})) = 0.000353$.

Sullivan 2018), during the product development stage. Second, efforts are also made to jointly allocate reliability–redundancy and maintenance policy (Bei et al. 2017) and joint allocation of reliability–redundancy and spare parts inventory (Jin et al. 2017). The third direction is to extend RRA to a more complex system structure with independent failure, such as network systems and failure corrections (Yeh and Fiondella 2017).

3.5 Failure-in-Time Based Design

3.5.1 Component Failure Rate Estimate

In the early design phase, a reliability prediction is often required for the components and systems in order to assess their future reliability and in many cases to meet customer specifications. For instance, IEEE issued a standard methodology (i.e. IEEE 1413) for the reliability prediction of electronic systems (IEEE 1413 1998; Pecht et al.

Table 3.7 Empirical and standard based prediction methods (ReliaSoft 2017).

Nation	Prediction method	Applications	Last updated
USA	MIL-HDBK-2017F and Notice 1 and 2	Military/Defense	1995
USA	Bellcore/Telcordia	Telecom	2006
France	RDF 2000	Telecom	2000
SAE USA	SAE Reliability Prediction	Automotive	2017
Japan	NTT Procedure	Telecom	1985
Germany	Siemens SN29500	Siemens Products	1999
China	China 299B	Military/Defense	1998
USA	PRISM	Military/Commercial	2000
USA	IEEE 1413	Electronics	2002

2002). The prediction model is built on the assumption of the exponential lifetime of components with independent failures. It is understood that the accuracy of the reliability prediction depends on two factors: (i) the appropriateness of the model to the component lifetime distributions and (ii) the completeness of the component information, such as derating, temperature, and other environmental stresses. Table 3.7 summarizes the empirical and standard-based prediction methods adopted in the US, China, France, Japan, and Germany.

Though various reliability prediction methods have been developed, the two standards mostly used by industry to forecast the new product reliability are MIL-HDBK-217F (US MIL-HDBK217 1995) and Bellcore TR332 (Bellcore 1995). Both standards predict the reliability of a new product by combining component failure rates and the bill of materials (BOMs). The Bellcore/Telcordia predictive method for an individual component is of the form

$$\lambda_o = \lambda_b \pi_Q \pi_T \pi_E \pi_O \qquad (3.5.1)$$

where

λ_b = the base failure rate or the nominal failure rate

π_Q = the quality factor

π_T = the temperature stress factor

π_E = the electrical stress factor

π_O = other factors such as humidity, usage, and vibrations

In the electronics industry, the nominal failure rate λ_b is usually estimated at 40 °C operating temperature with 50% of the electrical derating rate. Derating puts the operation of a device at lower than its rated maximum capability in order to extend its life. Typical electrical derating includes the voltage, current, and power. FIT is an alternative way to measure the product lifetime. It reports the number of failures per 10^9 hours (or 1 billion hours) of operation of a unit. For instance, a component with 100 FIT is equivalent to $10^9/100 = 10^7$ hours mean-time-between-failures (MTBF) for repairable system or mean-time-to-failure (MTTF) for non-repairable system. Therefore λ_o and λ_b in Eq. (3.5.1) can be expressed as either FIT or failures per unit time, depending on the application and industry preference.

Table 3.8 Nominal FIT and reliability factors of inverter components.

i	Component	Quantity	λ_b	π_Q	π_T	π_E	π_R
1	Transistor	6	100	1.5	0.8	2.5	1
2	Diode	6	50	1.2	0.7	3.2	1
3	Resistor	3	200	1.3	0.5	2.1	1

Example 3.7   A power inverter is an electronic device that changes direct current (DC) to alternating current (AC). It is widely used in solar photovoltaic (PV) systems by converting DC power into AC utility power. A three-phase inverter consists of six transistors, six diodes, and three power resistors. Failure of any component results in the malfunction of the inverter. Table 3.8 provides the baseline FIT and life factors of these components. Estimate: (1) the FIT of the inverter; (2) the inverter failure rate and MTBF; and (3) the inverter reliability by the end of 20 years assuming an exponential lifetime.

Solution:

1) The inverter can be treated as a series system. Taking into account the quality, temperature, application, and environmental factors, the FIT of the inverter is the sum of the FIT of individual components, that is

$$FIT = 6\lambda_{b1}\pi_{Q1}\pi_{T1}\pi_{E1}\pi_{R1} + 6\lambda_{b2}\pi_{Q2}\pi_{T2}\pi_{E2}\pi_{R2} + 3\lambda_{b3}\pi_{Q3}\pi_{T3}\pi_{E3}\pi_{R3}$$
$$= 6 \times 100 \times 1.5 \times 0.8 \times 2 + 6 \times 50 \times 1.2 \times 0.72 + 3 \times 200 \times 1.3 \times 0.52$$
$$= 3425.4$$

2) The failure rate λ_{inv} and MTBF of the inverter are estimated as follows:

$$\lambda_{inv} = \frac{FIT}{10^9} = \frac{3425.4}{10^9} = 3.4254 \times 10^{-6} \quad \text{failures/hour}$$

$$MTBF = \frac{1}{\lambda_{inv}} = \frac{1}{3.4254 \times 10^{-6}} = 291\,937 \quad \text{hours (or 33.3 years)}$$

3) Since the inverter lifetime is exponential, its reliability at the end of year 20 is

$$R(t = 20 \text{ years}) = \exp(-3.4254 \times 10^{-6} \times 20 \times 8760) = 0.549$$

Although the inverter has 33.33 years MTTF, the chance to survive for 20 years is only 54.9%. It is worth mentioning that λ_o in Eq. (3.5.1) is a point estimate. Uncertainties often exist that are associated with product design, manufacturing, and field operation. Therefore, reliability predictions obtained in the early development stage may exhibit a large discrepancy compared to the actual field data. In the next section, we develop a risk-averse reliability design method to cope with the uncertainties in design and operation.

3.5.2   Component with Life Data

A new design is often built upon the success or experience of legacy products with similar functions or operating conditions. Many technologies and components used in predecessor products are most likely to be reused in the new product. Therefore, reliability or failure rate of these components can be appropriately estimated from field failures of legacy products. This can be achieved through the following model:

$$\hat{\lambda} = \frac{\text{Total failures}}{\text{Cumulative operating time}} \tag{3.5.2}$$

where $\hat{\lambda}$ is the estimate of the component's true failure rate λ. This result from this model is rather accurate as long as the operating condition and electrical derating of the component in the new product are similar to the predecessors. We use the following numerical example to illustrate the estimation process.

Example 3.8 Assume that a type of 5 V relay is used in wireless communication systems. Each system contains 20 relays and total of 150 systems are deployed in the field. These systems operate for 24 hours a day and seven days a week. During the course of a year, four systems were returned for repair due to relay failures. Upon fix, these systems are returned to the customer for continuous operation. It is assumed that the system repair downtime is negligible compared to the uptime. Based on Eq. (3.5.2), the failure rate for the relay can be estimated by

$$\hat{\lambda} = \frac{4}{8760 \times 150 \times 20} = 1.903 \times 10^{-7} \text{ failures/hour} \tag{3.5.3}$$

If the manufacturer uses the same relay in the next generation product, the failure rate of the relay can be treated as 1.903×10^{-7} faults/hour provided the operating temperature, electrical derating, and application environment are similar to its predecessor.

3.5.3 Components without Life Data

For components without sufficient field data, their failure rate can be appropriately extrapolated based on the operating temperature and the derating rate. Component suppliers usually provide the base failure rate λ_b that represents the nominal reliability performance when they operate under the nominal conditions (e.g. 40 °C at 50% derating level for electronics). According to Eq. (3.5.1), the actual failure rate depends on π_Q, π_T, and π_E factors. Since π_Q is largely determined by the commodity supplier and π_Q is associated with the usage, the most effective way that a design engineer wants to prolong the component lifetime is to reduce the temperature stress, the derating rate, or both. Let us assume that $\pi_Q = \pi_R = 1$. Then Eq. (3.5.1) becomes

$$\lambda = \lambda_b \pi_T \pi_E \tag{3.5.4}$$

The temperature factor π_T can be estimated using the Arrhenius model, which is given by (Elsayed 2012)

$$\pi_T = e^{\frac{E_a}{k}\left(\frac{1}{T_0} - \frac{1}{T}\right)} \tag{3.5.5}$$

where

E_a = activation energy (eV)

k = Boltzman constant (8.62×10^{-5} eV/K)

T_0 = reference ambient temperature (313 K)

T = component ambient temperature (K)

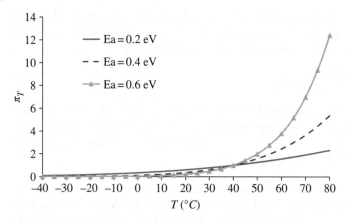

Figure 3.16 Ambient temperature versus the temperature factor.

In Figure 3.16 we plot the value of π_T as T increases from $-40\,°C$ to $80\,°C$ for $E_a = 0.2$, 0.4, and 0.6 eV, respectively. Two important observations are obtained. First, π_T is relatively insensitive to the activation energy E_a if $T < 40\,°C$. However, when T exceeds $40\,°C$, components with higher E_a fail much faster than those with lower E_a. Second, lowering the ambient temperature is critical to extending the component lifetime, especially for large E_a. For instance, a component with $E_a = 0.6$ eV, its π_T doubles whenever T increases by $10\,°C$, or equivalently cutting its lifetime by 50%.

According to Bellcore 332 (Bellcore 332 1995/Telcordia TR-332), the electrical derating factor π_E can be estimated by the following equation as

$$\pi_E = e^{m(p-p_0)} \tag{3.5.6}$$

where

p = actual electrical derating percentage

p_0 = reference derating with $p_0 = 50\%$

m = fitting parameter

The fitting value m ranges from 0.006 to 0.059 depending on the nature of the component and its composition materials. For details of values of m, one can refer to Telcordia TR-332 (Telcordia TR-332). Here we partially list the value of "m" for certain electronic devices, where m is in the range of [0.019, 0.024] (see Table 3.9). The following example illustrates the use of Eq. (3.5.6).

Example 3.9 A field programmable gate array (FPGA) is a programmable semiconductor device that is widely used in consumer electronics, avionics, and military systems. It is assumed that $E_a = 0.7$ eV, operating temperature $T = 60\,°C$ or 333 K, power

Table 3.9 Parametric value of m for certain electronic devices.

Component	1 MB EPROM	0.2 W 1% resistor	FPGA	Switch transistor	Optional amp	Clock driver
m	0.024	0.019	0.024	0.024	0.024	0.024

derated with $p = 80\%$, and fitting parameter $m = 0.024$. Given its $\lambda_b = 10^{-7}$ failures/hour, estimate its actual failure rate. Assume that 10 devices are used in the system. If the system lifetime is expected to be five years, what is the probability that the system fails because of FPGA failure?

Solution:

1) The temperature and derating factors are estimated as follows:

$$\pi_T = e^{\frac{0.7}{8.62 \times 10^{-5}}\left(\frac{1}{313} - \frac{1}{333}\right)} = 4.75$$

$$\pi_E = e^{0.024(0.8 - 0.5)} = 1.01$$

Hence, the actual failure rate of FPGA is (assuming $\pi_Q = \pi_R = 1$)

$$\lambda = \lambda_b \pi_T \pi_E = 10^{-7} \times 4.75 \times 1.01 = 4.785 \times 10^{-7} \text{ failures/hour}$$

2) Since 10 FPGA are used in each system, the system is functional provided no FPGA failure occurs in five years. Therefore, the functionality of 10 FPGA can be treated as a series system with the reliability

$$R = (\exp(-4.785 \times 10^{-7} \times 5 \times 8760))^{10} = 0.979^{10} = 0.81$$

Assuming other component types are reliable, the chance that an FPGA causes the system failure would be $1 - 0.81 = 0.19$.

3.5.4 Non-component Failure Rate

Besides the component (i.e. hardware) failures, there are many system-level failures caused by non-component issues. Typical non-component failures include design issues, manufacturing defects, software bugs, incorrect process, and no-fault-found (NFF). NFF happens when a part or field replacement unit is returned to the repair center for trouble shooting, but diagnostic tests are not able to duplicate the failure mode that occurred at the customer's site. More discussions about NFF will be provided in Chapter 4.

Non-component failure rates are usually more difficult to estimate than component failure rates. This is because non-component failure rates are influenced by a variety of factors, and some of them are human-related. For example, design errors and software bugs are correlated with the engineers' experience and the similarity to the predecessor product. Poor solder joints are closely associated with the complexity of the product and the reflow temperature. Process issues can be avoided by providing good training to employees or product users. Similar to the component failure rate, the non-component failure rate, denoted as λ, can be estimated as

$$\hat{\lambda} = \frac{\text{Non-component failures}}{\text{Cumulative operating time}} \qquad (3.5.7)$$

where $\hat{\lambda}$ represents the estimate of λ. Failures of non-component related issues include design errors, manufacturing defects, software bugs, process issues, and NFF. We use the following example to illustrate the application of Eq. (3.5.7).

Example 3.10   Assume each electric vehicle has two control boards that regulate the electricity flow between the battery and the motor. A car manufacturer sold 1000 cars in the last year, and 10 faulty control boards have been reported after one year. The root cause of these failure is poor solder joints that belong to the manufacturing failure. Assume the average mileage per car is 12 000 km/year. Estimate the failure rate of the manufacturing issue due to poor solder joints.

Solution:
The cumulative mileage of 1000 cars in a year is $12\,000 \times 1000 = 1.2 \times 10^7$ km. Since each car has two control boards, the cumulative mileage of all field control boards is $2 \times 1.2 \times 10^7 = 2.4 \times 10^7$ km. Based on Eq. (3.5.7), the failure rate of manufacturing issue of the control board is

$$\hat{\lambda} = \frac{10}{2.4 \times 10^7} = 4.167'10^{-7} \text{ faults/km} \tag{3.5.8}$$

which is different from component failures where the same failure mode may occur in different products. For example, a titanium capacitor is prone to burning when excessive voltage is applied. This failure mode may happen regardless of which titanium capacitor is used in products A or B. Non-component failure modes, however, usually do not have a high degree of duplication among different products. In fact, design errors and software bugs are often product-specific. Therefore useful non-component failure data are always scarce when applied to the reliability prediction of a new product.

The triangular distribution is often used as a subjective modeling of a population for which there is only limited failure data or no data available (Bowles and Pelaez 1995). Figure 3.17 depicts the triangular distribution represented by three parameters a, b, and c, where a denotes the smallest possible value of the failure rate, b denotes the largest possible value, and c is the mode representing the most likely value. The mathematical expression for $g(\lambda)$ is given as follows:

$$g(\lambda) = \begin{cases} \dfrac{2}{(c-a)(b-a)}(\lambda - a), & a \leq \lambda \leq c \\[3mm] \dfrac{2}{(c-b)(b-a)}(\lambda - b), & c < \lambda \leq b \\[3mm] 0, & \text{otherwise} \end{cases} \tag{3.5.9}$$

In reality, parameters a, b, and c are unknown and need to be estimated based on the field failure rate data. Assume the failure data do not contain any anomalous points, the minimum and the maximum points can be obtained by sorting the dataset in ascending

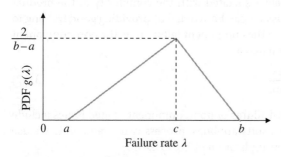

Figure 3.17 Triangular distribution for failure rate λ.

order and assigning the first and last points to a and b, respectively. Let $\bar{\lambda}$ be the sample mean of the dataset; then the mode c can be estimated by

$$c = 3\bar{\lambda} - b - a \tag{3.5.10}$$

The mean and the variance of the non-component failure rate estimate is given by

$$E[\lambda] = \frac{1}{3}(a + b + c) \tag{3.5.11}$$

$$Var(\lambda) = \frac{1}{18}(a^2 + b^2 + c^2 - ab - ac - bc) \tag{3.5.12}$$

Example 3.11 To estimate the manufacturing failure rate of newly designed PCB-D, we take data from three legacy products that are similar to PCB-D in terms of component type and quantity used, manufacturing technology, and total defects opportunity per board. Historical data shows that the failure rates due to manufacturing issue of the three predecessors are 1.2×10^{-6}, 1.4×10^{-6}, and 2.4×10^{-6} failures/hour. Estimate a, b, and c assuming the manufacturing failure rate follows the triangle distribution. What is the mean and the variance of the failure rate?

Solution:

$$a = \min\{1.2 \times 10^{-6}, 1.4 \times 10^{-6} \text{ and } 2.4 \times 10^{-6}\} = 1.2 \times 10^{-6} \text{ failures/hour}$$
$$b = \max\{1.2 \times 10^{-6}, 1.4 \times 10^{-6} \text{ and } 2.4 \times 10^{-6}\} = 2.4 \times 10^{-6} \text{ failures/hour}$$

To estimate c, we compute the sample mean which is

$$\bar{\lambda} = (1.2 \times 10^{-6} + 1.4 \times 10^{-6} + 2.3 \times 10^{-6})/3 = 1.6 \times 10^{-6} \text{ failures/hour}$$

Based on Eq. (3.5.10), we have

$$c = 3 \times 1.6 \times 10^{-6} - 2.4 \times 10^{-6} - 1.2 \times 10^{-6} = 1.3 \times 10^{-6} \text{ failures/hour}$$

Now substituting a, b, and c into Eqs. (3.5.11) and (3.5.12), we obtain $E[\lambda] = 1.63 \times 10^{-6}$ and $Var(\lambda) = 7.39 \times 10^{-14}$. Together with the component failure rates, the moments of non-component failure rates can be used as the basic inputs to model and optimize a risk-averse reliability design under the DFSS methodology.

3.6 Failure Rate Considering Uncertainty

3.6.1 Temperature Variation

It is quite common that multiple units of the same component type are used in a system. If the variation of the ambient temperature across these units is small, a single temperature reading T or the average temperature is adequate to calculate the temperature factor π_T in Eq. (3.5.5). When a large variation of temperature is observed, a single-point value of temperature becomes inadequate to estimate π_T. Figure 3.18 shows the temperature readings of 32 integrated devices called application-specific integrated circuits (ASICs) on a printed circuit board (PCB) module. The temperature varies from 65 °C to 88.2 °C with a maximum range of 23.2 °C when it is used in the field. This implies

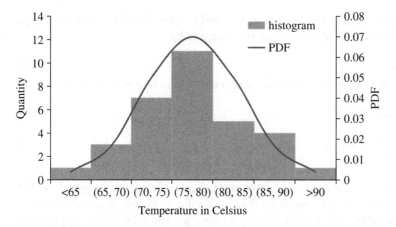

Figure 3.18 Temperature distribution of ASIC units in a PCB module.

that the devices under 88.2 °C will fail more than twice as fast as those under 65 °C. The temperature variation in this case fits the normal distribution with mean 76.3 °C and the standard deviation 5.7 °C.

To characterize the failure rate uncertainty that resulted from temperature variations, π_T should be treated as a random value. Based on the temperature distribution profile, the mean and the variance of π_T can be estimated as follows (Jin and Su 2005):

$$E[\pi_T] = \int_0^{+\infty} e^{\frac{E_a}{k}\left(\frac{1}{T_0} - \frac{1}{x}\right)} f_T(x) dx \qquad (3.6.1)$$

$$E[\pi_T^2] = \int_0^{+\infty} e^{\frac{2E_a}{k}\left(\frac{1}{T_0} - \frac{1}{x}\right)} f_T(x) dx \qquad (3.6.2)$$

$$Var(\pi_T) = E[\pi_T^2] - (E[\pi_T])^2 \qquad (3.6.3)$$

where $f_T(x)$ is the probability density function (PDF) of T. Modern electronic devices such as microprocessors, FPGA, and ASIC usually have built-in temperature sensors, and the real-time temperature can be directly measured, transferred and stored in a computer server for further analysis. To protect the outlier data, repeated sampling and multiple temperature readings are recommended for each component.

Example 3.12 In the design concept phase, the actual temperature distribution might not be available to the engineering team. Hence uniform distributions are used to model the component temperature variation based on the lower and upper bounds. If the ambient temperature of the 15 V capacitors falls in [40 °C, 80 °C], estimate the mean and the variance of π_T assuming $E_a = 0.5$ eV.

Solution:
The lower temperature limit in Kelvin is $273 + 40 = 313$ K and the upper limit is $273 + 80 = 353$ K. The uniform PDF of T is given by

$$f_T(x) = \begin{cases} \dfrac{1}{40}, & 313 \le T \le 353 \\ 0, & \text{otherwise} \end{cases} \qquad (3.6.4)$$

Based on Eqs. (3.6.1–3.6.3), the mean, the second moment, and the variance of π_T are

$$E[\pi_T] = \int_{313}^{353} e^{\frac{0.5}{8.62\times10^{-5}}\left(\frac{1}{313}-\frac{1}{x}\right)} \left(\frac{1}{40}\right) dx = 3.548 \tag{3.6.5}$$

$$E[\pi_T^2] = \int_{313}^{353} e^{\frac{2\times0.5}{8.62\times10^{-5}}\left(\frac{1}{313}-\frac{1}{x}\right)} \left(\frac{1}{40}\right) dx = 16.705 \tag{3.6.6}$$

$$Var(\pi_T) = 16.705 - 3.548^2 = 4.116 \tag{3.6.7}$$

Unfortunately, closed-form solutions to Eqs. (3.6.5) and (3.6.6) are not available, and $E[\pi_T]$ and $E[\pi_T^2]$ are obtained though numerical integration that can be implemented in Matlab or other computational tools.

3.6.2 Electrical Derating Variation

For the same component type in a PCB module, electrical derating such as power, voltage, or current may vary from one unit to other depending on the circuitry architecture and the functional requirements. Figure 3.19 shows the voltage derating percentage of 82 identical 10-V capacitors within a PCB module. The derated percentage varies from 50% to 85% with a maximum difference of 35%. Note that 50% derating equals 5 V and 85% derating corresponds to 8.5 V. The information about electrical derating can be obtained from the product design engineers who are responsible for determining the components and laying out the circuitry to meet the functional requirement. Once a prototype product is available, the range of p can be physically measured through multimeters or test-bed systems.

The variation of derating is manifested by parameter p in Eq. (3.5.6). The distribution of p can be appropriately extrapolated based on the actual derating of individual units. If p is uniformly distributed in a range of $[p_{min}, p_{max}]$, its PDF, denoted as $f_p(x)$, can be expressed by

$$f_p(x) = \begin{cases} \dfrac{1}{p_{max} - p_{min}}, & p_{min} \leq x \leq p_{max} \\ 0, & \text{otherwise} \end{cases} \tag{3.6.8}$$

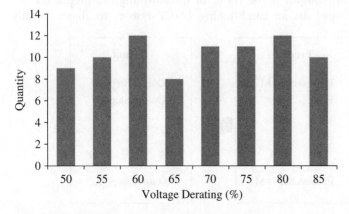

Figure 3.19 Distribution of voltage derating of capacitors.

Note that p_{min} and p_{max} represent the smallest and the largest derating percentages, respectively. Now the mean and variance of π_E can be estimated by

$$E[\pi_E] = \int_{p_{min}}^{p_{max}} e^{m(x-p_0)} f_p(x) dx = \frac{e^{m(p_{max}-p_0)} - e^{m(p_{min}-p_0)}}{m(p_{max} - p_{min})} \tag{3.6.9}$$

$$E[\pi_E^2] = \int_{p_{min}}^{p_{max}} e^{2m(x-p_0)} f_p(x) dx = \frac{e^{2m(p_{max}-p_0)} - e^{2m(p_{min}-p_0)}}{2m(p_{max} - p_{min})} \tag{3.6.10}$$

$$Var(\pi_E) = E[\pi_E^2] - (E[\pi_E])^2 \tag{3.6.11}$$

By considering the effects of temperature and derating variations in Eq. (3.5.4), the mean and variance of the component failure rate can be obtained as follows:

$$E[\widehat{\lambda}] = \lambda_b E[\pi_T] E[\pi_E] \tag{3.6.12}$$

$$Var(\widehat{\lambda}) = \lambda_b^2 (E[\pi_T^2] E[\pi_E^2] - (E[\pi_T])^2 (E[\pi_E])^2) \tag{3.6.13}$$

These moments along with the mean and variance of the non-component failure rate in Eqs. (3.5.11) and (3.5.12) constitute the basic building blocks in DFR under six-sigma criteria. This will be further elaborated in the case study of Section 3.9.

3.7 Fault-Tree Method

3.7.1 Functional Block Diagram

Building the functional black diagram (FBD) is the first step toward the analysis of the failure modes and root causes in the early product development phase. The FBD is a step-by-step procedure that describes the detailed functionality of a development process. As shown in Figure 3.20, the process consists of three sequential parts: Input, Process, and Output. Since the product is still in its early developmental phase, the steps identified under input, process, and output should not be too detailed. Each of these identified steps becomes a unique process that is later evaluated using a fault-tree analysis (FTA). For that reason, three to five steps are usually sufficient to describe any input, process, or output in FBD.

The FBD shows that any output is the result of transforming the inputs via the immediate process. We use an air-conditioning (A/C) system to illustrate this

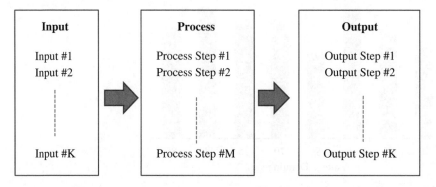

Figure 3.20 A functional block diagram with input, process, and output parts.

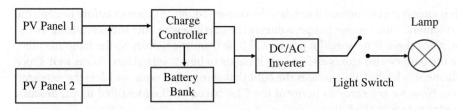

Figure 3.21 Schematic diagram of a PV system with battery storage.

transformation process. When you turn on the switch of an A/C system, the electricity drawn from the grid flows into the electric motor coils to generate the electromagnetic field. That magnetic field drives the rotor and generate the torque. The torque further pushes the compressor such that the working fluid leaves the compressor and flows into the condenser, where the ambient hot air will exchange the heat with fluid and convert the fluid to low pressurized liquid. As the output or result of the process, the room temperature drops.

Let us use a roof-top solar PV system in Figure 3.21 to illustrate how to construct the FBD and further define input, process, and output steps. The schematic in the figure shows the components that make up a typical roof-top PV system in a residential community. The system consists of a charger controller, battery bank, DC–AC inverter, light bulb (lamp), connecting wires, and two solar PV panels (more panels are installed in reality). The FBD begins with writing three high-level labels: Inputs, Process, and Outputs. Hence the schematic diagram in Figure 3.21 should be used as an aid to identify each of the significant processes involved.

Identifying the input, process, and output are the necessary steps toward the FTA, which will be discussed next. As shown in Figure 3.22, we use the input of "turn on the light switch" as an example to illustrate how to proceed and detail the process steps. By turning the light switch on, it closes the electric circuit. It is then expected that current flows from the PV panels to the lamp filament via the DC–AC inverter. Since PV

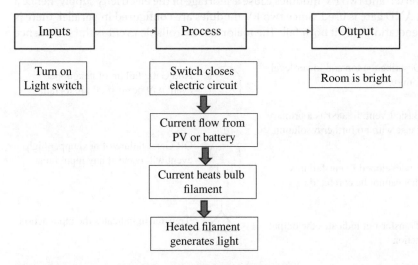

Figure 3.22 Functional block diagram of a roof-top PV system.

modules generate DC power, it needs to be converted into AC power before heating the lamp filament. This process happens during the daytime when the sun is available. In the night, since there is no sun, the power will flow from the battery to the lamb filament. Since the battery also stores DC energy, it needs to be converted into AC as well. Once the filament is heated, it generates the light that shines the room, which is the expected output. Now we are ready to perform the FTA based on the identified input, process, and output steps in the FBD.

3.7.2 Fault-Tree Analysis

The FTA is a logical, graphical diagram that describes the interactions between failure modes and their causes. The diagram is constructed upon the preliminary analysis of input, process, and output steps of the FBD. Particularly, the FTA diagram hierarchically shows all failures for a system, subsystem, assembly, and module down to an individual component. The FTA oftentimes starts from a consideration of the system failure effects, referred to as "top events," and the analysis proceeds by determining how these can be caused by individual or combined failure or events from lower-level subsystems or components. Standard logic symbols are often used to construct an FTA diagram to capture the logical connections between individual events or components. Basic logics symbols that are commonly used are shown in Figure 3.23. The results of the FTA can then be transferred to an FMECA spreadsheet with which the failure modes and their causes are assessed in terms of their effects on the system design. This section illustrates how to construct an FTA diagram for failure analysis, and FMECA will be discussed in Section 3.8.

Let us construct the FTA diagram based on the FBD of the solar PV system shown in the previous section. The FTA shown in Figure 3.24 has two PV panels that generate the power in parallel. The top event failures of this system can be caused by either no current from the PV panels or from the battery bank; hence a two-input OR gate captures this logic relation. At the next lower level (on the left side), any failure of the charge controller, DC–AC inverter, and two PV modules cause a shortage in the electricity supply; hence a three-input AND gate is used. Since two PV modules are configured in parallel, there is no power generation only if both fail. The major uncontrollable event is the sun. Hence

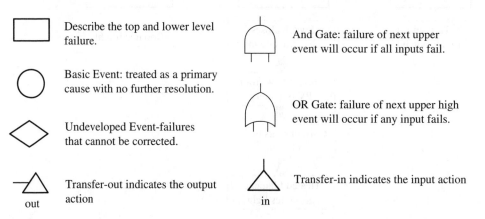

Figure 3.23 Standard logic symbols used in fault-tree analysis.

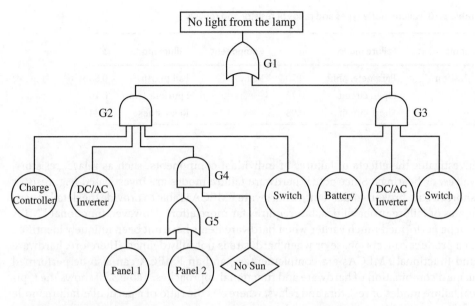

Figure 3.24 FTA for the roof-top PV system.

a three-input OR gate is applied to represent this relation. In case the power is supplied through the stored energy in the night (see the right side), it requires that battery bank, DC–AC inverter, and switch all to be working. Thus a three-input AND gate is used.

It is worth mentioning that each FTA diagram is constructed corresponding to a uniquely defined top event, which can be caused by different failure modes or logical connections between failure events. In the solar PV system, if the top event is "unsafe voltage," then it would be necessary for the DC–AC inverter's voltmeter to be available, and gate G1 would have to be changed to an OR gate along with the output of current G1.

In addition to depicting the logical connections between failure events in relation to top-level events, FTA can be used to estimate the top event probabilities, in the same way as in the reliability block diagram analysis in Chapter 2. Failure probabilities derived from reliability prediction models or simulation programs can be assigned to the failure events, and cut set and path set methods can be applied to evaluate system failure probability.

3.8 Failure Mode, Effect, and Criticality Analysis

3.8.1 Priority Risk Number

Failure mode, effect, and criticality analysis (FMECA) is perhaps the most widely used reliability analysis and improvement method in product design. The principle of FMECA is to analyze each failure mode of individual components and to determine the effects of system-level operation of each failure mode. FMECA may be performed at a functional level or on a piece part (i.e. hardware). Functional FMECA studies the effects of failure at the reliability block level, such as a power supply of a microprocessor. Hardware FMECA

Table 3.10 Failure mode types and ratios of resistor and replays.

Component	Failure mode	α	Component	Failure mode	α
Resistor	Parameter shift	0.70	Relay	Fail to trip	0.55
	Open circuit	0.27		Spurious trip	0.16
	Short circuit	0.03		Relay stuck	0.24

investigates the effects of failures of individual components, such as relays, resistors, valves, or bearings. Since actual hardware failure modes are investigated (e.g. resistor short circuit, broken solder joint), a hardware FMECA is able to provide the detailed estimates of failure probabilities, but requires far more effort. However, functional FMEA can be performed much earlier when hardware items have not been uniquely identified in a product concept phase or when hardware is not fully defined. Therefore, hardware and functional FMECAs are complementary, and an FMECA can also be performed using a combination of hardware and functional approaches. Table 3.10 shows the typical failure modes of resistors and relays, where α is the ratio of a particular failure mode of the component occurring in field products.

Figure 3.25 shows a typical FMECA worksheet taken from the Automotive Industry Action Group (ReliaSoft 2017). Method 101 is a non-quantitative method, which serves to highlight failure modes whose effects would be considered important in relation to severity, detectability, safety, or maintainability.

RAC CRTA-FMECA and MIL-HDBK-338 both identify Risk Priority Number (RPN) calculations as an alternate method to criticality analysis. The RPN is defined as follows:

$$\text{RPN} = \text{Detectability}(D) \times \text{Severity }(S) \times \text{Occurrence }(O) \tag{3.8.1}$$

where D, S, and O each is on a scale from 1 to 10. The highest RPN is $10 \times 10 \times 10 = 1000$, meaning that this failure is not detectable by inspection, is highly severe, and the occurrence is almost certain. If $O = 1$, the RPN decreases to 100. Hence the criticality analysis guides the focus on the highest failure risks.

Failure Modes, Effects and Cricality Analysis
FMECA #
Assembly #
Owner:
Date:
Team Member

No.	Failure Mode	Root Cause	Effects	Detectability	S	O	D	PRN	H	FRU	Recommendation	Who	When	Comments

Where: S = Severity, O = Occurance, D = Detectability, PRN = Risk Priority Number, H = Hazards, FRU = Field replacement unit

Figure 3.25 A typical FMECA spreadsheet.

3.8.2 Criticality Analysis

Criticality analysis includes consideration of the failure rate or probability, the failure mode ratio, and a quantitative assessment of criticality, in order to provide a quantitative rating of the criticality for the component or function. The failure mode criticality number is

$$C_m = \beta \alpha \lambda_p t \qquad (3.8.2)$$

where

β = conditional probity of losing function or mission
α = failure mode ratio (for an item sum of all $\alpha = 1$)
λ_p = part failure or hazard rate
t = operating or at-risk time of item

The expected number of failures $\lambda_p t$ can be replaced by the failure probability $(1 - \exp(-\lambda_p t))$. The item criticality number is the sum of the failure mode criticality numbers for the item under study.

In practice, when performing FMECA for quality and reliability assurance, interdependencies among various failure modes with uncertain or imprecise information are very difficult to incorporate into the assessment measure. Thus, methods like the Markov chain and fuzzy logistic are considered to be an effective tool to address the correlations of different failure modes or components (Xu et al. 2002; Xiao et al. 2011).

A system may be subject to a common cause failure (CCF), where the failure of multiple components is the result of a shared event or coupling factor (or mechanism). In fact, studies have shown that CCF events may account for between 20% and 80% of the unavailability of safety systems within nuclear reactors (Werner 1994). In addition, the CCF effect is a stochastic and time-varying process due to the influence of various degradation mechanisms, such as wear, corrosion, and fatigue (Fan et al. 2018). However, it is generally agreed that CCF does not include those multiple component failures because of a functional dependency that is modeled in a traditional fault tree or system reliability model (Wu et al. 2011). In particular on redundant systems, reliability dependency exists between components that are provided by the same company, or maintained by the same technician, or a reliability estimation derived from the same test sample.

3.9 Case Study: Reliability Design for Six Sigma

3.9.1 Principle of Design for Six Sigma

The design for Six Sigma (DFSS) emerged from the Six Sigma quality control and the Define-Measure-Analyze-Improve-Control (DMAIC) methodologies (ReliaSoft 2017). Though the primary goal of DFSS is to take proactive measures to reduce the number of non-conforming units at an early design and manufacturing stage, reliability also plays an important role in terms of achieving the MTBF target, meeting the safe requirement, and achieving customer satisfaction. In this section, we introduce a DFSS-based reliability design methodology to proactively address the long-term (post-installation) reliability issues in the early design and development phase. A generic system failure rate model that incorporates both component and non-component failures is given as follows:

$$\hat{\lambda}_{sys} = \hat{\lambda}_d + \hat{\lambda}_m + \hat{\lambda}_s + \hat{\lambda}_p + \hat{\lambda}_{nff} + \sum_{i=1}^{I} m_i \hat{\lambda}_i^{(n)} + \sum_{j=1}^{J} n_j \hat{\lambda}_j^{(o)} \tag{3.9.1}$$

where

$\hat{\lambda}_d$	=	failure rate of design issues
$\hat{\lambda}_m$	=	failure rate of manufacturing defects
$\hat{\lambda}_s$	=	failure rate of software bugs
$\hat{\lambda}_p$	=	failure rate of process issues
$\hat{\lambda}_{nff}$	=	failure rate of NFF issue
I	=	number of new component types
J	=	number of existing component types
$\hat{\lambda}_i^{(n)}$	=	failure rates of new component type i, for $i = 1, 2, ..., I$
$\hat{\lambda}_j^{(o)}$	=	failure rates of existing component type j, for $j = 1, 2, ..., J$
m_i	=	number of units of component type i used in the system
n_i	=	number of units of component type j used in the system

In Eq. (3.9.1), the first five terms represent the non-component failure rates, which include design, manufacturing, software, process, and NFF. The last two terms capture the aggregate failure rate of all hardware components. Particularly, $\hat{\lambda}_i^{(n)}$ represents the failure rate of new components used in the design and $\hat{\lambda}_j^{(o)}$ represents the failure rate of components used in predecessor products. Section 3.5.4 showed how to estimate failure rates of new and existing component types due to the different life data source.

Reliability uncertainties are caused by various factors, such as design changes, limited test time, temperature and electrical derating variations, and customer usage, among others. Though the exact distribution for $\hat{\lambda}_{sys}$ is difficult to obtain, the mean and the variance of $\hat{\lambda}_{sys}$ can be estimated as follows:

$$E[\hat{\lambda}_{sys}] = E[\hat{\lambda}_d] + E[\hat{\lambda}_s] + E[\hat{\lambda}_m] + E[\hat{\lambda}_p] + E[\hat{\lambda}_{nff}] + \sum_{i=1}^{I} m_i E[\hat{\lambda}_i^{(n)}] + \sum_{j=1}^{J} n_i E[\hat{\lambda}_j^{(o)}]$$

$$\tag{3.9.2}$$

$$Var(\hat{\lambda}_{sys}) = Var(\hat{\lambda}_d) + Var(\hat{\lambda}_s) + Var(\hat{\lambda}_m) + Var(\hat{\lambda}_p) + Var(\hat{\lambda}_{nff})$$

$$+ \sum_{i=1}^{I} m_i Var(\hat{\lambda}_i^{(n)}) + \sum_{j=1}^{J} n_i Var(\hat{\lambda}_j^{(o)}) \tag{3.9.3}$$

Both Eqs. (3.9.2) and (3.9.3) are derived by assuming that failures of components and non-components are mutually independent. Since $\hat{\lambda}_d$, $\hat{\lambda}_s$, $\hat{\lambda}_m$, $\hat{\lambda}_p$, $\hat{\lambda}_{nff}$, $\hat{\lambda}_i^{(n)}$, and $\hat{\lambda}_j^{(o)}$ are uniformly bounded, according to the central limit theorem (CLT), $\hat{\lambda}_{sys}$ tends to be normally distributed for large values of component numbers. The system failure rate with $(1 - \alpha) \times 100\%$ confidence can be estimated by

$$\lambda_{sys}^{\max} = E[\hat{\lambda}_{sys}] + Z_{1-\alpha}\sqrt{Var(\hat{\lambda}_{sys})} \tag{3.9.4}$$

where $Z_{1-\alpha}$ is the z-value of the standard normal distribution. Given the mean and variance of $\hat{\lambda}_{sys}$, the product reliability can also be specified by the MTBF with a pre-defined convergence. That is,

$$P\left\{\hat{\lambda}_{sys} \leq \frac{1}{MTBF}\right\} \geq 1 - \alpha \tag{3.9.5}$$

where *MTBF* is the target mean-time-between-failure of the system. Different confidence levels of the product MTBF are desirable before volume manufacturing begins. The lower bound of MTBF can be improved by using more reliable components, lowering the operating temperature, or reducing the derating percentage, or enhancing the design, software testing, manufacturing quality, and processes.

3.9.2 Implementation of Printed Circuit Board Design

In this section, the proposed DFSS model in Eq. (3.9.5) is applied to design a new printed circuit board (PCB) assembly. The board consists of 54 component types with a total of 759 parts. Seven new component types (with a total of 119 parts) are first used in the board. Since no field failure data are available for new components, their failure rates are extrapolated based on the methods described in Section 3.6. Table 3.11 presents the temperature and electrical derating parameters of seven new components. The activation energy E_a and fitting parameter m can be obtained from component suppliers or found in technical handbooks such as Bellcore SR-332. In industry, FIT is often used instead of failure rate. FIT is defined as the number of failures in 10^9 hours. For example, the FIT of 1 MB EPROM is 77.4, which is equivalent to a failure rate of 7.74×10^{-8} faults per hour (i.e. $77.4/10^9$). The other 47 component types have been used on predecessor products. Therefore their failure rates can be directly calculated using Eq. (3.5.2) and the results are summarized in Table 3.12.

Reliability of non-components is estimated from three predecessor products with similar complexity, namely PCB-A, PCB-B, and PCB-C, and the triangular distribution parameters $\{a, b, c\}$ are obtained and listed in Table 3.13. By combining data in Tables 3.11–3.13, we are able to calculate the mean and variance of PCB's failure rate using Eqs. (3.9.2) and (3.9.3), respectively. The $E[\lambda_{PCB}] = 1.08 \times 10^{-5}$ and $Var(\lambda_{PCB}) = 3.3 \times 10^{-12}$ are shown in Table 3.14. The detailed computation is automated

Table 3.11 Data for estimating failure rates of new components.

			Input data						Output data			
i	Component name	n_i	FIT	E_a	μ_T	σ_T	u	v	m	μ_{π_T}	σ_{π_T}	μ_{π_E}	σ_{π_E}
1	1 MB EPROM	14	37.4	1.4	43.7	1.7	60	80	0.024	1.90	0.53	1.63	1.04
2	0.2 W 1% Resistor	32	3.33	1.2	47.6	8.3	50	70	0.019	5.08	7.11	1.22	0.53
3	FPGA	26	6.1	0.7	56.2	4.1	60	85	0.024	3.74	1.16	1.74	1.18
4	Switch transistor	26	5.81	1.2	51.8	2.8	60	75	0.024	5.37	2.03	1.53	0.91
5	Op amp	3	9.00	1.2	46.6	5.2	55	75	0.024	3.17	2.46	1.45	0.83
6	Clock driver	14	0.823	1.2	65.1	1.9	65	80	0.024	27.86	6.51	1.73	1.13
7	Switch diode	4	11.25	1.0	63.2	6.4	60	80	0.029	15.74	11.08	1.81	1.25

Source: Reproduced with permission of IEEE.

Table 3.12 Data for failure rate of existing components.

i	Comp type	n_i	FIT	i	Comp type	n_i	FIT	i	Comp type	n_i	FIT
8	comp 1	16	130	24	comp 17	1	3.643	40	comp 33	2	0.823
9	comp 2	10	5.23	25	comp 18	1	3.643	41	comp 34	1	1.029
10	comp 3	10	10.00	26	comp 19	1	3.643	42	comp 35	1	1.00
11	comp 4	10	5.23	27	comp 20	1	3.33	43	comp 36	1	1.00
12	comp 5	1	10.00	28	comp 21	4	0.823	44	comp 37	1	0.823
13	comp 6	9	1.029	29	comp 22	4	0.823	45	comp 38	1	0.823
14	comp 7	1	9.00	30	comp 23	4	0.823	46	comp 39	1	0.823
15	comp 8	9	0.823	31	comp 24	4	0.823	47	comp 40	1	0.823
16	comp 9	2	3.643	32	comp 25	3	0.823	48	comp 41	1	0.823
17	comp 10	2	3.643	33	comp 26	3	0.823	49	comp 42	1	0.823
18	comp 11	8	0.823	34	comp 27	2	1.029	50	comp 43	1	0.823
19	comp 12	8	0.823	35	comp 28	2	1.00	51	comp 44	252	0.000 05
20	comp 13	6	1.029	36	comp 29	2	0.823	52	comp 45	24	0.0004
21	comp 14	7	0.823	37	comp 30	2	0.823	53	comp 46	177	0.000 05
22	comp 15	6	0.823	38	comp 31	2	0.823	54	comp 47	6	0.0002
23	comp 16	6	0.823	39	comp 32	2	0.823				

Source: Reproduced with permission of IEEE.

Table 3.13 Triangular distribution parameters for non-component failure rates.

	Input data (FIT)			Output data (faults/hour)		
	PCB-A	PCB-B	PCB-C	$a\,(\times 10^{-7})$	$b\,(\times 10^{-7})$	$c\,(\times 10^{-7})$
Design errors	148.5	423.6	855.0	1.49	8.55	4.24
Software bug	29.7	145.3	95.0	0.297	1.45	0.950
Manufacturing	207.9	242.1	285.0	2.08	2.85	2.42
Process	118.8	302.6	665.0	1.19	6.65	3.03
Others (NFF)	475.3	641.5	807.5	4.75	8.08	6.42

Source: Reproduced with permission of IEEE.

Table 3.14 Mean and variance of failure rates for the new PCB assembly.

Root case category	Components		Non-component					System
	New comps	Existing comps	Design	Software	Mfg	Process	NFF	PCB
mean	6.52×10^{-6}	2.42×10^{-6}	4.76×10^{-7}	9.0×10^{-8}	2.45×10^{-7}	3.62×10^{-7}	6.41×10^{-7}	1.08×10^{-5}
variance	2.42×10^{-12}	0	2.47×10^{-13}	8.66×10^{-15}	6.07×10^{-14}	1.44×10^{-13}	4.16×10^{-13}	3.30×10^{-12}

Source: Reproduced with permission of IEEE.

in Matlab. Based on Eq. (3.9.5), the lower bound of MTBF for the new design is also obtained as 63 000 hours with 99.7% confidence.

References

Bei, X., Chatwattanasiri, N., Coit, D.W., and Zhu, X. (2017). Combined redundancy allocation and maintenance planning using a two-stage stochastic programming model for multiple component systems. *IEEE Transactions on Reliability* 66 (3): 950–962.

Bellcore, (1995), Reliability Prediction Procedure for Electronic Equipment, SR-332, Issue 5, December 1995.

Bowles, J.B. and Pelaez, C.E. (1995). Application of fuzzy logic to reliability engineering. *Proceedings of the IEEE* 83 (3): 435–449.

Bulfin, R.L. and Liu, C.Y. (1985). Optimal allocation of redundancy components for large systems. *IEEE Transaction on Reliability, vol.* 34 (August): 241–247.

Chern, M.S. (1992). On the computational complexity of reliability redundancy allocation in a series system. *Operations Research Letter* 11 (6): 309–315.

Coit, D.W. and Liu, J. (2000). System reliability optimization with k-out-of-n subsystems. *International Journal of Reliability, Quality and Safety Engineering* 7 (2): 129–142.

Coit, D.W. and Smith, A.E. (1998). Redundancy allocation to maximize a lower percentile of system time-to-failure distribution. *IEEE Transactions on Reliability* 47 (1): 79–87.

Coit, D.W., Jin, T., and Wattanapongsakorn, N. (2004). System optimization considering component reliability estimation uncertainty: multi-criteria approach. *IEEE Transactions on Reliability* 53 (3): 369–380.

Elsayed, E. (2012). *Reliability Engineering*, 2e. Hoboken, NJ: Wiley and IEEE Press.

Fan, M., Zeng, Z., Zio, E. et al. (2018). A stochastic hybrid systems model of common-cause failures of degrading components. *Reliability Engineering and System Safety* 172: 159–170.

Fyffe, D.E., Hines, W.W., and Lee, N.K. (1968). System reliability allocation problem and a computational algorithm. *IEEE Transactions on Reliability* 17 (June): 64–69.

Ghare, P.M. and Taylor, R.E. (1969). Optimal redundancy for reliability in series system. *Operations Research* 17 (September): 838–847.

Guan, H., Chen, W.-R., Huang, N., and Yang, H.-J. (2010). Estimation of reliability and cost relationship for architecture-based software. *International Journal of Automation and Computing* 7 (4): 603–610.

Hadipour, H., M. Amiri, M. Sharifi, (2018) "Redundancy allocation in series-parallel systems under warm standby and active components in repairable subsystems," *Reliability Engineering and System Safety*, www.sciencedirect.com, available online 31 January 2018.

Heydari, M. and Sullivan, K.M. (2018). An integrated approach to redundancy allocation and test planning for reliability growth. *Computers and Operations Research* 92 (4): 182–193.

Holland, J.H., (1975), Adaptation in Natural and Artificial Systems, PhD Dissertation, University of Michigan Press, Ann Arbor, MI.

Huang, H.-Z., Liu, H.-J., and Murthy, D.N.P. (2007). Optimal reliability, warranty and price for new products. *IEEE Transactions* 39 (8): 819–827.

Hwang, C.L., Tillman, F.A., and Kuo, W. (1979). Reliability optimization by generalized lagrangian-function and reduced-gradient methods. *IEEE Transactions on Reliability* 28 (4): 316–319.

IEEE, (1998), IEEE Standard Methodology for Reliability Prediction and Assessment for Electronic System and Equipment. IEEE Standard 1413, New York.

Jia, H., Ding, Y., Peng, R., and Song, Y. (2017). Reliability evaluation for demand-based warm standby systems considering degradation process. *IEEE Transactions on Reliability* 66 (3): 795–805.

Jiang, R. and Murthy, D.N.P. (2009). Impact of quality variations on product reliability. *Reliability Engineering and System Safety* 94 (2): 490–496.

Jin, T., P. Su, (2005), "Minimize system reliability variability based on six-sigma criteria considering component operational uncertainties", in Proceedings of Annual Reliability and Maintainability Symposium, pp. 214–219.

Jin, T., Taboada, H., Espiritu, J., and Liao, H. (2017). Allocation of reliability-redundancy and spares inventory under Poisson fleet expansion. *IISE Transactions* 49 (7): 737–751.

Konak, A., Coit, D.W., and Smith, A.E. (2006). Multi-objective optimization using genetic algorithms: a tutorial. *Reliability Engineering and System Safety* 91 (9): 992–1007.

Kuo, W. and Zuo, M.J. (2003). *Optimal Reliability Modeling: Principles and Applications*, 1e. Hoboken, NJ: Wiley.

Kuo, W., Prassad, V.R., Tillman, F.A., and Hwang, C. (2001). *Optimal Reliability Design: Fundamental and Applications*, 1e. Cambridge, UK: Cambridge University Press.

Levitin, G., Xing, L., and Dai, Y. (2017). Optimization of component allocation/distribution and sequencing in warm standby series-parallel systems. *IEEE Transactions on Reliability* 66 (4): 980–988.

Liang, Y.-C. and Smith, A.E. (2004). An ant colony optimization algorithm for the redundancy allocation problem (RAP). *IEEE Transactions on Reliability* 53 (3): 417–423.

Mettas, A., (2000), "Reliability allocation and optimization for complex systems," in *Proceedings of Reliability and Maintainability Symposium*, pp. 216–221.

Öner, K.B., Kiesmüller, G.P., and van Houtum, G.J. (2010). Optimization of component reliability in the design phase of capital goods. *European Journal of Operational Research* 205 (3): 615–624.

Ozen, T.M. (1986). *Applied Mathematical Programming for Production and Engineering Management*. Prentice-Hall.

Pecht, M., Das, D., and Ramakrishnan, A. (2002). The IEEE standards on reliability program and reliability prediction methods for electronic equipment. *Microelectronics Reliability* 42 (2002): 1259–1266.

ReliaSoft (2017), FMECA, http://www.weibull.com/hotwire/issue46/relbasics46.htm (accessed on February 28, 2017).

Sayama, H., Kameyama, Y., Nakayama, J., and Sawaragi, Y. (1973). Iteration process in the Lagrange multipliers for multiplier method. *System and Control (Japan)* 17: 775–778.

Sinaki, G., (1994), Ultra-reliable fault tolerant inertial reference unit for spacecraft. *Advances in the Astronautical Sciences* 86: 239–248.

Sung, C.S. and Cho, Y.K. (1999). Branch-and-bound optimization for a series system with multiple-choice constraints. *IEEE Transactions on Reliability* 48 (2): 108–117.

Taboada, H.A., Espiritu, J.F., and Coit, D.W. (2008). MOMS-GA: a multi-objective multi-state genetic algorithm for system reliability optimization design problems. *IEEE Transactions on Reliability* 57 (1): 182–191.

Teng, J.T. and Thompson, G.L. (1996). Optimal strategies for general price-quality decision models of new products with learning production costs. *European Journal of Operational Research* 93 (3): 476–489.

US MIL-HDBK-217 (1995). Reliability prediction for electronic systems. In: *Available from National Technical Information Service*. VA: Springfield.

Werner, W. (1994). *Results of recent risk studies in France, Germany, Japan, Sweden and the United States*. Paris: OECD Nuclear Energy Agency, NEA/CSNI/R.

Wu, J., Yan, S., and Xie, L. (2011). Reliability analysis method of a solar array by using fault tree analysis and fuzzy reasoning Petri net. *Acta Astronautica* 69 (11/12): 960–968.

Xiao, N., Huang, H.-Z., Li, Y. et al. (2011). Multiple failure modes analysis and weighted risk priority number evaluation in FMEA. *Engineering Failure Analysis* 18 (4): 1162–1170.

Xu, K., Tang, L.C., Xie, M. et al. (2002). Fuzzy assessment of FMEA for engine systems. *Reliability Engineering and System Safety* 75 (1): 17–29.

Yeh, C.-T. and Fiondella, L. (2017). Optimal redundancy allocation to maximize multi-state computer network reliability subject to correlated failures. *Reliability Engineering and System Safety* 166 (October): 138–150.

Problems

Problem 3.1 Show that the cost function of Eq. (3.2.2) is an decreasing function with α, meaning $db(\alpha)/d\alpha > 0$. Also show that Eq. (3.2.2) is a convex function, meaning $d^2b(\alpha)/d\alpha^2 > 0$.

Problem 3.2 Show that $c(\alpha)$ in Eq. (3.2.3) is strictly decreasing and strictly convex in α.

Problem 3.3 Look at the data of Example 3.1. The analysis shows that Product A has a higher reliability gain than Product B if an additional budget of $5000 is allocated. If the total design budget is $25 000, do you prefer Product A or Product B?

Problem 3.4 Using the same data given in Example 3.2, do the following, assuming that the annual quantity of new product installation is $Q = 2000$/year: (1) find the optimal α^* for $T = 3$ years and (2) find the optimal α^* for $T = 5$ years (assuming that the annual discount factor is 10%).

Problem 3.5 (Reliability allocation) Given the total cost budget $C = 50\,000$, find the optimal component reliability to maximize the reliability of the following two designs, respectively. The component reliability and cost data are given in Table 3.2. Solve the problem using iteration.

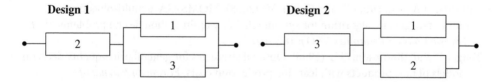

Problem 3.6 (Reliability allocation) Use the Lagrangian function to find the optimal component reliability in Problem 3.5 and compare the results obtained from the one based on iterations.

Problem 3.7 (Reliability allocation) Minimize the system cost in the two designs in Problem 3.4. For design 1, the minimum system reliability should not be less than 0.92. For design 2, the minimum system reliability should be higher than 0.88.

Problem 3.8 (Redundancy allocation) For a series system comprised of four subsystems, each subsystem is allowed to adopt parallel or redundant components to improve the overall reliability. The reliability and cost of each component choice is given in the table below. Given the total system cost budget is 15: (1) maximize the system reliability assuming only one component type is selected for each subsystem; (2) maximize the system reliability with no constraints on component types per subsystem. Component reliability and cost.

	Component type $j = 1$		Component type $j = 2$	
Subsystem i	Reliability (r_{ij})	Cost (c_{ij})	Reliability (r_{ij})	Cost (c_{ij})
1	0.90	1	0.95	2
2	0.88	2	0.92	3.5
3	0.86	1.5	0.91	3
4	0.85	0.5	0.93	2.5

Problem 3.9 (Redundancy allocation) For the same series system in Problem 3.8, we want to minimize the cost subject to the system reliability requirement. Assuming the system reliability must be above 0.97: (1) minimize the system cost assuming only one component type is selected for each subsystem; (2) minimize the system cost with no constraints on component types per subsystem.

Problem 3.10 Solve Example 3.5 using the heuristic method. The heuristic rule is given as follows: to maximize the reliability of a series system via redundant units, we want to maximize the reliability of each subsystem as the system reliability is the product of subsystem reliability.

Problem 3.11 (Allocation of k-out-of-n subsystems) Use the same reliability and cost data in Table 3.4 to perform the k-out-of-n redundancy allocation. The component failure rates for HSD, DC, Analog, and MW are 0.000 02, 0.000 025, 0.000 05 and 0.0001 failures/hour, respectively. The system reliability upon 1000 hours is required to be 0.8: find the optimal $\mathbf{x}^*$ that minimize the system cost. Solve this problem in two cases: (1) assuming all redundant components are in hot-standby mode; (2) redundant components are in cold-standby mode.

Problem 3.12 (Multicriteria design) Using the data in Table 3.5 and $C = 29$ and $W = 45$, find the Pareto optimal solutions for Model 3.8 to maximize the system reliability and minimize the reliability variance.

Problem 3.13 (Multicriteria design) Solve Model 3.8 again using the same data in Problem 3.11 on the condition that for each subsystem only one component type is chosen.

Problem 3.14 (Multicriteria design) Solve Model 3.8 again using the same data in Problem 3.11 on the condition that for each subsystem only one component type is chosen.

Problem 3.15 Assume that a solar park has 10 000 solar PV panels. Each panel is installed with one DC–AC inverter. In the last two years, 25 inverter failures are reported from the park. A failed inverter is replaced with a good unit so that the power generation is ensured. Assume that the system repair downtime is negligible compared to the uptime. Estimate the failure rate of the inverter.

Problem 3.16 The activation energy E_a of component A is 0.5 eV and the E_a of component B is 0.7 eV. If component B operates at 70 °C, at what temperature can component A operate such that it has the same π_T as component B?

Problem 3.17 Based on the fitting parameter m in Table 3.9, plot the electrical derating factor π_E for a 0.2 W 1% resistor and operational amplifier as p increases from 50% to 100%.

Problem 3.18 For a triangle distribution with parameters a, b, and c for the lowest, medium, and largest values, respectively, prove the mean and the variance are

$$E[\lambda] = \frac{1}{3}(a + b + c) \text{ and } Var(\lambda) = \frac{1}{18}(a^2 + b^2 + c^2 - ab - ac - bc)$$

Problem 3.19   Given the uniform PDF of p in Eq. (3.6.8), prove that Eqs. (3.6.9) and (3.6.10) are the actual mean and the second moment of π_E.

Problem 3.20   The mean failure rate of a PCB module is 1.5×10^{-5} failures/hour and the standard deviation of the failure rate is 4.5×10^{-6} failures/hour. Estimate the MTBF at the 90 and 99% confidence levels, respectively.

Problem 3.21   A simple solar PV system is shown below. Create the functional block diagram for this system and perform FTA based on the functional diagram.

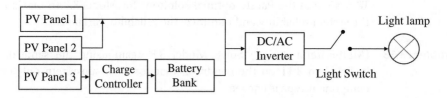

4

Reliability Growth Planning

4.1 Introduction

Reliability growth planning (RGP) aims to identify and eliminate critical failure modes during the design, manufacturing, and use of a new product. It consists of two processes: reliability growth test (RGT) and latent failure elimination (LFE). RGT is applied in the early development phase while LFE is implemented in volume production. Recently, a growing number of firms start to bundle the product with aftersales services by offering a lifecycle reliability growth program under performance-based maintenance. Therefore it is desirable to re-examine the conventional reliability growth strategy and develop new methodologies to meet the changing business environment. This chapter presents a life-cycle reliability growth methodology with which corrective actions (CAs) are integrated into product development, volume production, and field use. We investigate the inter-actions between the reliability testing time and CA effectiveness in two aspects. First, an optimal reliability growth testing model is presented to minimize the failure rate in the product development phase. Second, the reliability growth initiative is further extended to the after-sales market by continuously forecasting and eliminating latent failures via CAs. A case study is provided to demonstrate the application of the life-cycle reliability growth program to meet the time-to-market requirement.

4.2 Classification of Failures

Failures of a system can be classified into three categories: hardware failure, non-hardware failure, and human error. A hardware failure is associated with the malfunction or the defect of one or more components within the system, such as relays in a printed circuit board (PCB) or bearings in a gearbox. Non-hardware failures include software bugs, design weakness, manufacturing defects, process issue, and no-fault found (NFF). NFF is the kind of failure that happens at the customer site, but the failure mode cannot be duplicated at the repair shop. Human errors refer to the situations where the machine or equipment fails because of the operator's inappropriate use. This happens if the operators do not have sufficient training during the new product introduction, or the customer's actual usage profile exceeds the equipment design limit. Figure 4.1 graphically illustrates these types of failures that occur in engineering systems.

Reliability Engineering and Services, First Edition. Tongdan Jin.
© 2019 John Wiley & Sons Ltd. Published 2019 by John Wiley & Sons Ltd.
Companion website: www.wiley.com/go/jin/serviceengineering

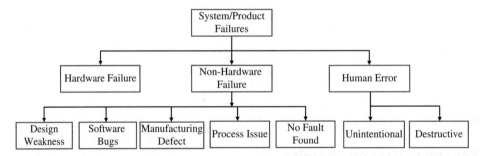

Figure 4.1 Classification of field failures.

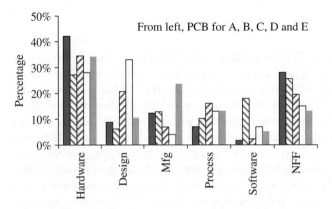

Figure 4.2 PCB failures breakdown by root-cause category.

A reliability growth model solely based on hardware failures may yield an optimistic prediction. This is because a field system could fail due to non-hardware or human errors as well, other than hardware failures. For instance, a desktop computer may fail either because of an overheated microprocessor or because of malicious malware introduced into the memory. In the electronic equipment industry, non-hardware failures sometimes dominate the field returns, especially during the introduction of new products. In Figure 4.2 failure data of four PCB products were collected during their first year of operation after the installation. All the failures are broken down by the root-cause categories: component, design, manufacturing, process, software, and NFF. It is interesting to see that component (or hardware) failures, though the largest in the Pareto chart, account for less than 45% of total failures across product types. In fact, component failures in PCB-B represent only 25% of all field failures. The second largest failure is either design or NFF, which accounts for 25% of all field returns on average for each. Therefore, non-component failures, especially the NFF issue, deserve great attention in new product design and introduction phases (Jin et al. 2006).

To compare how the non-hardware issues influence the reliability prediction, the mean-time-between-failure (MTBF) of PCB-A is computed by including: (i) component failures only; (ii) components, design weakness, and software bugs; and (iii) all field returns. The resulting MTBF corresponding to three prediction criteria are depicted in Figure 4.3. MTBF prediction solely based on component failures is also referred to as a parts count or bill of materials (BOM) method. The reliability target of this

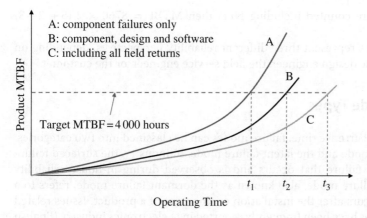

Figure 4.3 Actual MTBF versus predictions.

product is 40 000 hours MTBF. Based on the BOM estimate, the target reliability could be attained at t_1 when we only considered hardware failures, as shown in Line A. The MTBF perceived by the customer is indeed Line C, much lower than the BOM-based prediction. The product eventually attains the target MTBF at t_3. Line B shows the product MTBF typically viewed by the field service engineers who do not count the NFF as failures. This example shows that reliability prediction models incorporating both hardware and non-hardware issues will result in a more realistic estimation, especially during the new product introduction phase.

A distinction shall be made between failure mechanism and failure mode. Failure mechanism describes the physical, chemical, or radiative process that has led to a failure. For instance, water is leaking because of the crack in the ceramic tank. Hence, the failure mechanism is the ceramic crack. A cable fails to deliver the electric current because of a loose solder joint. Thus the loose solder joint is the failure mechanism. A failure mode is the way in which an item could fail to perform its required function. An item can fail in many different ways. For instance, a relay may fail in the closed or open state. Hence a failure mode is a description of a possible state of the item in which it has failed. However, the failure mode does not indicate why the relay failed in the closed or open state. A further experimental investigation shows that the corrosion of the contact causes the relay to fail in the open state. Now we can conclude that corrosion is a failure mechanism that causes the open mode failure. However, the relay could also fail in the open mode because of other failure mechanisms such as a broken spring or a cracked solder joint.

Example 4.1 Suppose a repairable system is installed in the field and operates in 24/7 mode. The system failed 10 times in two years, among which five were hardware issues, two were non-hardware issues, and the remaining three were NFF. Assuming the repair time is short, estimate the system MTBF.

Solution:
- Only consider hardware failures, MTBF = $8760 \times 2/5 = 3504$ hours.
- Both hardware and non-hardware failures are considered, MTBF = $8760 \times 2/(5+2) = 2503$ hours.

- All field returns are counted including NFF; then MTBF = $8760 \times 2/(5 + 2 + 3)$ = 1752 hours.
- These MTBF values represent three different reliability perspectives, depending on whether you are the design engineer, the field service engineer, or the customer.

4.3 Failure Mode Types

Depending on the occurrence time, failure modes can be classified into two categories: the surfaced failure mode and the latent failure mode. In industry, the surfaced failure mode is the type of failure that occurs and is observed during in-house reliability testing. The latent failure mode, also known as the dormant failure mode, refers to a failure mode that occurs after the installation or shipment of a product. Issues related to latent failure modes have been frequently reported in te electronics industry (English et al. 1995; Hokstad and Frøvig 1996; Jackson et al. 2002), robotics and mobile vehicles (Carlson and Mupthy 2005), and network servers (Grassi and Patella 2006). A latent failure mode, once it becomes a critical failure mode, often causes significant cost to the product manufacturer, not to mention the loss of customer goodwill and its quality reputation. Such expenses include escalated warranty costs, excessive spare parts inventory, and increased maintenance and repair overheads. Hence, it is desirable to understand the nature of latent failure so that countermeasures can be implemented prior to the plague of the issue.

Both the quantity and the intensity of latent failures are quite random in nature. We use the product failure data in Jin et al. (2010) to illustrate the first occurrence time of surfaced and latent failure modes for a type of electronic equipment. As shown in Table 4.1, there were 18 failure modes observed in 350 days for a fleet of systems since the first customer installation. In this particular case, the equipment manufacturer stipulates that a new failure mode belongs to the surfaced failure mode category if it occurs within 90 days of installation. New failure modes occurring thereafter fall into the latent failure mode category. According to this criterion, this product had six surfaced failure modes: open diode, no power supply, corrupted electrically erasable programmable read-only memory (EEPROM), cold solder, NFF, and flux contamination. The remaining 12 failure modes starting from the "SMC (system monitoring and control) limit table" belong to latent failure modes. Among the total 47 failures (or field returns) observed in 350 days, the number of returns caused by latent failure modes was 13, representing 38% of the total failures.

Assume that the fleet has 10 systems and each operates in 24/7 mode. The system MTBF can be estimated based on the failure occurrence times in Table 4.1. For comparison purposes, two estimations are made by considering: (i) surfaced failure modes only and (ii) surfaced and latent failures. As shown in Figure 4.4, the estimation using surfaced failure modes yields 2471 hours MTBF in day 350, but the actual MTBF including all failure modes is 1787 hours, which is 28% lower. Therefore it is imperative to monitor the trend of latent failure modes and incorporate them into the reliability prediction model; otherwise the resulting forecasting based on surfaced failure modes tends to be too optimistic.

There are many factors causing or inducing the occurrence of latent failures, including design weakness, software bugs, or improper usage (Jin et al. 2006). One particular reason that triggers a latent failure of electronic equipment is electrical static discharge

Table 4.1 The first occurrence time of different failure modes.

No		Week no. 1	2	3	12	15	23	24	30	33	38	41	45	49	50	
	Days	7	14	21	84	105	161	168	210	231	266	287	315	343	350	Sum
1	Open diode	1		1	4	1			2	6	1		1			17
2	No power supply		1			1	2					2		1		7
3	Corrupted EEPROM			2												2
4	Cold solder				1											1
5	NFF					1		1	1		1	2				6
6	Flux contamination					1										1
7	SMC limit table					1										1
8	Shorted capacitor						1									1
9	PPMU								1				1			2
10	Missing solder								1							1
11	Manufacturing defect								1							1
13	Bad ASIC chip										1					1
14	Blown fuse										1					1
15	Open trace												1			1
16	Bad op-amp												2			2
17	Timing generator												1			1
18	Solder short													1		1

Source: Reproduced with permission of IEEE.

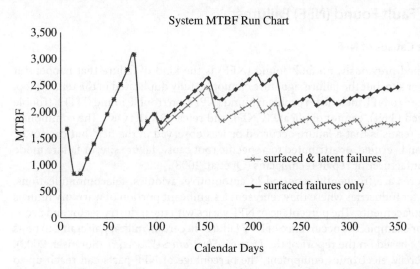

Figure 4.4 System MTBF with and without latent failures. Source: Reproduced with permission of IEEE.

(ESD), which is the sudden flow of electricity between two electrically charged objects. A build-up of static electricity can be caused by tribocharging or by electrostatic induction. ESD often damages the electronic components or microdevices during the assembly, handling, and installation of PCB items, but the failure symptoms did not show up until in field operation (Vinson and Liou 1998). Wearing protective devices like wrist straps, conductive shoes, or smocks helps to prevent the charge build-up. Besides, latent failures could be mitigated or avoided if the system health condition can be monitored using in-situ sensors or prognostics programs. Readers are referred to Si et al. (2011) for detailed discussions on relevant technologies, such as condition-based maintenance and prognostics and health management.

Example 4.2   Based on data in Table 4.1, estimate the MTBF of the product in days 100 and 200, respectively, by considering: (i) surfaced failure modes only and (ii) all field failures.

Solution:
Given that the fleet size is 10, the cumulative system operating hours by day 100 is $10 \times 100 \times 24 = 24\,000$ hours. A total of 10 failures occurred by day 100, and all belong to the surfaced failure modes: open diode, no power supply, corrupted EEPROM, cold solder, NFF, and flux contamination. Thus, MTBF = $24\,000/12 = 2000$ hours.

By day 200, the number of fleet operating hours doubles and reaches $10 \times 200 \times 24 = 48\,000$ hours. In the meantime, the cumulative failures reach 22, and 4 failures belong to latent failure modes. These are the SMC limit table, shorted capacitor, pin parametric measure unit (PPMU), and missing solder. Hence, the system MTBF = $48\,000/22 = 2182$ hours. If one ignores the latent failures, the projected MTBF would be $48\,000/18 = 2667$ hours, or 484 hours larger than the actual MTBF.

4.4   No Fault Found (NFF) Failures

4.4.1   The Causes of NFF

As mentioned previously, no fault found (NFF) is the kind of failure that happens at the customer site, but the failure signature cannot be fully duplicated in the repair shop. Terms related to NFF include no problem found (NPF), no trouble found (NTF), trouble not identified (TNI), cannot duplicate (CND), and retest OK (RTOK). The commonality of these terms is that a failure occurred or was observed in the field but cannot be replicated and verified, or attributed to a specific root cause, failure site, or failure mode by the manufacturer or repair technicians (Qi et al. 2008).

NFF failures are frequently reported in automotive, avionics, telecommunications, and computer industries where they represent a significant portion of warranty returns and field replacements. The percentage of NFF varies with the industry sector and product type. For example, the occurrence of NFF failures in certain military aircraft can be as high as 50%, based on the report of the *Defense Electronics Magazine* (Sorensen 2003). For commercial electronics equipment, the percentage of NFF parts can reach up to 30%, as shown in Figure 4.2 (Jin et al. 2006). When examining NFF by part or failure modes, the situation is often much graver. For example, NFF observations reported by commercial airlines and military repair depots have been found to be as high as 60–75%

(Burchell 2007). Though NFF is a notoriously difficult problem, there are three main factors responsible for the generation of NFF parts (Barkai 2014):

- In the course of the repair activity, the technician replaced multiple parts without exactly knowing which part was the actual root cause of the failure. This could be caused by poor service communications, lack of training, and suboptimal diagnostic techniques, but can also be encouraged or required, for example, when equipment availability is prioritized rather than the cost of the part's replacement in the field.
- The variation and diversity of customer usages push the equipment to operate in the design margin. From the design perspective, marginal design leads to "tolerance stacking" in which a part may fail under certain field conditions but not on the test-bed in the repair shop.
- Intermittent failures can be due to uncertainty in operating environment or customer usage. This refers to the malfunction of a machine or system that only occurs at irregular intervals with normal functionality at all other times. They can be defined as a loss of certain functions or electrical connections for a limited period of time, or can be minute breaks in a connection caused by vibration or through corrosions.

Increasing system complexities and operational environments have seen a rise in the number of NFF returns that are being reported during field service. There is strong evidence showing that a serviceable component was removed and attempts were made, but not successfully, to identify the root-cause. This chapter in particular, deals with the impact of NFF from an organizational culture and human factor point of view. A compressive review on the developments in NFF standards, its financial implications, and safety concerns is available in Khan et al. (2014). The authors emphasize that the NFF phenomenon cannot be treated as a single event, but rather as a sequence that results in a series of actions at various maintenance levels until a final decision is made to add the NFF label.

4.4.2 The Impact of NFF

NFF parts carry high direct and indirect costs, a bloated inventory, excessive warranty claims, and poor customer satisfaction. For example, in 1993 a British-based commercial airline operator formed a task force to investigate the 33% rate of NFF failures (Hockley and Phillips 2012). It was in fact 13.8% of all unscheduled removals that could be positively identified as NFF, costing £17.6 million (i.e. £1 = $1.5) per year. For equipment like aircraft and wind turbines with 20–30 years of life, the operating and support overhead typically account for between 60% and 80% of the lifecycle cost (LCC) of the equipment. The portion of LCC generated by NFF can be very significant and therefore has long been recognized in various industries as an area to mitigate the LCC of equipment. Costs driven by NFF occur at three different levels of support, namely equipment, repair shop, and spares inventory levels. These are further elaborated below.

4.4.2.1 Equipment Level of Support

At the equipment level there will be costs associated with investigations and diagnosis whether certain parts needs to be removed or replaced. These costs manifest themselves in three aspects. First, no failure is actually found in spite of lots of effort, but nothing is actually removed or changed and no affirmative solution is implemented. However, the same failure symposium then reoccurs subsequently in the next production period or soon after, and further investigation is required and necessary. Second, a wrong solution may be found and applied, which means the fault seems fixed, but it will reappear sooner or later in subsequent operations. The actual cost is then the sum of the cost of the first repair action (which was not a correct solution) and the cost of the subsequent failure and repair. Third, a replacement is performed speculatively, which might consist of exchanges or swapping of several parts to generate some positive outcomes. The equipment is declared to be serviceable, but creates one or more NFF opportunities in the next level (i.e. in the repair shop) as the parts were speculatively removed. Let C_{eq} be the NFF cost incurred by the equipment; then

$$C_{eq} = t_d c_l + t_d c_m + \sum_{i=1}^{n_s} c_i \tag{4.4.1}$$

where t_d is the equipment downtime, c_l the hourly labor cost c_m is the hourly production loss due to machine downtime, n_s is the number of parts replaced, and c_i is the unit cost of spare part i for $i = 1, 2, \ldots, n_s$.

4.4.2.2 At the Repair Shop Level of Support

At the repair shop level of support there will be wasted labor in unsuccessful or futile repairs. These include: (i) looking for failure modes that are not there as a result of speculative replacements; (ii) looking for intermittent failures that are unable to be duplicated because operating conditions (e.g., temperature, humidity, vibration, and location) that cause intermittent faults cannot be replicated on the test-bed; and (iii) implementing inadequate or inappropriate diagnostic tests such as the wrong procedures, or the test parameters are set at a level different from the ones at the customer site. Let C_{rs} be the NFF cost incurred to the repair shop; then

$$C_{rs} = t_r c_r + c_f \tag{4.4.2}$$

where t_r is the labor hours involved in testing and diagnosis, c_r is the hourly labor cost of a repair person, and c_f is the facility cost including the test bed, electricity, space, and other expenses associated with the repair.

4.4.2.3 In Spare Parts Inventory and Supply Chain

Throughout the spare parts supply chain there will be wasted transportation, labor hours, and materials: (i) stocking more spare items to keep up with demand because there are significant amounts of items in the repair pipeline that are indeed not faulty; (ii) purchasing more items to maintain the safety stock level due to large amounts of parts in the repair loop; and (iii) increasing the logistics and storage (or holding) cost due to the increased number of parts under repair when actually these items should not be in the loop at all. Since logistics and transportation can be considered as part of the holding cost, the supply chain cost due to NFF parts consists of holding cost and inventory capital. Let C_{sp} be the NFF cost incurred to the spare parts inventory and supply chain; then

$$C_{sp} = c_h n_s + c_p n_s \tag{4.4.3}$$

where c_h is the inventory holding cost per NFF part, n_s is the number of NFF parts, and c_p is the cost of purchasing a spare part for keeping the necessary safety stock level.

By aggregating the costs in Eqs. (4.4.1–4.4.3), the total cost of NFF returns can be estimated as follows:

$$C_{NFF} = C_{eq} + C_{rs} + C_{sp} \tag{4.4.4}$$

Example 4.3 A PCB module returned from a customer turns out to be an NFF part after being tested in the factory's repair shop. Estimate the NFF cost from the perspective of the customer and the factory. Related parameters are given as follows: $t_d = 4$ hours, $c_l = \$50$/hour, $c_m = \$4500$/hour, $n_s = 1$, $c_s = \$40\,000$/part, $t_r = 2$ hours, $c_r = \$25$/hour, $c_f = \$200$, $c_h = \$5000$/part/year, and $c_p = \$25\,000$/part.

Solution:
The NFF cost incurred to the customer consists of the payment of a spare part and the manufacturing losses in the downtime. Hence

$$C_{eq} = t_d c_l + t_d c_m + n_s c_s = 4 \times 50 + 4 \times 4500 + 1 \times 40\,000 = \$58\,200$$

If the PCB is still under the warranty period, there is no spare part cost incurred to the customer. However, the customer incurs the labor cost and the production losses, which are

$$C_{eq} = t_d c_l + t_d c_m + n_s c_s = 4 \times 50 + 4 \times 4500 = \$18\,200$$

The NFF cost incurred by the manufacturer consists of C_{rs} and C_{sp}. Based on Eqs. (4.4.2) and (4.4.3), we have

$$C_{rs} = t_r c_r + c_f = 2 \times 25 + 200 = \$250$$

$$C_{sp} = c_h n_s + c_p n_s = 5000 \times 1 + 25\,000 \times 1 = \$30\,000$$

Thus, the total cost incurred to the manufacturer is $\$250 + \$31\,000 = \$30\,250$. This example shows that an NFF part creates significant amounts of cost to the equipment user and the equipment provider.

4.5 Corrective Action Effectiveness

4.5.1 Engineering Change Order Versus Retrofit

Corrective action (CA) is the primary means to improve the reliability of a product by eliminating or mitigating the critical failure modes. Typical CA activities include redesign, redundancy allocation, use of more reliable components, software upgrade, manufacturing technology improvement, process change, and training of operators. Technically CA can be applied in any time during the product's design, development, manufacturing, and field use. For instance, CA is often used to eliminate the key failure modes that emerged from the in-house RGT. CA can also be implemented in the products that have been shipped and installed in the field. This can be done in the repair shop by applying the CA to the returned items. This process is also referred to as the engineering change order (ECO) in the equipment manufacturing industry. Due to the resource constraint, ECO is often applied to the critical failure modes that are responsible for the largest failure intensities. Finally, if the manufacturer decides to

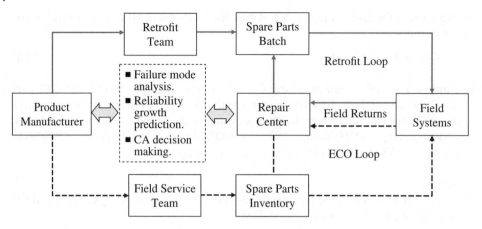

Figure 4.5 Implementation of ECO and retrofit post installation.

fully eliminate one or more failure modes in a short period, a common practice is to use the retrofit process. For a retrofit process, the manufacturer proactively replaces an in-service part that will fail owing to known failure modes.

Figure 4.5 depicts the flowchart of deploying ECO and retrofit processes for a repairable system. Both ECO and retrofit can only be applied to the field systems. In the automotive industry retrofit is also known as "recalls" when a fleet of vehicles are called back for fixing common or safety-related issues, such as brakes and fuel injection flaws (Trudell and Horie 2014). In terms of resource, ECO is a cost-effective approach to reliability growth, but the growth rate is slow and incremental. This is because a known failure mode will not be fully eliminated until all parts of the fleet fail and get fixed in the factory's repair shop. A retrofit process is often expensive because it requires a batch of spare parts and dedicated personnel for implementation, yet the benefit to the product reliability is immediate and profound.

4.5.2 Corrective Action Effectiveness

Corrective action effectiveness (CAE), also known as the fix effectiveness factor, measures the efficacy of the CA program against a specific, usually critical, failure mode. It measures the percentage of the failure reduction per failure mode. For instance, the monthly field returns of application specific integrated circuit (ASIC) units due to overheat is 30. If the operating temperature is reduced by 15 °C, the expected failures would be 10 per month. Then the value of CAE is $(30 - 10)/30 = 0.67$. The effectiveness of a CA depends on the nature of the failure mode and the fleet size. Take a PCB as an example. If a recurring failure is related to a particular component such as a relay or resistor, the failure mode could be completely eliminated by replacing all problem components with a better item. On the other hand, if an intermittent failure is associated with a solder joint problem, the full elimination of cold solder joints of field products is rather difficult because replacing all installed PCBs is too costly, though technically feasible. Gibson and Crow (1989) show that the CAE in real-world situations falls as [0.3, 0.7] when one considers the uncertainty in CA implementation. To optimize the resource distribution, it is necessary to define a cost model to characterize the effectiveness of a CA process

Figure 4.6 CA effectiveness under various b.

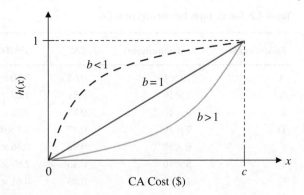

against the target failure mode. The following model is proposed to measure the CAE with respect to the resource investment (Jin et al. 2009):

$$h(x) = \left(\frac{x}{c}\right)^b, \quad \text{for } 0 \leq x \leq c \tag{4.5.1}$$

where x represents the amount of the cost (e.g. CA budget) associated with eliminating a particular failure mode. Both b and c are model parameters with positive values, and c is actually equal to the retrofit cost if all problem products associated with that failure mode are removed and replaced with updated units. If $x = 0$, it means there is no CA taken, and hence $h(x) = 0$. If $x = c$, it implies that retrofit is adopted and the effectiveness $h(x) = 1$. Thus the value of $h(x)$ varies between 0 and 1.

Depending on the value of b, Figure 4.6 depicts three types of CAE curves: linear, power, and rational. If $b = 1$, $h(x)$ is linearly proportional to x. This is applicable to most hardware failures for which the CA cost is proportional to the numbers of components being replaced. If $b > 1$, then $h(x)$ becomes a power function. This is applicable to product redesign as it often requires lots of effort and resources, but the benefit to the reliability growth is significant. For $b < 1$, Eq. (4.5.1) becomes a rational function, which implies that CAE levels off once x reaches a certain level. Such examples include software reliability growth and the solder joints reflow controls. In these cases, the initial improvement is always exciting, but the difficulties quickly increase if one wants to further reduce the failures.

Example 4.4   Nine critical failure modes are identified and assessed for CA. Their failure rates are summarized in Table 4.2. Based on the CAE, estimate the system failure rate and MTBF after the CA is implemented.

Solution:
To forecast the new system failure rate, we need to project the failure rate of individual failure modes in conjunction with CA activities. Without loss of generality, we use failure mode A in Table 4.2 to illustrate the estimation procedure. Let λ_A and $\lambda_A{}^{CA}$ be, respectively, the original failure intensity and the intensity after CA. Then

$$\lambda_A^{CA} = (1 - 0.63)\lambda_A = 0.37 \times 10 \times 10^{-5} = 3.70 \times 10^{-5} \text{ failure/hour}$$

Similarly, we can estimate the failure intensity of the remaining failure modes based on the CAE, and the results are summarized in the right column of Table 4.2. Note that

Table 4.2 Failure rates before and post CA.

Failure mode	λ (failure/hour)	CAE	λ^{CA} (failure/ hour)
A	10×10^{-5}	0.63	3.70×10^{-5}
B	9×10^{-5}	0.81	1.67×10^{-5}
C	8×10^{-5}	0.77	1.83×10^{-5}
D	7×10^{-5}	0.27	5.1×10^{-5}
E	6×10^{-5}	0.92	0.46×10^{-5}
F	5×10^{-5}	0.43	2.87×10^{-5}
G	4×10^{-5}	0.85	0.61×10^{-5}
H	3×10^{-5}	1.00	0.00×10^{-5}
I	2×10^{-5}	0.69	0.61×10^{-5}
Sum	5.4×10^{-4}	n/a	1.685×10^{-4}

the CAE of failure mode H is one, hence eliminating the entire failure mode. Now the projected failure intensity of the system is obtained by summing λ^{CA} from A to I, which yields

$$\lambda_{sys}^{CA} = \lambda_A^{CA} + \lambda_B^{CA} + \cdots + \lambda_I^{CA} = 16.85 \times 10^{-5} \quad \text{failure/hour}$$

Under the assumption of an exponential lifetime, the system MTBF upon CA is

$$MTBF = \frac{1}{\lambda_{sys}^{CA}} = \frac{1}{19.81 \times 10^{-5}} = 5935 \quad \text{hours}$$

Compared to the original MTBF, which is $1/(54 \times 10^{-5}) = 1852$ hours, it increases by 220%. Figure 4.7 graphically depicts the impacts of CAE on the failure intensity across all failure modes.

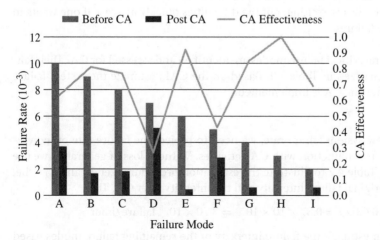

Figure 4.7 The impact of CA effectiveness on failure rate.

4.6   Reliability Growth Model

4.6.1   Duane Postulate

It is common that new products are less reliable in the prototype stage than later in the volume production and field use. This phenomenon was first analyzed by J.T. Duane who derived an empirical relationship between MTBF improvement and the cumulative system time based on a range of items used in aircraft (Duane 1964). Let θ_c be the cumulative MTBF defined as the total time divided by total failures. Duane observed that θ_c plotted against cumulative operating time in the log–log scale yields a straight line. That is,

$$\ln \theta_c = \ln \theta_0 + \alpha(\ln t_c - \ln t_0) \tag{4.6.1}$$

where θ_0 is the initial MTBF at the start of the reliability growth monitoring period; t_0 is the initial cumulative operating time; and t_c is the current cumulative operating time. If the product lifetime is exponential, the Duane model in Eq. (4.6.1) can also be expressed as

$$\ln \lambda_c = \ln \lambda_0 - \alpha(\ln t_c - \ln t_0) \tag{4.6.2}$$

where λ_c is the current failure rate and λ_0 is the initial failure rate of the system or item. Equation (4.6.2) can be easily derived from Eq. (4.6.1) by realizing $\lambda_c = 1/\theta_c$ and $\lambda_0 = 1/\theta_0$. The slope α gave an indication of the MTBF growth rate or the effectiveness of the CAs in eliminating the failure modes. Therefore, Eq. (4.6.1) can be rewritten as follows:

$$\theta_c = \theta_0 \left(\frac{t_c}{t_0}\right)^\alpha = \theta_0 \left(1 + \frac{t_c - t_0}{t_0}\right)^\alpha \tag{4.6.3}$$

The means that, given the growth rate α, we can predict the future MTBF based on the additional test time (i.e. $t_c - t_0$) to be spent. The value of α is in a range between 0.2 and 0.6 depending on the value of CAE. Table 4.3 summarizes the interdependency between

Table 4.3 Reliability growth rate under different CA effectiveness.

α	Level of CA effort	Stakeholders	Product stage
0.4–0.6	1) Weekly tracking and monitor the trend of failure modes. 2) Elimination of all critical failure modes is top priority, representing 30–50% failures.	Reliability, design, manufacturing, service, and market teams	Development, prototyping, early production, and new product introduction
0.2–0.4	1) Monthly tracking and monitor the trend of failure modes. 2) Selectively eliminate two to three critical failure modes, representing 15–25% of failures.	Reliability, manufacturing, and service teams	Volume production, shipment, and installation
0–0.2	1) Quarterly tracking and monitor the trend of failure modes. 2) Minor corrective actions are applied against known failure modes.	Reliability and service teams	Phase-out and decommission stage

the growth rate and the CA effectiveness in different product stages, including product design/development, early introduction, volume production, and phase-out periods.

The Duane model can also be used in principle to estimate how much test time is needed in order to attain the target MTBF. If the initial MTBF θ_0 is known at an early stage, the additional test time can be estimated based on the value of α that is appropriately assumed. The selection of α should be related to the expected CA effectiveness in detecting and eliminating the root causes of failures. Hence Table 4.3 serves as the guideline in choosing the appropriate value of α in order to figure out the additional testing time.

Sometimes, one is also interested in the instantaneous MTBF of the product. The instantaneous MTBF of the Duane postulate at the cumulative test time t_c during the test-fix-test development phase can also be derived and we use the derivation given by O'Connor (2012) to present the instantaneous MTBF. This can be obtained by realizing that

$$\theta_c = \frac{t_c}{n_c} \tag{4.6.4}$$

where n_c is the number of failures observed between $[0, t_c]$. Therefore

$$n_c = \frac{t_c}{\theta_c} = t_c^{(1-\alpha)} \left(\frac{t_0^\alpha}{\theta_0} \right) \tag{4.6.5}$$

Taking the derivative with respect to t_c in Eq. (4.6.5) yields

$$\frac{dn_c}{dt_c} = (1-\alpha)t_c^{-\alpha} \left(\frac{t_0^\alpha}{\theta_0} \right) = (1-\alpha)\frac{1}{\theta_0} \left(\frac{t_0}{t_c} \right)^\alpha \tag{4.6.6}$$

Now the instantaneous MTBF, denoted as θ_I, is obtained as follows:

$$\theta_I = \frac{dt_c}{dn_c} = \frac{\theta_0}{1-\alpha} \left(\frac{t_c}{t_0} \right)^\alpha = \frac{\theta_c}{1-\alpha} \tag{4.6.7}$$

Example 4.5 The initial MTBF of a new jet engine is $\theta_0 = 1000$ hours. Engineers would like to know how much additional test hours should be conducted to reach $\theta_c = 1500$ hours given $\alpha = 0.2$, 0.4, and 0.6, respectively. Note that the engine has accumulated 20 000 operating hours.

Solution:
From Eq. (4.6.3), we can express the additional test hours, $t_c - t_0$, as follows:

$$t_c - t_0 = t_0 \left[\left(\frac{\theta_c}{\theta_0} \right)^{1/\alpha} - 1 \right] \tag{4.6.8}$$

Therefore,

For $\alpha = 0.2$, we have $t_c - t_0 = 20\,000((15\,000/1000)^{1/0.2} - 1) = 54\,259$ hours
For $\alpha = 0.4$, we have $t_c - t_0 = 20\,000((15\,000/1000)^{1/0.4} - 1) = 18\,538$ hours
For $\alpha = 0.6$, we have $t_c - t_0 = 20\,000((15\,000/1000)^{1/0.6} - 1) = 10\,978$ hours

Given $\alpha = 0.6$, the number of additional test hours is about 50% of previously cumulated test hours. If we compare the additional test hours between $\alpha = 0.6$ and $\alpha = 0.2$, the time in the former is only 20% of the time in the latter. This shows that an aggressive reliability growth program can significantly accelerate the product's time-to-market.

4.6.2 Power Law Model

The failure sequence of a repairable system is often modeled as a counting process. Depending on the failure intensity rate, repairable systems can be classified into three categories: (i) reliability growth where critical failure modes are eliminated and the failure intensity declines over time; (ii) reliability deterioration due to the system aging; and (iii) constant failure intensity with no trend. Crow (1974) found that the reliability growth rate could be approximated by a non-homogenous Poisson process. His method became the well-known Crow/AMSAA (Army Materiel Systems Analysis Activity) model, which is also called the power law model. Since then, many studies pertaining to RGT implementation have been reported. For instance, Xie and Zhao (1996) developed a graphical tool to predict the reliability growth based on the Duane model. Recently, Peng et al. (2013) developed a Bayesian update model to assess the lifecycle reliability of new products. While Crow/AMSAA originally focuses on hardware, the RGT concept has been extended to software design, debugging, and performance verification (Inoue and Yamada 2007; Bai et al. 2008; Hwang and Pham 2009). In general, these models are quite effective in driving the reliability of new products when sufficient in-house testing time and resources are available.

The following power law intensity model, or Crow/AMSAA model, is widely used to characterize the failure intensity rate of a repairable system:

$$\lambda(t) = \alpha\beta t^{\beta-1} \tag{4.6.9}$$

where $\alpha > 0$ and $\beta > 0$ are the model parameters. Depending on whether $0 < \beta < 1$, $\beta = 1$, or $\beta > 1$, Eq. (4.6.9) is able to model three types of failure intensity: decreasing, constant, and increasing trend. This model is applicable to systems as well as to components that are repayable. Figure 4.8 depicts the failure intensity rate under different β with common α. It shows that if $\lambda(t)$ is decreasing over time, the time between two consecutive failures statistically increases over time. If $\lambda(t)$ is an increasing function, the time between two consecutive failures tends to be shorter over time. For systems with a constant intensity rate, the failure sequence can be modeled as a renewal process where successive repair actions render the system to an "as-good-as-new" state.

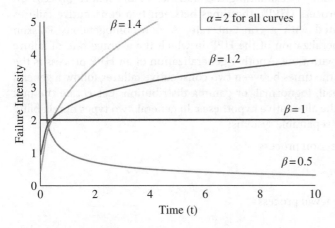

Figure 4.8 Crow/AMSAA failure intensity function.

Given the failure intensity rate, the cumulative failures during a time period, say $[t_1, t_2]$, can be estimated as

$$m(t_2, t_1) = \int_{t_1}^{t_2} \lambda(x)dx = \int_{t_1}^{t_2} \alpha\beta x^{\beta-1} dx = \alpha(t_2^\beta - t_1^\beta), \quad \text{for } 0 \le t_1 < t_2 \quad (4.6.10)$$

Example 4.6 The value of α and β of the Crow/AMSAA model of two repairable systems are given as follows: for System A, $\alpha_A = 0.05$ and $\beta_A = 2$ and for System B, $\alpha_B = 1$ and $\beta_B = 0.7$. Estimate the cumulative failures for $t_2 = 10$ and 20, respectively, assuming $t_1 = 0$.

Solution:
According to Eq. (4.6.10), the number of cumulative failures of System A at $t_2 = 10$ and 20 is estimated as

$$m_A(t_2 = 10) = 0.05 \times (10^2 - 0^2) = 5$$

$$m_A(t_2 = 20) = 0.05 \times (20^2 - 0^2) = 20$$

Similarly, the number of cumulative failures of System B at $t_2 = 10$ and 20 is obtained by

$$m_B(t_2 = 10) = 1 \times (10^{0.7} - 0^{0.7}) = 5$$

$$m_B(t_2 = 20) = 1 \times (20^{0.7} - 0^{0.7}) = 8.1$$

It is interesting to note that both systems have the same amount of failures in [0, 10]. Since the reliability of System A is degrading with $\beta_A = 2$, it generates an additional $20 - 5 = 15$ failures in the period of [10, 20]. On the contrary, the reliability of System B is growing with $\beta_A = 0.7$, and it only generate $8.1 - 5 = 3.1$ failures in the period of [10, 20]. Thus, Eq. (4.6.10) is capable of evaluating the effectiveness of CAs based on the number of failures in a given time period.

4.6.3 Trend Test Statistics

Reliability trend tests aim to determine whether the system reliability is growing, declining, or remains constant by examining the statistics of the failure process. In a homogeneous Poisson process (HPP), the times between two consecutive failures are exponentially distributed with a constant rate. A non-homogeneous Poisson process (NHPP) is the generalization of the HPP in which the average rate of failure arrivals is allowed to vary with time. Another generalization of an HPP process is the renewal process for which the times between two consecutive failures follow a general distribution, such as Weibull, lognormal, or gamma distribution. Let H_0 be the null hypothesis test and H_a be the alternative hypotheses. In general, two types of reliability trend tests are available for repairable systems:

Test 1: Trend test for the Poisson process
  H_0: HPP
  H_a: NHPP
Test 2: Trend test for the renewal process
  H_0: Renewal process
  H_a: Non-renewal process

Table 4.4 Trend test methods.

Test name	Test for what	Test statistics
Crow/AMSSA	NHPP versus HPP	Chi-square
Laplace test	NHPP versus HPP	Normal
PCNT	Renew versus non-renew	Normal
Lewis–Robinson test	Renew versus non-renew	Normal

Four quantitative trend tests are recommended: (i) Crow/AMSAA test; (ii) Laplace test; (iii) pair-wise comparison non-parametric test (PCNT); and (iv) Lewis–Robinson test. The Crow/AMSAA and Laplace tests are capable of testing NHPP versus HPP while the PCNT and Lewis–Robinson tests are often used to test renewal versus non-renewal processes.

Table 4.4 summarizes the test methods and the test statistics of these tools. Since the Crow/AMSAA test is probably the most widely used tool for testing the reliability growth, we will give a brief introduction here. For other test tools, readers can find the detailed discussions in the work by Wang and Coit (2005).

The Crow/AMSAA test is based on the assumption that a failure intensity in Eq. (4.6.9) is appropriate for a repairable system. When $\beta = 1$, the failure intensity reduces to $\lambda(t) = \alpha$, which means the failure process follows an HPP without a growth trend. If $\beta > 1$, the failure intensity increases with t, implying deterioration of reliability. If $\beta < 1$, the system reliability is growing with less failure frequency as time progresses. Therefore, the trend test involves whether the estimate of β is significantly different from 1. The hypothesis test is

$H_0: \beta = 1$ (HPP)
$H_a: \beta \neq 1$ (NHPP)

According to Crow (2015), for a system under test, the maximum likelihood estimate (MLE) is

$$\hat{\beta} = \frac{n}{\sum\limits_{i=1}^{n} \ln\left(\dfrac{t^*}{t_i}\right)} \tag{4.6.11}$$

where n is the number of observed failures, t_i is the ith failure arrival time, and t^* is the test termination time. The test statistics is $2n/\hat{\beta}$, which has a chi-squared distribution with degree $2n$. Given $100(1 - \gamma)\%$ confidence level, the rejection criteria of the null hypothesis is given by

$$\text{Reject } H_0 \text{ if } \quad \frac{2n}{\hat{\beta}} < \chi^2_{2n,1-\gamma/2} \quad \text{or} \quad \frac{2n}{\hat{\beta}} > \chi^2_{2n,\gamma/2} \tag{4.6.12}$$

Example 4.7 A prototype of a system was tested with corrective actions (CAs) incorporated during the in-house test. Table 4.5 presents the data collected over the entire test. Find the Crow/AMSAA parameters using MLE and conduct the hypothesis testing for the growth trend.

Table 4.5 Time-to-failure data.

Index i	Time-to-failure (hours)	Ln(t_i)	Index i	Time to failure (hours)	Ln(t_i)
1	2.6	0.948	11	154.9	5.043
2	9.6	2.261	12	157.0	5.056
3	14.8	2.697	13	233.2	5.452
4	35.4	3.566	14	295.4	5.688
5	60.5	4.102	15	279.0	5.631
6	46.6	3.842	16	348.6	5.854
7	85.7	4.451	17	384.3	5.951
8	109.4	4.695	18	352.5	5.865
9	146.7	4.988	19	382.0	5.945
10	140.1	4.942	20	487.6	6.189

Solution:
Substitute t_i for $i = 1, 2, \ldots, 20$ into Eq. (4.6.11) along with $n = 20$ and $t^* = 487.6$ hours. The MLE of β is given below:

$$\hat{\beta} = \frac{n}{\sum_{i=1}^{n} \ln\left(\frac{t^*}{t_i}\right)} = \frac{20}{30.62} = 0.65$$

$$\hat{\alpha} = \frac{n}{(t*)^{\hat{\beta}}} = \frac{20}{(487.6)^{0.65}} = 0.351$$

To perform the hypothesis test at $1 - \gamma = 90\%$ confidence level, we calculate the lower and upper limits of the chi-square values as follows:

$$\chi^2_{2n,\gamma/2} = \chi^2_{40,0.05} = 26.51 \quad \text{and} \quad \chi^2_{2n,1-\gamma/2} = \chi^2_{40,0.95} = 55.76$$

Notice that

$$\frac{2n}{\hat{\beta}} = \frac{2 \times 20}{0.65} = 61.24 > \chi^2_{40,0.95}$$

Therefore, the null hypothesis (i.e. HPP) is rejected and we accept the alternative hypothesis, which is NHPP. We conclude that the product reliability is growing with 90% confidence.

4.6.4 Bounded Failure Intensity Model

For both the loglinear (i.e. Duane postulate) and the power law models (Crow/AMSAA), the failure intensity goes up infinitely as the system's age increases. However, in some circumstances, the failure intensity of a repairable system is bounded as a result of minimal repairs and/or replacement of aging parts. Now the system is composed of parts with a mix of ages, some are old and some are relatively old. For systems with mixed component ages, the aggregate failure intensity tends to be constant as the operation age becomes sufficiently large. This phenomenon is often known as Drenick's theorem (Ascher and

Feingold 1984). Typical examples of such behavior include off-road vehicles, trucks, and automobiles, when their operating mileage goes beyond 100 000–150 000 km (Ascher 1986). Also, large subsystems within a complex system often show a similar trend of failure intensity. Considering NHPP with a bounded intensity rate, Pulcini (2001) proposed a two-parameter bounded intensity process (BIP) model with the following form:

$$\lambda(t) = \alpha \left(1 - e^{-\frac{t}{\beta}}\right), \quad \text{for } t \geq 0 \tag{4.6.13}$$

where α and β are non-negative parameters. Note that $\lambda(t)$ equals zero at $t = 0$. It increases over time and approaches an asymptote of α as t increases to infinity. Figure 4.9 plots $\lambda(t)$ with different values of α and β. As expected, the maximum intensity is capped by α as t increases. A large β implies that it takes a longer time for the intensity function to reach the upper bound.

Taking the derivative with respect to t in Eq. (4.6.13) yields

$$\lambda'(t) = \frac{d\lambda(t)}{dt} = \frac{\alpha}{\beta} e^{-\frac{t}{\beta}} \tag{4.6.14}$$

which is always positive for any $t \geq 0$ and for any α and β in the parameter space. Equation (4.6.14) tends to zero as t approaches infinity regardless of α and β values. The second derivative of $\lambda(t)$

$$\frac{d^2 \lambda(t)}{dt^2} = -\frac{\alpha}{\beta^2} e^{-\frac{t}{\beta}} \tag{4.6.15}$$

is always negative for any $t \geq 0$, indicating that $\lambda(t)$ is concave for any values of α and β. The expected number of failures between $[0, t]$ is given by taking the integration of $\lambda(t)$. That is,

$$m(t) = \int_0^t \lambda(x)dx = \int_0^t \alpha \left(1 - e^{-\frac{x}{\beta}}\right) dx = \alpha \left[t - \beta \left(1 - e^{-\frac{t}{\beta}}\right)\right] \tag{4.6.16}$$

The conditional probability density function of the generic ith failure time, say t_i, given that the previous failure occurs at t_{i-1}, is given by

$$f(t_i \mid t_{i-1}) = \lambda(t_i) \exp\left(-\int_{t_{i-1}}^{t_i} \lambda(x)dx\right)$$

$$= \alpha(1 - e^{-t_i/\beta}) \exp(-\alpha\{(t_i - t_{i-1}) - \beta[\exp(-t_{i-1}/\beta) - \exp(-t_i/\beta)]\}) \tag{4.6.17}$$

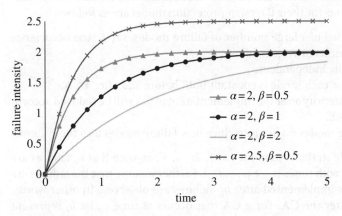

Figure 4.9 Failure intensity function with different α and β.

Maximum likelihood estimation. Let $t_1 < t_2 < \cdots < t_n$ denote the first n times to failure observed until $\tau > t_n$. According to Attardi and Pulcini (2005), the likelihood function relative to the failure data is given by

$$L(\widehat{\alpha}, \widehat{\beta}) = \prod_{i=1}^{n} f(t_i \mid t_{i-1})$$

$$= \widehat{\alpha}^n \prod_{i=1}^{n} [1 - \exp(-t_i/\widehat{\beta})] \times \exp(-\widehat{\alpha}\{\tau - \widehat{\beta}[1 - \exp(-\tau/\widehat{\beta})]\}) \qquad (4.6.18)$$

For the case of truncated sampling, we have $\tau = t_n$. Taking the logarithm of Eq. (4.6.18) and letting the first derivatives with respect to $\widehat{\alpha}$ and $\widehat{\beta}$ equal zero gives

$$\widehat{\alpha} = \frac{n}{\tau - \widehat{\beta}[1 - \exp(-\tau/\widehat{\beta})]} \qquad (4.6.19)$$

$$\frac{n[1 - \exp(-\tau/\widehat{\beta}) - (\tau/\widehat{\beta})\exp(-\tau/\widehat{\beta})]}{\tau/\widehat{\beta} - [1 - \exp(-\tau/\widehat{\beta})]} = \sum_{i=1}^{n} \frac{(t_i/\widehat{\beta})\exp(-t_i/\widehat{\beta})}{1 - \exp(-t_i/\widehat{\beta})} \qquad (4.6.20)$$

Solving Eqs. (4.6.19) and (4.6.20) yields $\widehat{\alpha}$ and $\widehat{\beta}$, which are the estimates of α and β, respectively.

4.6.5 Bayesian Projection Model

When projecting the reliability growth of a complex system, it is common to treat the system failure intensity as a sum of failure intensities of individual failure modes that are assumed to be statistically independent. Modeling the system as a combination of failure modes enables the use of CAE, which mathematically measures the fractional reduction in the failure intensity for a given failure mode upon the CA implementation. Wayne and Modarres (2015) present a Bayesian method for projecting the reliability growth of a complex repairable system based on the sum of mode intensity of individual failures. The model allows for arbitrary CA strategies and differs from other reliability growth projection models like those of Ellner and Wald (1995) and Crow (2004) in that it provides a complete inference framework via the posterior distribution on the failure intensity. Another unique feature of their approach compared to other Bayesian techniques is that the failure contributions from unobserved failure modes is also analytically expressed. The assumptions for their Bayesian projection model are as follows:

1) The system is comprised of a large number of failure modes, say K; the occurrence of any failure mode results in system failure.
2) All failures are mutually independent.
3) The failure intensity for each mode is constant both before and after a CA.
4) The resulting failure intensity after the implementation of CA will be reduced according to the assigned CAE.
5) CAs to existing failure modes do not introduce new failure modes into the system.

For the ith failure mode in the system for $i = 1, 2, \ldots, K$, assume that n_i failures are observed in test length T with times $(t_{i1}, t_{i2}, \ldots, t_{in_i})$. Further assume that the failures are divided such that a CA is implemented after n_{i1} failures are observed. In other words, $n_i - n_{i1}$ failures occur after the CA. For a CA that occurs at time v_i, let h_i represent the CAE resulting from the CA. For assumptions 2–4, the likelihood can then be expressed as

$$L(t_{i1}, t_{i2}, \cdots, t_{in_i}, n_i, n_{i1} \mid \lambda_i) = (1 - h_i)^{n_i - n_{i1}} \lambda_i^{n_i} \exp(-\lambda_i [v_i + (1 - h_i)(T - v_i)])$$

(4.6.21)

Note that the form of the likelihood function does not contain the actual failure times. The reason is due to the fact that they are replaced by the counts of failures before and after the CAs. For the prior on the individual failure model intensity, the gamma distribution is often adopted with the following form:

$$f_g(\lambda_i) = \frac{\lambda_i^{\alpha-1}}{\Gamma(\alpha)\beta^\alpha} \exp\left(-\frac{\lambda_i}{\beta}\right), \quad \text{for} \quad i = 1, 2, \ldots, K$$

(4.6.22)

where α is the shape parameter and β is the rate parameter, which is the inverse scale parameter. The choice of a gamma prior distribution is mathematically convenient because its conjugate distribution is still a gamma distribution. Ellner and Wald (1995) also have shown that the failure intensities for the collection of failure modes found in a complex system is adequately modeled as a random realization from a gamma distribution. In addition, using a gamma distribution recognizes what may be referred to as the vital few, whose numbers may be trivial but account for the majority of the K types of failure modes. This property indicates that each failure mode generates a different amount of contribution to the overall system failure intensity, yet only a relatively small number of failure modes are significant enough to be observed in the RGT. The posterior gamma distribution for the failure mode failure intensity is given as follows:

$$f_g(\lambda_i \mid t_{i1}, t_{i2}, \ldots, t_{in_i}, n_i) = \frac{\lambda_i^{\alpha+n_i-1}}{\Gamma(\alpha + n_i)\left(\frac{1}{\beta} + v_i + (1 - h_i)(T - v_i)\right)^{-(\alpha+n_i)}}$$

$$\exp\left(-\lambda_i\left(\frac{1}{\beta} + v_i + (1 - h_i)(T - v_i)\right)\right)$$

(4.6.23)

It is worth mentioning that n_{i1} in Eq. (4.6.21) only applies to the constant term $(1 - h_i)$ and is therefore canceled from the posterior distribution in Eq. (4.6.23). The distribution of the failure mode intensity of λ_i can be concisely expressed as

$$\lambda_i - Gamma\left((\alpha + n_i), \left(\frac{1}{\beta} + v_i + (1 - h_i)(T - v_i)\right)^{-1}\right)$$

(4.6.24)

Note that, if no CA is attempted, $h_i = 0$, and Eq. (4.6.24) reduces to the traditional gamma posterior distribution commonly found in many references (Martz and Waller 1982). From (4.6.24), the posterior distribution for the reduced failure intensity after the implementation of CA can also be deduced and the resulting distribution is

$$(1 - d_i)\lambda_i \sim Gamma\left((\alpha + n_i), (1 - h_i)\left(\frac{1}{\beta} + v_i + (1 - h_i)(T - v_i)\right)^{-1}\right)$$

(4.6.25)

By comparing Eq. (4.6.25) with Eq. (4.6.24), one finds that the shape parameter remains the same before and after the CAs, while the rate parameter is reduced to a fraction of $(1 - h_i)$. Also note that di is called effective failure reduction factor for mode i.

Assume m is the number of observed failure modes and $m \leq K$. Let $n = [n_1, n_2, ..., n_m]$ be the vector of the observed failure for each failure mode. One can also estimate the expected system failure intensity. The results are given as follows (Wayne and Modarres 2015):

$$E[\lambda_s \mid n] = \sum_{i=1}^{m} \left(\frac{(1 - h_i)\left(\frac{\lambda_B}{\beta K} + n_i\right)}{\left(\frac{1}{\beta} + v_i + (1 - h_i)(T - v_i)\right)} \right) + \left(\frac{1}{\beta} - \frac{m}{K\beta}\right)\left(\frac{\lambda_B}{\frac{1}{\beta} + T}\right) \qquad (4.6.26)$$

where

$$\lambda_B = K\alpha\beta \qquad (4.6.27)$$

The result for a complex system consisting of a large number of failure modes can be examined by taking the limit of Eq. (4.6.26) as it becomes large while holding the prior mean in Eq. (4.6.27) and the parameter constant. This approach yields the expression shown below:

$$E[\lambda_s \mid n] = \sum_{i=1}^{m} \left(\frac{(1 - h_i)n_i}{\left(\frac{1}{\beta} + v_i + (1 - h_i)(T - v_i)\right)} \right) + \frac{\lambda_B}{1 + \beta T} \qquad (4.6.28)$$

The variance of λ_s after the CA implementation can also be derived, and interested readers can refer to the Appendix of the paper by Wayne and Modarres (2015).

4.7 Reliability Growth and Demonstration Test

4.7.1 Optimal Reliability Growth Test

Reliability growth testing (RGT), also known as test-analyze-and-fix (TAAF), aims to identify the critical failure modes and implement corrective action (CA) in order to increase the reliability of the product before mass production (Krasich 2014). RGT is an in-house reliability growth strategy and is often carried out in the product development and prototype stage. The system could be tested in normal conditions or under accelerated or stressed conditions, such as temperature, humidity, vibration, voltage, and load. A good RGT program must take into account the resource and time constraints, such as materials, labor hours, and facility capacity. Therefore, determining how much test time is needed for a particular system is generally of great interest to the reliability management. This brings us the topic of RGT planning, which involves decisions, such as allocating the required test time, setting reliability goals, and designing a CA strategy for meeting those goals. The basis of planning an RGT procedure is the Duane postulate, discussed in Section 4.6, which is a model that incorporates the underlying realities faced during product development and can be used as a guideline in developmental testing (Table 4.6).

Under the given test time and budget, the RGT model is formulated to minimize the system failure rate at the product release time T_0. The expense of reliability growth consists of a fixed cost and a variable cost. The fixed cost pertains to the cost of the product

Table 4.6 Notation of RGT model.

Notation	Comments
m	Total number of subsystems (or component) within a new product
α_i	The growth rate of subsystem i, for $i = 1, 2, ..., m$
n_i	Planned number of test units for subsystem i
T_0	New product release time
$n_i^{(l)}$	Lower bound for the number of test units for subsystem i
$n_i^{(u)}$	Upper bound for the number of test units for subsystem i
t_i	Planned testing time for each subsystem i
T_i	Cumulative test time for subsystem i, and $T_i = n_i t_i$
$T_i^{(0)}$	Initial cumulative testing time for subsystem i
$c_i^{(f)}$	Fixed cost of subsystem i
$c_i^{(v)}$	Variable cost of testing one subsystem i
c_i	Total cost of subsystem i consisting of fixed cost and variable cost
C	Total budget for new product reliability grow test
$\lambda_i^{(0)}$	Initial failure rate of subsystem i
$\lambda_i(T_i)$	Achieved failure rate of subsystem i after testing time of T_i

itself and the variable cost is associated with implementing the reliability growth effects, including labor, test fixture, and CAs. The fixed cost is assumed to be proportional to the quantity of testing units and the variable cost is a function of the testing time of each subsystem.

The Duane postulate states that the logarithm of MTBF of the system increases linearly with the logarithm of the accumulative test time, as shown in Eq. (4.6.1), or equivalently the system failure rate decreases linearly with the total testing time in the log–log scale. In addition, it is assumed that the system is comprised of m subsystems connected in series. Subject to the resource constraints, an optimal RGT model to minimize the system failure rate is presented as follows (Jin and Li 2016).

Model 4.1

$$\text{Min}: \quad \lambda_{sys}(n, t) = \sum_{i=1}^{m} \lambda_i(T_i) \tag{4.7.1}$$

Subject to:

$$t_i \leq T_0, \quad \text{for} \quad i = 1, 2, ..., m \tag{4.7.2}$$

$$\sum_{i=1}^{m} c_i \leq C \tag{4.7.3}$$

$$n_i^{(l)} \leq n_i \leq n_i^{(u)} \tag{4.7.4}$$

where

$$T_i = \sum_{i=1}^{m} n_i t_i \tag{4.7.5}$$

$$c_i = n_i c_i^{(f)} + g_i(T_i), \quad \text{for} \quad i = 1, 2, \ldots, m \tag{4.7.6}$$

$$\ln \lambda_i(T_i) = \ln \lambda_i^{(0)} - \alpha(\ln T_i - \ln T_i^{(0)}), \quad \text{for} \quad i = 1, 2, \ldots, m \tag{4.7.7}$$

The decision variables in Model 4.1 is the number of testing units of each subsystem, $\mathbf{n} = [n_1, n_2, \ldots, n_m]$, and the testing time, $\mathbf{t} = [t_1, t_2, \ldots, t_m]$, for each subsystem. The objective function in (4.7.1) minimizes the sum of the subsystem failures at the end test, hence maximizing the system MTBF. There are three types of constraints in Model 4.1. First, constraint (4.7.2) prescribes that the maximum testing time for any subsystem should not exceed the product release time T_0. Second, the total testing cost should be confined to the available budget C_0 as shown in constraint (4.7.3). Third, constraint (4.7.4) states that the number of subsystems i under test must fall between $n_i^{(l)}$ and $n_i^{(u)}$ due to the test capacity limit. The evaluation of total testing time and the cost of each subsystem, testing units, and the Duane reliability growth model are provided in Eqs. (4.7.5–4.7.7), respectively.

In addition to the subsystem growth test, optimal allocation of both subsystem and system testing resources is also developed for the reliability growth of a series–parallel system. Heydari et al. (2014) seek to identify the testing strategy that maximizes the reliability growth attained at the system level. The mathematical model they proposed gives a framework for making an optimal decision on where and when growth testing is most effective at the component, subsystem, and system levels. They demonstrated that a possible trade-off is available to test planners by using two sample instances of the resource allocation problem and solving both of them using a simulated annealing heuristic.

Expert knowledge and subjective data are also valuable in planning RGT programs. Quigley and Walls (1999) introduced a reliability growth model that incorporates the concerns of experts about the likely faults in the initial system design with the information about faults that are realized on test. Their model distinguishes between the processes of detecting and removing faults. Thus the model captures the effectiveness of the test as well as the system developmental knowledge. Later on, Walls and Quigley (2001) described an elicitation process that was developed to ensure valid data are collected by suggesting how possible bias might be identified and managed. It provides a framework for managing the expert knowledge that will form a prior probability distribution for the relevant parameter in a Bayesian reliability growth projection.

4.7.2 Reliability Demonstration Test

The purpose of reliability demonstration testing (RDT) is to verify, with a pre-defined degree of statistical confidence, that the system (e.g. hardware or software products) meets the specified reliability requirement (Wang 2017). RDT is usually performed at the system level and is typically set up as a success test, but it can also be used at subsystem level depending on the reliability criticality. There exist four types of RDT, namely, cumulative binomial, non-parametric binomial, exponential chi-squared, and non-parametric Bayesian. The principles of these RDT methods are elaborated below and numerical examples are also provided to illustrate their applications.

4.7.2.1 Cumulative Binomial

This is used when the test duration is different from the time of the required reliability. This methodology requires the underlying distribution of the product's lifetimes to be known. Other required information includes: (i) the target reliability; (ii) the reliability confidence level of the demonstration test; (iii) the acceptable number of failures; and (iv) either the number of available test units or the amount of available test time. The form of the cumulative binomial appears as

$$\sum_{k=0}^{f} \binom{n}{k} (1 - R_T)^k R_T^{n-k} = 1 - CL \tag{4.7.8}$$

where

CL = the required confidence level

f = the allowable number of failures in test

n = the total number of test units

R_T = the system reliability on test

Since CL and f are the required inputs to the RDT process and R_T has been estimated, the number of test units n is the only unknown parameter that needs to be determined from Eq. (4.7.8). We use the example below to illustrate how to determine the value of n subject to available time.

Example 4.8 The reliability of a system is required to be 0.90 at $t_D = 200$ hours with 95% conference if no failures occur during the test. The available testing capability is only $t_T = 150$ hours. We must now determine the number of test units for 150 hours of testing with no failures in order to demonstrate the reliability goal. Assume the system lifetime follows the Weibull distribution with shape parameter $\beta = 2$.

Solution:

First we estimate the Weibull scale parameter. From $R(t) = \exp(-(t/\eta)^\beta)$, we have

$$\eta = t_D[-\ln(R_D)]^{-1/\beta} = 200 \times (-\ln 0.9)^{-1/2} = 616.2$$

Next we estimate the system reliability at $t_D = 150$ hours, that is,

$$R_T = \exp(-(t_T/\eta)^\beta) = \exp(-(150/616.2)^2) = 0.9425$$

The last step is to substitute $\eta = 616.2$ and $R_T = 0.9425$ into Eq. (4.7.8), which for the Weibull reliability appears as

$$1 - 0.95 = \sum_{k=0}^{0} \frac{n!}{k!(n-k)!} (1 - 0.9425)^k 0.9425^{n-k} = 0.9425^n$$

Solving the equation yields $n = 50.53$. Since the fractional value must be rounded up to the next integer, we set $n = 51$ units as the final result.

4.7.2.2 Exponential Chi-Squared

This process aims at designing tests for products that have an assumed constant failure rate based on the chi-squared distribution. This method only returns the required accumulated test time for a demonstrated reliability or mean-time-to-failure (MTTF).

The accumulated test time is equal to the total amount of time applied to all of the units under test. Assuming n units are subject to an equal length of test time, then the chi-squared equation of test time is

$$nt_T = T_a = \frac{MTTF\chi^2_{1-CL,2f+2}}{2} \tag{4.7.9}$$

where

$\chi^2_{1-CL,2f+2}$ = chi-squared distribution with degree of freedom $2f+2$

T_a = cumulative test time of n units

All other parameters carry the same meaning as in the cumulative binomial RDT. Since this method is only applicable to a constant failure rate with an exponential lifetime, the testing statistics in Eq. (4.7.9) can be rewritten as

$$T_a = \frac{t_D \chi^2_{1-CL,2f+2}}{2(-\ln(R_D))} \tag{4.7.10}$$

The result is obtained by realizing that $MTTF = t/(-\ln R)$ for an exponential distribution where R is the product reliability under test.

Example 4.9 In this example, the chi-squared method is applied to demonstrate a product reliability of 0.85 at $t_D = 300$ hours at CL = 90%. Maximum two failures are allowed. The question is to determine the cumulative test time T_a. What if no failure is allowed; then what is the required cumulative test time?

Solution:
The chi-squared value at CL = 90% is given as

$$\chi^2_{1-CL,2f+2} = \chi^2_{1-0.9,2\times2+2} = \chi^2_{0.1,6} = 10.6446$$

Substituting the above value into the test statistics in Eq. (4.7.10), we obtain

$$T_a = \frac{(300)(10.6446)}{2(-\ln(0.85))} = 7155 \text{ hours}$$

This means that a total of 16 374 hours of test time needs to be accumulated with no more than two failures in order to demonstrate the required reliability. If only one failure is allowed instead of two, then the chi-squared value

$$\chi^2_{1-CL,2f+2} = \chi^2_{1-0.9,2\times0+2} = \chi^2_{0.1,2} = 4.605$$

Substituting the value into the test statistics in Eq. (4.7.10), we obtain

$$T_a = \frac{(300)(7.779)}{2(-\ln(0.85))} = 3096 \text{ hours}$$

This example shows that if there is no failure observed by 3096 hours, the product can be claimed to meet the specified reliability.

In addition to analytical methods, simulation is also a useful tool to help the reliability engineer determine the sample size, test duration, and expected number of failures. Since simulation usually does not have to make assumptions such as the lifetime distribution and model parameters, it could be more flexible or convenient than the analytical method, especially for tests with a small sample size.

4.8   Lifecycle Reliability Growth Planning

4.8.1   Reliability Growth of Field Systems

To gain the market window, the development cycle of many new products continues to shrink in order to meet the time-to-market goal. This phenomenon is prevailing in electronics and consumer products, such as cell phones, semiconductor devices, home appliances, and automobiles. This creates grand challenges in the implementation of the RGT process. First, manufacturers have to meet the product delivery deadline in order to increase the market share; hence RGT needs to be conducted in a compressed time window. Second, in today's distributed manufacturing environment, many subsystems and components are designed or made by subcontract manufacturers. Outsourcing like off-shore manufacturing makes it difficult to perform a system-level test in the early development stage. Third, under performance-based maintenance contracting, a growing number of firms start to bundle the product with the service by offering a lifecycle reliability warranty to their customers. These new business and service strategies call for the integration of reliability growth into the product design, manufacturing, installation, and field use, instead of being limited to the in-house stage.

RGP has emerged as a new methodology to facilitate the design, introduction, and maintenance of new products in the distributed manufacturing environment. In an RGP program, the growth initiative begins with a product early design and development phase, and is extended to volume shipment and post-installation phases. It differs from the traditional RGT in two aspects. First, RGP enables the manufacturer to drive the reliability growth in product development stage as well as in after-sales market, while RGT is usually confined to early developmental period. Second, RGP represents a holistic reliability solution with which repair capacity, CAs, and spare parts inventory are jointly coordinated. In addition, a traditional RGT process usually concentrates on eliminating surfaced failure modes, while RGP is able to track the latent failures arising from field use, and further launches CA programs.

Figure 4.10 graphically depicts the RGP process that spans the product design, prototyping, volume shipment, and field operation in the after-sales market. RGP can be divided into two phases: reliability growth testing and LFE. Technically RGT enables the manufacturer to achieve the MTBF goal through an in-house test. During the installation and field use, there are many uncertain factors causing the latent failures, such as customer usage, uncertain operating environment, software bugs, electric static discharge, and mishandling of the product (Greason 2007; Vinson and Liou 1998). To ensure the reliability of field products, the goal of RGP in the post-installation phase is to monitor, analyze, and remove critical latent failure modes. The following trade-offs need to be made during the planning and execution of an RGP program.

- Allocating more subsystems/systems for RGT in the development phase benefits the reliability growth rate, yet a prolonged in-house test may postpone a new product release time.
- Allocating more resources (e.g. repair capacity and spare parts) to the post-installation phase accelerates the product release time. However, low product reliability (e.g. MTBF) results in a larger amount of field returns. Consequently, both the repair and the warranty costs will significantly increase due to CAs, product recalls, and spares inventory.

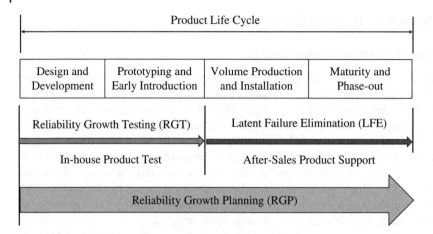

Figure 4.10 A holistic approach to reliability growth planning.

Therefore, an effective reliability growth planning (RGP) program should balance the resource distribution between RGT and LFE processes by considering the following factors: (i) product release time; (ii) target reliability; (iii) number and intensity of latent failure modes; (iv) CA capacity; (v) spare parts inventory; and (vi) warranty cost and customer satisfaction. Among these factors, forecasting and eliminating latent failure modes plays a vital role in designing an effective RGP program in the post-installation phase.

4.8.2 Prediction of Latent Failure Modes

In a classical RGT process, the reliability growth prediction is often made based on the surfaced failure modes that appeared during the in-house test. Unlike a surfaced failure mode, a latent failure mode has a dormant status and does not occur until the system has operated in the field for months or even a year. If failure occurrence rates are relatively stable, Markov chain models can be used to predict the latent failures (Somani et al. 1993; Susova and Petrov 1997). Quite often the intensity rate of the failure mode during the new product introduction is time-varying because of the on-going CAs and uncertain customer usage. We present a time-varying failure intensity model to predict latent failure modes by taking into account the usage uncertainties. More detailed discussions about the model are available in Jin et al. (2010). The notation pertaining to the prediction model is listed in Table 4.7.

Let us use Figure 4.11 to illustrate how latent failure modes emerge and evolve over time. Assume that a new product has undergone an in-house GRT process in $[0, t_1]$, and two surfaced failure modes, $\mu_1(t)$ and $\mu_2(t)$, remain at the time when the product is released to the market at t_1. After the product operates for a time period T_1, two new failure modes, $\mu_3(t)$ and $\mu_4(t)$, emerged in the field. Both are designated as latent failure modes because they do not occur prior to t_1. Unless CAs are initiated, both $\mu_3(t)$ and $\mu_4(t)$ will remain thereafter. As time evolves and enters $[t_2, t_3]$, additional latent failure modes, $\gamma_1(t)$ and $\gamma_2(t)$, are observed in the field, as shown in Figure 4.11. The prediction model needs to address two questions: (i) how to estimate the number of new latent failure modes as time evolves and (ii) how to predict the system failure intensity

Table 4.7 Notation for latent failure prediction model.

Notation	Explanation
t_1	Product release time
t_2	Current time
t_3	Future time instance
T_1	Time period between t_1 and t_2
T_2	Time period between t_2 and t_3
m	Number of surfaced failure modes that occurred
k_1	Number of latent failure modes that occurred in $[t_2, t_1]$
k	Number of new latent failure modes that will occur in $[t_2, t_3]$
$\mu_i(t)$	Failure mode i observed prior to t_2 for $i = 1, 2, ..., m$
$\gamma_j(t)$	Latent failure mode j expected to occur in $[t_2, t_3]$ for $j = 1, 2, ..., k$
$\mu_s(t\|t_2)$	System or subsystem failure intensity at time t for $t_2 \leq t \leq t_3$
$\Gamma(t)$	Aggregate intensity of latent failure modes in $[t_2, t_3]$
$\hat{\Gamma}(t), \hat{k}$	The estimate of $\Gamma(t)$ and k, respectively

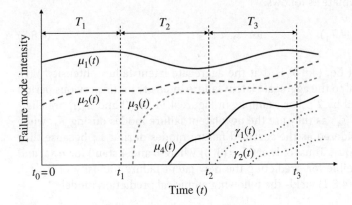

Figure 4.11 Surfaced and latent failure intensity function.

rate subject to both surfaced and uncertain latent failure modes. Both questions will be addressed in this section.

Assume t_2 is the current time. Our objective is to predict the system failure intensity in $[t_2, t_3]$. Let $\mu_s(t|t_2)$ be the system failure intensity to be predicted. Then

$$\mu_s(t \mid t_2) = \sum_{i=1}^{m} \mu_i(t) + \sum_{j=1}^{k} \gamma_j(t), \quad \text{for } t_2 < t \leq t_3 \tag{4.8.1}$$

where the first sum in Eq. (4.8.1) captures the intensity of surfaced failure modes by t_2 and the second sum represents the projected intensity of latent failure modes in $[t_2, t_3]$. For instance $m = 4$ in Figure 4.11 by time t_2. Similarly, k is the number of latent failure modes that will occur in $[t_2, t_3]$ and $\gamma_j(t)$ is the intensity of the jth latent failure mode for $j = 1, 2, ..., k$. Next we give an empirical method to estimate the model parameters.

There are three unknown parameters in Eq. (4.8.1), namely $\mu_i(t)$, k, and $\gamma_j(t)$, and they must be estimated. Since $\mu_i(t)$ is the intensity of the ith surface failure mode, it can be estimated based on the historical failure data by fitting appropriate intensity models, such as the Crow/AMSSA model or the bounded intensity model. Intuitively, the value of k depends on the length of $T_2 = t_3 - t_2$. The longer the T_2, the higher is the possibility of occurrence of the latent failure mode. Meanwhile, k is also correlated with the number of new latent failure modes that occurred during T_1. If the number of new latent failure modes occurred in T_1 is high, it is likely that the number of new latent failure modes emerging in T_2 will be high as well. We then propose an empirical estimate of k as follows:

$$\hat{k} = \left\lfloor \frac{k_1(t_3 - t_2)}{t_2 - t_1} \right\rfloor = \left\lfloor \frac{k_1 T_2}{T_1} \right\rfloor, \qquad \text{for } t_2 < t \le t_3 \qquad (4.8.2)$$

Note that $\lfloor \bullet \rfloor$ means the round-down to the integer and k_1 is the number of latent failure modes occurring in T_1. For instance, $k_1 = 2$, $T_2 = 100$ hours, and $T_1 = 150$ hours; then $\hat{k} = (2 \times 100/150) = 1$. On the other hand, if $T_1 = T_2$, then $k_2 = k_1$.

Estimating $\lambda_j(t)$ for each j in Eq. (4.8.1) is possible if not difficult, but there is no necessity to do so. Rather we are more interested in the aggregate intensity rate of latent failure modes during T_2. Hence the sum of $\gamma_j(t)$ for all j defined as $\Gamma(t) = \sum_{j=1}^{k} \gamma_j(t)$ can be approximated by the estimate as follows:

$$\hat{\Gamma}(t) = \frac{T_2}{T_1} \sum_{j \in L} \hat{\mu}_j(t - T_1), \qquad \text{for } t_2 < t \le t_3 \qquad (4.8.3)$$

The rationality behind Eq. (4.8.3) is that the aggregate latent failure intensity, $\Gamma(t)$, during T_2 is proportional to the aggregate latent failure intensity in the previous period T_1, yet shifted or delayed by T_1. For example, in Figure 4.11, $\mu_3(t)$ and $\mu_4(t)$ are eligible for estimating $\hat{\mu}_j(t - T_1)$ as they are the new latent failure modes during T_1, while $\mu_1(t)$ and $\mu_2(t)$ are considered as the surfaced failure modes during T_1 because they started to appear prior to t_1. Thus, only set $L = \{3, 4\}$ where 3 and 4 stand for $\mu_3(t)$ and $\mu_4(t)$, respectively, is eligible for predicting the new latent failure intensity in $[t_2, t_3]$. Substituting (4.8.3) into (4.8.1) yields the following empirical prediction model:

$$\hat{\mu}_s(t \mid t_n) = \sum_{i=1}^{m} \hat{\mu}_i(t) + \frac{T_{n+1}}{T_n} \sum_{j \in L_n} \hat{\mu}_j(t - T_n), \qquad \text{for } t_n < t \le t_{n+1} \qquad (4.8.4)$$

where t_n for $n = 1, 2, \dots$ is the nth time instance and $T_n = t_{n+1} - t_n$ with $t_0 = 0$. Note that L_{n-1} is the set containing the latent failure modes occurring in T_{n-1}. Now we summarize the procedure for predicting the system failure intensity subject to uncertain latent failure modes as follows:

Step 1: Estimating $\mu_i(t)$ for surfaced failure mode i based on reliability growth models, such as the Crow/AMSAA model in (4.6.9) or the bounded intensity model (4.6.13).

Step 2: Estimating k and $\Gamma(t)$ using Eqs. (4.8.2) and (4.8.3), respectively.

Step 3: Predicting the failure intensity for the interval of $[t_2, t_3]$ using Eq. (4.8.4).

Step 4: Updating the parameters, i.e. $t_{n+1} \to t_n$, and $T_{n+1} \to T_n$, and moving the prediction window to the next period.

During the prediction process, $\mu_s(t|t_n)$ will be updated sequentially as time evolves to the next period. For instance, when t_3 becomes the current time and t_4 is the future time, then $T_2 = t_3 - t_2$ and $T_3 = t_4 - t_3$ in Eq. (4.8.4). Latent failure modes occurring during the previous period become surfaced failure modes. New $\mu_s(t|t_3)$ is obtained by repeating Steps 1 to 4.

4.8.3 Allocation of Corrective Action Resource

The failure intensity model discussed thus far has not explicitly connected the failure corrections with the CA budget. To manage the correction actions, Crow (2004, 2006) proposes to classify all surfaced failure modes into Type A and Type B modes. Type A failure modes are all modes such that if seen during the test no CA will be taken. Type B failure modes are all modes such that if seen during the test a CA will be taken. Such decisions are often made based on the cost-effectiveness of correction or the amount of failures per mode generated from the systems. Figure 4.12 plots the failure mode Pareto chart based on the failure data collected over three months across a fleet system. Among the six failure modes, relay and resistor failures represent 65% of field returns. In traditional CA allocation, priorities are often given to the failure mode with the highest failure quantities in the Pareto chart. CA resources will be first applied to the top critical failure modes: relays and resistors. If surplus resources are available, NFF will be the one to be addressed as well. Such CA resource allocation is justifiable if the cost of eliminating a single failure is the same and independent of failure modes.

In reality, the CA cost often varies and is dependent upon the underlying failure mode. Let us use Figure 4.12 as an example. Assume it is more costly to eliminate an NFF issue than removing and replacing an op-amp device in the PCB. The manufacturer aims to fix these failure modes at relatively low cost, even if the NFF ranks third in the Pareto chart. This is because the net reliability growth in terms of MTBF are identical provided the same amount of failures are eliminated. Jin and Wang (2009) propose a generic CA allocation model with which CA resources could be potentially allocated to any surfaced failure modes regardless of their intensity rate. The system failure intensity

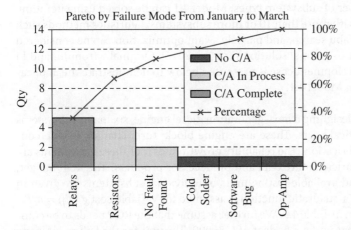

Figure 4.12 Failure mode Pareto chart (C/A = corrective action).

Table 4.8 The format of $\hat{\mu}_i(t)$ for a surfaced failure mode in Eq. (4.8.5).

Model	Failure intensity	Parameters	Model
1	Constant	$\alpha > 0, \beta = 0$	Crow/AMSAA
2	Increasing	$\alpha > 0, \beta > 1$	Crow/AMSAA
3	Decreasing	$\alpha > 0, 0 < \beta < 1$	Crow/AMSAA
4	Bounded	$\alpha > 0, \beta > 0$	Bounded failure intensity

after incorporating the CA effectiveness function becomes

$$\hat{\mu}_{s,CA}(\mathbf{x}; t_n) = \sum_{i=1}^{m}(1 - h_i(x_i))\hat{\mu}_i(t) + \frac{T_{n+1}}{T_n}\sum_{j \in L}\hat{\mu}_j(t - T_n)$$

$$= \sum_{i=1}^{m}g_i(x_i)\hat{\mu}_i(t) + \frac{T_{n+1}}{T_n}\sum_{j \in L}\hat{\mu}_j(t - T_n) \tag{4.8.5}$$

where $\mathbf{x} = [x_1, x_2, \ldots, x_m]$ is the decision variable representing the amount of CA budget allocated to the ith surfaced failure mode. Here, $g_i(x_i) = 1 - h(x_i)$ is the CA ineffectiveness for the ith surfaced failure mode and $h_i(x_i)$ is given in Eq. (4.5.1). An assumption behind (4.8.5) is that an ongoing CA neither induces nor eliminates any latent failure modes. Table 4.8 summarizes available failure intensity models that accommodate four types of intensity profiles, namely, constant, increasing, decreasing, and bounded rates.

4.9 Case Study

4.9.1 Optimizing Reliability Growth Test of Diesel Engines

Model 4.1 is an example of non-linear mixed integer programming, and this type of problem is difficult to solve because of the existence of local optima. For a small problem with a limited number of subsystem types, Model 4.1 can be solved using iteration. For large sized systems with more than 10 or more subsystems, heuristics methods such as a genetic algorithm, Tabu search, and particle swam optimization can be considered to find the optimal or near-optimal solution (Li et al. 2016). Originally from Jin and Li (2016), a new engine development case is presented below to demonstrate the application of the RGP based on Model 4.1.

Example 4.10 When designing a next-generation diesel engine, six major subsystems are identified as the critical parts. These are engine block, turbocharger, engine control, cooling, and fuel lubrication. The reliability growth rates α of these subsystems are estimated based on historical failure data and predecessor products. The RGP budget, product release time, and available testing units of individual subsystems are given in Table 4.9. For simplicity, a quadratic function is assumed for variable cost $g_i(T_i) = b_i T_i^2$ and the value of b_i is given in Table 4.9. We further assume that the RGP needs to be completed within $T_0 = 0.8$ year under a budget of $C = 900$. The unit for the failure intensity rate is failures/year and the time unit is year.

Table 4.9 Input parameters to the engine RGP.

i	Subsystem	α_i	$\lambda_i^{(0)}$	$n_i^{(i)}$	$n_i^{(u)}$	$c_i^{(f)}(\times\$10^3)$	$b_i(\times\$10^3)$	$T_i^{(0)}$
1	Engine block	0.4	1.38	3	6	60	3	1
2	Turbocharger	0.4	0.4	2	4	45	3	1
3	Engine control	0.3	0.8	4	8	80	3	1
4	Cooling	0.3	0.2	4	8	20	3	1
5	Fuel	0.2	0.25	4	8	30	3	1
6	Lubricating	0.2	0.3	5	10	20	3	1

Source: Reproduced with permission of IEEE.

Table 4.10 Optimal testing units and time.

i	Subsystem	n_i^*	t_i^*	T_i	$c_i(\times\$10^3)$	$\lambda_i(T_i)$
1	Engine block	3	0.30	0.90	182	1.123
2	Turbocharger	2	0.45	0.90	92	0.143
3	Engine control	4	0.18	0.74	322	0.634
4	Cooling	4	0.19	0.74	82	0.034
5	Fuel	4	0.14	0.55	121	0.162
6	Lubricating	5	0.11	0.55	101	0.212

Source: Reproduced with permission of IEEE.

Based on the given RGP input parameters, the optimal testing units and times of each subsystem can be solved based on Model 4.1. Since there are only 12 design variables in this example, Excel Solver can solve this type of small optimization problem; the results are presented in Table 4.10. The optimal system failure intensity rate is $\lambda_{sys}^* = 2.307$, reduced by 31% compared with the initial value of 3.33 failures/year.

4.9.2 Multiphase Reliability Growth Strategy

In addition to the Duane and AMSAA/Crow models, the failure intensity in RGT can also be assumed to be dynamic and changing and is modeled as a Weibull distribution. Awad (2016) emphasizes that using Weibull is more realistic and expands the applicability of the RGT model in real-life systems. The GRT model can also be formulated to minimize the variability of failure intensity along with the mean failure intensity. Coit (1998) proposed a multiobjective optimization to allocate subsystem RGT time in order to minimize the failure intensity and the variance of intensity for all subsystems under the test.

Modeling and planning reliability growth for a repairable system is a very complex task as it involves many uncertain factors, such as design, manufacturing, testing, operation, and maintenance, and some are them are indeed mutually dependent. Growing the reliability of a new product often involves multiple objectives, such as cost, test time, and time-to-market. Quite often these objectives are conflicting due to the limited resources and time. Therefore, equipment industries often adopt a multiphase and multiobjective approach to planning the CAs and allocating the resources for reliability growth. Each phase involves two iterative steps: reliability prediction and CA resource allocation. These steps are repeated until the system reliability reaches the design target. This iterative process allows the manufacturers to assess the effectiveness of an ongoing CA program and also to adjust or redistribute the resources against new or latent failure modes in the next phase. During the introduction of a product, this reliability growth strategy is particularly beneficial to the manufacturer given the uncertainties in market size and customer usage. As the reliability increases along with the customer shipment, more sales revenue is generated, which gives the manufacturer the incentive to implement more rigorous CA programs. Interested readers are referred to the works by Jin et al. (2013) and Li et al. (2016) where reliability growth is planned under a multiphase and multiobjective framework.

References

Ascher, H. (1986). Reliability models for repairable systems. In: *Reliability Technology: Theory and Applications* (ed. J. Moltoft and F. Jensen), 177–185. Elsevier Science.

Ascher, H. and Feingold, H. (1984). *Repairable Systems Reliability: Modeling, Misconceptions and Their Causes*. New York, NY: Marcel Dekker.

Attardi, L. and Pulcini, G. (2005). A new model for repairable systems with bounded failure intensity. *IEEE Transactions on Reliability* 54 (4): 572–582.

Awad, M. (2016). Economic allocation of reliability growth testing using Weibull distributions. *Reliability Engineering and System Safety* 152: 273–280.

Bai, C.-G., Cai, K.-Y., Hu, Q.-P., and Ng, S.-H. (2008). On the trend of remaining software defect estimation. *IEEE Transactions on Systems, Man and Cybernetics, Part A* 38 (5): 1129–1142.

Barkai, J., (2014), "Method to reduce no fault found rates," Aerospace, Automotive, Aviation, Reliability, Service, Service Lifecycle Management (SLM), Service Technology, available at: http://joebarkai.com/method-reduce-no-fault-found-rates (accessed on September 29, 2016).

Burchell, B., (2007), "Untangling no fault found", *Aviation Week*, 9 February, 2007, available at: http://www.aviationweek.com/aw/generic/story_generic.jsp?channelom&id=news/om207cvr.xml (accessed 5 December 2011).

Carlson, J. and Murphy, R.R. (2005). How UGVs physically fail in the field. *IEEE Transactions on Robotics and Automation* 21 (3): 423–437.

Coit, D.W. (1998). Economic allocation of test times for subsystem-level reliability growth testing. *IIE Transactions* 30 (12): 1143–1151.

Crow, L.H. (1974). Reliability analysis for complex, repairable systems. In: *Reliability and Biometry* (ed. F. Proschan and R.J. Serfling), 379–410. Philadelphia, PA: SIAM.

Crow, L. H., (2004), "An extended reliability growth model for managing and assessing corrective actions," in *Proceedings of Annual Reliability and Maintainability Symposium*, Los Angeles, CA, 2004, pp. 73–80.

Crow, L. H., (2006), "Useful metrics for managing failure mode corrective action," in *Proceedings of Annual Reliability and Maintainability Symposium*, Los Angeles, CA, 2006, pp. 247–252.

Crow, L. H., (2015), "Reliability growth planning curves based on multi-phase projections," in *Proceedings of the Annual Reliability and Maintainability Symposium*, pp. 1–6.

Duane, T.J. (1964). Learning curve approach to reliability monitoring. *IEEE Transactions on Aerospace* 2: 563–566.

Ellner, P. M., L. C. Wald, (1995), "AMSAA maturity projection model," in *Proceedings of IEEE Annual Reliability and Maintainability Symposium*, pp. 174–181.

English, J. R., L. Yan, T. L. Landers, (1995), "A modified bathtub curve with latent failures," in *Proceedings of Annual Reliability and Maintainability Symposium*, pp. 217–222.

Gibson, G.J., L. H Crow, (1989), "Reliability fix effectiveness factor estimation," in *Proceedings of Reliability and Maintainability Symposium*, pp. 171–177.

Grassi, V. and Patella, S. (2006). Reliability prediction for service-oriented computing environments. *IEEE Internet Computing* 10 (3): 43–49.

Greason, W.D. (2007). Review of charge and potential control of electrostatic discharge (ESD) in microdevices. *IEEE Transactions on Industry Applications* 43 (5): 1149–1158.

Heydari, M. H., Sullivan, K.M., Pohl, E. A., (2014), "Optimal allocation of testing resources in reliability growth," In *Proceedings of IIE Annual Conference*, pp. 3423–3428.

Hockley, C. and Phillips, P. (2012). The impact of no fault found on through-life engineering services. *Journal of Quality in Maintenance Engineering* 18 (2): 141–153.

Hokstad, P. and Frøvig, A.T. (1996). The modeling of degraded and critical failures for components with dormant failures. *Reliability Engineering and System Safety* 51 (2): 189–199.

Hwang, S. and Pham, H. (2009). Quasi-renewal time-delay fault-removal consideration in software reliability modeling. *IEEE Transactions on Systems, Man and Cybernetics, Part A* 39 (1): 200–209.

Inoue, S. and Yamada, S. (2007). Generalized discrete software reliability modeling with effect of program size. *IEEE Transactions on Systems, Man and Cybernetics, Part A* 37 (2): 170–179.

Jackson, D.S., Pant, H., and Tortorella, M. (2002). Improved reliability-prediction and field-reliability-data analysis for field-replaceable units. *IEEE Transactions on Reliability* 51 (1): 8–16.

Jin, T., Z. Li, (2016), "Reliability growth planning for product-service integration," in *Proceedings of Annual Reliability and Maintainability Symposium*, pp. 1–6.

Jin, T., H. Wang, (2009), "A multi-objective decision making on reliability growth planning for in-service systems," in *Proceedings of IEEE Conference on Systems, Man and Cybernetics*, San Antonio, TX, pp. 2–6.

Jin, T., P. Wang, Q. Huang, (2006), "A practical MTBF estimate for PCB design considering component and non-component failures", in *Proceedings of Annual Reliability and Maintainability Symposium*, pp. 604–610.

Jin, T., Y. Yu, F. Belkhouche, (2009), "Reliability growth planning using retrofit or engineering change order: A budget based decision making," in *Proceedings of Industrial Engineering Research Conference*, pp. 2152–2157.

Jin, T., Liao, H., and Kilari, M. (2010). Reliability growth modeling for in-service systems considering latent failure modes. *Microelectronics Reliability* 50 (3): 324–331.

Jin, T., Yu, Y., and Huang, H.-Z. (2013). A multiphase decision model for system reliability growth with latent failures. *IEEE Transactions on Systems, Man and Cybernetics, Part A* 43 (4): 958–966.

Khan, S., Phillips, P., Jennions, I., and Hockley, C. (2014). No fault found events in maintenance engineering. Part 1: current trends, implications and organizational practices. *Reliability Engineering and System Safety* 123 (3): 183–195.

Krasich, M., (2014), "Reliability growth testing, what is the real final result? Accelerated test methodology for reliability growth," in *Proceedings of Annual Reliability and Maintainability Symposium (RAMS)*. Colorado Springs, CO.

Li, Z., Mobin, M., and Keyser, T. (2016). Multi-objective and multi-stage reliability growth planning in early product-development stage. *IEEE Transactions on Reliability* 65 (2): 769–781.

Martz, H.F. and Waller, R.A. (1982). *Bayesian Reliability Estimation*. New York: Wiley.

O'Connor, P. (2012). *Practical Reliability Engineering*, Chapter 12, 4e. West Sussex, UK: Wiley.

Peng, W., Huang, H.-Z., Li, Y. et al. (2013). Life cycle reliability assessment of new products – a Bayesian model updating approach. *Reliability Engineering and Systems Safety* 112: 109–119.

Pulcini, G. (2001). A bounded intensity process for the reliability of repairable equipment. *Journal of Quality Technology* 33 (4): 480–492.

Qi, H., Ganesan, S., and Pecht, M. (2008). No-fault-found and intermittent failures in electronic products. *Microelectronics Reliability* 48 (5): 663–674.

Quigley, J. and Walls, L. (1999). Measuring the effectiveness of reliability growth testing. *Quality Reliability Engineering International* 15 (2): 87–93.

Si, X.S., Wang, W., Hu, C.H., and Zhou, D.H. (2011). Remaining useful life estimation – a review on the statistical data driven approaches. *European Journal of Operational Research* 213 (1): 1–14.

Somani, A.K., S. Palnitkar, T. Sharma, (1993), "Reliability modeling of systems with latent failures using Markov chains," in *Proceedings of Annual Reliability and Maintainability Symposium*, pp. 120–125.

Sorensen, B., (2003), "Digital averaging – the smoking gun behind 'No-Fault-Found'," *Aviation Today*, February 24, 2003, available at: http://www.aviationtoday.com/regions/ sa/Digital-Averaging-The-Smoking-Gun-Behind-\stquoteNo-Fault-Found\stquote_ 2120.html (accessed on September 30, 2016).

Susova, G.M., A.N. Petrov, (1997), "Markov model-based reliability and safety evaluation for aircraft maintenance-system optimization," in *Proceedings of Annual Reliability and Maintainability Symposium*, pp. 29–36.

Trudell, C., M. Horie, (2014), "Toyota recalls 1.75 million vehicles worldwide for fuel, brake flaws," *Insurance Journal*, October 15, 2014. Available at: http://www.insurancejournal .com/news/international/2014/10/15/343502.htm (accessed on October 24, 2015).

Vinson, J.E. and Liou, J.J. (1998). Electrostatic discharge in semiconductor devices: an overview. *Proceedings of the IEEE* 86 (2): 399–420.

Walls, L. and Quigley, J. (2001). Building prior distributions to support Bayesian reliability growth modelling using expert judgement. *Reliability Engineering and System Safety* 74 (2): 117–128.

Wang, W. (2017), "Planning reliability demonstration test with performance requirements," in *Proceedings of Reliability and Maintainability Symposium*, pp. 1–65.

Wang, P., D. W. Coit, (2005), "Repairable systems reliability trend tests and evaluation", in *Proceedings of Annual Reliability and Maintainability Symposium*, Alexandria, VA, pp. 416–421.

Wayne, M. and Modarres, M. (2015). A Bayesian model for complex system reliability growth under arbitrary corrective actions. *IEEE Transactions on Reliability* 64 (1): 206–220.

Xie, M. and Zhao, M. (1996). Reliability growth plot: an underutilized tool in reliability analysis. *Microelectronics and Reliability* 36 (6): 797–805.

Problems

Problem 4.1 Based on the failure information in Table 4.1, estimate and plot the product MTBF run chart between [0, 350 days] assuming 50 systems are installed on day 1 and these systems operate 12 hours in day time.

Problem 4.2 Based on the failure data in Table 4.1, plot the failure Pareto chart in the following cases: (i) all failures between [1, 90]; (ii) between [1, 200]; and (iii) between [1, 350]. Does the pattern change?

Problem 4.3 Assuming 50 systems operating in the field 12 hours/day. Based on the failure rate in Table 4.1, estimate the failure intensity rate of the following failure modes: (i) open diode; (ii) NFF; and (iii) PPMU.

Problem 4.4 Explain the reasons that lead to the NFF failures?

Problem 4.5 What are the measures one can adopt to mitigate the NFF failures or returns?

Problem 4.6 Every year there are 200 products returned from the field that turn out to be NFF. Estimate the NFF cost from the perspective of the customer and the factory. Related parameters are given as follows: $t_d = 3$ hours, $c_l = \$40$/hour, $c_m = \$3500$/hour, $n_s = 1$, $c_s = \$30\,000$/part, $t_r = 2.5$ hours, $c_r = \$30$/hour, $c_f = \$300$, $c_h = \$6000$/part/year, and $c_p = \$35\,000$/part.

Problem 4.7 Explain the difference between an engineering change order and a retrofit process, though both are commonly used to improve the reliability of field products.

Problem 4.8 The initial MTBF of a new car is $\theta_0 = 2000$ miles. The management would like to know how much additional miles should be run to reach $\theta_c = 3500$ hours given $\alpha = 0.2$, 0.4, and 0.6, respectively. Note that the car has accumulated 10 000 miles during the in-house test.

Problem 4.9 Three critical failure modes are identified for receiving corrective actions, shown in the table below. If the total CA budget is $20 000, how do you allocate the budget to each failure mode such that the aggregate failure intensity rate is lowest upon CA actions?

Failure mode	λ ($\times 10^{-5}$ failure/hour)	CA effectiveness	CA cost ($)
A	7.5	0.6	10 000
B	6.3	0.8	15 000
C	5.1	0.7	12 000

Problem 4.10 Prove that the MLE in Eq. (4.6.18) is correct:

$$L(\alpha, \beta) = \prod_{i=1}^{n} f(t_i \mid t_{i-1})$$

$$= \alpha^n \prod_{i=1}^{n} [1 - \exp(-t_i/\beta)]$$

$$\times \exp(-\alpha\{\tau - \beta[1 - \exp(-\tau/\beta)]\})$$

where $0 < t_1 < t_2 < \cdots < t_n \leq \tau$ denote the first n times to failure observed in period $[0, \tau]$.

Problem 4.11 The reliability of a system is required to be 0.95 at $t_D = 300$ hours with 90% confidence that no failure will occur during the test. The available testing capability is only $t_T = 250$ hours. We must now determine the number of test units for 250 hours testing with no failures in order to demonstrate the reliability goal. Assume the system lifetime follows the Weibull distribution with shape parameter $\beta = 2.5$.

Problem 4.12 The chi-squared method is applied to demonstrate a product reliability of 0.90 at $t_D = 400$ hours at CL = 95%. A maximum of two failures are allowed. The question is to determine the cumulative test time T_a. What if no failure is allowed? What then is the required cumulative test time?

5

Accelerated Stress Testing and Economics

5.1 Introduction

In accelerated stress testing (AST), the product undergoes higher-than-normal operating conditions in an effort to extrapolate the product reliability in use conditions by precipitating faults in a compressed period. Typical accelerating stresses include environmental, electrical, mechanical, and chemical factors. The choice of accelerating stress levels and the allocation of available test units to these levels are the key considerations in executing accelerated test experiments. This chapter begins with the introduction of the concept of AST that include a highly accelerated life test (HALT) and highly accelerated stress screening (HASS). The reliability extrapolation is often justified on the basis of physical and chimerical laws, or a combination with statistical models derived from the lifetime data. The chapter describes one-stress Arrhenius law and multistress Eyring models, both of which are physics-experimental-based models. We also present three types of statistics-based AST models: a scale and use rate acceleration model, a non-parametric model, and semi-parametric models that include a well-known proportional hazard model (PHM). Both HALT and HASS are costly to implement due to the consumption of equipment, materials, and labor. Economic models are developed to guide the product manufacturer to realize the cost savings in the deployment of HASS. This chapter concludes by presenting a case study of implementing HASS and environmental stress screening (ESS) to improve the product reliability in a subcontract manufacturing facility.

5.2 Design of Accelerated Stress Test

5.2.1 HALT, HASS, and ESS

Accelerated life stress testing or AST can be simply defined as: applying high levels of stresses to a device under test (DUT) for a short period of time to extrapolate the lifetime of the device in the use condition assuming it will exhibit the same failure mechanisms. Also known as accelerating variables, such stresses include temperature, humidity, voltage, power, speed, mechanical force, and torque, among others. The key here is to understand the root causes of failures and their relation to the applied stresses, either environmental, mechanical, or electrical stresses. The main purpose of AST is to accelerate the reliability growth in the product development stage such that the reliability

Reliability Engineering and Services, First Edition. Tongdan Jin.
© 2019 John Wiley & Sons Ltd. Published 2019 by John Wiley & Sons Ltd.
Companion website: www.wiley.com/go/jin/serviceengineering

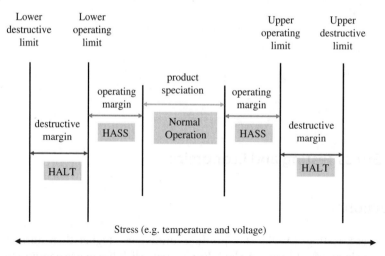

Figure 5.1 ALT and HASS and product design margins.

performance at the time to shipment meets the design requirement. Depending on the stress levels and testing time, two types of AST are generally adopted in industry: HASS and HALT. We elaborate on both techniques below.

As shown in Figure 5.1, the overall operating range of a product can be divided into three areas: product specification, operating margin, and the destructive margin. The product specifications represent the normal stress levels that the product is expected to use in the field. The operating margin formed between the lower (upper) operating limit and the lower (upper) specification limit is the safety area where the product is still able to operate normally, but with a higher failure rate. This is also the operating area of implementing the HASS process, a technique that uses stresses beyond product specifications in order to detect infant mortality failures and shorten the time to corrective actions in mass production.

The destructive margin formed between the lower (upper) destructive limit and the lower (upper) operating limit represents the area where the product operates under a much higher than usual stress, and will fail quickly. This is the testing domain of the HALT process that aims to precipitate the failure modes in a compressed time window. Sometimes HALT is also referred to as the test, analyze, and fix (TAAF) during the product design and development stage.

If a product operates beyond the lower or upper destructive limits, failure will occur immediately because the product suffers from failure mechanisms that may not happen if operating within the destructive margins. Although Figure 5.1 depicts a two-side (i.e. upper and lower) specification, many products like electronic devices are more susceptible to failures under high temperature and voltage rather than at the low stresses. Mechanical components like bearings and gears are prone to failure only if they are subject to high rotating speeds or torques. Hence both HASS and HALT processes in these circumstances is single-sided (right) testing rather than two-sided experimentation.

Both HALT and HASS are discovery testing as compared to compliance testing, like the reliability demonstration test. However, the following distinctions should be made between HASS and HALT. First, HASS is a reliability growth technique incorporated in the production phase to identify manufacturing defects that could cause early infant mortality, while the purpose of HALT is to identify the design flaws or weakness in the early development and prototyping phase. The result of HALT can be applied to the robust product design, like the design of the integrated chip wire-bonding process that is less susceptible to environmental stresses (Yang and Yang 1999). Second, HASS is a quality control process and applicable to all finished products, while HALT is implemented on sampled products, usually in prototype forms. Third, HASS in general is more economical than HALT because the use of equipment and materials, such as chambers, shakers, instruments, power, and cooling facility (e.g. liquid nitrogen), is less intensive in HASS.

In certain applications where the operating limits are already extreme (i.e. operating margin and product specification overlaps), ESS is often preferred over the HASS process. The key difference is that the stresses applied by ESS are within a product's non-destructive operating range used in the field. Like HASS, the ESS program is also implemented in post-production in which 100% of produced units are subjected to more severe stresses than in normal service. However, each method achieves the goal differently, depending on application-specific timelines, costs, and stress levels.

Other types of filtering processes exist to precipitate early infant mortality, such as burn-in, thermal cycling, power cycling, and thermal shocks. For instance, Yan and English (1997) proposed a modified bathtub curve that integrates the concept of latent failures and obsolescence for microdevice manufacturing. The idea is to construct an integrated cost model used to determine both optimal burn-in and ESS times. Ye et al. (2013) design an optimal burn-in process to minimize the warranty cost of a new product. Like ESS, the stress levels of these techniques are typically set in the boundary between the operating margin and the product specification. Table 5.1 summarizes the

Table 5.1 Summary of HALT, HASS, ESS, and other screening techniques.

Test type	Application	Stress level, timeframe	Impacts	Purpose	Cost ($/item)
HALT	Design, development, and prototyping	Destructive margin, very short time	Small sample	Identify design defect and weakness	High
HASS	Production stage	Operating margin	100% products	Remove latent failure or infant mortality	Medium to low
ESS	Production stage	Extreme limits of product specification	100% products	Remove latent failure, infant mortality, and manufacturing defects	Low
Burn-in, thermal cycling, thermal shock, power cycling, voltage margining	Production stage	Extreme limits of product specification	100% products	Remove latent failure, infant mortality, and manufacturing defects	Low

applications, stress levels, and impact units of different accelerated testing and screening techniques.

5.2.2 Types of Accelerating Stresses

In an effective accelerated test, the reliability expert chooses one or more stress types that cause the product to fail under normal operating conditions. The stresses are then applied at various accelerated levels and the time-to-failure and time-to-degradation for the units under test are recorded. For example, a product normally operates at 30 °C ambient temperature with 40% relative humidity (RH). If high temperature and humidity cause the product to fail more quickly, the product can be tested under 60 °C with RH = 70% or 100 °C with RH = 90% in order to accelerate the units to fail more rapidly. In this example, the stress type is temperature and humidity and the accelerated stress levels are 60 and 100 °C for temperature and 70% and 90% for RH. Depending on the nature of product materials and the operating conditions, stresses used to accelerate the failure can be classified into three categories, as shown in Table 5.2. These are environmental, electrical, mechanical, and chemical stresses.

5.2.2.1 Environmental Stresses

They are the factors that are closely related to the surroundings of the operating product. Temperature, humidity, and thermal cycling are the typical environmental stresses. It is important to determine the critical environmental stresses and assign appropriate stress levels that do not induce different failure modes other than the ones at the use condition. For instance, the life of semiconductor devices are sensitive to the operating temperature and RH. However, if the same devices are used in satellites or space stations orbiting outside the atmosphere, beta radiation and gamma rays have the ability to ionize semiconductor materials, which results in new failure effects: (i) producing additional electron-hole pairs and (ii) creating high-energy charges to be injected into silicon dioxide regions, causing the degradation and failure of transistors. While the humidity emerges as a key environmental stress for electronic devices used in vapor-intensive tropical areas or rainforest regions, cosmic radiation becomes one of the primary environmental stresses when they are used in space engineering systems.

5.2.2.2 Electrical Stress

Electrical stress is applied to exercise a product near or at its electrical limits. Examples of electrical stress tests include simulating junction temperatures on semiconductors

Table 5.2 Classification of stress types and their factors.

Stress type	Stress factors
Environmental	Temperature, thermal cycle, humidity, thermal shock, vibration, sand and dust, nuclear and cosmos radiation, altitude
Electrical	Voltage, current, power, frequency, electrical field, power cycling
Mechanical	Force, friction, torque, fatigue, vibration, pressure
Chemical	Corrosions, diffusions

and testing the insulation of circuit breakers of high-voltage transmission. Two basic types of electrical stress tests are available: voltage margining and power cycling. Voltage margining pertains to varying input current or voltage above and below nominal operating limits. A subset of voltage margining is frequency margining which is often used in stressing the speed (or the clock cycle) of microprocessors like the central processing unit. Other types of voltage margining include electric field. Power cycling consists of turning a machine's power on and off at specified levels with predetermined time intervals. It is often used to induce the solder joint failure by creating thermal fatigue when the temperature of solder joints increases and decreases cyclically with the on–off power. Electrical stress alone is not able to expose the number of defects commonly found under the vibration test or temperature cycling. However, it is often economical to implement the electrical stress along with other stresses to increase the overall effectiveness of ALT or HASS programs. This is because it is often required to supply electric power to products under test in order to stimulate soft or hard failures induced by mechanical or environmental factors.

5.2.2.3 Mechanical Stress

This type of stress can be induced by force, torque, vibration, and thermal shocks, among others. The effort would be caused by internal or external factors. For instance, a solder joint crack in a circuit board is often induced by repetitive thermal cycling, which is an internal factor. The disconnection of a universal serial bus connector in a computer could be caused by frequent insertion and extraction operations, belonging to an external factor. The breakage of wind turbine blades is induced by material fatigue due to vibration and wind shocks repeatedly applied to the blade. Therefore, failure of mechanical systems is largely associated with external factors.

5.2.2.4 Chemical Stress

Corrosion and diffusion are two basic failure mechanisms induced by chemical stresses. Corrosion is a natural process converting a refined metal to a more chemically stable form, such as its oxide, hydroxide, or sulfide. It is the gradual destruction of materials by chemical and/or electrochemical reaction with their environment. Diffusion is the net movement of molecules or atoms from a high concentration region to a low concentration region as a result of random motion of tiny particles. Diffusion is driven by a gradient in the chemical potential of the diffusing species.

5.2.3 Stress Profiling

Different types of load profile are available in AST and their application depends on the stress type, availability of the test bed, and the product's operating condition. Based on the stress amplitude and its variation frequency, stress loading profiles can be classified into five categories: (i) constant stress, (ii) sinusoid stress, (iii) step stress, (iv) ramp-up and dwell, and (v) zigzag and cyclic stress.

Figure 5.2 depicts the loading profile of a constant stress. As the name implies, the stress level remains constant during the entire test period. For ALT and HASS processes subject to a constant stress level, the decision variables are the level of the stress A_1 and the duration of the test period t_1. For instance, Yang (1994) proposed an optimal design of four-level constant-stress ALT plans that chose the stress levels, test units of

Figure 5.2 Constant stress level.

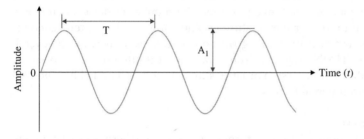

Figure 5.3 Sinusoid stress.

each stress, and censoring times to minimize the asymptotic variance of the maximum likelihood estimators of the mean life. Stresses like temperature, humidity, voltage, or power can be set at a fixed level during the accelerated testing period.

A sinusoid stress profile is constantly used in fatigue testing of mechanical systems. As shown in Figure 5.3, the stress profile is characterized by two parameters: the amplitude A_1 and the cycle period T. The mathematical expression of the sinusoid stress is

$$S(t) = A_1 \sin(2\pi(1/T)t) = A_1 \sin(2\pi ft) \qquad (5.2.1)$$

where f is the frequency of stress and $f = 1/T$ is in units of Hertz. Obviously, the frequency and the amplitude are the key factors that determine the severity of the stresses imposed on the testing units.

Figure 5.4 shows a step-stress loading profile where a constant stress is applied for a period of time and then the stress is escalated to a new level for another period of time. This process is repeated until it reaches the end of the test time or all the stress levels have been applied. The decision variables in a step-stress test include the number of stress steps and the amplitude and the duration of each step stress. The mathematical formula is

$$S_i(t) = A_i(u(t - t_{i-1}) - u(t - t_i)), \quad \text{for } t_{i-1} < t \le t_i \qquad (5.2.2)$$

where $t_0 = 0$ and $u(t)$ is a standard step function with $u(t) = 1$ for $t \ge 0$ or $u(t) = 0$ for $t < 0$. For example, for the three-step stress test in Figure 5.4, the values of A_1, A_2, A_3, t_1, t_2, and t_3 need to be determined prior to the execution of the ALT or HASS test. As an example, Miller and Nelson (1983) designed an optimum plan for two-step stress tests where all units are run to failure and the goal is to minimize the asymptotic variance of the maximum likelihood estimator of the mean life. In electrical stress testing, it is relatively easy for stresses like voltage, power, and electric field to be transitioned from one level to another instantaneously. Hence the step stress loading profile is commonly adopted in an electrical test.

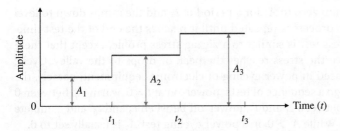

Figure 5.4 Step stress (or stair stress).

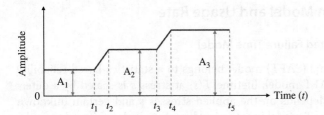

Figure 5.5 Dwell and ramp-up stress.

Figure 5.5 depicts the loading profile of a dwell and ramp-up stress where the transition from the low stress level to the upper level is not instantaneous, but it takes an amount of time, i.e. $t_2 - t_1$ and $t_4 - t_3$, before reaching level A_2 and A_3, respectively. Hence, dwell and ramp-up profiles are usually applied in environmental tests where different levels of stresses like temperature and humidity cannot be reached immediately due to the limitation of the test equipment.

There exist other types of stress profiles such as the zigzag stress in Figure 5.6 and the cyclic stress in Figure 5.7. These can be treated as the variation of the dwell and ramp-up stress in Figure 5.5. For instance, in the zigzag stress, there is no dwell period

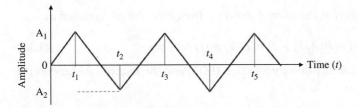

Figure 5.6 Zigzag stress profile.

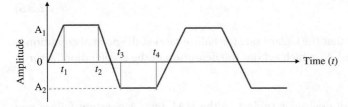

Figure 5.7 Cyclic stress profile.

and the stress ramps up from zero to A_1 for a period of t_1 and the ramps down to level A_2 between t_1 and t_2. This process is repeated until it reaches the end of the test time. The pattern of a cyclic stress test is similar to a zigzag stress profile, except that there exist dwelling periods once the stress reaches the peak or drops to the valley. Cyclic stress profiles are widely used in power cycling of electronics equipment, namely, the products under test undergo a sequence of tests: power-on at $t = 0$, warm-up between 0 and t_1, normal operation between t_1 and t_2, power-off between t_2 and t_3, and complete shut-down from t_3 and t_4. While $A_1 > 0$, in a power cycling test A_2 is usually set to 0.

5.3 Scale Acceleration Model and Usage Rate

5.3.1 Exponential Accelerated Failure Time Model

A scale accelerated failure time (SAFT) model belongs to a statistics-based reliability modeling approach. In an SAFT model, lifetime $T(s)$ at stress s is scaled by a deterministic number that often depends on the applied stresses **s** and certain unknown parameters. Therefore, the SAFT model in statistical literature is also referred to as the accelerated failure time (AFT) model. Let T_n and T_s be, respectively, the lifetime under the normal use condition and the accelerated stress condition. Their relation is governed by the acceleration factor (AF) as follows:

$$T_s = \frac{T_n}{A_f} \tag{5.3.1}$$

where A_f is the acceleration factor and T_n and T_s are the lifetime at normal and the stressed condition, respectively. Lifetime is accelerated when $A_f > 1$ and decelerated if $A_f < 1$. For example, if the lifetime at the stressed level is exponentially distributed with failure rate λ_s, then the reliability at stress s is

$$R_s(t) = P\{T_s > t\} = e^{-\lambda_s t} \tag{5.3.2}$$

Let $R_n(t)$ be the reliability at the normal stress s_n. Then it can be extrapolated as

$$R_n(t) = P\{T_n > t\} = P\{A_f T_s > t\} = P\{T_s > (t/A_f)\} = e^{-\frac{\lambda_s}{A_f} t} \tag{5.3.3}$$

Furthermore, the probability density function (PDF) and the hazard rate function at s_n can also be obtained as

$$f_n(t) = \frac{\lambda_s}{A_f} e^{-\frac{\lambda_s}{A_f} t} \tag{5.3.4}$$

$$\lambda_n = \frac{f_n(t)}{R_n(t)} = \frac{\lambda_s}{A_f} \tag{5.3.5}$$

Equation (5.3.5) shows that the hazard rate (or failure rate) at the normal condition is scaled down by a factor of A_f, which echoes the definition of the acceleration factor in Eq. (5.3.1).

Example 5.1 A group of devices is subject to the HAL test. A constant failure rate $\lambda_s = 0.1$ failures/hour is derived based on the testing data. Assuming the acceleration factor $A_f = 5$, what is the reliability at $t = 10$ hours in stressed and normal conditions, respectively?

Solution:

Based on Eq. (5.3.3), the reliability function in the normal use condition is

$$R_n(t) = \exp\left(-\frac{\lambda_s}{A_f}t\right) = \exp\left(-\frac{0.1}{5}t\right) = \exp(-0.02t)$$

For $t = 10$ hours, $R_n(t = 100) = \exp(-0.02 \times 10) = 0.819$.

Let us compute the reliability under the stressed level at $t = 10$ hours. That is

$$R_s(t = 100) = \exp(-\lambda_s t) = \exp(-0.1 \times 10) = 0.368$$

Figures 5.8 and 5.9 plot the reliability function and PDF function under $A_f = 1, 2$, and 5, respectively, based on λ_s in the example. Note that reliability with $A_f = 1$ corresponds to the normal condition.

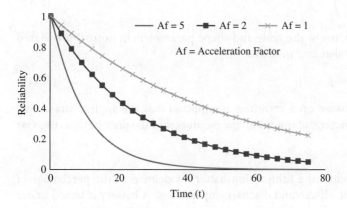

Figure 5.8 Reliability with different A_f values.

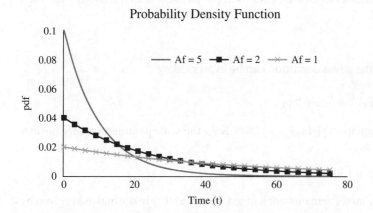

Figure 5.9 Probability density with different A_f values.

5.3.2 Weibull AFT Models

In this section, we extend the exponential AFT model to the two-parameter Weibull lifetime distributions. Let θ_s and β_s be the Weibull scale and shape parameters under the accelerated condition. Then the reliability function under the stress condition is

$$R_s(t) = P\{T_s > t\} = e^{-\left(\frac{t}{\theta_s}\right)^{\beta_s}} \tag{5.3.6}$$

Given $R_s(t)$, we can derive the reliability, PDF, and hazard rate in normal use; they are given as follows:

$$R_n(t) = P\{T_s > (t/A_f)\} = e^{-\left(\frac{t}{A_f \theta_s}\right)^{\beta_s}} \tag{5.3.7}$$

$$f_n(t) = \frac{\beta_s}{A_f \theta_s}\left(\frac{t}{A_f \theta_s}\right)^{\beta_s - 1} e^{-\left(\frac{t}{A_f \theta_s}\right)^{\beta_s}} \tag{5.3.8}$$

$$h_n(t) = \frac{\beta_s}{A_f \theta_s}\left(\frac{t}{A_f \theta_s}\right)^{\beta_s - 1} \tag{5.3.9}$$

Let θ_n and β_n be, respectively, the scale and shape parameters in normal use. Based on Eqs. (5.3.7–5.3.9), it is also easy to realize that

$$\theta_n = A_f \theta_s \quad \text{and} \quad \beta_n = \beta_s \tag{5.3.10}$$

The above results are based on a common assumption that the applied stress only influences the scale parameter, but not the shape parameter of the distribution (Escobar and Meeker 2006).

Example 5.2 The reliability of a lithium-ion battery is defined as the percentage of capacity it has upon cyclic charge and discharge operations. A battery is tested under accelerated stress conditions by applying deep charge-and-discharge cycles, and the reliability reaches 0.7436 after 1000 cycles. Assume $A_f = 1.5$ and $\beta_s = 3$. Estimate the battery reliability function under a normal use condition. What is the actual reliability for 1000 cycles in normal use?

Solution:
The reliability under the stress condition can be expressed by

$$R_s(t = 1000) = e^{-\left(\frac{1000}{\theta_s}\right)^3} = 0.744$$

Solving the above equation yields $\theta_s = 1500$. Now the scale parameter under normal use can be estimated as

$$\theta_n = A_f \theta_s = 1.5 \times 1500 = 2250$$

Since the shape parameter remains unchanged, the reliability in normal use is given by

$$R_n(t = 1000) = e^{-\left(\frac{t}{2250}\right)^3} = 0.916$$

5.3.3 Lognormal AFT Models

Lognormal distribution is widely used for modeling and estimating the failure times of electronics components subject to thermal or electric stresses. These include temperature, voltage, power, and the electric-magnetic field. Another application of lognormal distribution is to model the fracture of the substrate of integrated circuits. The root cause of this failure mechanism is power cycling. It makes the device junction temperature fluctuate because of the differences between the coefficients of thermal expansion of device packaging materials. Let μ_s and σ_s be the parameters of the lognormal distribution under the stressed condition; then the PDF is

$$f_s(t) = \frac{1}{\sigma_s t \sqrt{2\pi}} e^{-\frac{(\ln t - \mu_s)^2}{2\sigma_s^2}}, \qquad t \geq 0 \tag{5.3.11}$$

The AF condition in Eq. (5.3.11) in logarithmic scale can be expressed as

$$\ln T_s = \ln T_n - \ln A_f \tag{5.3.12}$$

By substituting Eq. (5.3.12) into Eq. (5.3.11), the PDF at the normal use condition is

$$f_n(t) = \frac{1}{\sigma_s t \sqrt{2\pi}} e^{-\frac{(\ln t - \mu_s - \ln A_f)^2}{2\sigma_s^2}} = \frac{1}{\sigma_n t \sqrt{2\pi}} e^{-\frac{(\ln t - \mu_n)^2}{2\sigma_n^2}} \tag{5.3.13}$$

where μ_n and σ_n are the parameters of the lognormal distribution under normal use conditions. The following relations held for the lognormal AFT model:

$$\mu_n = \mu_s + \ln A_f \quad \text{and} \quad \sigma_n = \sigma_s \tag{5.3.14}$$

The lognormal distribution belongs to the location-scale distribution family where μ_n (μ_s) are called the location parameters that are dependent on stress level s, and σ_n (σ_s) are called the scale parameters that are independent of s. Other location-scale distributions include normal, uniform, and Cauchy distributions.

Example 5.3 Show that the PDF in Eq. (5.3.13) is correct.

Proof:

$$
\begin{aligned}
F_n(t) &= P\{T(s_n) < t\} = P\{T(s) < (t/A_f)\} \\
&= \int_0^{t/A_f} \frac{1}{\sigma_s x \sqrt{2\pi}} e^{-\frac{(\ln x - \mu_s)^2}{2\sigma_s^2}} dx \;\;\overset{x = t/A_f}{=}\;\; \int_0^{t} \frac{1}{\sigma_s A_f x \sqrt{2\pi}} e^{-\frac{(\ln x - \mu_s - \ln A_f)^2}{2\sigma_s^2}} A_f dx \\
&= \int_0^{t} \frac{1}{\sigma_s x \sqrt{2\pi}} e^{-\frac{(\ln x - (\mu_s + \ln A_f))^2}{2\sigma_s^2}} dx \\
&= \int_0^{t} f_n(x) dx \tag{5.3.15}
\end{aligned}
$$

Hence we complete the proof of the PDF in Eq. (5.3.13) via defining $x = t/A_f$.

5.3.4 Linear Usage Acceleration Model

While the product is placed in the normal operating environment, increasing the usage can be an effective approach to accelerating the life as well. This differs from HALT and HASS techniques where operating conditions such as temperature, voltage, and vibrations are escalated to higher than the normal operating condition. Usage acceleration can be applied to products subject to intermittent or non-continuous operations, such as relays, switches, bearings, motors, gearbox, vehicle tires, washing machines, and air conditioners. The basic assumption of usage acceleration models is that the product useful life should not be affected by the increased rate or cycles of operations during the short time period. This is important because cycling simulates the actual use and if the cycling frequency is low enough, the test units can return to the steady state prior to the start of the next cycle. As such, the time-to-failure distribution is independent of the cycling rate or there is no reciprocity effect. The implies that there exists a linear relation between the accelerated lifetime and the lifetime under normal use, namely

$$A_f = \text{(Life at normal usage rate)}/\text{(Life at accelerated usage rate)} \qquad (5.3.16)$$

The model in Eq. (5.3.16) is also called the SAFT because the lifetime is proportional to the usage rate. For example, Johnston et al. (1979) observed that the cycles-to-failure of a type of insulation material was shortened with the increased alternating current (AC) frequency. The acceleration factor can be estimated as $A_f(412) = 412/60 \approx 6.87$ when the AC frequency in voltage endurance tests was increased from 60 to 412 Hz by keeping the voltage at the same level.

Ideally increasing the usage rate should not significantly change the actual use condition of the product. Hence in accelerated usage rate tests, other relevant factors should be identified and controlled to reflect the actual use environment. If the cycling rate is too high, it can cause reciprocity breakdown (Escobar and Meeker 2006). For example, in a power cycling test excessive heat may build up on test units (e.g. microprocessor chips) if the time interval between two consecutive power cycles are too short. This induces the reciprocity breakdown because the cycles-to-failure distribution depends on the cycling rate. Thus, it is necessary to let the test units "cool down" between the cycles of consecutive operations.

5.3.5 Miner's Rule under Cyclic Loading

Most mechanical systems often endure repeated loads that are constantly applied to the object, and the magnitude of the loads may exceed the upper or lower limits of the material strength. The stresses above or below the limits are called critical stresses. Typically examples include wind turbine blades, suspension spring, aircraft wings, and gearbox, among others. Fatigue is generated by cyclic stresses beyond the critical values, especially the upper limit. The system eventually breaks down as the result of the accumulation of fatigues.

Figure 5.10 shows a typical cyclic load profile that fluctuates between the upper and lower limits. It is assumed that there is no fatigue effect as long as the stress does not exceed the critical levels. The fatigue lifetime of an item subject to varying stress can be estimated using Miner's rule. This is expressed as

$$\frac{n_1}{N_1} + \frac{n_2}{N_2} + \cdots + \frac{n_k}{N_k} = 1 \qquad (5.3.17)$$

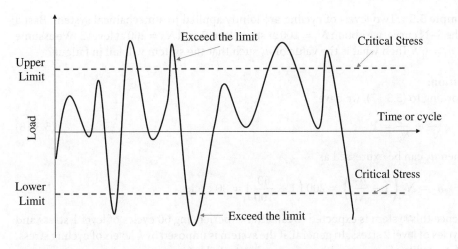

Figure 5.10 A typical cyclic load profile for mechanical systems.

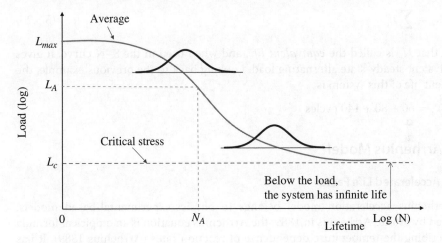

Figure 5.11 S–N curve.

where

n_i = the number of cycles at a specific load level (above the fatigue limit)
N_i = the median number of cycles to failure at that level
k = the number of the load levels or profiles

The values of N_i for $i = 1, 2, ..., k$ can be obtained from the so-called stress-cycle curve, or S–N curve. Figure 5.11 depicts the empirical relationship between stress and cycles to fatigue. An important assumption is that the life is infinite below the fatigue limit stress L_c, while the system fails immediately if the maximum stress L_{max} is imposed. When the applied stress varies between L_c and L_{max}, it induces material fatigue, leading to a failure as a result of cumulative damages.

Example 5.4 If $k = 1$ in Miner's rule, meaning that we only have one level of load (or stress), then the cycles leading to the failure is simply given by $n_1 = N_1$.

Example 5.5   Two levels of cycling are jointly applied to a mechanical system. Based on the S–N curve, we obtain $N_1 = 100$ at stress level 1 and $N_2 = 200$ at level 2. We assume that $n_1 = 60$. Then what is the value of n_2 such that the system will fail in fatigue?

Solution:
According to (5.3.17), we have

$$\frac{n_1}{N_1} + \frac{n_2}{N_2} = 1 \tag{5.3.18}$$

Then n_2 can be expressed as

$$n_2 = N_2 \left(1 - \frac{n_1}{N_1}\right) = 200 \left(1 - \frac{60}{100}\right) = 80 \text{ cycles} \tag{5.3.19}$$

Hence this system is expected to fail after experiencing 60 cycles of level 1 stress and 80 cycles of level 2 stress. In general, if the system is imposed by k levels of cycling stress, the fatigue life of an item with a mean-value-to-failure is

$$N_e = \sum_{i=1}^{k} n_i \tag{5.3.20}$$

Note that N_e is called the *equivalent life*, and when used in the S–N curve, it gives an equivalent steady-state alternating load. Looking back to the previous example, the equivalent life of this system is

$$N_e = 60 + 80 + 140 \text{ cycles}$$

5.4   Arrhenius Model

5.4.1   Accelerated Life Factor

The Arrhenius acceleration model belongs to physics-experimental-based models. Proposed by Svante Arrhenius in 1889, the Arrhenius equation is an empirical formula for describing the temperature dependence of reaction rates (Arrhenius 1889). It has been recognized as one of the earliest and most effective acceleration models to predict how the time-to-failure changes with the imposed temperature stress. The model has been widely used for failure mechanisms that depend on diffusion processes, migration processes or chemical reactions. Hence, it covers many of the non-mechanical (or non-material) fatigu failure modes that are responsible for the dysfunction of electronic circuits or devices. The Arrhenius model takes the following form:

$$t_f(T) = A \exp\left(\frac{E_a}{kT}\right) \tag{5.4.1}$$

where
$T =$ the temperature measured in degrees Kelvin (K)
$E_a =$ the activation energy in units of electronvolts (eV)
$k =$ Boltzmann's constant with $k = 8.617 \times 10^{-5}$ eV /K

Parameter A is a non-thermal constant called the scaling factor. The value of E_a depends on the materials of the product and the failure mechanism. Typically it is in a

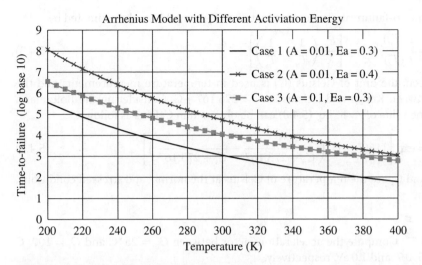

Figure 5.12 Time-to-failure with different A and E_a.

range between 0.3 and 0.4 eV, but can go up to 1.5 eV or higher. The Arrhenius model argues that for reactants to transform the product into a chemical process, they must first acquire a minimum amount of energy, called the activation energy E_a. The concept of activation energy explains the exponential relationship between the reaction rate and the fraction of molecules having kinetic energy larger than E_a. The latter can be calculated from statistical mechanics.

Under different A and E_a values, Figure 5.12 plots three cases to show how the time-to-failure decreases with the elevation of temperature. Note that the vertical axis represents $t_f(T)$ in logarithmic scale with base 10. Assume Case 1 as the baseline with $A = 0.01$ and $E_a = 0.3$. By comparing Cases 1 with 2, it is found that activation energy plays a critical role in determining the time-to-failure (i.e. reliability). If E_a increases from 0.3 to 0.4 with the same $A = 0.01$, the value of $t_f(T)$ in Case 2 increases over 300 times at $T = 200$ K and 20 times at $T = 400$ K as opposed to Case 1.

Example 5.6 Suppose an electronic device operates at $T = 333$ K continuously for $t_f = 1.44$ years prior to its failure. Estimate E_a of this device assuming $A = 0.05$ hour.

Solution:
From Eq. (5.4.1), the activation energy can be expressed as

$$E_a = kT(\ln t_f(T) - \ln A) \tag{5.4.2}$$

To use Eq. (5.3.1), we will convert $t_f = 1.44$ years into hours, namely $t_f = 1.44 \times 8760$ hours/year $= 14\,047$ hours. Hence,

$$E_a = kT(\ln t_f(T) - \ln A) = 8.617 \times 10^{-5} \times 333 \times [\ln(14\,047) - \ln(0.05)] = 0.36 \text{ eV}$$

Let T_n and T_s be the normal and the stressed temperatures, respectively, and the unit is Kelvin. The acceleration factor A_f is defined as the ratio of the time-to-failure at T_n

over the time-to-failure at T_s. Assume that $T_s > T_n$; then AF can be estimated by

$$A_f = \frac{t_f(T_n)}{t_f(T_s)} = \exp\left[\frac{E_a}{k}\left(\frac{1}{T_n} - \frac{1}{T_s}\right)\right] \tag{5.4.3}$$

Quite often, the unit of Celsius (C) is used in temperature measurements and the relation between Kelvin and Celsius is $K = 276.167 + C$. Considering the Boltzmann constant, the value of A_f in Eq. (5.4.3) can be obtained in Celsius as follows:

$$A_f = \exp\left[11605 \times E_a\left(\frac{1}{C_n + 273.167} - \frac{1}{C_s + 273.167}\right)\right] \tag{5.4.4}$$

where C_n and C_s are the temperature of Celsius at the normal and stressed conditions, respectively.

Example 5.7 Compute the acceleration factor between $C_n = 25\,°C$ and $C_s = 100\,°C$ for $E_a = 0.5, 0.7$, and $1.0\,eV$, respectively.

Solution:
Based on Eq. (5.4.4), we have

$$A_f(E_a = 0.5) = \exp\left[11\,605 \times 0.5 \times \left(\frac{1}{25 + 273.16} - \frac{1}{100 + 273.16}\right)\right] = 50$$

$$A_f(E_a = 0.7) = \exp\left[11\,605 \times 0.7 \times \left(\frac{1}{25 + 273.16} - \frac{1}{100 + 273.16}\right)\right] = 240$$

$$A_f(E_a = 1.0) = \exp\left[11\,605 \times 1.0 \times \left(\frac{1}{25 + 273.16} - \frac{1}{100 + 273.16}\right)\right] = 2518$$

It is interesting to see that the A_f increases over 50 times (i.e. $2518/50 = 50.4$) if the value of E_a doubles from 0.5 to 1.0 eV.

Figure 5.13 plots the AF value by increasing the temperature from 25 to 100 °C under $E_a = 0.3$ and 0.4 eV, respectively. It is observed that for $E_a = 0.3$, the device lifetime reduces approximately by 50% whenever the operating temperature goes up 20 °C. For $E_a = 0.4$, the lifetime reduces roughly by 50% whenever the temperature increases by

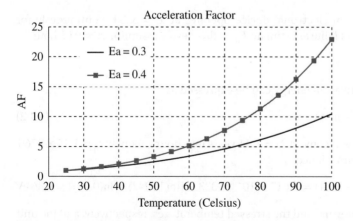

Figure 5.13 Acceleration factor increases with temperature.

15 °C. Since E_a of most electronic devices is within $[0.3, 0.4]$, a rule of thumb in electronics design is that the lifetime of the electronic product doubles whenever its operating temperature is reduced by 15–20 °C. Equivalently, the device lifetime shrinks by a half whenever the temperature increases by 15–20 °C. Various techniques are available for lowering the temperature of electronics systems in the use condition, including air fans, liquid cooling, and heat sinks. For instance, the power density for typical data centers reaches $60 \text{ kW}/\text{m}^2$, and almost a half of the electric energy is used to maintain a low-temperature environment for computing clusters (Pakbaznia and Pedram 2009).

5.4.2 Other Units for Activation Energy

When dealing with the degradation of organic materials, such as food, plastics, pharmaceuticals, paints, and coatings, it is more common to replace the Boltzmann constant k with the universal gas constant R. Then the Arrhenius model becomes

$$t_f(T) = A \exp\left(\frac{E_a}{RT}\right) \tag{5.4.5}$$

For example, $R = 8.314\,46\,\text{J}/(\text{mol K})$ is commonly given in units of joules per mole Kelvin. In this formula, E_a is the activation energy in units of joules per mole (J/mol). According to Escobar and Meeker (2006), the corresponding Arrhenius acceleration factor is

$$A_f = \frac{t_f(T_n)}{t_f(T_s)} = \exp\left[0.120\,27 E_a \left(\frac{1}{T_n} - \frac{1}{T_s}\right)\right] \tag{5.4.6}$$

where T_n and T_s are the temperature in Kelvin in the use and the stressed condition, respectively.

It is worth mentioning that the Arrhenius model is not applicable to all thermal acceleration problems. A main reason is because both the activation energy E_a and the rate constant k are experimentally determined. They represent macroscopic parameters that are not truly related to threshold energy and the success of individual molecular collisions. Rather, macroscopic measurements are the result of a group of molecular collisions with different parametric scenarios. Nevertheless, the Arrhenius model has been proven to be satisfactorily in many different applications.

5.5 Eyring Model and Power Law Model

5.5.1 Eyring Model

The Eyring model has its theoretical basis in chemical reactions and quantum mechanics. It is a physics-experimental-based acceleration model. If an aging process, such as chemical reaction, corrosion, diffusion, or metal migration, is driving the reliability degradation toward failure, the Eyring model is able to capture how the rate of degradation varies with the imposed stress. Unlike the Arrhenius model, the Eyring model can be used to model life acceleration under multiple stresses, includes temperature, voltage, current, and other relevant stresses. A two-stress Eyring model accommodating temperature and one non-thermal stress takes the following form (NIST 2017):

$$t_f(T, S) = A \times T^m \exp\left[\frac{E_a}{kT} + S\left(B + \frac{C}{T}\right)\right] \tag{5.5.1}$$

where

T = temperature stress in degrees Kelvin
S = non-thermal stress in forms of voltage or current (or a function of voltage or current), or any other relevant stress
$A, B, C,$ and m = model parameters

As with the Arrhenius model, k is the Boltzmann constant and E_a is the activation energy. Parameters A, B, and m determine the acceleration effect between stressed and use conditions, and C determines the combined effect of stresses T and S. Applications in the literature have typically use a fixed value of m ranging from $m = 0$ (Boccaletti et al. 1989), $m = 0.5$ (Klinger 1991a,b), to $m = 1$ (Mann et al. 1974; Nelson 1990).

If one compares Eq. (5.5.1) with Eq. (5.4.1), the temperature terms are very similar between them, explaining why the Arrhenius model has been so effective in establishing the connection between the parameter E_a and the quantum theory concept of "activation energy needed to cross an energy barrier and initiate a reaction" (NIST 2017). The general Eyring model accommodating temperature and n non-thermal stress factors can be expressed as

$$t_f(T, S) = AT^m \exp\left[\frac{E_a}{kT} + \sum_{i=1}^{n} S_i\left(B_i + \frac{C_i}{T}\right)\right] \tag{5.5.2}$$

where $\mathbf{S} = \{S_1, S_2, \ldots, S_n\}$ is the non-thermal stress vector and B_i and C_i are the parameters associated with stress S_i for $i = 1, 2, \ldots, n$. In the general Eyring model there exist terms characterizing the interactions between the temperature and non-thermal stresses. This means that the effect of changing temperature on the lifetime depend on the levels of other stresses. In models with no stress interaction, the acceleration factors can be computed separately for each stress and then multiplied together. This would not be the case if the interaction terms are necessary and required for the underlying physical mechanism, like the temperature and other stresses in the general Eyring model.

5.5.2 Inverse Power Law Model

The inverse power law (IPL) model is commonly used for non-thermal accelerated stress tests and possesses the following form (Yang 2007; Elsayed 2012):

$$L(V) = \frac{1}{KV^n} \tag{5.5.3}$$

where

L = a quantifiable life measure, such as mean life, characteristic life, median life
V = the stress level such as voltage
K and n = model parameters to be determined, with $K > 0$ and $n > 0$

The IPL in Eq. (5.5.3) can also be expressed in logarithmic scale as follows:

$$\ln L = -\ln K - n \ln V \tag{5.5.4}$$

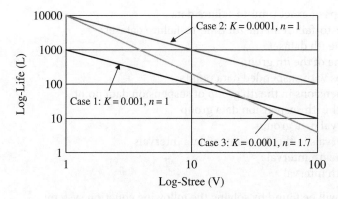

Figure 5.14 Plot of inverse power law model in log base 10.

The linear relation between $\ln V$ and $\ln L$ is appealing because $-\ln (K)$ now become the intercept and n is the slope. Both $\ln (K)$ and n can be directly estimated from the log-scale plot. In Figure 5.14, the IPL model is plotted in log-scale with base 10 in three cases: $\{K = 0.001, n = 1\}$, $\{K = 0.0001, n = 1\}$, and $\{K = 0.0001, n = 1.7\}$, respectively. A common observation is that lifetime decreases with the increased stress level. Given the same initial life at normal condition, a larger n implies that the product life decreases faster (see Cases 2 and 3). The value of K determines the product life time at the normal condition (see Cases 1 and 2) and a smaller K implies a longer life.

Let V_n and V_s be the stresses in normal and accelerated conditions, respectively. The acceleration factor for the IPL is

$$A_f = \frac{L(V_n)}{L(V_s)} = \left(\frac{V_s^n}{V_n^n}\right) \tag{5.5.5}$$

The PDF of the IPL-exponential model takes the following form:

$$f(t; V) = KV^n e^{-KV^n t} \tag{5.5.6}$$

Note that this is a two-parameter $\{K, n\}$ distribution model. The reliability function and the failure rate of the model are also obtained as follows:

$$R(t; V) = \int_t^\infty f(x; V)dx = e^{-KV^n t} \tag{5.5.7}$$

$$h(t) = \frac{f(t; V)}{R(t; V)} = KV^n \tag{5.5.8}$$

The failure rate is indeed constant (time-invariant). This result is coincident with the expected lifetime model in Eq. (5.5.3) because $h(t) = 1/L(V)$ under exponential distribution. Both K and n can be estimated through the maximum likelihood estimation (MLE) method. The MLE function in the logarithmic scale is given as follows:

$$L(K, n) = \sum_{i=1}^{F_e} N_i \ln(KV_i^n e^{-KV_i^n T_i}) - \sum_{i=1}^{S} N_i' KV_i^n T_i' + \sum_{i=1}^{FI} N_i'' \ln(R_{Li}'' - R_{Ri}'') \tag{5.5.9}$$

where

$$R_{Li}'' = e^{-T_{Li}'' KV_i^n}$$

$$R_{Ri}'' = e^{-T_{Ri}'' KV_i^n}$$

F_e = the number of groups of exact time-to-failure data
N_i = the number of time-to-failure in the ith time-to-failure dataset
V_i = the stress level of the ith dataset
T_i = the exact failure time of the ith group
S = the number of groups with suspended data points
N_i' = the number of suspensions in the ith group of suspension data points
T_i' = the testing time of the ith suspension data group
F_I = the number of interval data groups
N_i'' = the number of intervals in the ith group of data intervals
T_{Li}'' = the beginning of the ith interval
T_{Ri}' = the ending of the ith interval

The solution to K and n will be found by solving the following equation system:

$$\begin{cases} \dfrac{\partial}{\partial K} L(K, n) = 0 \\[2mm] \dfrac{\partial}{\partial n} L(K, n) = 0 \end{cases} \qquad (5.5.10)$$

5.6 Semiparametric Acceleration Models

5.6.1 Proportional Hazard Model

Cox (1972) proposed a proportional hazard model (PHM) in 1972. It is a semiparametric model and has been widely accepted for analysis of failures with covariates. In statistics, a semiparametric model is a statistical model that has parametric and non-parametric components. PHM is built upon the hazard rate function, assuming that the hazard rate $h(t; \mathbf{z})$ under the covariate $\mathbf{s}$ is the multiplication of an unspecified baseline hazard rate $h_0(t)$ and a relative weight, $\exp(\boldsymbol{\beta}^T \mathbf{s})$, where $\boldsymbol{\beta}$ is the regression coefficient vector. A covariate in accelerated life testing represents the stresses imposed, such as temperature, voltage, and force. The PHM model can be expressed as:

$$h(t; \mathbf{s}) = h_0(t) \exp(\boldsymbol{\beta}^T \mathbf{s}) \qquad (5.6.1)$$

where

$\mathbf{s} = [s_1, s_2, \ldots, s_n]$ is the covariate vector with n being the number of covariates
$\boldsymbol{\beta} = [\beta_1, \beta_2, \ldots, \beta_n]$ is the regression coefficient vector

The model has been successfully used for survival analysis in medical areas (O'Quigley 2008) and reliability forecasting in accelerated life testing (Elsayed and Chan 1990; Elsayed and Jiao 2002). If we are interested in the time-to-failure T, the cumulative distribution function (CDF) function is given by

$$F(t) = 1 - \exp\left(-\int_0^t h_0(u) \exp(\boldsymbol{\beta}^T \mathbf{s}) du\right), \quad \text{for } t \geq 0 \qquad (5.6.2)$$

In this semiparametric model, the parameter has both a finite-dimensional component and an infinite-dimensional component. Note that $\{\boldsymbol{\beta}, h_0(u)\}$ are model parameters, where $\boldsymbol{\beta}$ is finite-dimensional and is of interest and $h_0(u)$ is an unknown non-negative function of time. The collection of possible candidates for $h_0(u)$ is infinite-dimensional.

If the analytical form of $h_0(t)$ is known, the conventional MLE can be employed to estimate the regression coefficient vector $\boldsymbol{\beta}$. On the other hand, what makes the PHM so attractive is that $\boldsymbol{\beta}$ can also be estimated by maximizing the corresponding partial likelihood function without specifying $h_0(t)$. Cox (1972) developed a non-parametric method called partial likelihood estimation to estimate the covariate parameters. The partial likelihood estimator is given as

$$L(\boldsymbol{\beta}) = \prod_{i=1}^{d} \left(e^{(\beta^T s(i))} \left(\sum_{r}^{i_r} e^{(\beta^T s(r))} \right) \right) \tag{5.6.3}$$

or, expressed in the logarithmic form,

$$\ln L(\boldsymbol{\beta}) = \sum_{i=1}^{d} \beta^T s(i) - \sum_{i=1}^{d} \ln \left(\sum_{r}^{i_r} e^{(\beta^T s(r))} \right) \tag{5.6.4}$$

where $s(i)$ is the regressor variable associated with the testing samples or items that failed at $t_{(i)}$. The index r refers to the units under test at $t_{(i)}$. Estimation of the parameter values is then obtained by use of the maximum partial likelihood estimation (MPLE). The partial likelihood function is derived by taking the product of the conditional probability of a failure at time t_i, given the number of items that are at risk of failure at that time. Below we use a two-level stress experiment to illustrate how to estimate the regression coefficients based on Eq. (5.6.3).

Example 5.8　To extrapolate the lifetime of lighting emitting diodes (LEDs), 10 units are chosen for a reliability test under accelerated voltage stress. Five items have been applied with $s_1 = 15$ V and the other five items are subjected to $s_2 = 20$ V. The test ends at $t = 60$ days with only one LED surviving at $s_1 = 15$ V. Detailed failure time and censored data are listed in Table 5.3. Note that a censored unit is the one that is removed from the test with no failure.

Let $\phi = \exp(\boldsymbol{\beta}^T \mathbf{s}) = \exp(\beta_1 s_1 + \beta_2 s_2 + \cdots + \beta_n s_n)$. We present the condition probability for testing items at failure risk at any time t. Then at time $t = 7$ days, Item 1 failed at $s_2 = 20$ V, and the remaining nine items are still functioning; hence the failure probability of an item at level s_2 at $t = 7$ is

$$\frac{e^{\beta_2 s_2}}{5e^{\beta_1 s_1} + 5e^{\beta_2 s_2}} \tag{5.6.5}$$

At time $t = 15$ days, Item 2 failed at s_2, and the remaining eight items are still functioning; hence the probability of failure of an item at s_2 and at $t = 15$ is

$$\frac{e^{\beta_2 s_2}}{5e^{\beta_1 s_1} + 4e^{\beta_2 s_2}} \tag{5.6.6}$$

At time $t = 21$ days, Item 3 failed at s_2. Between $28 \leq t \leq 30$ days, Items 4 and 5 have been censored at $s_1 = 15$ V. Hence, prior to $t = 36$ there still exist five items that are still functioning. At $t = 36$, Item 6 failed at $s_1 = 15$ V. Hence the probability of failure of that item is

$$\frac{e^{\beta_1 s_1}}{3e^{\beta_1 s_1} + 2e^{\beta_2 s_2}} \tag{5.6.7}$$

Table 5.3 The experimental outcome of 10 LED devices.

Item no.	Duration t (days)	Voltage (V)	Censored or failure
1	7	20	Failed
2	15	20	Failed
3	21	20	Failed
4	28	15	Censored
5	30	15	Censored
6	36	15	Failed
7	45	20	Censored
8	46	15	Failed
9	51	20	Failed
10	60	15	Censored

This process continues as time evolves. At $t = 51$, only two items are subject to failure risk, namely Item 9 is at $s_2 = 20\,\text{V}$ and Item 10 at $s_1 = 15\,\text{V}$. Hence the probability of failure of Item 9 at $t = 51$ is

$$\frac{e^{\beta_2 s_2}}{e^{\beta_1 s_1} + e^{\beta_2 s_2}} \tag{5.6.8}$$

If we consider all failure instances of $t = 7, 15, 21, 36, 46,$ and 51, the partial likelihood function is obtained as follows:

$$L(\beta_1, \beta_2) = \left(\frac{e^{\beta_2 s_2}}{5e^{\beta_1 s_1} + 5e^{\beta_2 s_2}}\right)\left(\frac{e^{\beta_2 s_2}}{5e^{\beta_1 s_1} + 4e^{\beta_2 s_2}}\right)\left(\frac{e^{\beta_2 s_2}}{5e^{\beta_1 s_1} + 3e^{\beta_2 s_2}}\right)$$
$$\times \left(\frac{e^{\beta_1 s_1}}{3e^{\beta_1 s_1} + 2e^{\beta_2 s_2}}\right)\left(\frac{e^{\beta_1 s_1}}{2e^{\beta_1 s_1} + e^{\beta_2 s_2}}\right)\left(\frac{e^{\beta_2 s_2}}{e^{\beta_1 s_1} + e^{\beta_2 s_2}}\right) \tag{5.6.9}$$

or it can be expressed in the logarithmic format

$$\ln L(\beta_1, \beta_2) = 2\beta_1 s_1 + 4\beta_2 s_2 - \ln(5e^{\beta_1 s_1} + 5e^{\beta_2 s_2})$$
$$- \ln(5e^{\beta_1 s_1} + 4e^{\beta_2 s_2}) - \ln(5e^{\beta_1 s_1} + 3e^{\beta_2 s_2})$$
$$- \ln(3e^{\beta_1 s_1} + 2e^{\beta_2 s_2}) - \ln(2e^{\beta_1 s_1} + e^{\beta_2 s_2}) - \ln(e^{\beta_1 s_1} + e^{\beta_2 s_2}) \tag{5.6.10}$$

To find the covariate coefficient, we take the partial derivatives with respect to β_1 and β_2, and set them to zero, that is

$$\frac{\partial \ln L(\beta_1, \beta_2)}{\partial \beta_1} = 2s_1 - \frac{5z_1 e^{\beta_1 s_1}}{5e^{\beta_1 s_1} + 5e^{\beta_2 s_2}} - \frac{5z_1 e^{\beta_1 s_1}}{5e^{\beta_1 s_1} + 4e^{\beta_2 s_2}} - \frac{5z_1 e^{\beta_1 s_1}}{5e^{\beta_1 s_1} + 3e^{\beta_2 s_2}}$$
$$- \frac{3z_1 e^{\beta_1 s_1}}{3e^{\beta_1 s_1} + 2e^{\beta_2 s_2}} - \frac{2z_1 e^{\beta_1 s_1}}{2e^{\beta_1 s_1} + e^{\beta_2 s_2}} - \frac{z_1 e^{\beta_1 s_1}}{e^{\beta_1 s_1} + e^{\beta_2 s_2}} \tag{5.6.11}$$

$$\frac{\partial \ln L(\beta_1, \beta_2)}{\partial \beta_2} = 4s_2 - \frac{5s_2 e^{\beta_2 s_2}}{5e^{\beta_1 s_1} + 5e^{\beta_2 s_2}} - \frac{4s_2 e^{\beta_2 s_2}}{5e^{\beta_1 s_1} + 4e^{\beta_2 s_2}} - \frac{3s_2 e^{\beta_2 s_2}}{5e^{\beta_1 s_1} + 3e^{\beta_2 s_2}}$$
$$- \frac{2s_2 e^{\beta_2 s_2}}{3e^{\beta_1 s_1} + 2e^{\beta_2 s_2}} - \frac{s_2 e^{\beta_2 s_2}}{2e^{\beta_1 s_1} + e^{\beta_2 s_2}} - \frac{s_2 e^{\beta_2 s_2}}{e^{\beta_1 s_1} + e^{\beta_2 s_2}} \tag{5.6.12}$$

Given $s_1 = 15$ V and $s_2 = 20$ V, a Matlab program is developed to search for the solution based on the above equation system, and results are $\beta_1 = 0.013$ and $\beta_2 = 0.023$.

5.6.2 PH Model with Weibull Hazard Rate

In certain semi-parametric acceleration models, the form of degradation path or lifetime distribution is specified or at least partially specified (Kobbacy et al. 1997, Wang and Kececioglu 2000). Depending on the actual distributions, these models can be classified as: (i) Weibull PHM; (ii) logistic regression model; and (iii) log-logistic regression model. These models will be elaborated in the next three sections.

When $h_0(t)$ in Eq. (5.6.1) resumes a Weibull baseline hazard function, the PM model is referred to as the Weibull proportional hazard model (WPHM). The mathematical form is resumed as follows (Jardine et al. 1987):

$$h(t; s) = \frac{\beta}{\eta} \left(\frac{t}{\eta} \right)^{\beta-1} \exp(\beta^T s) \tag{5.6.13}$$

By maximizing this likelihood function, regression coefficients β and baseline hazard rate parameters $\{\beta, \eta\}$ in the model are estimated. Gorjian et al. (2009) discussed the advantages and limits of the Weibull PH model. The key advantages include:

1) Explanatory variables have a multiplicative effect, rather than an additive effect, on the baseline hazard function. Thus it is a more realistic and reasonable assumption.
2) The model can be used to investigate the effects of various explanatory variables on the life length of assets or items that possess increasing, decreasing, or constant failure rates.

According to Gorjian et al. (2009), two main limitations of the WPHM are:

1) Due to multicollinearity, estimated values of regression coefficients (β) are sensitive to omission, misclassification, sample size, and time dependence of explanatory variables.
2) Proportionality assumption imposes a severe limitation. This means survival curves for assets and individual components can never cross, even if they have different covariates.

5.6.3 Logistic Regression Model

The logistic regression model, also called a logit model, is an ordinal regression model that is a special case of a proportional odds model (POM). It is a regression model for ordinal dependent variables and was first considered by McCullagh (1980). A logistic regression model is usually adopted to relate the probability of an event to a set of covariates. This concept can be used in degradation analysis. If the current degradation features are $s(t)$, the odds ratio between the reliability function $R(t; s(t;))$ and the cumulative distribution function can be defined as follows (Liao et al. 2006):

$$\frac{R(t, s)}{1 - R(t, s)} = \exp(a + \beta s) \tag{5.6.14}$$

where both $\alpha > 0$ and β are the model parameters to be estimated. Therefore, the reliability function can be expressed as

$$R(t; s) = \frac{\exp(a + \beta s)}{1 + \exp(a + \beta s)} \tag{5.6.15}$$

The maximum likelihood function for the model parameters can be obtained by maximizing the log-likelihood function using the Nelder–Mead algorithm.

The advantages and limits of the logistic PH model are summarized as given by Gorjian et al. (2009). The key advantage is that compared with the PHM model, it requires less computation effort to estimate the parameter of the logistic regression model based on its likelihood function. There are two limitations of the model: (i) unlike POM, the model in Eq. (5.6.14) assumes a specific distribution; and (ii) to estimate parameters and evaluate the reliability function, this model takes into account only the current covariates, whereas PHM incorporates both current and historical covariates.

5.6.4 Log-Logistic Regression Model

The log-logistic regression model is a special case of POM when a log-logistic distribution is assumed for the failure times. The log-logistic regression model is described in which the hazard for separate samples converges with time. Therefore, this provides a linear model for the log odds on survival by any chosen time. This model is developed to overcome some shortcomings of the Weibull distribution in the modeling of failure time data, including time-varying failure rates with an up-and-down profile.

The distribution used frequently in the modeling of survival and failure time data is the Weibull distribution. However, its application is limited by the fact that its hazard must be monotonic (either increasing or decreasing), whatever the values of its parameters. Bennett (1983) claims that the Weibull distribution may be inappropriate where the course of the failure (e.g. disease in individuals) is such that mortality reaches a peak after some finite period and then slowly declines. To characterize this type of non-monotonic pattern, the following hazard function of a log-logistic regression model is proposed by Bennett (1983):

$$h(t; s) = \frac{\delta}{t(1 + t^{-\delta} \exp(-\beta s))} \tag{5.6.16}$$

where δ is a measure of precision and β is a measure of location. The hazard is assumed to be increasing first and then decreasing with a change at the time, which has its maximum value at

$$t = 1 - \delta \exp(-\beta s)^{1/\delta} \tag{5.6.17}$$

The parameters of the log-logistic regression model can be estimated by maximizing the likelihood function. The ratio of the hazard for a covariate s takes two values s_1 and s_2 is given by Bennett (1983):

$$\frac{h(t; s_1)}{h(t; s_2)} = \frac{1 + t^{-\delta} \exp(-\beta_1 s_1)}{1 + t^{-\delta} \exp(-\beta_2 s_2)} \tag{5.6.18}$$

If we let t increase, it converges to unity. Gorjian et al. (2009) summarize the advantages and limits of the log-logistic PH model. The key advantages are: (i) it is more suitable to apply in the analysis of survival data rather than a lognormal distribution; (ii) the model extremely fits the applications where hazard reaches a peak after some finite period, and then slowly declines; and (iii) unlike the Weibull PH model, the hazard for different samples is not proportional for a log-logistic regression model – rather their ratio trends to unity for large t, as shown in Eq. (5.6.18). Thus, this property implies that the survival probabilities of different groups of asset/components become more similar as t increases. The main limitation of the model is that it assumes a specified distribution.

5.7 Highly Accelerated Stress Screening Testing

5.7.1 Reliability with HASS Versus Non-HASS

HASS stands for highly accelerated stress screening, and it is a treated as the production equivalent of HALT. It is a post-production test activity with coverage of 100% finished units. The stress intensity is typically half that used in accelerated life testing. The stress levels in HASS are near the operating limits, creating enough fatigue to precipitate latent or hidden failures owing to poor workmanship issue. As such it prevents faulty units from entering the higher level of assembly or end customer. Meanwhile, HASS can be treated as an aggressive version of ESS in which the thermal and vibration stresses are typically set near the limit of product specification (see Figure 5.1).

Manufacturers dealing with "fast-to-market" products are often under tremendous pressure in order to gain the market share, ward off competition, and meet reliability requirements. In the product development and prototype stage, an intensive and prolonged reliability growth test (e.g. HALT) is infeasible because of a compressed design schedule. HASS can be considered as an alternative approach to improving the reliability under fast-to-market pressure. To implement HASS, it is important to convince the management team that the return-on-investment of HASS, though it may extend product delivery time, could be larger than the products with no HASS. Below we present a framework to estimate the benefit of HASS in terms of reliability growth and financial savings of a new product introduction.

Figure 5.15 shows a typical service flow diagram for repairable systems during the product introduction period. New products are produced in the factory and are

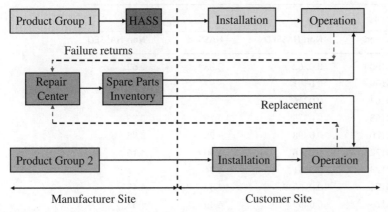

Figure 5.15 HASS versus non-HASS product flow.

shipped and installed at customer sites. To assess the effectiveness of HASS, all shipped products are divided into two groups, one that is subjected to the HASS process and the other without HASS (referred to as "Non-HASS" hereinafter). Two groups of products are continuously released into the market under these categories. Further assume that the manufacturer negotiates with the customers to achieve a desired mean-time-between-failure (MTBF) over a promised period of time (e.g. six months or one year). Failures are returned and repaired in the factory's repair shop. Meanwhile a good unit from the inventory is delivered to replace the field failure. Let suffixes "*H*" and "*NH*" indicate HASSed and non-HASSed groups. Assuming that the failure intensity rate is the one in the Crow/AMSAA model, we have

$$u_H(t) = \lambda_1 \beta_1 t^{\beta_1 - 1} \tag{5.7.1}$$
$$u_{NH}(t) = \lambda_2 \beta_2 t^{\beta_2 - 1} \tag{5.7.2}$$

where $\{\lambda_1, \beta_1\}$, and $\{\lambda_2, \beta_2\}$ are the model parameters corresponding to two groups, respectively. Based on the respective failure intensity rates of product groups with and without HASS, the rate difference is obtained as follows:

$$\Delta u(t) = u_{NH}(t) - u_H(t) = \lambda_1 \beta_1 t^{\beta_1 - 1} - \lambda_2 \beta_2 t^{\beta_2 - 1} \tag{5.7.3}$$

Considering that HASS eliminates infant mortality, we would expect the difference between HASSed and non-HASSed groups to be positive, i.e. $\Delta u(t) > 0$ in Eq. (5.7.3).

Example 5.9 Two batches of printed circuit board (PCB) modules are produced and shipped to customers. The size of the batches are $n_1 = 250$ and $n_2 = 100$ in which Group 1 experienced HASS and Group 2 did not. The Crow/AMSS model is applied to estimate the failure intensity of both production batches, and the results are given as: $\{\lambda_1 = 0.012, \beta_1 = 1.05\}$ and $\{\lambda_2 = 0.01, \beta_2 = 1.2\}$ for the two groups. Estimate how many failures are expected from each group from $t = 50$ to 700?

Solution:
We use Eqs. (5.7.1) and (5.7.2) to compute the failure intensity of a single PCB with HASS and without HASS, and the results are presented in Table 5.4. The intensity without

Table 5.4 Failure intensity of HASSed and non-HASSed products.

Time (t)	Single PCB		Group PCB	
	HASSed	Non-HASSED	HASSed	Non-HASSED
50	0.0153	0.0262	3.83	2.62
100	0.0159	0.0301	3.97	3.01
200	0.0164	0.0346	4.11	3.46
300	0.0168	0.0375	4.19	3.75
400	0.0170	0.0398	4.25	3.98
500	0.0172	0.0416	4.30	4.16
600	0.0173	0.0431	4.34	4.31
700	0.0175	0.0445	4.37	4.45
800	0.0176	0.0457	4.40	4.57
900	0.0177	0.0468	4.43	4.68
100	0.0159	0.0301	3.97	3.01

HASS is 1.7–2.5 times that of the HASSed product. Next we compute the group failure intensity by multiplying $u_H(t)$ with n_1, so do $u_{NH}(t)$ and n_2; the results are listed in the left two columns. Though Group 1 possesses a higher failure intensity for $50 \leq t \leq 600$, Group 2 eventually exceeds Group 1 from $t = 700$. This is to consider the fact that the size of Group 1 is 1.5 times larger than Group 2. Therefore, HASS significantly reduces the field returns, especially for a medium or large installed base of products.

5.7.2 Financial Justification of HASS

In general all business activities are profit-driven and accordingly implementing HASS is not different from any business decisions. As such it must be financially justifiable from the management team perspective. It is relatively easy to estimate the cost of the HASS process, and the comparisons can be made between HASSed products and non-HASSed products in terms of cost savings. Assumptions for developing the financial model are made as follows:

- HASS will eliminate infant mortality by identifying and remedying the causes for early failures due to design weakness, quality defects, and process issues. See Silverman (2000) and Doganaksoy (2001).
- Field failure data and repair analysis reports are used to revise the product design and manufacturing processes to iron out all causes for no fault found (NFF) failures (Janamanchi and Jin 2010). The investigation by Jones and Hayes (2001) indicates that the complexity of electronic equipment, customer usages, and the operating environment are the potential causes for NFF, yet none of them can be effectively eliminated by HASS.
- Products with or without HASS are manufactured using the continually improving and evolving product design and production processes. This assumption actually occurs in the manufacturing industries where information systems such as the Failure Reports and Corrective Actions System (FRACAS) are employed to systematically track failure modes, identify critical issues, and allocate resources to improve the reliability of field products.

There are four major cost items associated with HASS implementation: labor, material, facility, and opportunity costs. Material costs refer to the consumables (e.g. liquid nitrogen, electricity), replacement components, and other incidental materials used. Investment in testing beds like environmental chambers and its depreciation (or lease or rentals) are all assumed to be part of the facility costs. Opportunity cost is referred to as the loss of goodwill because a HASS test likely will postpone the product delivery time. Other expenses such as training and documentation costs may also be significant, but for the model simplicity these costs are considered to be insignificant and as such are not included. Therefore, we have

$$C_{HASS} = C_l + C_m + C_f + C_o$$

where

C_l = labor cost per item
C_m = material cost per item
C_f = facility cost per item
C_o = opportunity cost per item

The manufacturer incurs the cost of field failures that take place within the warranty period. There are three major cost components of field failures: repair costs (C_r), shipping/logistics costs (C_s), and inventory holding costs (C_h). Repair costs typically consist of material cost, labor cost, and overhead cost at the repair shop. Shipping costs include the delivery of spare parts to the customer sire for replacement and the returning of defective items. Inventory holding costs include the cost of holding the spare parts in the stockroom and the defective items in the repair pipeline. During the time interval $[t, t + \tau]$, the cost savings C_{SAV} on account of HASS can be expressed as

$$
\begin{aligned}
C_{SAV}(\tau; t) &= \int_t^{t+\tau} (C_r + C_s + C_h)\Delta u(x)dx \\
&= \int_t^{t+\tau} (C_r + C_s + C_h)(\lambda_2\beta_2 x^{\beta_2-1} - \lambda_1\beta_1 x^{\beta_1-1})dx \\
&= (C_r + C_s + C_h)(m_2(t, t + \tau) - m_1(t, t + \tau))
\end{aligned}
\tag{5.7.4}
$$

where

$$
m_k(t, t + \tau) = \int_t^{t+\tau} \lambda_k\beta_k x^{\beta_k-1}dx = \lambda_k((t + \tau)^{\beta_k} - t^{\beta_k}), \qquad \text{for } k = 1 \text{ or } 2
\tag{5.7.5}
$$

Notice that $k = 1$ is for HASSed and $k = 2$ is for non-HASSed products. If during the $[t, t + \tau]$, τ_0 time ($\tau_0 \le \tau$) is used for HASS, then to justify the continuation of HASS, we need the factory repair cost to be higher than HASS costs for the relevant period. By comparing Eqs. (5.7.4) and (5.7.5), we have

$$
C_{SAV}(\tau; t) > \tau_0(C_l + C_m + C_f + C_o)
\tag{5.7.6}
$$

It then follows that, if C_f and C_l are incurred more as fixed costs rather than variable costs, then they have to be accounted for accordingly and the use of ($\tau_0 \le \tau$) has no meaning. In other words, idle time of resources set aside for repairing field failures needs to be absorbed in the factory repairs costs. In effect, we would be using τ as a factor instead of τ_0 to compute the variable cost component of repairs, while accounting the prorated fixed costs for the relevant period of τ as follows:

$$
C_{SAV}(\tau; t) > \tau(C_l + C_m + C_f)
\tag{5.7.7}
$$

Example 5.10 Two batches of products are shipped and installed in the field. One is subject to HASS and the other is not. Each batch has 10 items. The failure intensity post-HASS is $\lambda_1 = 0.0001$ failures/hour and $\beta_1 = 1.2$. The item without HASS is $\lambda_2 = 0.0001$ failures/hour and $\beta_2 = 1.4$. The costs associated with HASS are $C_l = \$150$/item, $C_m = \$120$/item, $C_f = \$250$/item, and $C_o = \$5000$. The costs associated with repairing a failed item are $C_r = \$250$/item, $C_s = \$160$/item, and $C_h = \$150$/item. Assuming that the HASS is implemented at $t = 0$, estimate the breakeven point of τ such that HASS becomes more economical than non-HASSed products.

Solution:
The HASS cost for one item of product is
$$C_{HASS} = 150 + 120 + 250 + 5000 = \$5520/\text{item}$$
Therefore, the total HASS cost for 10 items is $10 \times C_{HASS} = \$55\,200$.

Next we estimate the failure reduction between $[0, \tau]$ as a result of HASS. According to Eq. (5.7.5), we have

$$\Delta m = m_2(t, t + \tau) - m_1(t, t + \tau) = \lambda_2 t^{\beta_2} - \lambda_1 t^{\beta_1} = 0.0001\tau^{1.4} - 0.0001\tau^{1.1}$$

where Δm is the amount of field failures avoided during $[0, \tau]$ for $t = 0$. According to Eq. (5.7.4), the aggregate cost saving is

$$C_{SAV}(\tau; t) = (C_r + C_s + C_h) \times \Delta m = (250 + 160 + 150) \times 0.0001 \times (\tau^{1.4} - \tau^{1.1})$$

Setting $C_{SAV}(\tau; t) = \$55\,200$, we solve for $\tau = 3927$ hours, meaning if the warranty period is longer than 3927 hours, it is more economical to implement the HASS test because the savings due to reduced field returns are higher than the repair costs under no HASS.

For new products that have not been released in the market, manufacturers can predicts the warranty cost based on the lifetime extrapolated from the HALT process, instead of using field data. Recently Zhao and Xie (2017) present a framework to predict the warranty cost and risk under a one-dimensional warranty policy by analyzing HALT data. Two sources of variability typically arise: the uncertainty of estimated parameters and the variation in field conditions. The consideration of these variabilities is necessary to avoid the underestimation of the warranty cost, especially during the new product introduction phase.

5.8 A Case Study for HASS Project

5.8.1 DMAIC in Six Sigma Reliability Program

Company ABC designs and markets high-end testing equipment for wafer probing and device test in semiconductor manufacturing sector. A system is usually configured with 30–40 PCB modules depending on the functional requirements. While ABC designs the PCB, the manufacturing of these boards are subcontracted to external suppliers. Upon receiving the PCB by ABC's assembly factory, each board undergoes optional testing, system configuration, and system test before customer shipment and installation. If a PCB fails in any one of the in-house processes, it is returned to the repair shop for root-cause analysis. When fixed, the board is routed to "Incoming Stock" again. After the system is shipped and installed in the field, the system will be tested for one more time. If a PCB failed in the field test, it is returned to the repair shop. When fixed, it is routed to the "Incoming Stock" as well. The flowchart in Figure 5.16 shows the subcontract manufacturing process, system configuration, in-house testing, field installation, and final test of new systems.

Jin et al. (2011) proposed a closed-loop HASS program with the objective to implement reliability growth initiatives in a distributed manufacturing and service chain. As shown in Figure 5.17, the proposed program consists of six functional blocks to form a Six Sigma reliability control scheme through DMAIC: Define, Measure, Analyse, Improve, and Control. DMAIC is a data-driven improvement strategy to optimize and stabilize the business processes of a new product introduction. Tang et al. (2007) emphasize the importance of incorporating operations research and management science techniques for enhancing the effectiveness of DMAIC methodology. It is the core tool to orchestrate and guide Six Sigma projects in manufacturing industries.

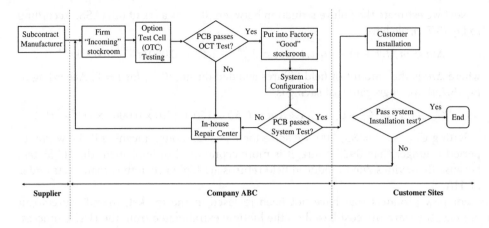

Figure 5.16 Process map from subcontractor, in-house testing to field installation.

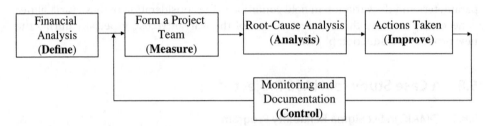

Figure 5.17 DMAIC in closed-loop Six-Sigma reliability growth (Jin et al. 2011).

5.8.2 Define – Financial Analysis and Project Team

5.8.2.1 Financial Analysis

The Six Sigma program was motivated by customer satisfaction, but the potential cost savings resulted from the improved process is the actual incentive. The resources involved in a repair process include shipping logistics, labor, materials, and testing facilities. To assess the cost benefit of the Six Sigma project, a Monte Carlos simulation program was developed to estimate the cost savings assuming that the infant mortality rate (IMR) could be reduced from the current 14 % to 7% or 3%, and the results are summarized in Table 5.5.

For example, if IMR is down to 7%, the average saving over three months would be $358 209, and $1 429 820 over 12 months. The simulation data allows us to estimate the mean and standard deviation of the savings. At 90% confidence level, the optimistic savings (OSs) and the pessimistic savings (PSs) over the three-month period are given as

$$OS = 358\,209 + Z_{1-0.05} \times 63\,692 = \$462\,972 \tag{5.8.1}$$

$$PS = 358\,209 - Z_{0.05} \times 63\,692 = \$253\,445 \tag{5.8.2}$$

The upper and lower cost savings are very important because they comprehend the uncertainties of the SixSigma program in actual implementations. Similar interpretations can be applied to the potential cost savings under $IMR_a = 3\%$.

Table 5.5 Potential cost savings of reducing infant mortality rates (IMR) (unit: $).

Target rate	Time (month)	$T = 3$	$T = 6$	$T = 9$	$T = 12$
$IMR_a = 7\%$	Average	358 209	716 613	1 074 189	1 429 820
	Standard deviation	63 692	90 500	110 737	129 480
	Optimistic	462 972	865 473	1 256 336	1 642 796
	Pessimistic	253 445	567 754	892 043	1 216 845
$IMR_a = 3\%$	Average	564 734	1 128 324	1 696 137	2 259 437
	Standard deviation	102 049	147 199	180 477	206 932
	Optimistic	732 590	1 370 445	1 992 996	2 599 810
	Pessimistic	396 878	886 202	1 399 278	1 919 064

Source: Reproduced with permission of Elsevier.

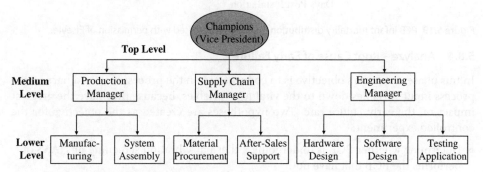

Figure 5.18 Formation of a cross-functional team.

5.8.2.2 Forming a Cross-Functional Team

Effective implementation of Six-Sigma programs requires the formation of a cross-functional team spanning engineering, operations, supply, marketing, and field services. Figure 5.18 shows the hierarchical structure in which members from different departments are selected both vertically and horizontally to form the Six-Sigma team. For example, the members from engineering are responsible for eliminating design weakness, correcting software bugs, and updating the software versions. Operations engineers are responsible for manufacturing, handling, and installation issues. Elimination of NFF failures is challenging because various reasons cause NFF returns. One approach is that field engineers work closely with the customer to collect the onsite failure signature, and then send it to the repair center along with the defective part. Finally, the market engineer is able to design the financial metrics to gauge the cost benefit based on the savings upon the implementation of the Six-Sigma program.

5.8.3 Measure – Infant Mortality Distribution

Based on the system installation dates and the defective PCB return time, Pareto charts are generated to visualize the early PCB failures, and the results are shown in Figure 5.19. The Pareto chart shows that among all the failures, 34% occurred within 30 days of installation, and 18% of the failures occurred within 10 days of installation. The failure trend strongly supports the hypothesis that the current PCB manufacturing and testing process needs to be re-examined so as to reduce the IMR.

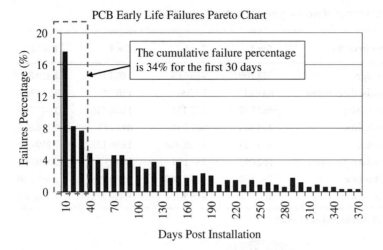

Figure 5.19 PCB infant mortality distribution. Source: Reproduced with permission of Elsevier.

5.8.4 Analyze – Root Cause of Early Failures

In this phase, the team's objective is to concentrate on the process map and narrow the process input variables down to the vital few variables, because they have the greatest impact on the early failure rate. Two hypotheses are created as the objective for the controlled experiment:

- Determine whether PCB is undergoing a thermal/vibration screening or HASS outperforms the PCB that have not.
- Determine whether PCB is undergoing multiple power cycling tests or outperforms PCB that have not.

Two controlled batches of PCB with 100 each were used in the controlled experiment. The first batch of PCB underwent the thermal/vibration screening and power cycling before they were delivered to the system configuration. The second batch of PCB, after being received from subcontract manufacturers, go directly into the system integration without going through HASS and power cycling tests. To minimize the customer usage effects, these boards were mixed randomly and configured into new systems being shipped to different customers. The reliability of these PCB can be easily tracked by the board's unique series number.

After the 30-day field operation, failure reports of both batches revealed that there were eleven failures among 100 non-HASS and non-Power Cycle boards (11%). Only three failures were observed from 100 HASSed and Power Cycled boards (3%). A two-proportion hypothesis test (Gupta 2004) was performed comparing the two batches of PCB populations. The hypothesis of the test is given below:

$$H_0 : p_1 = p_2 \text{ or } H_1 = p_1 > p_2 \tag{5.8.3}$$

The test statistics are

$$Z = \frac{\hat{p}_1 - \hat{p}_2}{\sqrt{\hat{p}(1 - \hat{p})(1/n_1 + 1/n_2)}} \tag{5.8.4}$$

$$\hat{p} = \frac{n_1\hat{p}_1 + n_2\hat{p}_2}{n_1 + n_2} \tag{5.8.5}$$

Here p_1 and p_2 represents the failure rate for both batches, respectively. In addition, $\hat{p}_1 = x_1/n_1$ and $\hat{p}_2 = x_2/n_2$ are the corresponding estimates. In our example, $x_1 = 11$, $x_2 = 3$, $n_1 = n_2 = 100$. The two-proportion hypothesis test revealed that we could conclude with 98% confidence that the HASSed and Power Cycled Tested population performed much better than those without HASS and Power Cycled during the first 30-day operational period.

5.8.5 Improve – Action Taken

Based on the result of hypothesis test, the recommendations were given by the team to implement a one-day HASS process followed by automated power cycling tests in subcontract manufacturers. Specifically, the improvement plan highlights the following action items:

- Establish an ad-hoc team consisting of mangers and engineers between the host company and each subcontract manufacturer to standardize the improvement process. Agreement should be reached among cost sharing and benefits resulting from the new process.
- Implement HASS and power cycling tests in all subcontract manufacturers within three to five weeks.
- Develop an automated database visible to relevant managers and engineers of the host company and the subtract manufacturers to track the implementation progress.

5.8.6 Control – Monitoring and Documentation

A control plan is developed to monitor and respond to any issues arising from the key inputs and outputs of the new process. The plan included two core components:

- Process Implementation Qualification Report (PIQR) – This report is used to evaluate the HASS and Power Cycling tests conducted by the subcontract manufacturers. The report documents the physical observations and quantitative data related to the process inputs and outputs. If variations arise from the process, the team will take corrective actions along with the subcontract manufacturers.

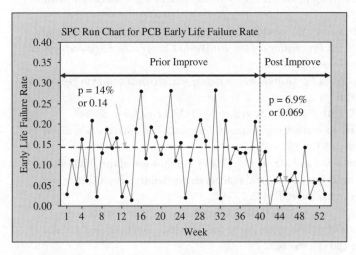

Figure 5.20 Tracking the infant mortality rate. Source: Reproduced with permission of Elsevier.

- Generate and update a statistical process control (SPC) chart for the PCB IMRs – The purpose of the SPC chart is to monitor the reliability performance of the PCB coming from the subcontract manufacturers in a time sequence. The SPC chart in Figure 5.20 shows that the IMR is reduced from 14% to 6.9% in week 40, and remains at the low level thereafter showing the effectiveness of the DMAIC procedure.

References

Arrhenius, S.A. (1889). Über die dissociationswärme und den einfluß der temperatur auf den dissociationsgrad der elektrolyte. *Zeitschrift Für Physikalische Chemie* 4: 96–116. doi: 10.1515/zpch-1889-0108.

Bennett, S. (1983). Log-logistic regression models for survival data. *Applied Statistics* 32 (2): 165–171.

Boccaletti, G., Borri, F.R., D'Esponosa, F., and Ghio, E. (1989). Accelerated tests. In: *Microelectronic Reliability II. Integrity Assessment and Assurance* (ed. E. Pollino). Chapter 11. Norwood, MA: Artech House.

Cox, D.R. (1972). Regression models and life tables (with discussion). *Journal of the Royal Statistical Society, Series B* 34: 187–220.

Doganaksoy, N. (2001). HALT, HASS and HASA explained: accelerated reliability techniques by Harry McLean. *Technometrics* 43 (4): 489–490.

Elsayed, E. (2012). *Reliability Engineering*, Chapter 6, 2e. Hoboken, NJ: Wiley.

Elsayed, E. and Chan, C.K. (1990). Estimation of thin-oxide reliability usin g proportional hazards model. *IEEE Transactions on Reliability* 39 (3): 329–335.

Elsayed, E. and Jiao, L. (2002). Optimal design of proportional hazards based accelerated life testing plan. *International Journal of Materials and Product Technology* 17 (5/6): 411–424.

Escobar, L.A. and Meeker, W.Q. (2006). A review of accelerated test model. *Statistical Science* 21 (4): 552–577.

Gorjian, N., Ma, L., Mittinty, M., Yarlagadda, P., Sun, Y., (2009), "A review on reliability models with covariates," in *Proceedings of the 4th World Congress on Engineering Asset Management*, Athens, Greece, pp. 1–6.

Gupta, P. (2004). *The Six Sigma Performance Handbook: A Statistical Guide to Optimizing Results*, 1e. McGraw-Hill.

Janamanchi, B. and Jin, T. (2010). Reliability growth vs. HASS cost for product manufacturing with fast-to-market requirement. *International Journal of Productivity and Quality Management* 5 (2): 152–170.

Jardine, A.K.S., Anderson, P.M., and Mann, D.S. (1987). Application of the Weibull proportional hazards model to aircraft and marine engine failure data. *Quality and Reliability Engineering International* 3 (2): 77–82.

Jin, T., Janamanchi, B., and Feng, Q. (2011). Reliability deployment in distributed manufacturing chain via closed-loop six sigma methodology. *International Journal of Production Economics* 130 (1): 96–103.

Johnston, D.R., LaForte, J.T., Podhorez, P.E., and Galpern, H.N. (1979). Frequency acceleration of voltage endurance. *IEEE Transactions on Electrical Insulation* 14 (3): 121–126.

Jones, J. and Hayes, J. (2001). Investigation of the occurrence of: no-faults-found in electronic equipment. *IEEE Transactions on Reliability* 50 (3): 289–292.

Klinger, D. J. (1991a). "On the notion of activation energy in reliability: Arrhenius, Eyring and thermodynamics," in *Proceedings of Annual Reliability and Maintainability Symposium*, New York, NY, pp. 295–300.

Klinger, D.J. (1991b). Humidity acceleration factor for plastic packaged electronic devices. *Quality and Reliability Engineering International* 7 (5): 365–370.

Kobbacy, K.A.H., Fawzi, B.B., Percy, D.F., and Ascher, H.E. (1997). A full history proportional hazards model for preventive maintenance scheduling. *Quality and Reliability Engineering International* 13 (4): 187–198.

Liao, H., Zhao, W., Guo, H., (2006), "Predicting remaining useful life of an individual unit using proportional hazards model and logistic regression model," in *Proceedings of Annual Reliability and Maintainability Symposium*, pp. 127–132.

Mann, N.R., Schafer, R.E., and Singpurwalla, N.D. (1974). *Methods for Statistical Analysis of Reliability and Life Data*. New York, NY: Wiley.

McCullagh, P. (1980). Regression models for ordinal data. *Journal of the Royal Statistical Society, Series B* 42 (2): 109–142.

Miller, R. and Nelson, W. (1983). Optimum simple step-stress plans for accelerated life testing. *IEEE Transactions on Reliability* R-32 (1): 59–65.

Nelson, W. (1990). *Accelerated Testing Statistical Models: Test Plans and Data Analyses*. New York, NY: Wiley.

NIST (National Institute of Standards and Technology), *Engineering Statistics Handbook*, Chapter 8. Available at: http://www.itl.nist.gov/div898/handbook/apr/section1/apr152 .htm (accessed on September 2017)

O'Quigley, J. (2008). *Proportional Hazard Regression*. New York, NY: Springer.

Pakbaznia, E., Pedram, M., (2009), "Minimizing data center cooling and server power costs," In *ISLPED'09 Proceedings of the 2009 ACM/IEEE International Symposium on Low Power Electronics and Design*, pp. 145–150, San Fancisco, CA, August 19–21, 2009.

Silverman, M. (2000), "HASS development method: screen development, change schedule, and re-prove schedule," in *Proceedings of the Annual Reliability and Maintainability Symposium*, pp. 245–247.

Tang, L.C., Goh, T.N., Lam, S.W., and Zhang, C.W. (2007). Fortification of Six Sigma: expanding the DMAIC toolset. *Quality and Reliability Engineering International* 23 (1): 3–18.

Wang, W. and Kececioglu, D.B. (2000). Fitting the Weibull log-linear model to accelerated life-test data. *IEEE Transactions on Reliability* 49 (2): 217–223.

Yan, L. and English, J.R. (1997). Economic cost modeling of environmental-stress-screening and burn-in. *IEEE Transactions on Reliability* 46 (2): 275–282.

Yang, G. (2007). *Life Cycle Reliability Engineering*, Chapter 7, 1e. Hoboken, NY: Wiley.

Yang, G.B. (1994). Optimum constant-stress accelerated life-test plans. *IEEE Transactions on Reliability* 43 (4): 575–581.

Yang, K. and Yang, G. (1999). Robust reliability design using environmental stress testing. *Quality and Reliability Engineering International* 14 (6): 409–416.

Ye, Z.-S., Murthy, D.N.P., Xie, M., and Tang, L.-C. (2013). Optimal burn-in for repairable products sold with a two-dimensional warranty. *IIE Transactions* 45 (2): 164–176.

Zhao, X. and Xie, M. (2017). Using accelerated life tests data to predict warranty cost under imperfect repair. *Computers and Industrial Engineering* 107 (5): 223–234.

Problems

Problem 5.1 State the difference between testing specifications of HALT and HASS.

Problem 5.2 List the common stresses that are used in HALT or HASS tests.

Problem 5.3 Across the product lifecycle, which ones are subject to HAL and HASS tests: early design and prototype, new product introduction period, volume production, and end of product life.

Problem 5.4 Is ESS a type of HASS test? What are the main differences between ESS and HASS?

Problem 5.5 Burn-in, thermal cycling, thermal shock, power cycling, and voltage margining all belong to the HASS process. Give one application (i.e. example) for each test where they are used in product development or manufacturing.

Problem 5.6 The following stress profiles are commonly applied: sinusoid, step-stress, dwell and ramp-up, and cyclic stress. If the stress types are temperature, please explain which of these stress profiles are more applicable to temperature in reality? What about other stress types like vibration, voltage, power cycling, and force?

Problem 5.7 Plot the PDF of a lognormal distribution at $A_f = 10$ and 100 given $\mu_n = 10$ and $\sigma_n = 3$.

Problem 5.8 The failure time of the FPGA (field programmable gate array) follows the gamma distribution and the PDF is given by

$$f(t) = \frac{\lambda e^{-\lambda t}(\lambda t)^{\theta-1}}{\Gamma(\theta)}, \qquad \text{for } t \geq 0$$

where λ is the scale parameter and θ is the shape parameter. Assuming that the FPGA is subject to the HALT test and the relation between the normal and stressed levels are governed by $E[T_n] = A_f E[T_s]$, where T_n and T_s are the normal and stressed life and A_f (for $A_f > 1$) is the acceleration factor, do the following:

1) Let λs be the scale parameter at the stress level. Estimate λn, the scale parameter at the normal condition, assuming that θ remains constant.
2) Find the hazard rate function in the normal condition.

3) Given $\lambda_s = 0.01$ failure/hour, $\theta = 2.5$, and $A_f = 200$. If the FPGA has tested for 20 hours at the stressed level, what is its equivalent operating hours at the normal condition?

Problem 5.9 The lifetime of a bearing can be modeled as a two-parameter Weibull distribution $\{\theta, \beta\}$. The bearing is tested under an accelerated stress condition by applying accelerated forces and vibration stresses, and the reliability reaches 0.65 after 100 000 rotations. Assuming that $\theta_n = 4.5\theta_s$, where θ_n and θ_s are the scale parameters in the normal and stressed conditions, respectively, do the following, assuming that $\beta = 3.2$ does not change under the stress:

(1) Estimate the PDF of the bearing life under normal operating conditions.

(2) What is the reliability upon 100 000 cycles in normal use?

(3) Estimate the acceleration factor.

Problem 5.10 Reliability of electronics devices are sensitive to their operating temperature. Under the normal operating temperature of 30 °C, the mean-time-to-failure (MTTF) of a microprocessor is 10 years and the standard deviation is three years. The manufacturer wants to design a HALT test to verify this MTTF value. It is known that the relation between the acceleration factor and temperature is as follows:

$$A_f = 2^{(T_s - T_n)/15}$$

where T_s and T_n are the temperature at the stressed and normal conditions, respectively. Do the following:

1) If the HALT test must be completed in 30 days, what is the minimum temperature stress applied?

2) If the device lifetime follows the lognormal distribution, find the PDF under the normal condition and the stressed condition, respectively.

3) If the customer requires that the device must survive for 15 years at 90% confidence, determine the highest operating temperature under which this device can run.

Problem 5.11 Briefly describe the three most common causes of strength degradation of mechanical components. Give an example of each and also provide methods to prevent or reduce the chance of failures.

Problem 5.12 Miner's law is used to predict the expected time-to-failure of mechanical components in fatigue. A component was tested under three stress levels and their mean cycles to failure are listed in the table below. Do the following:

Stress level ($\times 10^6$ N/m^2)	6.5	8	10
Mean cycles to failure ($\times 10^5$)	13	11.5	7.4

(1) The proportion of using these stress level in the field would be 0.7, 0.2, and 0.1, respectively. What is the equivalent life of the component?

(2) Can you estimate the mean cycles to failure under the stress level of $7 \times 10^6 \, \text{N/m}^2$?

(3) If the component operates at a single stress level and the measured cycles to failure is 9.5×10^5, what is the stress level?

Problem 5.13   Suppose an electronic device operating at $T = 40\,°C$ continuously for $t_f = 2.5$ years prior to its failure. Based on the Arrhenius law, estimate E_a of this device assuming $A = 0.1$ hour.

Problem 5.14   For the Arrhenius model, plot the relation between time-to-failure and activation energy at $T = 25\,°C$ and $50\,°C$, respectively, assuming $A = 0.01$.

Problem 5.15   A two-stress Eyring model involves temperature T and voltage V. Assuming $E_a = 0.4 \, \text{eV}$, $B = 1$, and $C = 1$, do the following:

(1) Plot the relationship between the mean lifetime and T for given $A = 0.1$, $m = 0.3$ and $S = 10 \, \text{V}$ and $15 \, \text{V}$.

(2) Plot the relationship between the mean lifetime and S (voltage) for given $A = 0.1$, $m = 0.3$ and $T = 25\,°C$ and $70\,°C$.

(3) Since there is no interaction between T and S, the Eyring model in Eq. (5.5.1) can be expressed as

$$\ln t_f(T, S) = \ln A + m \ln T + \frac{E_a}{kT} + S\left(B + \frac{C}{T}\right)$$

Re-plot subproblems (1) and (2) in logarithmic scale.

Problem 5.16   Consider the following times-to-failure data at two different stress levels.

Stress (V)	Times to failure
20	11.1, 16.3, 18.5, 24.1
36	2.2, 3.1 3.6, 4.7, 4.9, 5.8, 6.1, 6.8, 7.4, 8.2, 9.1, 10.4, 10.7, 12.2 14.3, 15.2

Using the maximum likelihood function, use the Weibull inverse power law to model the parameter (also see http://reliawiki.com/index.php/Inverse_Power_Law_Relationship).

Problem 5.17   The PHM belongs to the statistics-based (or non-parametric) HAT model. Describe the concept of the PHM. State two major differences between the statistical HAL model and physical-experimental based models, such as the Arrhenius and Eyring models.

Problem 5.18   In an accelerated life test, 20 devices are placed in an environmental chamber for constant temperature stressing at $T_1 = 150\,°C$. Two devices failed at $t = 900$ and 960 hours, and the test is terminated at $t = 1000$ hours with 18 devices surviving. Another test is performed on 4 devices under a higher temperature of $T_2 = 310\,°C$. Three devices failed at $t = 450$, 630, and 850 hours. The last device is removed from the test at $t = 700$ hours. Do the following:

(1) Define the covariate of this HAL test and present the PHM.
(2) Formulate the partial likelihood function and solve for the coefficients of corresponding covariates.
(3) Assuming the Weibull baseline hazard rate function with $\theta_o = 1200$ and $\beta = 3.5$ and estimate the hazard rate function at $T_1 = 150\,°C$ and $T_2 = 310\,°C$, respectively.
(4) What is the hazard rate function at temperature $T = 200\,°C$?

Problem 5.19 A group of 20 products spend between 0 and 2.75 days in HASS. Note that "0" means the product failed in three months of field use and "1" means passes through three months. How does the number of hours affect the probability that the product will survive the field during the first three months? Use the logistics regression model to solve this.

The table shows the reliability data of 20 samples in the first three-month field operation.

Days	0.5	0.75	1	1.25	1.5	1.75	1.75	2	2.25	2.5	0.5
Pass	0	0	0	0	0	0	1	0	1	0	0
Days	2.75	3	3.25	3.5	4	4.25	4.5	4.75	5	5.5	2.75
Pass	1	0	1	0	1	1	1	1	1	1	1

Problem 5.20 Two batches of PCB modules will be shipped and installed in the field. The size of the batches are $n_1 = 30$ and $n_2 = 20$ and group 1 products undergo HASS and group 2 do not. The table below lists the field returns of failures after one year installation. Note that hours represent the cumulative operating time of that returned item.

Three HASSed products: 1050, 3700, 7900
Seven non-HASSed products: 330, 540, 890, 1200, 3400, 5700, 8200
Do the following:

(1) Estimate $\{\lambda_1, \beta_1\}$, and $\{\lambda_2, \beta_2\}$ for both product groups using the Crow/AMSS model.
(2) What is the intensity difference between two groups at $t = 100$, 200, and 300 days?
(3) How many failures would be expected from both groups within a year if $n_1 = n_2 = 100$?

Problem 5.21 Two batches of products are installed in the field. One is subject to HASS and the other is not. Each batch has 20 items. The failure intensity post HASS is $\lambda_1 = 0.001$ failures/hour and $\beta_1 = 1.1$. The item without HASS is $\lambda_2 = 0.0012$ failures/hour and $\beta_2 = 1.3$. The costs associated with HASS are $C_l = \$250/item$, $C_m = \$220/item$, $C_f = \$350/item$, and

$C_o = \$7000$. The costs for repairing a failed item are $C_r = \$370/\text{item}$, $C_s = \$260/\text{item}$, and $C_h = \$250/\text{item}$. Assuming that the HASS implemented at $t = 0$, estimate the breakeven point of τ such that HASS is more economical than non-HASSed products.

6

Renewal Theory and Superimposed Renewal

6.1 Introduction

The number of renewals provides valuable information for spares inventory management, queuing analysis, maintenance planning, and warranty services. In this chapter, we introduce some basic theories and models about the renewal process and its applications. There are two different approaches to solving renewal function (RF): the analytical method and numerical computation. In Sections 6.2–6.4, we adopt Laplace transform to search the closed-form solutions when the inter-renewal times follow exponential, generalized exponential, or Erlang distribution. In Sections 6.5 and 6.6, we propose several approximation methods to compute the Weibull and gamma renewals based on the mixture of exponential functions and the sinc function. In Sections 6.7 and 6.8 we extend the renewal function to situations where the installed base of field products varies and are time-dependent. Both the aggregate fleet renewal function and the lead-time renewal model are explicitly derived under deterministic and stochastic fleet expansion. These models are built upon the superimposed renewal processes. In Section 6.9 we present a case study from the wind industry to illustrate the application of the superimposed renewal process.

6.2 Renewal Integral Equation

6.2.1 Overview of Renewal Solution Methods

The renewal theory was originally proposed by Feller (1941) in 1940s. Later it was further developed by Smith (1958) and Cox (1962). Since then, the renewal theory has been widely used for inventory management (Kumar and Knezevic 1998; Moors and Strijbosch 2002), queuing analysis (Cox and Ishman 1980; Elsayed 2012), telecommunication networks (Brown et al. 1981), and warranty services (Jaquette 1972; Murthy and Blischke 1992). The number of renewals is often used to evaluate the product performance such as reliability trend, the mean-time-between-failures (MTBF), spare parts demand, and maintenance cost. In general, the computation of the renewal function (RF) is not an easy task for many lifetime distributions. Analytical solutions are available only for certain types of failure probabilities such as exponential or Erlang distributions. For most empirical distributions, like Weibull, gamma, and normal distributions, approximation methods have to be used to estimate the actual RF values. For instance, Cheng et al.

Reliability Engineering and Services, First Edition. Tongdan Jin.
© 2019 John Wiley & Sons Ltd. Published 2019 by John Wiley & Sons Ltd.
Companion website: www.wiley.com/go/jin/serviceengineering

(2004) proposed a closed form of the Weibull generation function, but the function structure is highly complicated, making it difficult to derive the closed-form RF solution. For a comprehensive review on the RF computation and the approximation, refer to the discussions by Cui and Xie (2003) and Tortorella (2005).

Existing RF approximations can be classified into two categories: numerical computation and analytical approximation. The former calculates the RF data iteratively using computer programming. The latter approximates the RF based on the mathematical structure of the underlying renewal function. Numerical computation methods include the direct Laplace–Stieltjes integration (Xie 1989) and Monte-Carlo simulations (Brown et al. 1981; Liao et al. 2008). Analytical approximations include cubic/spine curves (Cleroux and McConalogue 1976; McConalogue and Pacheco 1981), moment methods (Kambo et al. 2012), generating functions (Smeitink and Dekker 1990; Cui and Xie 2003; Jiang 2008, 2010), and power series functions (Chaudhry 1995; Garg and Kalagnanam 1998). The classification of computational methods of RF is shown in Figure 6.1.

Due to the advancement of computer technology, it is now relatively easy to apply numeral methods to find the very accurate RF solution if the interrenewal time distribution is known explicitly (Jaquette 1972). In fact, studies have shown that numerical methods can achieve accurate approximations up to four decimals by appropriately choosing the grid size (Tortorella 2005; Xie 1989). Existing algorithms used to compute the numerical RF typically involve $O(n^2)$ complexity and also require a large data storage. However, these algorithms could become extremely expensive as t becomes large.

6.2.2 Generic Renewal Function

In many applications, it is still necessary to evaluate the probabilistic quantity directly based on the analytical RF expression. As pointed by Osaki (2002), an analytical function or an approximation model usually makes it more convenient for decision makers to determine multi-item problems simultaneously rather than using numeral solutions. For instance, if one considers the mean interdeparture time in a $G1/G/1$ loss system, then the expected number of renewals in a random interval has to be evaluated analytically.

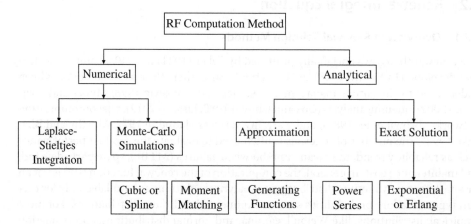

Figure 6.1 Computational methods of renewal functions.

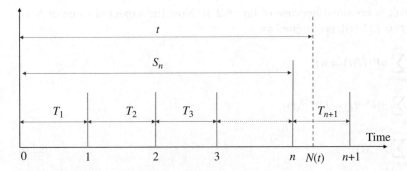

Figure 6.2 Failure sequences of a renewal process.

Hence, it will be meaningful to approach the renewal solutions either numerically or analytically.

Figure 6.2 depicts a sequence of failures that are observed over a period of time. The system is installed and starts to run at $t = 0$. Assume that the downtime is short and negligible compared with the uptime. The first failure occurs at t_1 and the system is restored to an as-good-as-new condition. Upon recovery, the system continues to operate before it fails again at t_2. Then the system is repaired and restored to the as-good-as-new state again for the next operation. Here t_i for $i = 1, 2, 3, \ldots$ are called the failure arrival times and T_i for $i = 1, 2, 3, \ldots$ are called the inter-renewal times. The relation between T_i and t_i are given as follows:

$$T_i = t_i - t_{i-1}, \qquad \text{for } i = 1, 2, 3, \ldots \tag{6.2.1}$$

All T_i in a renewal process are assumed to be identically and independently distributed. Before we present the renewal equation, the following notation is defined:

$N(t) =$ the number of renewals (or failures) by time t, a random variable
$F(t)$ and $f(t) =$ CDF and PDF for T_i for $i = 1, 2, \ldots, N(t)$
$S_n =$ cumulative time for n renewals, a random variable, which is defined as

$$S_n = T_1 + T_2 + \cdots + T_n = \sum_{i=1}^{n} T_i \tag{6.2.2}$$

The probability that the number of failures (renewals) $N(t) = n$ is the same as the probability that t falls in between the nth and $(n+1)$th failures. That is,

$$P\{N(t) < n\} = P\{N(t) \le n - 1\} = P\{S_n > t\} = 1 - F^{(n)}(t) \tag{6.2.3}$$

where $F^{(n)}(t)$ is the cumulative distribution function (CDF) of S_n. One also realizes that

$$P\{N(t) < n\} + P\{N(t) = n\} + P\{N(t) > n\} = 1 \tag{6.2.4}$$

Hence

$$
\begin{aligned}
P\{N(t) = n\} \\
&= 1 - P\{N(t) < n\} - P\{N(t) > n\} \\
&= P\{N(t) < n + 1\} - P\{N(t) < n\} \\
&= P\{S_{n+1} > t\} - P\{S_n > t\} \\
&= F^{(n)}(t) - F^{(n+1)}(t) \tag{6.2.5}
\end{aligned}
$$

The final result is obtained because of Eq. (6.2.3). Now the expected value of $N(t)$, denoted as $M(t) = E[N(t)]$, is obtained as

$$M(t) = \sum_{n=0}^{\infty} nP\{N(t) = n\}$$

$$= \sum_{n=0}^{\infty} n[F^{(n)}(t) - F^{(n+1)}(t)]$$

$$= \sum_{n=1}^{\infty} F^{(n)}(t) \tag{6.2.6}$$

where $F^{(n)}(t)$ is the nth-fold convolution of the underlying failure time distribution $F(t)$, which is recursively defined by

$$\begin{cases} F^{(n)}(t) = \int_0^t F^{(n-1)}(t-x)dF(x), & 0 \le x \le t, \ for \ n = 2, 3, \cdots \\ F^{(1)}(t) = F(t), & t \ge 0, \ for \ n = 1 \end{cases} \tag{6.2.7}$$

Equation (6.2.6) is referred to as renewal equation or RF, which can also be rewritten as

$$M(t) = F^{(1)}(t) + \sum_{n=1}^{\infty} F^{(n+1)}(t)$$

$$= F^{(1)}(t) + \sum_{n=1}^{\infty} \int_0^t F^{(n)}(t-x)dF(x)$$

$$= F^{(1)}(t) + \int_0^t \sum_{n=1}^{\infty} F^{(n)}(t-x)f(x)dx$$

$$= F(t) + \int_0^t M(t-x)f(x)dx \tag{6.2.8}$$

The last step is obtained by realizing $F^{(1)}(t) = F(t)$ and $f(t) = dF(t)/dt$. Since the integral involves the convolution between $M(t)$ and $f(t)$, Eq. (6.2.8) is also called the renewal integral equation.

Let $m(t)$ be the density function of $M(t)$. By taking the derivative with respect to t in Eq. (6.2.8), we have

$$m(t) = \frac{dM(t)}{dt} = \frac{dF(t)}{dt} + M(t-t)f(t) + \int_0^t \left(\frac{dM(t-x)}{dt} \right) f(x)dx$$

$$= f(t) + \int_0^t m(t-x)f(x)dx \tag{6.2.9}$$

The above result is obtained by realizing that $M(t-t) = M(0) = 0$ for $t \ge 0$. Furthermore, given the RF solution $M(t)$, the variance of the RF and its asymptotic value are also available as follows:

$$Var(N(t)) = \sum_{k=1}^{\infty} (2k - 1)F^{(k)}(t) - (M(t))^2 \tag{6.2.10}$$

$$\lim_{t \to \infty} M(t) = \frac{t}{\mu} + \frac{1}{2} \left(\frac{\sigma^2}{\mu^2} - 1 \right) = At + B \tag{6.2.11}$$

with

$$A = \frac{1}{\mu} \quad \text{and} \quad B = \frac{1}{2}\left(\frac{\sigma^2}{\mu^2} - 1\right) \tag{6.2.12}$$

Notice that μ and σ^2 are the mean and the variance of the underlying inter-renewal distribution function $F(t)$. For example, if the inter-renewal times are normally distributed with $\mu = 10$ and $\sigma^2 = 100$, then we have $A = 1/10 = 0.1$ and $B = 0$ based on Eq. (6.2.12). The asymptotic RF is $M(\infty) = 0.1t$. More discussions about $m(t)$, $Var(N(t))$, and the asymptotic value of $M(t)$ are also available in Cox and Isham (1980) and Elayed (2012).

6.2.3 Renewal in Laplace Transform

From Eqs. (6.2.8), it is realized that the renewal integral equation involves the convolution between $M(t)$ and $f(t)$. The integral involving the convolution usually makes the analysis and computation much more complicated than regular integration. However, it is relatively easy to use Laplace transform to handle the convolution in the frequency domain or the s-domain. This is because the convolution of two functions in the t-domain becomes the multiplication in the s-domain. Given any continuous function $f(t)$ for $t \geq 0$, the Laplace transform and the inverse Laplace transform are defined as

$$\text{Laplace transform: } f_L(s) = L\{f(t)\} = \int_0^\infty f(t)e^{-st}dt \tag{6.2.13}$$

$$\text{Inverse Laplace transform: } f(t) = L^{-1}\{f_L(s)\} = \int_{-\infty}^\infty f_L(s)e^{st}ds \tag{6.2.14}$$

The subscript "L" represents the Laplace transform of the original function. $L\{\cdot\}$ is the operator used to denote the operation of the Laplace transform. Similarly, $L^{-1}\{\cdot\}$ is the operator for the inverse Laplace transform. Next we review two basic properties of the Laplace transform as they will be used in the development of the RF solution.

$$\text{Convolution: } f_1(t) * f_2(t) \Leftrightarrow f_{L1}(s)f_{L2}(s) \tag{6.2.15}$$

$$\text{First derivative: } f_L(s) = sF_L(s) - F(0) = sF_L(s) \tag{6.2.16}$$

Here "$*$" represents the convolution between two functions. The result in Eq. (6.2.16) is due to the fact that $F(0) = 0$ for any lifetime CDF. Recall from Eq. (6.2.8) that the RF in the t-domain can be expressed in the form of the convolution as follows:

$$M(t) = F(t) + \int_0^t M(t-x)f(x)dx = F(t) + M(t) * f(t) \tag{6.2.17}$$

Applying the Laplace transform to the above equation yields

$$M_L(s) = F_L(s) + M_L(s)f_L(s) \tag{6.2.18}$$

where $M_L(s)$, $F_L(s)$, and $f_L(s)$ are the corresponding Laplace transforms of $M(t)$, $F(t)$, and $f(t)$, respectively. After rearranging Eq. (6.2.18), the RF in the s-domain is obtained as

$$M_L(s) = \frac{F_L(s)}{1 - f_L(s)} = \frac{F_L(s)}{1 - sF_L(s)} = \frac{f_L(s)}{s(1 - f_L(s))} \tag{6.2.19}$$

The last two terms in $M_L(s)$ are obtained by using the first-order derivative property in Eq. (6.2.16), namely, $f_L(s) = sF_L(s)$.

To obtain the actual RF solution, $M_L(s)$ needs to be converted back to the t-domain. The inverse Laplace transform is often performed through the partial fraction expansion that reduces the degree of the numerator and the denominator of the original rational function. An s-domain RF upon partial fraction expansion can be expressed as

$$M_L(s) = \sum_{i=1}^{m} \frac{a_i}{s^i} + \sum_{j=1}^{n} \frac{b_j}{(s + \lambda_j)} \tag{6.2.20}$$

where a_i, b_j, and λ_j are the model coefficients or parameters to be determined. Now each faction term in $M_L(s)$ can be converted into a t-domain function accordingly. Therefore, the t-domain RF solution is obtained as

$$M(t) = \sum_{i=1}^{m} a_i t^i + \sum_{j=1}^{n} b_j e^{-\lambda_j t} \tag{6.2.21}$$

Example 6.1 Suppose the PDF of the inter-renewal time is given as $f(t) = 0.5e^{-t} + e^{-2t}$ for $t \geq 0$. Use the Laplace transform to find the renewal function $M(t)$.

Solution:
First we find the Laplace transform of $f(t)$, which is given as

$$f_L(s) = L\{f(t)\} = \frac{1}{2(s + 1)} + \frac{1}{s + 2} = \frac{1.5s + 2}{(s + 1)(s + 2)} \tag{6.2.22}$$

Substituting Eq. (6.2.22) into Eq. (6.2.19) and performing the partial fraction expansion yields

$$M_L(s) = \frac{1.5s + 2}{s^2(s + 1.5)} = \frac{4}{3s^2} + \frac{1}{9s} - \frac{1}{9(s + 1.5)} \tag{6.2.23}$$

Taking the inverse Laplace transform, we obtain RF in the t-domain, namely

$$M(t) = \frac{4}{3}t + \frac{1}{9}(1 - e^{-1.5t}) \tag{6.2.24}$$

The renewal density function $m(t)$ and the asymptotic renewal are also obtained as

$$m(t) = \frac{\partial M(t)}{\partial t} = \frac{4}{3} - \frac{1}{6}e^{-1.5t} \tag{6.2.25}$$

$$\lim_{t \to \infty} M(t) = \lim_{t \to \infty} \left[\frac{4}{3}t + \frac{1}{9}(1 - e^{-1.5t}) \right] = \frac{4}{3}t + \frac{1}{9} \tag{6.2.26}$$

6.2.4 Geometric and Geometric-Type Renewal

The geometric process is a meaningful generalization of a renewal process. In contrast to a perfect renewal process, the geometric process can be useful to model an imperfect repair in which its cycles are independently, but not identically, distributed (Cha and Finkelstein 2018). However, the durations of the cycle follow the same distribution family. The corresponding distribution in the nth repair, denoted as $F_n(t)$, is defined by the underlying distribution $F(t)$ as (Lam 2007)

$$F_n(t) = F(a^{n-1}t), \quad \text{for } n = 1, 2, \ldots \tag{6.2.27}$$

where a is a positive constant. Then the time sequence $\{S_n\}$ for $n \geq 1$ is called a *geometric process*. Equivalently, the corresponding counting process $N(t)$ for $t \geq 0$ is defined as $N(t) = \sup\{n: S_n \leq t\}$, for $t \geq 0$. Lam (1988) introduced the geometric process for modeling the maintenance planning and spare parts provisioning. Finkelstein (1993) considered some generalizations of Eq. (6.2.27) to non-linear scale transformations. Wang and Pham (2006) called a similar process a quasi-renewal process.

When $a = 1$, a geometric process reduces to a perfect renewal process. As in the case of a perfect renewal process, an important feature of a geometric process is that it is also governed by one underlying distribution $F(t)$. For $a > 1$, the cycles of this process are stochastically decreasing as follows:

$$F(a^n t) > F(a^{n-1}t), \quad \text{for } n = 1, 2, \ldots \tag{6.2.28}$$

which implies that $T_{n+1} < T_n$ for $n \geq 1$. Therefore, this process is suitable for modeling an imperfect repair action when a system lifetime becomes worse after each repair. On the other hand, when $a < 1$, a system's reliability is improving with each repair, which is often seen during the reliability growth test in practice. Define the mean and variance of the time-to-failure T_1 in the first cycle as follows:

$$E[T_1] = \mu \quad \text{and} \quad Var(T_1) = \sigma^2 \tag{6.2.29}$$

It follows from Cha and Finkelstein (2018) that

$$E[T_n] = \frac{\mu}{a^{n-1}} \quad \text{and} \quad Var(T_n) = \frac{\sigma^2}{a^{2(n-1)}} \tag{6.2.30}$$

where $E[T_n]$ and $Var(T_n)$ are the mean and variance of the time-to-failure of the nth cycle. The corresponding density function and the corresponding failure rates are

$$f_n(t) = a^{n-1}f(a^{n-1}t), \quad \text{for } n = 1, 2, 3, \ldots \tag{6.2.31}$$
$$h_n(t) = a^{n-1}h(a^{n-1}t), \quad \text{for } n = 1, 2, 3, \ldots \tag{6.2.32}$$

where $f(t)$ and $h(t)$ stand for the density and the failure rate of the underlying distribution $F(t)$, respectively. Therefore, for $a > 1$, the sum of expectations is converging to

$$\sum_{n=1}^{\infty} E[T_n] = \frac{a\mu}{(a-1)}, \quad \text{for } a > 1 \tag{6.2.33}$$

Finally, by referring to Eqs. (6.2.8) and (6.2.9) under a perfect renewal condition, the following renewal-type equations with a convolution in the right-hand side are derived:

$$M(t) = F(t) + \int_0^t M(a(t-x))f(x)dx \tag{6.2.34}$$

$$m(t) = f(t) + \int_0^t m(a(t-x))f(x)dx \tag{6.2.35}$$

6.2.5 Generalized Renewal Process

Repairable systems can be brought to one of possible states following a repair. These states are: "as good as new," "as bad as old," "better than old but worse than new," "better than new," and "worse than old." Most probabilistic models used to estimate the expected number of failures account for the first two states, but they are not properly applicable to the last three states, which are more realistic in practice. The seminal papers by Kijima and Sumita (1986) and Kijima (1989) model imperfect repair using the generalized renewal process (GRP) with the idea of estimating the "virtual age" of a repairable system. This spurred a tremendous growth in the imperfect maintenance literature. As of today both the concept and theory of the GRP approach has been well established. The main reason for the ubiquitous popularity is that GRP can incorporate all of the five after-repair states in the same mathematical framework. The GRP framework is comprised of two distinctive approaches for modeling imperfect repair, namely, the arithmetic reduction of age (ARA) and the arithmetic reduction of intensity (ARI) (Doyen and Gaudoin 2004; Tanwar et al. 2014). In ARA models, the effect of repair is expressed by a reduction of the system virtual age. In the ARI approach, the repair effect is incorporated by the failure intensity change before and after failure.

For the ARA modeling accept, Kijima (1989) introduced the notion of the virtual age of the system as follows. Let V_n be the virtual age of the system immediately after the nth repair; the system will experience an $(n+1)$th repair at an actual time T_{n+1}, according to the following probability:

$$P\{T_{n+1} \le t \mid V_n = y\} = \frac{F(t+y) - F(y)}{R(y)} \tag{6.2.36}$$

where $F(t+y)$ and $F(y)$ is the CDF and $R(y)$ is the reliability function of the system. Also T_{n+1} is the time between the nth and $(n+1)$th repair, namely, the interarrival time between two successive failures. Define a partial sum

$$S_n = \sum_{k=1}^{n} T_k, \quad \text{with } S_0 = 0 \tag{6.2.37}$$

This represents the real age of the system at the time of the nth repair (or failure); namely, S_n is the actual cumulative operating time elapsed since its installation. Let a_n represent the repair quality factor of the nth repair. It is assumed that the nth repair can remove the failures incurred during the period of T_n. In other words, a_n is selected such that $T_n \ge a_n T_n$; then the virtual age of the system after the nth repair becomes

$$V_n = V_{n-1} + a_n T_n, \quad \text{for } n = 1, 2, \ldots \tag{6.2.38}$$

where V_{n-1} is the virtual age right after the $(n-1)$th repair with $V_0 = 0$. In a special case with $a_k = a$ for $k = 1, 2, \ldots, n$ for all the repairs, then the virtual age of the system upon the nth repair is given by

$$V_n = a \sum_{k=1}^{n} T_k \tag{6.2.39}$$

Several extensions of Kijima models have been developed by various scholars. Finkelstein (1989, 2008) discovered that the term "virtual age" conceived by Kijima was for a specific model of imperfect repair with the underlying assumption that the repair action does not change the baseline CDF or $F(t)$, or the baseline failure rate $\lambda(t)$. The only change is the "initial time" after each repair. Therefore, the CDF of a lifetime after repair is defined as a remaining lifetime distribution $F(t/x)$. Other extensions of Kijima models such as proportional age reduction and proportional age setback models have been developed by various authors (Zhou et al. 2007; Martorell et al. 1999; and Sanchez et al. 2009). By realizing that a closed-form solution of the GRP equation is often unavailable, which makes the respective statistical estimation challenging, Kaminskiy and Krivtsov (1998) proposed a Monte Carlo approach that can be applied to estimate parameters of Kijima's Models I and II (Kijima 1989). In the operation and regulation of complex engineering systems, such as those in the space and process industries, it requires the use of sound models for predicting failures based on the past performance of the systems. The GRP solution proposed by Yanez et al. (2002) is a promising and efficient approach for such performance-based applications.

6.3 Exponential and Erlang Renewal

6.3.1 Exponential Renewal

For an exponential renewal process, the times between two consecutive failures, denoted as T_i for $i = 1, 2, 3, \ldots, N(t)$, are exponentially distributed with a constant failure rate λ. The cumulative distribution function is

$$F(t) = 1 - e^{-\lambda t}, \qquad \text{for } t \geq 0 \qquad (6.3.1)$$

Thus an exponential renewal process is equivalent to a Poisson process from the perspective of counting the number of failure events. We are interested in how many failures will occur between $[0, t]$. We use the Laplace transform to find $M(t)$. Performing the Laplace transform on Eq. (6.3.1) yields

$$F_L(s) = \int_0^\infty (1 - e^{-\lambda t}) e^{-st} dt = \frac{1}{s} - \frac{1}{s + \lambda} \qquad (6.3.2)$$

Substituting Eq. (6.3.2) into Eq. (6.2.19), we obtain $M_L(s)$ as follows:

$$M_L(s) = \frac{F_L(s)}{1 - sF_L(s)} = \frac{\lambda}{s^2} \qquad (6.3.3)$$

After performing the inverse Laplace transform to Eq. (6.3.3), we have the t-domain RF solution for the exponential renewal process. That is,

$$M(t) = \int_{-\infty}^\infty \frac{\lambda}{s^2} e^{st} ds = \lambda t \qquad (6.3.4)$$

Equation (6.3.4) shows that the number of failures under exponential renewal is a linear function of t with a positive slope of λ. For instance, if $\lambda = 0.1$ failures/month, for $t = 30$ and 50 months, we have $M(10) = 3$ and 5 failures, respectively.

6.3.2 Erlang Renewal

The Erlang distribution belongs to the gamma distribution family provided that the shape parameter is always an integer. If inter-renewal times follow the Erlang distribution, the PDF is given as

$$f(t) = \frac{\lambda^k t^{k-1} \exp(-\lambda t)}{\Gamma(k)} \qquad (6.3.5)$$

where k is a positive integer and λ is called the rate. The Laplace transform is

$$L\{f(t)\} = f_L(s) = \frac{\lambda^k}{(s+\lambda)^k} \tag{6.3.6}$$

substitute Eqs. (6.3.6) into (6.2.19) yields

$$M_L(s) = \frac{\lambda^k}{s[(s+\lambda)^k - \lambda^k]} \tag{6.3.7}$$

The partial fraction method can be applied to obtain the inverse Laplace transform, which leads to the following result:

$$M_L(s) = \frac{\lambda^k}{s((s+\lambda)^k - \lambda^k)} = \frac{\lambda^k}{a_k s^{k+1} + a_{k-1} s^k + \cdots + a_1 s^2}$$

$$= \frac{b_k}{s^2} + \frac{b_{k-1}}{s} + \sum_{i=1}^{k-2} \frac{b_i}{s+p_i} \tag{6.3.8}$$

where $\mathbf{a} = \{a_k, a_{k-1}, \ldots, a_1\}$ is the coefficient of the polynomial function of s and $\mathbf{b} = \{b_k, a_{k-1}, \ldots, b_1\}$ is the coefficient in the expansion series. By applying the inverse Laplace transform, the renewal function in the t-domain under the Erlang distribution is obtained as

$$M(t) = b_k t + b_{k-1} + \sum_{i=1}^{k-2} b_i e^{-p_i t} \tag{6.3.9}$$

The following example is provided to illustrate how to calculate $\mathbf{a}$ and $\mathbf{b}$ and further derive $M(t)$ under the Erlang distribution.

Example 6.2   The inter-renewal times follow the Erlang distribution with $\lambda = 1$ and shape parameter $k = 2$. Find the Laplace transform of $f(t)$ and $F(t)$, and obtain $M(t)$.

Solution:
Step 1: From Eq. (6.3.5), the Erlang PDF with $\lambda = 1$ and $k = 2$ is given as follows:

$$f(t) = te^{-t}, \qquad \text{for } t \geq 0 \tag{6.3.10}$$

According to the definition of the Laplace transform, we have $L\{f(t)\} = f_L(s)$ as follows:

$$f_L(s) = \int_0^\infty te^{-t}e^{-st}dt = \int_0^\infty te^{-(s+1)t}dt = \frac{1}{s+1}\int_\infty^0 tde^{-(s+1)t}d(s+1)t$$

$$= \frac{1}{s+1} te^{-(s+1)t}\Big|_\infty^0 - \frac{1}{s+1}\int_\infty^0 e^{-(s+1)t}dt$$

$$= \frac{1}{(s+1)^2}\int_\infty^0 de^{-(s+1)t} = \frac{1}{(s+1)^2}e^{-(s+1)t}\Big|_\infty^0 = \frac{1}{(s+1)^2} \tag{6.3.11}$$

Step 2: Based on the property in Eq. (6.2.16), the Laplace transform for $F(t)$ is

$$F_L(s) = \frac{f_L(s)}{s} = \frac{1}{s(s+1)^2} \tag{6.3.12}$$

Step 3: By substituting Eq. (6.3.12) into (6.2.19), and further carry out partial fraction expansion, the Erlang renewal in the *s*-domain is obtained as

$$M_L(s) = \frac{1}{s^3 + 2s^2} = \frac{0.5}{s^2} - \frac{0.25}{s} + \frac{0.25}{s+2} \tag{6.3.13}$$

Performing the inverse Laplace transform, we have

$$M(t) = 0.5t + 0.25(e^{-2t} - 1) \tag{6.3.14}$$

Realizing $A = 0.5$ and $B = -0.25$, the asymptotic renewal function is

$$\lim_{t\to\infty} M(t) = 0.5t - 0.25 \tag{6.3.15}$$

6.4 Generalized Exponential Renewal

6.4.1 Generalized Exponential Distribution

The generalized exponential function belongs to the so-called exponentiated-Weibull distribution (EWD). It was first proposed by Mudholkar and Srivastava (1993) as a simple generalization of the well-known Weibull family by introducing one more shape parameter. Later Gupta and Kundu (1999) simplified the EWD model as the generalized exponential distribution. It was noticed that the generalized exponential distribution can be used to approximate many empirical distributions including the Weibull and the gamma distributions (Gupta and Kundu 2001, 2007). The CDF and the PDF for the generalized exponential functions are given as

$$F_g(t) = (1 - e^{-bt})^n, \qquad t \geq 0 \tag{6.4.1}$$

$$f_g(t) = bne^{-bt}(1 - e^{-bt})^{n-1}, \qquad t \geq 0 \tag{6.4.2}$$

The generalized exponential function has two parameters: the scale parameter b and the shape parameter n. Both are positive parameters and could be integers or non-integers.

Figure 6.3 depicts the probability density functions (PDFs) for the generalized exponential functions for different scale and shape parameters. For example, when $n = 1$, it becomes the ordinary exponential distribution. It is also equivalent to the Weibull distribution with $\beta = 1$. For $0 < n < 1$, its PDF is a decreasing function and behaves like Weibull distributions with shape parameter $0 < \beta < 1$. Similarly, when $n > 1$, its PDF exhibits the bell-shape curve, which is similar to the Weibull distributions with $\beta > 1$ or gamma distributions with shape parameter $k > 1$. Hence, the generalized exponential distribution can be used to approximate both Weibull and gamma distributions.

The mean and the variance of the generalized exponential function were also obtained by Gupta and Kundu (1999) and the results are

$$\mu = E[T] = \frac{1}{\lambda}(\psi(n+1) - \psi(1)) \tag{6.4.3}$$

$$\sigma^2 = Var(T) = -\frac{1}{\lambda^2}(\psi'(n+1) - \psi'(1)) \tag{6.4.4}$$

where $\psi(.)$ denotes the digamma function, which is defined as the logarithmic derivative of the gamma function, namely, $\psi(n) = -d\ln[\Gamma(n)]/dn$. Also, $\psi'(.)$ denotes the

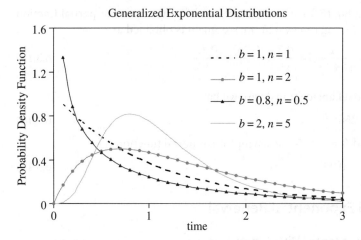

Figure 6.3 PDF for generalized exponential distributions.

derivative of $\psi(.)$. The hazard rate function of the generalized exponential function is given as follows:

$$h_g(t) = \frac{f_g(t)}{1 - F_g(t)} = \frac{bne^{-bt}(1 - e^{-bt})^{n-1}}{1 - (1 - e^{-bt})^n}, \qquad t \geq 0 \qquad (6.4.5)$$

Figure 6.4 plots $h_g(t)$ for different pairs of b and n. Depending on the value of n, the hazard rate function $h_g(t)$ may decrease, increase, or remain constant. This observation further indicates that the generalized exponential function possesses similar properties to the Weibull hazard rate function. Taking the limit of $h_g(t)$ yields

$$\lim_{t \to \infty} h_g(t) = \lim_{t \to \infty} \frac{bne^{-bt}(1 - e^{-bt})^{n-1}}{1 - (1 - e^{-bt})^n} = b \qquad (6.4.6)$$

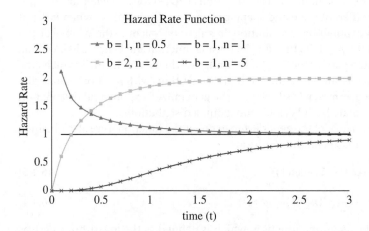

Figure 6.4 Hazard rate of generalized exponential function.

The limit of $h_g(t)$ is b for a decreasing hazard rate, and the upper limit of $h_g(t)$ is also b for an increasing hazard rate. For instance, given $b = 2$ and $n = 2$, the upper limit of $h_g(t) = 2$ as $t \to \infty$.

In summary, the generalized exponential distribution is a good candidate for the approximation of Weibull or gamma distributions. It is capable of modeling decreasing, constant, and increasing failure rates. Another reason is that it is relatively convenient to perform the Laplace and the inverse Laplace transforms compared with Weibull and gamma distributions.

6.4.2 Renewal in Laplace Transform

In this section, we present a closed-form solution for a generalized exponential renewal function based on the Laplace transform, which was originally developed by Jin and Gonigunta (2009). For the distribution with integer n, Eq. (6.4.1) can be expanded as the sum of $n + 1$ binomial terms as follows:

$$F_g(t) = (1 - e^{-bt})^n = \sum_{k=0}^{n} \binom{n}{k} (-1)^k e^{-kbt} = \sum_{k=0}^{n} (-1)^k C_k^n e^{-bkt} \qquad (6.4.7)$$

where $C_k^n = n!/(k!(n-k)!)$. The Laplace transform of $F_g(t)$ with the integer shape parameter n can now be readily obtained as

$$F_{gL}(s) = \sum_{k=0}^{n} \frac{(-1)^k C_k^n}{s + bk} = \sum_{k=0}^{n} \frac{d_k}{s + c_k} \qquad (6.4.8)$$

where

$$d_k = (-1)^k C_k^n, \qquad \text{for } k = 0, 1, \dots, n \qquad (6.4.9)$$

$$c_k = bk; \qquad \text{for } k = 0, 1, \dots, n \qquad (6.4.10)$$

By substituting Eq. (6.4.8) into (6.2.19), the generalized exponential renewal function $M_{gL}(s)$ in the s-domain is obtained as

$$M_{gL}(s) = \frac{\displaystyle\sum_{k=0}^{n} \frac{d_k}{s + c_k}}{1 - s \displaystyle\sum_{k=0}^{n} \frac{d_k}{s + c_k}} = \frac{\displaystyle\sum_{k=0}^{n} \frac{d_k}{s + c_k}}{1 - \frac{d_0 s}{s + c_0} - s \displaystyle\sum_{k=1}^{n} \frac{d_k}{s + c_k}} \qquad (6.4.11)$$

Realizing $d_0 = 1$ and $c_0 = 0$, Eq. (6.4.11) can be further simplified and expressed as

$$M_{gL}(s) = -\frac{\displaystyle\sum_{k=1}^{n} \frac{d_k}{s + c_k} + \frac{1}{s}}{s \displaystyle\sum_{k=1}^{n} \frac{d_k}{s + c_k}} = -\frac{1}{s^2 \displaystyle\sum_{k=1}^{n} \frac{d_k}{s + c_k}} - \frac{1}{s} = -\frac{1}{s^2 Y(s)} - \frac{1}{s} \qquad (6.4.12)$$

where

$$Y(s) = \sum_{k=1}^{n} \frac{d_k}{s + c_k} \qquad (6.4.13)$$

This means that $M_{gL}(s)$ essentially can be expressed as a polynomial function in which all these coefficients are determined by n, d_k, and c_k. Notice that n is the shape parameter of the original distribution function $F_g(t)$ while d_k and c_k are given by Eqs. (6.4.9) and (6.4.10), respectively. Therefore, the explicit expression for $M_{gL}(s)$ is known for a given distribution $F_g(t)$. This also implies that the value of c_k and d_k are known given $F_g(t)$. Let us further expand $Y(s)$ as follows:

$$Y(s) = \sum_{k=1}^{n} \frac{d_k}{s + c_k} = \frac{\sum_{k=0}^{n-1} q_k s^k}{\sum_{k=0}^{n} p_k s^k} = \frac{Q(s)}{P(s)} \tag{6.4.14}$$

where

$$Q(s) = \sum_{k=0}^{n-1} q_k s^k \tag{6.4.15}$$

$$P(s) = \sum_{k=0}^{n} p_k s^k \tag{6.4.16}$$

Equation (6.4.14) shows that $Y(s)$ in the denominator of $M_{gL}(s)$ can be represented by the quotient of two polynomial functions $Q(s)$ with degree $n - 1$ and $P(s)$ with degree n. In other words, both q_k (for $k = 0, 1, \ldots, n - 1$) and p_k (for $k = 1, \ldots, n$) are uniquely determined by d_k and c_k. After substituting Eqs. (6.4.15) and (6.4.16) into Eq. (6.4.12), we obtain $M_{gL}(s)$ as follows:

$$M_{gL}(s) = -\frac{1}{s^2 Y(s)} - \frac{1}{s} = -\frac{P(s)}{s^2 Q(s)} - \frac{1}{s} \tag{6.4.17}$$

Further applying the partial fraction expansion to $M_{gL}(s)$ yields

$$M_{gL}(s) = \frac{C_1}{s^2} + \frac{C_2}{s} - \frac{1}{s} + \frac{R(s)}{Q(s)} = \frac{C_1}{s^2} + \frac{C_2 - 1}{s} + \frac{R(s)}{Q(s)} \tag{6.4.18}$$

where

$$C_1 Q(s) + C_2 s Q(s) + s^2 R(s) = -P(s) \tag{6.4.19}$$

$$R(s) = \sum_{k=0}^{n-2} r_k s^k \tag{6.4.20}$$

It is noticed that C_1 and C_2 are constant and $R(s)$ is a polynomial function to be determined. By comparing Eq. (6.4.18) with Eq. (6.4.11) and also realizing that the Laplace transform $L[t] = 1/s^2$ and $L[1] = 1/s$, the following inference can be immediately obtained:

$$C_1 = \frac{1}{\mu}, \qquad C_2 = \frac{1}{2}\left(\frac{\sigma^2}{\mu^2} + 1\right) \tag{6.4.21}$$

where μ and σ^2 are the mean and variance of $F_g(t)$ and can be estimated from Eqs. (6.4.3) and (6.4.4), respectively. Next we need to determine r_k, i.e. coefficients of $R(s)$, in Eq. (6.4.20). From Eqs. (6.4.15) and (6.4.16), it is known that the degrees of $Q(s)$ and $P(s)$ are $n - 1$ and n, respectively. Hence the degree of $R(s)$ cannot exceed $n - 2$ in order to maintain the degree balance in Eq. (6.4.19). Substituting Eqs. (6.4.15), (6.4.16), and (6.4.20) into Eq. (6.4.18) and rearranging the equation, we obtain the following:

$$C_1 \sum_{k=0}^{n-1} q_k s^k + C_2 \sum_{k=0}^{n-1} q_k s^{k+1} + \sum_{k=0}^{n-2} r_k s^{k+2} = -\sum_{k=0}^{n} p_k s^k \tag{6.4.22}$$

$$r_k = -p_{k+2} - C_1 q_{k+2} - C_2 q_{k+1}, \qquad \text{for } k = 0, 1, 2, \ldots, n-2 \tag{6.4.23}$$

Obviously, coefficient r_k can be determined iteratively based on C_1, C_2, p_k, and q_k. Now we can rewrite Eq. (6.4.18) as

$$M_{gL}(s) = \frac{C_1}{s^2} + \frac{C_2 - 1}{s} + G(s) \tag{6.4.24}$$

where

$$G(s) = \frac{R(s)}{Q(s)} = \frac{r_{n-2} s^{n-2} + \cdots + r_1 s + r_0}{q_{n-1} s^{n-1} + q_{n-2} s^{n-2} + \cdots + q_1 s + q_0} \tag{6.4.25}$$

Equation (6.4.24) represents the s-domain renewal function for the generalized exponential distribution $F_g(t)$. Notice that C_1 and C_2 are known constants given by Eq. (6.4.21). Procedures to estimate the coefficients of $Q(s)$ and $R(s)$, i.e. q_k and p_k, are also given by Eqs. (6.4.15) and (6.4.16). Hence the explicit formula of $M_{gL}(s)$ can be derived given the original function $F_g(t)$.

6.4.3 Inverse Laplace Transform

The partial fraction expansion can be applied to Eq. (6.4.24) for obtaining $M_g(t)$ in the t-domain. Assuming $Q(s)$ in Eq. (6.4.25) has l real roots and m pairs of complex roots (i.e. $l + 2m = n - 1$). By applying the partial fraction expansion, $G(s)$ can be decomposed and expressed as the sum of the real parts and the complex parts as follows:

$$G(s) = \sum_{k=1}^{l} \frac{a_k}{s + \lambda_k} + \sum_{k=1}^{m} \frac{g_k s + h_k}{(s + \omega_k)^2 + \varphi_k^2} \tag{6.4.26}$$

Coefficients a_k, g_k, h_k, λ_k, ω_k, and φ_k are the natural results of the partial fraction expansion. Many off-the-shelf software tools are available for conducting the partial fraction expansion. For instance, Matlab provides the function called "residue" to obtain the expansion series. Upon obtaining Eq. (6.4.26), the corresponding t-domain function of $G(s)$ is obtained as follows:

$$g(t) = L^{-1}[G(s)] = \sum_{k=1}^{l} a_k e^{-\lambda_k t} + \sum_{k=1}^{m} g_k e^{-\omega_k t}(\cos \varphi_k t + \eta_k \sin \varphi_k t) \tag{6.4.27}$$

where

$$\eta_k = \frac{h_k - \omega_k g_k}{\varphi_k g_k} \tag{6.4.28}$$

Finally, combining Eqs. (6.4.27) with (6.4.24), the t-domain expression $M_g(t)$ is obtained as

$$M_g(t) = C_1 t + (C_2 - 1) + g(t) \tag{6.4.29}$$

Here we summarize the procedures to solve the generalized exponential renewal equation based on Laplace transform. Given $F_g(t)$, we do the following:

Step 1: Based on Eq. (6.4.8), perform the Laplace transform to obtain d_k and c_k using Eqs. (6.4.9) and (6.4.10).

Step 2: Determine C_1 and C_2 using Eqs. (6.4.21).

Step 3: Compute p_k and q_k based on Eqs. (6.4.14)–(6.4.16).

Step 4: Estimate r_k using Eq. (6.4.20).

Step 5: Perform partial fractions to get $G(s)$ in Eq. (6.4.26).

Step 6: Obtain $g(t)$ by performing an inverse Laplace transform using Eq. (6.4.27). Finally obtain $M_g(t)$ based on Eq. (6.4.29).

Example 6.3   Take the example of the Weibull distribution $F_W(t)$ with $\alpha = 1$ and $\beta = 2$, and the mean $\mu = 0.886$ and the variance $\sigma^2 = 0.215$. The corresponding approximation distribution is $F_g(t)$ with $b = 2.28$ and $n = 4$. Substituting b and n into Eq. (6.4.8) gives the Laplace transform

$$F_{gL}(s) = \frac{1}{s} - \frac{4}{s + 2.28} + \frac{6}{s + 4.56} - \frac{4}{s + 6.84} + \frac{1}{s + 9.12}$$

Here $\mathbf{d} = [1, -4, 6, -4, 1]$ and $\mathbf{c} = [0, 2.28, 4.56, 6.84, 9.12]$. Using Eqs. (6.4.15) and (6.4.16), both $P(s)$ and $Q(s)$ can be determined as

$$P(s) = s^4 + 22.8s^3 + 181.9s^2 + 592.6s + 648.6$$

$$Q(s) = -2s^3 - 36.5s^2 - 239.1s - 663.7$$

Furthermore, C_1, C_2, and $R(s)$ can be determined based on Eqs. (6.4.21) and (6.4.20) and the resultant values are given as

$C_1 = 1.13$

$C_2 = 0.637$

$$R(s) = -0.36s^2 - 7.16s - 40.39$$

Now $G(s)$ is obtained by substituting $R(s)$ and $Q(s)$ into Eq. (6.4.26). Then the partial fraction expansion is applied to $G(s)$, which yields

$$G(s) = \frac{0.159}{s - 1.14} + \frac{2.475s + 3.632}{(s - 5.7)^2 + 4.415^2}$$

and finally the t-domain expression for Weibull RF is given as

$$M(t) = 1.13t - 0.363 + 0.159e^{-1.14t} + 2.475e^{-5.7t}(\cos 4.415t + 1.468 \sin 4.415t)$$

where

$$g(t) = 0.159e^{-1.14t} + 2.475e^{-5.7t}(\cos 4.415t + 1.468 \sin 4.415t)$$

6.5   Weibull Renewal with Decreasing Failure Rate

6.5.1   Approximation by Mixed Exponential Functions

Let α and β be the Weibull scale and shape parameters, respectively. Depending on the value of β, the Weibull distribution is capable of modeling products or systems with decreasing $(0 < \beta < 1)$, constant $(\beta = 1)$, and increasing $(\beta > 1)$ failure rates. For $\beta = 1$, the Weibull distribution is reduced to the exponential distribution. The Weibull PDF and reliability function are given as

$$f(t) = \alpha\beta(\alpha t)^{\beta-1}\exp(-(\alpha t)^\beta), \qquad t \geq 0 \qquad (6.5.1)$$

$$R(t) = \exp(-(\alpha t)^\beta), \qquad t \geq 0 \qquad (6.5.2)$$

If the inter-renewal time follows the Weibull distribution, in theory $M(t)$ can be derived by substituting the Laplace transform of the Weibull PDF into Eq. (6.2.19). Unfortunately, the analytical form of $f_L(s)$ is not available when β is a non-integer. If β is an integer, the closed-form solution for $f_L(s)$ is available, but the analytical expression is very complicated with the involvement of the so-called Meijer G-function (Cheng et al. 2004). In theory, the G-function can be substituted into Eq. (6.2.19) to obtain $M_L(s)$, but the partial fraction technique is still difficult to perform in order to obtain $M(t)$ in the t-domain.

Approximation can be considered as an alternative approach to estimating Weibull $M(t)$. The approximation function must be chosen such that it is mathematically accurate to approximate the original Weibull distribution, yet computationally convenient for performing Laplace and inverse Laplace transforms. Common approximation functions include exponential functions and polynomial functions, because Laplace transform and inverse Laplace transforms can be conveniently carried out. Below we propose a mixed exponential function to approximate the Weibull distribution for $0 < \beta < 0$. In Section 6.6, we propose a different approach to approximating the Weibull RF for $\beta > 1$.

Feller (1971) and Jewell (1982) found that a Weibull distribution with a shape parameter β can be expressed as a mixture of Weibull distributions with a fixed shape parameter γ as long as $\gamma > \beta$. This means any Weibull distribution with a smaller shape parameter can be represented by the linear combination of Weibull distributions having larger shape parameters. That is,

$$\exp(-(\alpha t)^\beta) = \sum_{i=1}^{n} b_i \exp(-(\lambda_i t)^{\gamma_i}) \qquad (6.5.3)$$

where $\beta < \gamma_i$ for $i = 1, 2, \ldots, n$ and $\sum_{i=1}^{n} b_i = 1$. It is noticed that some or all values of γ_i could be the same as long as they are larger than β.

Equation (6.5.3) implies that for $0 < \beta < 1$, the Weibull distribution can be represented by a sum of the exponential function with $\gamma = 1$. Obviously the simplest mixture model to represent the Weibull reliability $R(t)$ is the linear combination of two exponential functions (i.e. $n = 2$) as follows:

$$\exp(-(\alpha t)^\beta) = b_1 \exp(-\lambda_1 t) + b_2 \exp(-\lambda_2 t) + e(t)$$
$$= R_a(t) + e(t) \qquad (6.5.4)$$

where

$$R_a(t) = b_1 \exp(-\lambda_1 t) + b_2 \exp(-\lambda_2 t) \qquad (6.5.5)$$

Note that $R_a(t)$ is called the mixture of exponential functions, which is a linear combination of two exponential functions. Here $e(t)$ represents the approximation error. It is desirable to find the best parameters $\{b_1, b_2, \lambda_1, \lambda_2\}$ such that the approximation error is minimized for $t \geq 0$. To that end, Jin and Gonigunta (2010) proposed the following optimization model to minimize the mean squared error (MMSE) between $R(t)$ and $R_a(t)$.

Model 6.1

$$\text{Min: } v(b_1, b_2, \lambda_1, \lambda_2) = \frac{1}{m} \sum_{i=1}^{m} \left(e^{-(\lambda t_i)^\beta} - b_1 e^{-\lambda_1 t_i} - b_2 e^{-\lambda_2 t_i} \right)^2 \tag{6.5.6}$$

$$\text{Subject to: } \quad b_1 + b_2 = 1 \tag{6.5.7}$$

$$0 < b_1 < 1, \ \ 0 < b_2 < 1 \ \text{ and } \ \lambda_1, \lambda_2 > 0 \tag{6.5.8}$$

The objective function (6.5.6) aims to minimize the sum of squared errors between the actual Weibull reliability and the mixture model. Note that t_i for $i = 1, \ldots, m$ is the time point located in the scope of the interested horizon. It is recommended to choose t_m such that $R(t_m) \leq 0.05$, or in other words it covers at least 95% of the reliability range. A constant time step Δt is used for the increment from t_{i-1} to t_i, but a variable Δt is also applicable. The incremental step size is defined as $\Delta t = t_m/m$ for given m. Many optimization software are available to search for the optimal values of b_1, b_2, λ_1, and λ_2, including Excel solver, AMPL, and Matlab. The following example is presented to illustrate the performance of this approximation method.

Example 6.4 Assume $\alpha = 1$ for the Weibull distribution and β decreases from 0.9 to 0.2. Based on Model 6.1, parameters $\{b_1, b_2, \lambda_1, \lambda_2\}$ are optimized and listed in Table 6.1.

Figure 6.5 depicts the actual Weibull reliability function versus its mixture exponential model base on the parameters listed in Table 6.1. As β increases and approaches 1, the approximation error (i.e. MMSE) becomes smaller. This is because the Weibull distribution with larger β is closer to the classical exponential function. Nevertheless, even if $\beta = 0.1$, the MMSE is only 7.6×10^{-5} using $m = 21$ sample points. The approximation quality can be further improved by increasing t_m or reducing the step size Δt for smaller β.

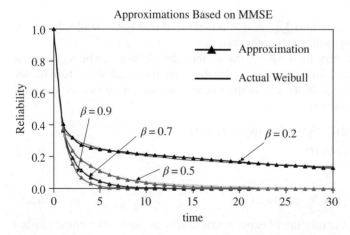

Figure 6.5 Weibull reliability approximation under different shape parameters. Source: Jin (2010) Reproduced with the permission of Taylor & Francis Ltd.

Table 6.1 Optimal solutions for Weibull reliability with $\alpha = 1$.

β	b_1	b_2	λ_1	λ_2	t_m	Δt	m	MMSE
0.95	0.16	0.84	1.94	0.89	11	1	11	8.8×10^{-9}
0.9	0.72	0.28	0.79	1.92	11	1	11	5.6×10^{-8}
0.8	0.44	0.56	1.93	0.62	11	1	11	5.4×10^{-7}
0.7	0.54	0.46	1.95	0.47	11	1	11	3.3×10^{-6}
0.6	0.39	0.61	0.33	1.93	11	1	11	9.9×10^{-6}
0.5	0.35	0.65	0.23	1.95	11	1	11	2.3×10^{-5}
0.4	0.28	0.72	0.12	1.8	21	1	21	3.4×10^{-5}
0.3	0.27	0.73	0.06	1.76	31	1	31	4.5×10^{-5}
0.2	0.275	0.725	0.025	1.275	42	2	21	6.7×10^{-5}
0.1	0.26	0.74	0.0014	0.6	42	2	21	7.6×10^{-5}

Source: Jin (2010) Reproduced with the permission of Taylor & Francis Ltd.

6.5.2 Laplace and Inverse Laplace Transform

After the original Weibull $R(t)$ is approximated by $R_a(t)$, the computation of the Laplace transform of RF function becomes straightforward. We quickly review this computational procedure below. The Weibull PDF $f(t)$ can be approximated by

$$f(t) \approx f_a(t) = -\frac{d}{dt} R_a(t) = b_1 \lambda_1 \exp(-\lambda_1 t) + b_2 \lambda_2 \exp(-\lambda_2 t) \qquad (6.5.9)$$

where $f_a(t)$ is the PDF of the mixture model. The Laplace transform of $f_a(t)$ is given as

$$f_{aL}(s) = L\{f_a(t)\} = \frac{b_1 \lambda_1}{(s + \lambda_1)} + \frac{b_2 \lambda_2}{(s + \lambda_2)} \qquad (6.5.10)$$

Substituting Eq. (6.5.10) into Eq. (6.2.19), the approximation to the Weibull renewal function, denoted as $M_{aL}(s)$, is obtained as

$$\begin{aligned} M_{aL}(s) &= \frac{(b_1 \lambda_1 + b_2 \lambda_2)s + (b_1 \lambda_1 \lambda_2 + b_2 \lambda_1 \lambda_2)}{s^3 + (\lambda_1 + \lambda_2 - b_1 \lambda_1 - b_2 \lambda_2)s^2} \\ &= \frac{c_1}{s^2} + \frac{c_2}{s} + \frac{c_3}{s + c_4} \end{aligned} \qquad (6.5.11)$$

where $\{c_1, c_2, c_3, c_4\}$ are uniquely determined by $\{b_1, b_2, \lambda_1, \lambda_2\}$. Performing an inverse Laplace transform to $M_{aL}(s)$, the approximation to the original Weibull renewal is obtained as

$$M_a(t) = c_1 t + c_2 + c_3 e^{-c_4 t} \qquad (6.5.12)$$

Here we summarize the approximation method in six steps:

Step 1: Given the Weibull reliability function $R(t)$, determine t_m such that $R(t_m) \le 0.05$.
Step 2: Define the incremental step $\Delta t = t_m / \Delta t$ by appropriately choosing m.
Step 3: Solve Model 6.1 and find the optimal values of $\{b_1, b_2, \lambda_1, \lambda_2\}$.
Step 4: Perform the Laplace transform on $f_a(t)$ using Eq. (6.5.10).

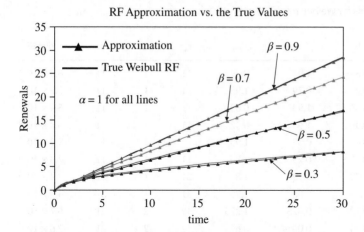

Figure 6.6 Weibull renewal and it approximation for $0 < \beta < 1$ and $\alpha = 1$. Source: Jin (2010) Reproduced with the permission of Taylor & Francis Ltd.

Step 5: Obtain $M_{aL}(s)$ based on Eq. (6.5.11).
Step 6: Finally, obtain $M_a(t)$, which is the approximation of the original Weibull renewal.

To demonstrate the performance of the mixed exponential function, we compute the true Weill RF and its approximation by varying beta from 0.3 to 0.9; the results are plotted in Figure 6.6. It is found that the approximation overlaps the actual RF very well, especially for $\beta > = 0.5$.

Example 6.5 Let us continue Example 6.4 by computing $M_{aL}(t)$. Take $\beta = 0.7$ as an example. We substitute $b_1 = 0.54$, $b_2 = 0.46$, $\lambda_1 = 1.95$, and $\lambda_2 = 0.47$ into Eq. (6.5.11) and perform the particle fraction expansion, which yields

$$M_{aL}(s) = \frac{1.248s + 0.855}{s^3 + 1.103s^2} = \frac{0.7752}{s^2} + \frac{0.4287}{s} - \frac{0.4287}{s + 1.103} \tag{6.5.13}$$

In this case, the partial fraction expansion is obtained by directly calling the Matlab function "residue." Finally, by applying the inverse Laplace transform to $M_{aL}(s)$, the renewal function in the t-domain is obtained as

$$M_a(t) = 0.7752t + 0.4287(1 - e^{-1.103t}) \tag{6.5.14}$$

This is the approximation model for the actual Weibull renewal function with $\alpha = 1$ and $\beta = 0.7$.

6.6 Weibull Renewal with Increasing Failure Rate

6.6.1 Transient Renewal Function

Two criteria are often used to evaluate the merits of an analytical RF approximation: computational simplicity and approximation accuracy. Although various analytical models have been proposed, very few of them can achieve both merits at the same time. To achieve high approximation accuracy, it is often necessary to increase the complexity of the model structure. A common approach is to increases the number of model parameters. This often leads to the growth of the computational load, making the model less applicable in practice. In addition, research has often been carried out in parallel between analytical approximation and numerical computation. Few attempts have been made to combine the numerical method with the analytical model to obtain a simple yet accurate RF model.

This section introduces an analytical model to approximate the Weibull RF for $\beta > 1$. It has a simple structure and only two model parameters need to be determined. The new method takes advantage of computational accuracy of the numerical method and integrates it into the analytical model. First, a generic model is created to approximate the actual Weibull RF data, but the model coefficients need to be determined. Now the problem becomes how to determine the coefficients such that the approximation error is minimized. The idea is to compute partial RF values at selected time points to estimate the model parameters. For $\beta > 1$, the Weibull RF data exhibits so-called oscillatory behavior. This phenomenon is captured by the sinc function of which the parameters are estimated based on the partial numerical analysis of the transient renewal.

Figure 6.7 plots the Weibull renewals under different values of $\alpha > 0$ and $\beta > 1$. These renewal data are computed using the Steljet–Lapance numerical method (Xie 1989). An interesting observation is that the scale parameter α determines the slope of the renewal curve. For instance, given t, the number of renewals of $M_1(t)$, is 1.5 times that of $M_2(t)$, then $M_2(t)$ is always twice that of $M_3(t)$. This means that the average number of renewals at a specific t is proportional to α. The book by Fausett (2007) is a good reference for implementing numerical analysis in the Matlab environment.

Another observation is that, when t is small, the Weibull renewal exhibits the so-called oscillation behavior for large β. The oscillation diminishes and merges into the asymptotic line as t becomes large. However, less oscillation phenomenon is observed for small β such as $\beta = 1.5$, yet the oscillation aptitude increases with β (e.g. $\beta = 5$ and 10). Because of the oscillation phenomenon, it is not easy to find an analytical model to approximate the Weibull $M(t)$ for $\beta > 1$. To cope with oscillation, the Weibull $M(t)$ can be treated as the superposition of a steady-state renewal function and a transient renewal function.

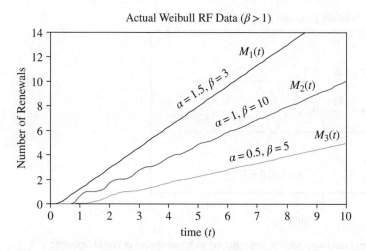

Figure 6.7 Weibull renewal data with different $\beta > 1$.

That is,

$$M(t) = At + B + g(t)$$
$$= G(t) + g(t) \tag{6.6.1}$$

where $g(t)$ describes the oscillation behavior and $g(t) = 0$ as $t \to \infty$. Note that $G(t) = At + B$ represents the steady-state renewal. Indeed, $G(t)$ is the same as the asymptotic renewal in Eq. (6.2.11). Based on Eq. (6.6.1), the transient renewal $g(t)$ can be expressed as

$$g(t) = M(t) - (At + B) \tag{6.6.2}$$

If the actual Weibull $M(t)$ is known, $g(t)$ can be computed directly by subtracting $At + B$ from $M(t)$. Unfortunately, $M(t)$ is unknown and itself needs to be determined. Obviously Eq. (6.6.2) cannot be directly used to estimate $g(t)$. Though the actual $M(t)$ is not known, the discrete values of $M(t)$, denoted as $\hat{M}(t)$, can be obtained through a numerical method like the Laplace–Stieltjes integration method. If we replace $M(t)$ with $\hat{M}(t)$ in Eq. (6.6.2), it becomes

$$g(t) \cong \hat{M}(t) - (At + B) \tag{6.6.3}$$

For instance, we can examine the characteristics of $g(t)$ under different α and β based on numerical computation of $\hat{M}(t)$, and then determine the structure of $g(t)$. Based on Eq. (6.6.2), transient renewals with three different values of $\{\alpha, \beta\}$ are computed and plotted in Figure 6.8. Note that A and B are estimated from Eq. (6.2.12) and $\hat{M}(t)$ is obtained using the Laplace–Stieltjes integration method (Xie 1989).

As expected, there are few oscillations observed when β is small (i.e. $\beta = 1.5$). The oscillation becomes phenomenal as β exceeds 1.5 and becomes large. Another observation is that the scale parameter α controls the oscillation cycle period. For instance, T_2 is the upper-and-down cycle for $\alpha = 0.5$, and is almost twice the cycle time T_1 for $\alpha = 1$. Regardless of α and β, the initial value of $g(t)$ is always equal to B, i.e. $g(0) = B$. Since $g(t)$ exhibits different types of oscillation patterns, we will present two approximation models for $g(t)$ corresponding to $1 < \beta \le 1.5$ and $\beta > 1.5$, respectively.

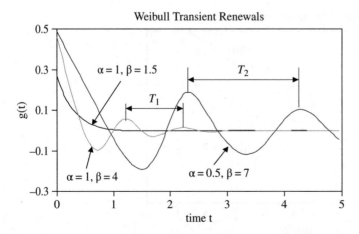

Figure 6.8 Weibull transient renewals. Source: Reproduced with permission of World Scientific Publishing.

6.6.2 Approximation without Oscillation

This section focuses on the derivation of the approximation model for $g(t)$ for $1 < \beta \leq 1.5$. It has been shown in a previous section that $g(t)$ has no or almost no oscillations when $0 < \beta \leq 1.5$. Figure 6.9 depicts a typical $g(t)$ for that case. It is observed that the transient renewal decreases from B to 0 monotonically. This motivates the idea of using an exponential function to approximate $g(t)$ as follows (Jin et al. 2014):

$$g(t) \cong g_a(t) = B \exp(-t/T) \tag{6.6.4}$$

where $g_a(t)$ represents the approximation of $g(t)$. This model is attractive because only one unknown parameter T needs to be estimated and the value of B is given in Eq. (6.2.12). In the following, a detailed procedure will be provided to estimate T and the error between $g_a(t)$ and $g(t)$ will be discussed in terms of approximation accuracy.

Several approaches are proposed to estimate T including the MMSE method (Garg and Kalagnanam 1998). MMSE intends to find the best T that minimizes the sum of mean squared errors between $g(t)$ and $g_a(t)$. Although MMSE ensures a high level of approximation accuracy, the computation is a little bit complicated with the involvement of optimization. Below we propose a simple, yet quite accurate, algebraic method to estimate T. To ensure the approximation accuracy, $g_a(t)$ satisfies the following condition:

$$g_a(t_0) = B \exp(-t_0/T) = g(t_0) \tag{6.6.5}$$

where t_0 is the time when $g(t)$ drops to certain percentages of B. To ensure the approximation accuracy, t_0 is often chosen such that $g(t_0) \leq 0.1B$. Based on this criterion, T can be determined as

$$T = -\frac{1}{t_0} \ln\left(\frac{g(t_0)}{B}\right) = \frac{2.3}{t_0} \tag{6.6.6}$$

Based on this criterion, a set of T was obtained for various values of α and β, and the results are listed in Table 6.2.

The data in Table 6.2 shows that T increases with β regardless of the value of α. On the other hand, for given β, the value of T is proportional to α. For instance, when $\alpha = 1$

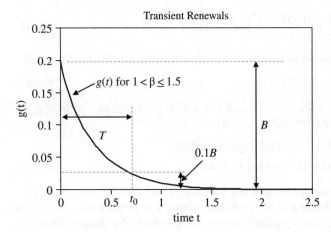

Figure 6.9 Estimation T for Weibull $0 < \beta \leq 1.5$. Source: Reproduced with permission of World Scientific Publishing.

Table 6.2 Values of T for various α and β based on $g_a(t_0) = 0.1B$.

β	$T\,(\lambda = 0.5)$	$T\,(\lambda = 1)$	$T\,(\lambda = 2)$
1.01	0.922	1.844	3.695
1.1	1.107	2.214	4.444
1.2	1.305	2.611	5.198
1.3	1.492	2.983	5.935
1.4	1.669	3.338	6.715
1.5	1.828	3.657	7.360

Source: Reproduced with permission of World Scientific Publishing.

and $\beta = 1.3$, we have $T = 2.983$, and $T = 5.935$ for $\alpha = 2$ and $\beta = 1.3$. Based on the observations in Table 6.2, the following regression model is obtained to extrapolate the value of T for any pairs of $\{\alpha, \beta\}$ for $0 < \beta \le 1.5$:

$$T = \frac{1}{\alpha}(7.965 - 11.697e^{-\beta} - 5.063\beta e^{-\beta}) \tag{6.6.7}$$

Example 6.6 For instance, given Weibull $\alpha = 0.75$ and $\beta = 1.35$, find the approximate transient renewal $g_a(t)$ using Eq. (6.6.4).

Solution:
Substituting $\alpha = 0.75$ and $\beta = 1.35$ into Eq. (6.6.7) to estimate T gives

$$T = \frac{1}{0.75}(7.965 - 11.697e^{-1.35} - 5.063 \times 1.35e^{-1.35}) = 4.2143 \tag{6.6.8}$$

For the Weibull distribution, the mean and variance are $\mu = 1.223$ and $\sigma^2 = 0.838$. Now A and B are obtained from Eq. (6.2.12), that is

$$A = \frac{1}{\mu} = \frac{1}{1.223} = 0.8179 \tag{6.6.9}$$

$$B = \frac{1}{2}\left(\frac{\sigma^2}{\mu^2} - 1\right) = \frac{1}{2}\left(\frac{0.838}{1.223^2} - 1\right) = -0.2197 \tag{6.6.10}$$

Now $g_a(t)$ is obtained by substituting Eqs. (6.6.8) and (6.6.10) into Eq (6.6.4); that is

$$g_a(t) = -0.2197 \exp(-t_0/4.2143) \tag{6.6.11}$$

Finally, the analytical $M_a(t)$ for $\alpha = 0.75$ and $\beta = 1.35$ is obtained by combining the steady-state renewal $G(t) = At + B$ and $g_a(t)$; that is

$$M_a(t) = 0.8179t + 0.2197 - 0.2197\exp(-t/4.2143)$$
$$= 0.8179t + 0.2197(1 - \exp(-t/4.2143)) \tag{6.6.12}$$

Table 6.3 compares the values between the actual $g(t)$ and its approximation $g_a(t)$ when β varies in [1.1, 1.5]. Since α is the scale parameter, it is set to unity. The actual $g(t)$ is obtained from the Laplace–Stieltjes integration method. The results show that $g_a(t)$ converges to $g(t)$ as t increases. However, the convergence is much faster for large β as opposed to a small one.

Table 6.3 Approximations versus actual $g(t)$ for $\alpha = 1$ and $1 < \beta \leq 1.5$.

t	$\beta = 1.1$		$\beta = 1.2$		$\beta = 1.3$		$\beta = 1.4$		$\beta = 1.5$	
	$g(t)$	$g_a(t)$	$g(t)$	$g_a(t)$	$g(t)$	$g_a(t)$	$g(t)$	$g_a(t)$	$g(t)$	$g_a(t)$
0	0.086	0.086	0.150	0.150	0.199	0.199	0.238	0.238	0.270	0.270
0.1	0.061	0.069	0.106	0.115	0.141	0.148	0.168	0.170	0.190	0.187
0.2	0.047	0.055	0.080	0.089	0.104	0.110	0.122	0.122	0.136	0.129
0.3	0.038	0.044	0.062	0.069	0.078	0.082	0.089	0.087	0.096	0.090
0.4	0.030	0.035	0.048	0.053	0.058	0.061	0.064	0.062	0.067	0.062
0.5	0.024	0.028	0.038	0.041	0.044	0.045	0.046	0.045	0.046	0.043
0.6	0.020	0.023	0.029	0.031	0.033	0.033	0.033	0.032	0.031	0.030
0.7	0.016	0.018	0.023	0.024	0.025	0.025	0.023	0.023	0.020	0.021
0.8	0.014	0.015	0.018	0.019	0.018	0.018	0.016	0.016	0.012	0.014
0.9	0.011	0.012	0.014	0.014	0.013	0.014	0.011	0.012	0.007	0.010
1.0	0.009	0.009	0.011	0.011	0.010	0.010	0.007	0.008	0.003	0.007
1.1	0.008	0.007	0.009	0.009	0.007	0.008	0.004	0.006	0.002	0.005
1.2	0.006	0.006	0.007	0.007	0.005	0.006	0.003	0.004	0.002	0.003
1.3	0.005	0.005	0.005	0.005	0.004	0.004	0.001	0.003	0.001	0.002
1.4	0.004	0.004	0.004	0.004	0.003	0.003	0.001	0.002	0.001	0.002
1.5	0.004	0.003	0.003	0.003	0.002	0.002	0.000	0.002	0.001	0.001
1.6	0.003	0.002	0.003	0.002	0.001	0.002	0.000	0.001	0.000	0.001
1.7	0.003	0.002	0.002	0.002	0.001	0.001	0.000	0.001	0.000	0.001
1.8	0.002	0.002	0.002	0.001	0.000	0.001	0.000	0.001	0.000	0.000
1.9	0.002	0.001	0.001	0.001	0.000	0.001	0.000	0.000	0.000	0.000
2.0	0.002	0.001	0.001	0.001	0.000	0.001	0.000	0.000	0.000	0.000

Source: Reproduced with permission of World Scientific Publishing.

6.6.3 Approximation with Oscillation

The procedure is more involved in finding an approximation model of $g(t)$ for $\beta > 1.5$ because of oscillation in the transient renewals. Figure 6.10 depicts a typical transient renewal curve for $\beta > 1.5$. It is observed that $g(t)$ possesses a sinusoidal wave, yet diminishes as t becomes large. The time duration between two adjacent peaks is called the cycle time, which is denoted as T.

Careful observation reveals that $g(t)$ actually contains two different phases. Phase I covers the time period of $[0, t_1]$ and Phase II encompasses the period of $[t_1, \infty)$. The cycle time in Phase I is twice of the cycle time of Phase II, though Phase I lasts only a half of T. Thus, the sinusoidal function is not appropriate to approximate $g(t)$ because a sine wave has a fixed cycle time or frequency for $0 \leq t < \infty$.

The sinc function is widely used in signal processing and wireless communication theory (Barry et al. 2009). The full name is "sine cardinal," but it is commonly referred to by its abbreviation, "sinc." Two properties of the sinc function resemble the Weibull transient renewal: (1) a sinc function also has two phases with $T = 2t_1$ for $0 \leq t < \infty$ and (2) a sinc function oscillates sinusoidally with a decaying envelope as $t \to \infty$. By observing

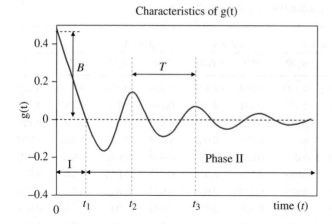

Figure 6.10 Characteristic terms of transient renewal. Source: Reproduced with permission of World Scientific Publishing.

this dual oscillation frequency, Jin et al. (2014) proposed to use the sinc function to approximate the actual $g(t)$ as follows:

$$g(t) \cong g_a(t) = B\frac{\sin(2\pi t/T)}{2\pi t/T} \tag{6.6.13}$$

where $B = g(0)$ for $t = 0$ and T is the cycle time in Phase II. By substituting Eq. (6.6.13) into Eq. (6.6.1) and rearranging terms, the approximation for the Weibull RF, denoted as $M_a(t)$, is obtained as

$$M_a(t) = At + B\left(1 - \frac{\sin(2\pi t/T)}{2\pi t/T}\right) \tag{6.6.14}$$

In Eq. (6.6.14), A and B are known constants for a given Weibull distribution. Thus only T needs to be determined, which will be discussed in the following.

We compute the transient renewal using the Laplace–Stieltjes integration (Xie 1989) and obtain a set of T as β varies from 1 to 10. The results of T are summarized in Table 6.4. The value of β is confined to (1.5, 10] simply because this is the shape parameter range most likely to be used for Weibull applications. Table 6.4 shows that T decreases as β increases from 1.6 to 6, and then remains at a relatively stable level even if β continues to increase from 6.5 to 10. Another observation is, for a given β, that the cycle time T is proportional to $1/\alpha$. For instance, for $\beta = 4$, $T = 0.915$ if $\alpha = 1$ and $T = 0.455$ if $\alpha = 2$. The cycle time of the latter is a half of the former. If $\alpha = 0.5$, then $T = 1.825$, which doubles compared to $T = 0.915$ with $\alpha = 1$. This observation is similar to the Weibull renewal for $0 < \beta \leq 1.5$. Based on the data in Table 6.4, the following regression model is obtained for estimating T for given any pairs of $\{\alpha, \beta\}$. The regression model is

$$T = \frac{1}{\alpha}(0.942 + 10.566e^{-\beta} - 3.575\beta e^{-\beta}) \tag{6.6.15}$$

Equation (6.6.15) is valid for any Weibull distributions as long as $\beta > 1.5$. For instance, when $\alpha = 0.35$ and $\beta = 4.7$, the cycle time T would be 2.531 based on Eq. (6.6.15).

Table 6.4 Values of T for $g(t)$ for various α and $\beta \in (1.5, 10]$.

β	$T\ (\alpha = 0.5)$	$T\ (\alpha = 1)$	$T\ (\alpha = 2)$
1.6	3.958	1.978	0.989
2.0	2.615	1.305	0.655
2.5	2.115	1.055	0.525
3.0	1.935	0.965	0.485
3.5	1.855	0.925	0.465
4.0	1.825	0.915	0.455
4.5	1.815	0.905	0.455
5.0	1.805	0.905	0.455
5.5	1.815	0.905	0.455
6.0	1.815	0.905	0.455
6.5	1.825	0.915	0.455
7.0	1.835	0.915	0.455
7.5	1.845	0.925	0.465
8.0	1.845	0.925	0.465
8.5	1.855	0.925	0.465
9.0	1.865	0.935	0.465
9.5	1.875	0.935	0.465
10.0	1.875	0.935	0.465

Source: Reproduced with permission of World Scientific Publishing.

Example 6.7 Based on the procedure described in previous sections, various examples are used to demonstrate the performance of the approximation model for $\beta \in (1.5, 10]$. Without loss of generality, the scale parameter is set to be $\alpha = 1$. The actual $g(t)$ is obtained using the Laplace–Stieltjes integration and further compared with $g_a(t)$.

Solution:
Step 1: Given Weibull parameters λ and β, we estimate the A and B based on Eqs. (6.2.12).
Step 2: Estimate the cycle time T using Eq. (6.6.15).
Step 3: Construct the transient renewal $g_a(t)$ based on Eq. (6.6.13).
Step 4: Obtain the approximation model $M_a(t)$ based on Eq. (6.6.14).

We compute the actual Weibull $M(t)$ using the Laplace–Stieltjes integration and compare it with $M_a(t)$. Table 6.5 lists the actual RF data $M(t)$ and the approximation $M_a(t)$ for $\beta = 2, 4, 6$, and 8, all under $\alpha = 1$. When $\beta = 4$ or smaller, there is almost no approximation error up to two decimals. The error, namely $M(t) - M_a(t)$, slightly increases as β becomes larger, especially for small t. However, $M_a(t)$ quickly converges to the true RF data as t becomes larger regardless of the value of β.

Next we examine whether λ influences the approximation accuracy of $M_a(t)$. We set $\alpha = 0.5, 1$, and 2, respectively. Figure 6.11 provides the comparisons between the results from $M_a(t)$ and $M(t)$ for $\beta = 3$ under different α. It is observed that the approximation model can generate very accurate approximations that almost overlap the actual renewal curve. Similar observations can be made from Figures 6.12 and 6.13, which correspond

Table 6.5 Comparison between $M(t)$ and $M_a(t)$ for different β with $\alpha = 1$.

t	$\beta = 2$		$\beta = 4$		$\beta = 6$		$\beta = 8$		$\beta = 10$	
	M(t)	$M_a(t)$	M(t)	$M_a(t)$	M(t)	$M_a(t)$	M(t)	$M_a(t)$	M(t)	$M_a(t)$
0.0	0.00	0.00	0.00	0.00	0.00	0.00	0.00	0.00	0.00	0.00
0.5	0.23	0.23	0.06	0.04	0.02	0.01	0.00	0.02	0.00	0.03
1.0	0.75	0.75	0.64	0.66	0.63	0.66	0.63	0.65	0.63	0.64
1.5	1.33	1.33	1.19	1.19	1.09	1.09	1.04	1.03	1.01	1.00
2.0	1.89	1.89	1.75	1.75	1.71	1.71	1.71	1.69	1.71	1.68
2.5	2.46	2.46	2.30	2.30	2.18	2.19	2.10	2.12	2.05	2.07
3.0	3.02	3.02	2.85	2.85	2.78	2.77	2.76	2.73	2.76	2.71
3.5	3.59	3.59	3.40	3.40	3.27	3.28	3.17	3.20	3.10	3.15
4.0	4.15	4.15	3.95	3.95	3.85	3.84	3.81	3.78	3.80	3.74
4.5	4.71	4.71	4.50	4.50	4.36	4.37	4.24	4.28	4.16	4.21
5.0	5.28	5.28	5.06	5.06	4.92	4.91	4.86	4.83	4.84	4.78
5.5	5.84	5.84	5.61	5.61	5.44	5.45	5.32	5.35	5.22	5.28
6.0	6.41	6.41	6.16	6.16	5.99	5.99	5.91	5.89	5.87	5.82
6.5	6.97	6.97	6.71	6.71	6.52	6.52	6.39	6.41	6.29	6.33
7.0	7.54	7.54	7.26	7.26	7.07	7.06	6.96	6.95	6.91	6.87
7.5	8.10	8.10	7.81	7.81	7.60	7.60	7.46	7.47	7.36	7.39
8.0	8.66	8.66	8.37	8.37	8.14	8.14	8.02	8.01	7.95	7.92
8.5	9.23	9.23	8.92	8.92	8.68	8.68	8.53	8.54	8.42	8.44
9.0	9.79	9.79	9.47	9.47	9.22	9.22	9.07	9.07	8.99	8.97
9.5	10.36	10.36	10.02	10.02	9.76	9.76	9.60	9.60	9.48	9.49
10.0	10.92	10.92	10.57	10.57	10.30	10.30	10.13	10.13	10.03	10.02

Source: Reproduced with permission of World Scientific Publishing.

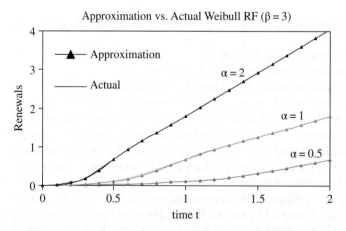

Figure 6.11 Approximations versus actual for $\beta = 3$. Source: Reproduced with permission of World Scientific Publishing.

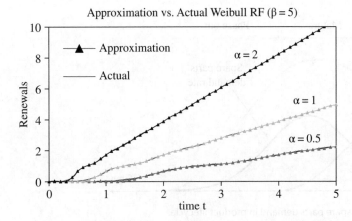

Figure 6.12 Approximations versus actual for $\beta = 5$. Source: Reproduced with permission of World Scientific Publishing.

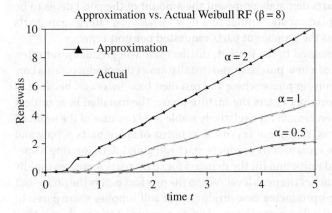

Figure 6.13 Approximation versus actual for $\beta = 8$. Source: Reproduced with permission of World Scientific Publishing.

to $\beta = 5$ and 8, respectively. In summary, the sinc function in Eq. (6.6.13) is shown to be effective to approximate the actual Weibull transient renewal behavior.

6.7 Renewal under Deterministic Fleet Expansion

6.7.1 Superimposed Exponential Renewal

The lifecycle of a product can be divided into three phases: (i) new product introduction; (ii) volume manufacturing and installation; and (iii) phase-out or decommissioning. This lifecycle pattern is commonly observed in capital goods, such as wind turbines (WTs), semiconductor equipment, airplanes, telecommunication equipment, and defense systems, among others. Figure 6.14 graphically shows how the installed base of a product changes in three phases. An installed base, also known as the product fleet, represents the cumulative number of products operating in the field. A new installation rate represents the number of products that are purchased by customers and is often

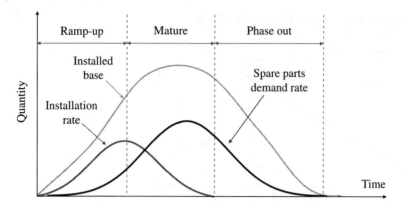

Figure 6.14 Installation and spare parts demand in product lifecycle.

measured in installed systems or products per time period, such as 10 systems/month or 5 systems/week. Spare parts demands represent the amount of the good items to be provided for field failure replacements. Similar to a new installation rate, a spare parts demand is often measured as the number of parts requested per unit time.

When a new product is released to the market, the fleet size will gradually, yet often randomly, increase because of a new purchase and installation of the products. This corresponds to the product ramp-up phase where the installed base increases because of the market growth. As the product enters the mature phase, the installed base or fleet size reaches the peak and then remains at a relatively stable level because of the reduced installation rate. Nevertheless, this phase is critical in terms of spare parts supply and maintenance support for two reasons: (i) products start aging and deteriorating, especially for those early installed units, and (ii) the demand for spare parts increases rapidly because the installed base reaches the peak level. When the product enters the phase-out stage, the manufacturer stops producing new products, yet still supplies spare parts. In this post-production phase the fleet size starts to decline because of the product retirement. However, spare parts supply in this phase is crucial to ensure the system operation and availability because the majority of field systems are aging.

We intend to estimate the renewal of an installed base during the ramp-up phase. Figure 6.15 shows an installation scenario of a new product along with subsequent failures during the introduction phase. Following the initial installation at $t = 0$, new systems are shipped and installed at time $w_1, w_2, \ldots, w_{n(t)}$ where $n(t)$ is the number of systems installed between $(0, t]$. The actual fleet size is $n(t) + 1$, including the system installed at $t = 0$. For repairable systems, each installed system will independently generate failures and maintenance requests. For instance, system "1" is installed at w_1 and two failures occur between $[w_1, t]$, requiring two spare parts for failure replacement. For simplicity, we consider single-item systems and each failure corresponds to one item renewal.

The system is brought to as good-as-new state upon the replacement of the spare part. Therefore, the number of aggregate failures of the fleet in $[0, t]$ is equivalent to the number of renewals generated by the installed base. Given the number of renewals in $[0, t]$, the manufacturer or service provider can mitigate or minimize the stockout probability by appropriately allocating the safety stock level or stocking sufficient amounts of

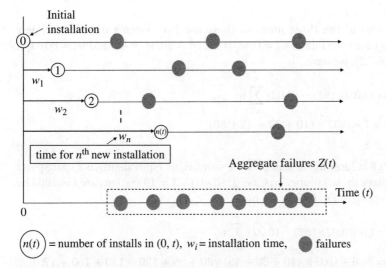

$\boxed{n(t)}$ = number of installs in $(0, t)$, w_i = installation time, ● failures

Figure 6.15 Spare part demand under a single-item system fleet.

spare parts. By referring to Figure 6.15, the aggregate fleet renewals between $[0, t]$ can be treated as a superposition of $n(t) + 1$ independent renewal processes, each corresponding to one installed system. Let $Z(t)$ be the aggregate fleet renewals between $[0, t]$. Then

$$Z(t) = M(t) + M(t - w_1) + \cdots + M(t - w_{n(t)})$$

$$= M(t) + \sum_{i=1}^{n(t)} M(t - w_i) \tag{6.7.1}$$

where $M(t - w_i)$ represents the number of renewals for the system that is installed at w_i for $i = 1, 2, \ldots, n(t)$. When the inter-renewal times are exponentially distributed with constant rate λ, the closed-form renewal solution is shown in Eq. (6.3.4). By substituting $M(t) = \lambda t$ into Eq. (6.7.1), the aggregate fleet renewal is obtained as

$$Z(t) = \lambda t + \sum_{i=1}^{n(t)} \lambda(t - w_i) = (n(t) + 1)\lambda t - \lambda \sum_{i=1}^{n(t)} w_i \tag{6.7.2}$$

Equation (6.7.2) is derived for a single-item system fleet. If a system consists of m identical items, each fails independently. Then the aggregate fleet renewal under multi-item systems is

$$Z_m(t) = m\lambda t + \sum_{i=1}^{n(t)} m\lambda(t - w_i) = m(n(t) + 1)\lambda t - m\lambda \sum_{i=1}^{n(t)} w_i \tag{6.7.3}$$

Example 6.8 Ten new products were released and installed worldwide over the past six months. The sales data shows that these systems were installed on the following dates: 0, 10, 30, 45, 80, 95, 120, 130, 160, and 175 days. Assume an exponential lifetime of the product with $\lambda = 0.02$ failures/day. Estimate the number of failures generated by the fleet in two cases: (1) at the end of three months (91 days) and (2) at the end of six months (183 days).

Solution:

In Case 1, at the end of the three months, there are five systems installed, namely $n(t) + 1 = 5$. In this case $t = 91$ days, $w_0 = 0$, $w_1 = 10$, $w_2 = 30$, $w_3 = 45$, and $w_4 = 80$ days. According to Eq. (6.7.2), we have

$$Z(t = 91) = (5)(0.02)(91) - (0.02) \sum_{i=1}^{4} w_i$$

$$= 9.1 - 0.02 \times (10 + 30 + 45 + 80)$$

$$= 5.8 \tag{6.7.4}$$

Thus we expect 5.8 failures by the end of three months, or, equivalently, 5.8 spare parts are required to replace these failures. In Case 2, all $n(t) + 1 = 10$ systems are installed by $t = 183$ days. According to Eq. (6.7.2), we have

$$Z(t = 183) = (10)(0.02)(183) - (0.02) \sum_{i=1}^{9} w_i$$

$$= 36.6 - 0.02 \times (10 + 30 + 45 + 80 + 95 + 120 + 130 + 160 + 175)$$

$$= 19.7 \tag{6.7.5}$$

Thus there are 19.7 failures by the end of the six months from all 10 installed products.

6.7.2 Superimposed Erlang Renewal

Not all the inter-renewal times are exponential. Many products or components exhibit an increased failure rate between two consecutive renewals. Then Erlang distribution can be considered as a useful tool to model this type of renewal. Suppose the inter-renewal time follows the Erlang distribution with $k = 2$ and its PDF is given by

$$f(t) = \frac{\lambda^k t^{k-1} \exp(-\lambda t)}{(k-1)!} = \lambda^2 t e^{-\lambda t} \tag{6.7.6}$$

The reliability function with $k = 2$ is

$$R(t) = 1 - F(t) = 1 - \sum_{i=0}^{k-1} \frac{1}{i!} e^{-\lambda t} (\lambda t)^i = (1 + \lambda t) e^{-\lambda t} \tag{6.7.7}$$

where $F(t)$ is the CDF and the hazard rate function is obtained by dividing $f(t)$ by $R(t)$, that is

$$h(t) = \frac{f(t)}{R(t)} = \frac{\lambda^2 t e^{-\lambda t}}{(1 + \lambda t) e^{-\lambda t}} = \frac{\lambda^2 t}{1 + \lambda t} \tag{6.7.8}$$

Note that $h(t)$ increases with t with the initial $h(0) = 0$. It levels off at $h(\infty) = \lambda$ for $t \to \infty$. Figure 6.16 shows Erlang hazard rates with different values of λ.

The Laplace transform is

$$L\{f(t)\} = f_L(s) = \frac{\lambda^2}{(s + \lambda)^2} \tag{6.7.9}$$

By substituting Eq. (6.7.9) into Eq. (6.2.19), the renewal function in the s-domain and in the t-domain are obtained, respectively, as follows:

$$M(s) = \frac{\lambda^2}{s^2(s + 2\lambda)} = \frac{0.5\lambda}{s^2} - \frac{0.25}{s} + \frac{0.25}{s + 2\lambda} \tag{6.7.10}$$

$$M(t) = 0.5\lambda t + 0.25(e^{-2\lambda t} - 1) \tag{6.7.11}$$

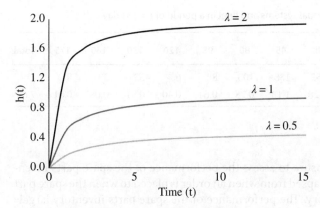

Figure 6.16 Erlang hazard rate function with $k = 2$.

Example 6.9 Assume that the sales data are the same as in Example 6.8, but the inter-renewal time shows a degradation trend that fits the Erlang distribution with $k = 2$ and $\lambda = 0.02$ failures/day. Estimate the number of failures generated by the fleet in two cases: (1) at the end of 3 months (91 days) and (2) at the end of 6 months (183 days).

Solution:
In Case 1, at the end of $t = 91$ days, a total of five systems are installed and $n = 5 - 1 = 4$. The installation dates are $w_0 = 0$, $w_1 = 10$, $w_2 = 30$, $w_3 = 45$, and $w_4 = 80$ days. According to Eq. (6.7.11), we have

$$M_0(t = 91) = (0.5)(0.02)(91) + 0.25(e^{-(2)(0.02)(91)} - 1) = 0.67$$

$$M_1(t = 91) = (0.5)(0.02)(91 - 10) + 0.25(e^{-(2)(0.02)(91-10)} - 1) = 0.57$$

$$M_2(t = 91) = (0.5)(0.02)(91 - 30) + 0.25(e^{-(2)(0.02)(91-30)} - 1) = 0.38$$

$$M_3(t = 91) = (0.5)(0.02)(91 - 45) + 0.25(e^{-(2)(0.02)(91-45)} - 1) = 0.25$$

$$M_4(t = 91) = (0.5)(0.02)(91 - 80) + 0.25(e^{-(2)(0.02)(91-80)} - 1) = 0.02$$

Thus, the aggregate fleet renewals by the end of $t = 91$ days is

$$M(t = 90) = 0.67 + 0.57 + 0.38 + 0.25 + 0.02 = 1.89$$

In Case 2, at the end of $t = 183$ days, a total of 10 systems are installed and $n = 10 - 1 = 9$. Since the inter-renewal is not exponential, we cannot directly use Eq. (6.7.2); rather, we substitute Eq. (6.7.11) into Eq. (6.7.1), which is the generic fleet renewal, giving

$$Z(t) = 0.5\lambda t + 0.25(e^{-2\lambda t} - 1) + \sum_{i=1}^{n} \left(0.5\lambda(t - w_i) + 0.25(e^{-2\lambda(t-w_i)} - 1)\right) \quad (6.7.12)$$

where w_i is the installation date of the $(i + 1)$th system for $i = 1, 2, \ldots, n$, and the first system is installed at $w_0 = 0$. Table 6.6 summarizes the Erlang renewals of 10 systems. The aggregate fleet renewals reach 7.69 at the end of six months, four times higher than $M(t = 91)$. This is because new systems are installed between 91 and 183 days, causing more failures as a result of the fleet expansion.

Table 6.6 Erlang renewals of individual systems and fleet in a period of $t = 183$ days.

w_i (days)	0	10	30	45	80	95	120	130	160	175	Total
$t - w_i$ (days)	183	173	153	138	103	88	63	53	23	8	
$M(t - w_i)$	1.58	1.48	1.28	1.13	0.78	0.64	0.40	0.31	0.08	0.01	7.69

6.7.3 Lead-Time Renewal

The lead time is a critical measure to assess the performance of the spare parts inventory. It is defined as the time elapsed from when an order is placed to when the spare part actually arrives in the inventory. The performance of the spare parts inventory largely depends on how much the safety stock is allocated against the backorders and how quickly the inventory can be replenished during the lead time. Lead-time renewal also allows the product manufacturer to optimize the warranty policy by appropriately allocating free replacement parts (Free 1986). Hence understanding the lead-time demand enables the inventory planner to optimize the safety stock, ensure service quality, and lower the parts cost.

Given the lead time l, a replacement order placed at time t for $t > 0$ will arrive to the inventory at $t + l$. According to Eq. (6.7.1), the number of aggregate fleet renewals in $[0, t]$ is $Z(t)$, and is $Z(t + l)$ in $[0, t + l]$. Then the demands for the spare parts during the lead-time period between $[t, t + l]$ can be estimated by

$$
\begin{aligned}
D(l; t) &= Z(t + l) - Z(t) \\
&= M(t + l) + \sum_{i=1}^{n(t+l)} M(t + l - w_i) - M(t) - \sum_{i=1}^{n(t)} M(t - w_i) \\
&= [M(t + l) - M(t)] + \sum_{i=1}^{n(t)} [M(t + l - w_i) - M(t + l - w_i)] \\
&\quad + \sum_{i=n(t)+1}^{n(t+l)} M(t + l - w_i)
\end{aligned}
\tag{6.7.13}
$$

This shows that $D(l; t)$ can be treated as a sum of two independent renewal streams, namely, the renewals from the systems that are installed between $[0, t]$ and renewals from the systems that are installed during $(t, t + l]$. If the inter-renewal time is exponential, Eq. (6.7.13) can be further simplified by replaying $M(t)$ with λt, that is

$$
\begin{aligned}
D(l; t) &= \lambda(t + l) + \sum_{i=1}^{n(t+l)} \lambda(t + l - w_i) - \lambda t - \sum_{i=1}^{n(t)} \lambda(t - w_i) \\
&= \lambda l + \sum_{i=1}^{n(t)} [\lambda(t + l - w_i) - \lambda(t - w_i)] + \sum_{i=n(t)+1}^{n(t+l)} \lambda(t + l - w_i) \\
&= \lambda l (1 + n(t)) + \sum_{i=n(t)+1}^{n(t+l)} \lambda(t + l - w_i) \\
&= D_1 + D_2
\end{aligned}
\tag{6.7.14}
$$

Under exponential inter-renewal times, Eq. (6.7.14) indicates that the lead-time renewal is the sum of two independent demand steams:

$D_1 = \lambda l[1 + n(t)]$ that represent the renewals from the systems installed prior to t and

$$D_2 = \sum_{i=n(t)+1}^{n(t+l)} \lambda(t + l - w_i) \text{ are the renewals from systems installed in } [t, t+l].$$

6.8 Renewal under Stochastic Fleet Expansion

6.8.1 Aggregate Exponential Renewal

The aggregate fleet renewal models studied thus far assume that the system installation time w_i is known or deterministic. In reality, the manufacturer usually does not have the advance information about when and how many new systems will be purchased by the potential customers; hence the actual system installation time during the new introduction phase is uncertain. Even if the installation rate can be estimated, the exact installation time of each new system is still unknown until the customer has placed the purchase order. In such an uncertain market, the Poisson counting process has been used to forecast the new product adoption scenarios (Farrel and Saloner 1986). The basic assumption is that the time to the next installation is exponentially distributed. It is possible that spikes of product orders may occur in the ramp-up phase, as shown in Figure 6.14, but the product installation rate, especially for capital equipment, is rarely sky rocketing. The main reasons are because of capacity constraints ranging from manufacturing facilities, tools, skilled labor, materials, and available subcontractors. Assuming the installed base expands under a Poisson process, we present an analytical model to estimate the aggregate fleet renewal under Poisson fleet expansion. The following assumptions are made on the installation rate and replacement process:

- After the initial system is installed at $t = 0$, the installed base increases following the Poisson counting process with a rate of λ. The number of installations in $(0, t]$, denoted as $N(t)$, has the following mass probability function:

$$P\{N(t) = n\} = \frac{(\lambda t)^n e^{-\lambda t}}{n!}, \qquad \text{for } n = 0, 1, 2, \dots \qquad (6.8.1)$$

- For single-item systems, each system is configured with one line replaceable unit (LRU). Upon failure, the defective LRU is replaced with a good part, and the system is restored to an "as good as new" state. The downtime for replacement is negligible compared with the uptime.

We use Figure 6.15 to illustrate the system installation and aggregate renewal process under stochastic fleet expansion. Two mutually independent processes are involved: the Poisson installation process and the system renewal process. The increment of the installed base is modeled as a homogeneous Poisson process with a rate of λ. After the initial installation at time 0, new systems are installed at random points in time $W_1, W_2, \dots, W_{N(t)}$ in $(0, t]$. Note that both $N(t)$ and W_i for all i are random variables. This differs from the deterministic fleet expansion in Section 6.7.1 where the installation time w_i and $n(t)$ are assumed to be known or fixed. The aggregate fleet renewals in $[0, t]$ can be

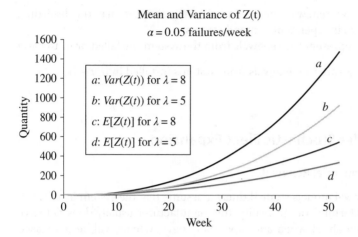

Figure 6.17 Mean and variance of $Z(t)$ under different α and λ.

treated as a superposition of $N(t) + 1$ independent renewal processes. Let $Z(t)$ be the aggregate fleet renewals in $[0, t]$. Then

$$Z(t) = M(t) + \sum_{i=1}^{N(t)} M(t - W_i), \qquad \text{for } t \geq 0 \tag{6.8.2}$$

where $M(t - W_i)$ represents the number of renewals for the ith system installed at W_i. Note that $Z(t)$ is a random number due to the stochastic nature of $N(t)$ and W_i. If the inter-renewal time is exponential with a constant failure rate α, then $M(t) = \alpha t$ and Eq. (6.8.2) can be simplified as

$$Z(t) = \alpha t + \sum_{i=1}^{N(t)} \alpha(t - W_i) = \alpha[1 + N(t)] - \alpha \sum_{i=1}^{N(t)} W_i \tag{6.8.3}$$

It is usually difficult, if not impossible, to derive the underlying distribution of $Z(t)$, but the mean and the variance of $Z(t)$ can be obtained explicitly based on the statistical property of $N(t)$ and W_i. The detailed derivation is available in Jin and Liao (2009) and the results are presented below:

$$E[Z(t)] = \alpha t + \frac{1}{2}\alpha \lambda t^2 \tag{6.8.4}$$

$$Var(Z(t)) = \alpha t + \frac{1}{2}\alpha \lambda t^2 + \frac{1}{3}\alpha^2 \lambda t^3 \tag{6.8.5}$$

Note that both $E[Z(t)]$ and $Var(Z(t))$ are time-varying functions in quadratic and cubic formats, respectively. Assuming $\alpha = 0.05$ faults/week, Figure 6.17 shows the spare parts demand profile for a period of 52 weeks under two different installation rates. Lines a and b represent $E[Z(t)]$ and $Var(Z(t))$ under $\lambda = 10$ systems/week. Lines c and d correspond to a lower installation rate of $\lambda = 5$ systems/week. Obviously, a smaller λ results in a lower mean and variation of fleet renewals.

Let $z(t)$ be the aggregate failure intensity rate. By taking the derivatives of Eqs. (6.8.4) and (6.8.5) with respect to t, the mean and the variance of $z(t)$ are

$$E[z(t)] = \alpha + \alpha\lambda t \qquad (6.8.6)$$

$$Var(z(t)) = \alpha + \alpha\lambda t + \alpha^2\lambda^2 t^2 \qquad (6.8.7)$$

This proves that the aggregate spare parts demand stream is a non-stationary process for which the mean and the variance are the increasing functions of time, respectively. This justifies the fact that the aggregate fleet renewal, i.e. the demand for spare parts, is a non-stationary process.

For less costly products, a nonlinear or exponential installation process rather than a Poisson counting process could be more suitable to model the installation process because of the occurrence of demand spikes. For instance, for systems that are less capital intensive (typically below \$0.5 M/system), the manufacturer can ramp-up the production to meet the demand surge by expanding the production capacity. They can do it because less investment in facilities, raw materials, and personnel is required compared to multimillion dollar capital products such as aircraft engines, WTs, and semiconductor lithography equipment.

6.8.2 Lead-Time Renewal

Under one-for-one inventory replenishment, the amount of lead-time demand is equivalent to the number of fleet renewals that occurred in the lead time. This section presents the lead-time renewal under stochastic fleet expansion, and the results presented here were originally from Jin et al. (2017).

Let l be the lead time starting from when a spare part is ordered at t till the part arrives to the inventory at $t+l$. Let $D(l; t)$ be the lead-time renewal during $[t, t+l]$. Based on Eq. (6.8.2), we first estimate the lead-time renewal by conditioning on $N(t+l)$ and $\mathbf{W}_{t+l}$, that is,

$$E[D(t) \mid N(t+l), \mathbf{W}_{t+l}]$$
$$= E[Z(t+l) \mid N(t+l), \mathbf{W}_{t+l}] - E[Z(t) \mid N(t), \mathbf{W}_t]$$
$$= \alpha l(1 + n_{t+l}) - \alpha \sum_{k=1}^{n_l} w_k \qquad (6.8.8)$$

where $Z(t+l)$ and $Z(t)$ represent the aggregate fleet renewals in $[0, t+l]$ and $[0, t]$, respectively. Note that $E[N(t+l)] = n_{t+l}$ and $E[N(l)] = n_l$, respectively, and $\mathbf{W}_{t+l} = [W_1 = w_1, W_2 = w_2, \ldots, W_{N(t+l)} = w_{n_{t+l}}]$. A simple proof of Eq. (6.8.8) is provided as follows:

Proof: From Eq. (6.8.2), we have

$$E[Z(t+l) \mid N(t+l), \mathbf{W}_{t+l}] = M(t+l) + \sum_{i=1}^{n_{t+l}} M(t+l-w_i) \qquad (6.8.9)$$

By substituting Eqs. (6.8.9) and (6.8.2) into Eq. (6.8.8), we have

$$E[D(t) \mid N(t+l), \mathbf{W}_{t+l}] = E[Z(t+l) \mid N(t+l), \mathbf{W}_{t+l}] - E[Z(t) \mid N(t), \mathbf{W}_t]$$
$$= M(t+l) + \sum_{i=1}^{n_{t+l}} M(t+l-w_i) - M(t) + \sum_{i=1}^{n_t} M(t-w_i) \qquad (6.8.10)$$

Equation (6.8.10) captures the lead-time renewal under general lifetime distributions. If the inter-renewal is exponential with constant failure rate α, we have $M(t+l) = \alpha(t+l)$ and $M(t) = \alpha t$. Then Eq. (6.8.10) can be further simplified as

$$E[D(t) \mid N(t+l), W_{t+l}] = \alpha(t+l) + \sum_{k=1}^{n_{t+l}} \alpha(t+l-w_k) - \alpha t - \sum_{k=1}^{n_t} \alpha(t-w_k)$$

$$= \alpha l + \sum_{k=1}^{n_t} \alpha(t+l-w_k) + \sum_{k=n_t+1}^{n_{t+l}} \alpha(t+l-w_k) - \sum_{k=1}^{n_t} \alpha(t-w_k)$$

$$= \alpha l + \sum_{k=1}^{n_t} \alpha l + \sum_{k=1}^{n_t} \alpha(l-w_k) = \alpha l(1 + n_t + n_l) - \sum_{k=1}^{n_l} w_k$$

$$= \alpha l(1 + n_{t+l}) - \alpha \sum_{k=1}^{n_l} w_k \tag{6.8.11}$$

Next we remove the condition on $N(t)$. For a Poisson installation with rate λ, we have $E[N(t)] = Var(N(t)) = \lambda t$. By removing the conditions on $N(t+l)$ and W_{t+l} in Eq. (6.8.11), the mean and the variance of $D(t)$ can now be expressed as

$$E[D(t)] = \alpha(1 + \lambda t)l + \frac{1}{2}\alpha\lambda l^2 \tag{6.8.12}$$

$$Var(D(t)) = \alpha(1 + \lambda t)l + \left(\frac{1}{2}\alpha\lambda + a^2\lambda t\right)l^2 + \frac{1}{3}a^2\lambda l^3 \tag{6.8.13}$$

Since $E[D(t)]$ and $Var(D(t))$ are quadratic and cubic functions of l, the lead-time renewal is a non-stationary process for which the mean and variance are time-dependent. Given the mean and variance of $D(t)$, different types of inventory control policies, such as (Q, r), $(s-1, s)$, or (s, S), can be designed to meet the required performance requirements, such as the stockout probability, fill rate, and parts availability. This will be further discussed in Chapters 9 and 10.

6.9 Case Study

6.9.1 Installed Base of Wind Turbines in the USA

Wind power emerged as a clean and sustainable energy resource to meet the growing electricity needs in the twenty-first century. At the close of 2016, the US wind fleet totaled 82 GW, enough to power 24 million American homes (US DOE 2015). There are now more than 52 000 individual wind turbines (WTs) in 41 states plus Guam and Puerto Rico (AWEA 2017). According to the US Department of Energy, by 2030, the total wind power capacity in the US will reach 300 GW, comprising 20% of the electricity market as opposed to 5.5% in 2017 (US DOE 2008). To meet this target, it is anticipated that every day 17–19 new WTs will be installed in continental USA. Figure 6.18 shows the trend line of the cumulative installed WT systems and the average turbine capacity. It is interesting to see that the capacity per turbine increases from 0.5 MW in 1998 to the present 1.6 MW/turbine. Over the past 18 years, the total installation reached 52 000 turbines. This implies that the average installation rate is 2889 WTs/year (or 28 WTs/week).

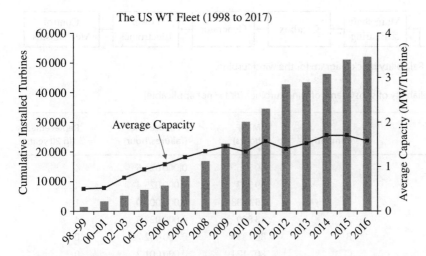

Figure 6.18 Current WT fleet size and installed capacity.

The rapid deployment of WTs creates tremendous challenges in equipment design, maintenance, repair, and overhaul services. The actual situation could be worse as wind farms are often located in remote regions or offshore areas, making it difficult to access these giant machines and perform regular maintenance. Given 52 000 WTs operating in the US by 2016, every day 125 turbines will fail based on current 10 000 hours of MTBF. In other words, every 12 minutes there is a WT failure in the field. The issue will be exacerbated as the installed base continues to grow and reach 300 GW capacity in 2030. The capacity of modern WTs are typically in the range between 1.5 MW and 3 MW. Assume the average capacity is 3 M per turbine. The number of turbines would be 300 GW/3 MW = 100 000 in 2030. If the average capacity is 1.5 MW, the fleet size could be 200 000 in 2030. The actual installed base most likely falls between both numbers by considering the factor that the early installations usually have a smaller capacity.

WTs are complex aerodynamic, electromechanical systems incorporating multiple subsystems, including blades, rotor, gearbox, power electronics, and control modules, among others. Gearbox failures typically account for the largest amount of downtime costs because of complexity in replacement and repair logistics. It is estimated that when a $1500 bearing fails unnoticed in a gearbox, it can lead to power interruption and revenue loss, including an unscheduled replacement of a $100 000 gearbox and an unscheduled crane cost of up to $70 000 to access the failed components. These costly failures can consume 15–20% of the price of the turbine itself. Therefore, timely supply and delivery of spare parts is a key to supporting the maintenance and operation of a WT during its 20–25 years of field service. The purpose of this case study is to analyze the reliability of the WT subsystem and predict the expected number of failures during its lifetime.

6.9.2 Spare Parts Prediction under Fleet Expansion

As shown in Figure 6.19, a WT can be modeled as a series system interconnected by the blades/rotor, the main shaft/bearing (MS/B), the gearbox, the generator, power electronics, and control mechanisms, among others. Upon failure, the defective component

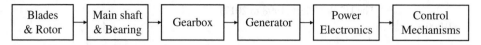

Figure 6.19 Reliability block diagram for the wind turbine.

Table 6.7 Reliability of subsystems of wind turbines (n/a = not applicable).

Subsystem	Company	MTBF (hours)	Failure rate (failures/hour)	Replacements (in 20 years)
Blades	A	40 745	0.000 024 5	4
	B	39 297	0.000 025 4	4
	C	252 033	0.000 004 0	1
Main S/B	A	43 800	0.000 022 8	4
	B	365 339	0.000 002 7	0
	C	807 174	0.000 001 2	0
Gearbox	A	39 371	0.000 025 4	4
	B	87 174	0.000 011 5	2
	C	218 871	0.000 004 6	1
Generator	A	54 750	0.000 018 3	3
	B	73 234	0.000 013 7	2
	C	365 534	0.000 002 7	0
Power electronics	A	22 902	0.000 043 7	8
	B	39 205	0.000 025 5	4
	C	175 561	0.000 005 7	1
WT system	A	7 424	0.000 134 7	n/a
	B	12 688	0.000 078 8	n/a
	C	54 923	0.000 018 2	n/a

is removed and replaced by a good part, and the system can be restored to production immediately.

According to the US Department of Energy, by 2030, the total wind power capacity in the US will reach 300 GW, comprising 20% of the electricity market as opposed to 5.5% in 2017 (US DOE 2008). Relevant reliability data of individual subsystems are collected and presented in Table 6.7. There exist large MTBF variations between different subsystem types. For instance, MTBF of the Main S/B are generally high while blades, gearbox, and power electronics have a lower MTBF across all three manufacturers. Within the same subsystem, MTBF also varies between different manufacturers. For instance, the MS/B MTBF varies from 43 800 to 807 174 hours across three different manufacturers. Though life data directly from WT manufacturers are generally not available, readers can refer to studies by Vittal and Teboul (2005) and Tavner et al. (2007) to gain additional information about WT reliability. Practically, the former investigates small and medium WTs and the latter analyses large WT systems.

We plan to deploy one wind farm with the plan of installing 60 WTs over the next five years. This means each month a WT will be installed and connected to the grid for power

Table 6.8 Aggregate fleet renewals of components.

Component	Company	Failure rate (failures/month)	Renewals (3 years)	Renewals (5 years)	Renewals (10 years)	Renewals (20 years)
Blades	A	0.0176	33.3	93.7	284.2	665.2
	B	0.0183	11.5	32.4	98.2	229.9
	C	0.0029	1.8	5.1	15.5	36.2
Main S/B	A	0.0164	10.3	29.1	88.2	206.3
	B	0.0019	1.2	3.4	10.4	24.4
	C	0.0009	0.5	1.5	4.6	10.9
Gearbox	A	0.0183	11.5	32.4	98.2	229.9
	B	0.0083	5.2	14.7	44.5	104.1
	C	0.0033	2.1	5.9	17.8	41.6
Generator	A	0.0132	8.3	23.3	70.8	165.6
	B	0.0099	6.2	17.5	53	124
	C	0.0019	1.2	3.4	10.4	24.4
Power electronics	A	0.0315	19.8	55.7	169	395.5
	B	0.0184	11.6	32.5	98.6	230.8
	C	0.0041	2.6	7.3	22	51.6

generation. Assume that it always takes exactly 30 days to complete the installation of one WT. Based on the reliability data in Table 6.7, our objective is to estimate the fleet renewals of each component types for a period of 5, 10, and 20 years corresponding to WTs of companies A, B, and C, respectively. The results are summarized in Table 6.8.

For instance, to estimate the aggregate feel renewals for the gearbox of Company A during the first three years, or $t = 36$ months, we use Eq. (6.7.2) to perform the prediction of renewals. Note that $\lambda = 0.000\,025\,4$ failures/hour $= 0.018\,3$ failures/month, $w_i = 1, 2, \ldots, 36$, and $n(t) = 36$ as one WT is installed per month:

$$Z_{gearbox}(t = 36) = (n(t) + 0)\lambda t - \lambda \sum_{i=1}^{n(t)} w_i = 36 \times 0.0183 \times 36 - 0.0183 \sum_{i=1}^{36} i = 11.5$$

(6.9.1)

It is worth mentioning that the initial system is installed at the end of month 1; hence there is no system at $t = 0$. Similarly, the aggregate fleet renewals of a gearbox at $t = 60$, 120, and 240 months can also be obtained using Eq. (6.7.2). For $t = 120$ months (i.e. 10 years), we have $w_i = 1, 2, \ldots, 60$ and $n(t) = 60$. Then

$$Z_{gearbox}(t = 120) = (n(t) + 0)\lambda t - \lambda \sum_{i=1}^{n(t)} w_i$$

$$= 60 \times 0.0183 \times 120 - 0.0183 \sum_{i=1}^{60} i = 98.2.$$

(6.9.2)

Since a WT system has three blades, Eq. (6.7.3) needs to be used to estimate the aggregate fleet failures of blades by realizing $m = 3$. For $t = 36$ months, the failure rate of the blade is $\lambda = 0.0176$ failures/month for Company A. Thus

$$Z_{blade}(t = 36) = m(n(t) + 0)\lambda t - m\lambda \sum_{i=1}^{n(t)} w_i$$

$$= 3 \times 36 \times 0.0176 \times 36 - 3 \times 0.0176 \sum_{i=1}^{36} i = 33.3. \tag{6.9.3}$$

For $t = 120$ months (i.e. 10 years), we have $w_i = 1, 2 \ldots, 60$ and $n(t) = 60$. Then the aggregate fleet renewals of blades are

$$Z_{blade}(t = 120) = 3 \times 60 \times 0.0176 \times 120 - 3 \times 0.0176 \sum_{i=1}^{60} i = 284.2 \tag{6.9.4}$$

References

AWEA, (2017), "Near-record growth propels wind power into first place as America's largest renewable resource," available at: https://www.awea.org/MediaCenter/pressrelease.aspx?ItemNumber=9812 (accessed on 20 September 2017).

Barry, J.R., Lee, E.A., and Messerschmitt, D.G. (2009). *Digital Communication*, 3e. Norwell, MA: Kluvwer Academic Publisher.

Brown, M., Solomon, H., and Stephens, M.A. (1981). Monte Carlo simulation of the renewal function. *Journal of Applied Probability* 18 (2): 426–434.

Cha, J.H. and Finkelstein, M. (2018). Generalizations of renewal process. In: *Point Processes for Reliability Analysis*, 211–246. UK: Springer.

Chaudhry, M.L. (1995). On computations of the mean and variance of the number of renewals: a unified approach. *Journal of the Operational Research Society* 44: 1352–1364.

Cheng, J., Tellambura, C., and Beaulieu, N. (2004). Performance of digital linear modulations on Weibull slow-fading channels. *IEEE Transactions on Communications* 52 (8): 1265–1268.

Cleroux, R. and McConalogue, D.J. (1976). A numerical algorithm for recursively-defined convolution integrals involving distribution functions. *Management Science* 22 (10): 1138–1146.

Cox, D.R. (1962). *Renewal Theory*. London, UK: Methuen.

Cox, D.R. and Isham, V. (1980). *Point Processes*. Chapman and Hall.

Cui, L. and Xie, M. (2003). Some normal approximations for renewal function of large Weibull shape parameters. *Communications in Statistics B: Simulation and Computers* 32 (1): 1–16.

Doyen, L. and Gaudoin, O. (2004). Classes of imperfect repair models based on reduction of failure intensity or virtual age. *Reliability Engineering and Systems Safety* 84: 45–56.

Elayed, E. (2012). *Reliability Engineering*, Chapter 7, 2e. Hoboken, NJ: Wiley.

Farrel, J. and Saloner, G. (1986). Installed base and compatibility: innovation, product preannouncements and predation. *The American Economic Review* 76 (5): 940–955.

Fausett, L.V. (2007). *Applied Numerical Analysis Using Matlab*, 2e. Prentice Hall.

Feller, W. (1941). On the integral equation of renewal theory. *Annals of Mathematical Statistics* 12: 243–267.

Feller, W. (1971). *An Introduction to Probability Theory and its Applications*, vol. II. New York, NY: Wiley.

Finkelstein, M. (1989). Perfect, minimal and imperfect repair (in Russian). *Reliability and Quality Control* 3: 17–21.

Finkelstein, M.S. (1993). A scale model of general repair. *Microelectronics Reliability* 33 (1): 41–46.

Finkelstein, M. (2008). Failure rate modeling for reliability and risk. In: *Springer Series in Reliability Engineering*. London, UK: Springer-Verlag, London Limited.

Free, E.W. (1986). Warranty analysis and renewal function estimation. *Naval Research Logistics Quarterly* 33 (3): 361–372.

Garg, A. and Kalagnanam, K. (1998). Approximations for the renewal functions. *IEEE Transactions on Reliability* 47 (1): 66–72.

Gupta, R.D. and Kundu, D. (1999). Generalized exponential distribution. *Australian and New Zealand Journal of Statistics* 41: 901–916.

Gupta, R.D. and Kundu, D. (2001). Generalized exponential distribution: different method of estimation. *Journal of Statistical Computation and Simulation* 69: 315–337.

Gupta, R.D. and Kundu, D. (2007). Generalized exponential distribution: existing results and some recent developments. *Journal of Statistical Planning and Inference* 137 (11): 3537–3547.

Jaquette, D. (1972). Approximation to the renewal function m(t). *Operations Research* 20 (3): 722–727.

Jewell, N. (1982). Mixtures of exponential distributions. *The Annals of Statistics* 10 (2): 479–484.

Jiang, R. (2008). A gamma-normal series truncation approximation for computing the Weibull renewal function. *Reliability Engineering and System Safety* 93 (4): 616–626.

Jiang, R. (2010). An analytically tractable approximation to the renewal function. *International Journal of Performability Engineering* 6 (2): 109–121.

Jin, T. and Gonigunta, L. (2009). Weibull and gamma renewal approximation using generalized exponential functions. *Communications in Statistics – Simulation and Computation* 38 (1): 154–171.

Jin, T. and Gonigunta, L. (2010). Exponential approximation to Weibull renewal functions for decreasing failure rate. *Journal of Statistical Computation and Simulation* 80 (3): 273–285.

Jin, T. and Liao, H. (2009). Spare parts inventory control considering stochastic growth of an installed base. *Computers and Industrial Engineering* 56 (1): 452–460.

Jin, T., F. Sun, H. Taboada, J. Espiritu, (2014), "A hybrid approximation for Weibull renewal with increasing failure rate," in *Proceedings of ISSAT Conference*. August 7–9, Seattle, WA, pages 1–6.

Jin, T., Taboada, H., Espiritu, J., and Liao, H. (2017). Allocation of reliability – redundancy and spares inventory under Poisson fleet expansion. *IISE Transactions* 49 (7): 737–751.

Kambo, N.S., Rangan, A., and Hadji, E.M. (2012). Moments-based approximation to the renewal function. *Communications in Statistics – Theory and Methods* 41 (5): 851–868.

Kaminskiy, M.P. and Krivtsov, V.V. (1998). A Monte Carlo approach to repairable system reliability analysis. *Probabilistic Safety Assessment and Management* 1063–1068.

Kijima, M. (1989). Some results for repairable systems with general repair. *Journal of Applied Probability* 26 (1): 89–102.

Kijima, M. and Sumita, U. (1986). A useful generalization of renewal theory: counting processes governed by non-negative Markovian increments. *Journal of Applied Probability* 23 (1): 71–88.

Kumar, U.D. and Knezevic, J. (1998). Availability based spare optimization using renewal process. *Reliability Engineering and System Safety* 59 (2): 217–223.

Lam, Y. (1988). Geometric process and the replacement problem. *Acta Mathematicae Applicatae Sinica* 4 (4): 366–382.

Lam, Y. (2007). *The Geometric Process and Applications*. Singapore: World Scientific.

Liao, H., P. Wang, and T. Jin, (2008), "Spare parts management considering new sales," in *Proceedings of Annual Reliability and Maintainability Symposium*, January 28–31, Las Vegas, NV.

Martorell, S., Sanchez, A., and Serradell, V. (1999). Age-dependent reliability model considering effects of maintenance and working conditions. *Reliability Engineering and Systems Safety* 64: 19–31.

McConalogue, D.J. and Pacheco, A. (1981). Numerical treatment of convolution integrals involving distributions with densities having singularities at the origin. *Communications in Statistics* 10 (3): 265–280.

Moors, J.J.A. and Strijbosch, L.W.G. (2002). Exact fill rates for (R, s, S) inventory control with gamma distributed demand. *Journal of Operations Research Society* 53 (11): 1268–1274.

Mudholkar, G.S. and Srivastava, D.K. (1993). Exponentiated Weibull family for analyzing bathtub failure-rate data. *IEEE Transactions on Reliability* 42: 299–302.

Murthy, D.N.P. and Blischke, W.R. (1992). Product warranty management – III: a review of mathematical models. *European Journal of Operational Research* 62: 1–34.

Osaki, S. (2002). *Stochastic Models in Reliability and Maintenance*, 1e. Springer.

Sanchez, A., Carlos, S., Martorell, S., and Villanueva, J.F. (2009). Addressing imperfect maintenance modeling uncertainty in unavailability and cost based optimization. *Reliability Engineering and Systems Safety* 94: 22–32.

Smeitink, E. and Dekker, R. (1990). A simple approximation to the renewal function. *IEEE Transactions on Reliability* 39 (1): 71–75.

Smith, W.L. (1958). Renewal theory and its ramifications. *Journal of Royal Statistical Society, Series B* 20: 243–302.

Tanwar, M., Rai, R.N., and Bolia, N. (2014). Imperfect repair modeling using Kijima type generalized renewal process. *Reliability Engineering and System Safety* 124 (4): 24–31.

Tavner, P.J., Xiang, J., and Spinato, F. (2007). Reliability analysis for wind turbines. *Wind Energy* 10 (1): 1–18.

Tortorella, M. (2005). Numerical solution of renewal-type integral equations. *INFORMS Journal of Computing* 17 (1): 66–74.

US DOE, (2008), "*20% Wind Energy by 2030: Increasing Wind Energy's Contribution to U.S. Electricity Supply*," available at: https://www.nrel.gov/docs/fy08osti/41869.pdf (accessed on 20 September 2017).

US DOE, (2015), *Wind Vision, A New Era for Wind Power in the United States*, Chapter 2, available at: https://www.energy.gov/sites/prod/files/wv_chapter2_wind_power_in_the_united_states.pdf (accessed on September 2017).

Vittal, S., M. Teboul, (2005), "Performance and reliability analysis of wind turbines using Monte Carlo methods based on system transport theory," in *Proceedings of the 46th AIAA/ASME/ASCE/AHS/ASC Structures, Structural Dynamics and Materials Conference*, 2005, pp. 1–8.

Wang, H.Z. and Pham, H. (2006). *Reliability and Optimal Maintenance*. London, UK: Springer.

Xie, M. (1989). On the solution of renewal-type integral equations. *Communications in Statistics – B: Simulation and Computation* 18 (1): 281–293.

Yanez, M., Joglar, F., and Modarres, M. (2002). Generalized renewal process for analysis of repairable systems with limited failure experience. *Reliability Engineering and System Safety* 77: 167–180.

Zhou, X., Xi, L., and Lee, J. (2007). Reliability-cantered predictive maintenance scheduling for a continuously monitored system subject to degradation. *Reliability Engineering and Systems Safety* 92: 530–534.

Problems

Problem 6.1 Please find the renewal function given that the inter-renewal time is exponential with the following failure rates (failures/week): (1) $\lambda = 0.1$; (2) $\lambda = 2$; and (3) $\lambda = 3.5$.

Problem 6.2 Equation (6.3.2) represents the sum of the inter-renewal times of n failures. If T_i for $i = 1, 2, \ldots, n$ are independently and identically exponential distributions with rate λ, show that S_n is a gamma distribution with parameters $\{\lambda, n\}$.

Problem 6.3 Prove the variance of renewal $N(t)$ in Eq. (6.2.10).

Problem 6.4 Given the inter-renewal times are Weibull with the following scale and shape parameters, obtain the asymptotic renewals: (1) $\alpha = 1$, $\beta = 0.5$; (2) $\alpha = 1$, $\beta = 1.5$; (3) $\alpha = 1$, $\beta = 5$; (4) $\alpha = 0.5$, $\beta = 0.5$; (5) $\alpha = 0.1$, $\beta = 4$; (5) $\alpha = 2.5$, $\beta = 0.5$; and (6) $\alpha = 25$, $\beta = 3.5$.

Problem 6.5 Suppose the PDF of the inter-renewal time is given as $f(t) = 2e^{-2t}$ for $t \geq 0$. Use the Laplace transform to find the renewal function $M(t)$.

Problem 6.6 Suppose the PDF of the inter-renewal time is given as $f(t) = e^{-2t} + 2e^{-4t}$ for $t \geq 0$. Use the Laplace transform to find the renewal function $M(t)$.

Problem 6.7 The inter-renewal times follow the Erlang distribution with $\lambda = 1$ and shape parameter $k = 3$. The probability density function is given as $f(t) = 0.5t^2e^{-t}$. Find the renewal function $M_L(s)$ in the s-domain and the $M(t)$ in the t-domain.

Problem 6.8 Find the asymptotic renewal function of the following generalized exponential distributions and also plot their probability density function: (1)

$F_g(t) = (1 - e^{-t})^2$, (2) $F_g(t) = (1 - e^{-3t})^2$, (3) $F_g(t) = (1 - e^{-0.5t})^2$, and (4) $F_g(t) = (1 - e^{-t})^4$.

Problem 6.9 Find the renewal function in the s-domain and then find the t-domain function given the following generalized exponential distribution function: (1) $F_g(t) = (1 - e^{-t})^2$, (2) $F_g(t) = (1 - e^{-2t})^2$, (3) $F_g(t) = (1 - e^{-0.5t})^2$, and (4) $F_g(t) = (1 - e^{-t})^3$.

Problem 6.10 Using the minimum mean squared error to find the mixture of exponential functions for the application of the following Weibull distribution function with the scale parameter α and shape parameter β. These are: (1) $\alpha = 1, \beta = 0.2$; (2) $\alpha = 1, \beta = 0.5$; and (3) $\alpha = 1, \beta = 0.9$.

Problem 6.11 Using the minimum mean squared error to find the mixture of exponential functions for the following Weibull distributions with the scale parameter α and shape parameter β. These are: (1) $\alpha = 0.1, \beta = 0.1$; (2) $\alpha = 0.1, \beta = 0.5$; and (3) $\alpha = 0.1, \beta = 0.9$.

Problem 6.12 Based on the result from the previous problem, find the corresponding approximated Weibull renewal function for these distributions.

Problem 6.13 Using the minimum mean squared error to find the mixture of exponential functions for the following Weibull distribution function with the scale parameter α and shape parameter β. These are: (1) $\alpha = 10, \beta = 0.1$; (2) $\alpha = 10, \beta = 0.5$; and (3) $\alpha = 10, \beta = 0.9$.

Problem 6.14 Based on the result from previous problem, find the corresponding approximated Weibull renewal function for these distributions.

Problem 6.15 Find the transient renewal function $g(t)$ for the following Weibull distribution function with the scale parameter α and shape parameter β. These are: (1) $\alpha = 1, \beta = 1.1$; (2) $\alpha = 1, \beta = 1.3$; and (3) $\alpha = 1, \beta = 1.5$.

Problem 6.16 Based on the transient renewal from the previous problem, find the corresponding renewal function and then plot each Weibull distribution in a chart.

Problem 6.17 Find the transient renewal function $g(t)$ for the following Weibull distribution function with the scale parameter α and shape parameter β. These are: (1) $\alpha = 0.1, \beta = 1.1$; (2) $\alpha = 0.01, \beta = 1.3$; and (3) $\alpha = 0.001, \beta = 1.5$.

Problem 6.18 Based on the transient renewal from the previous problem, find the corresponding renewal function and then plot each Weibull distribution in a chart.

Problem 6.19   Find the transient renewal function $g(t)$ for the following Weibull distribution function with the scale parameter α and shape parameter β. These are: (1) $\alpha = 10$, $\beta = 1.1$; (2) $\alpha = 100$, $\beta = 1.3$; and (3) $\alpha = 1000$, $\beta = 1.5$.

Problem 6.20   Based on the transient renewal from the previous problem, find the corresponding renewal function and then plot each Weibull distribution in a chart.

Problem 6.21   Find the transient renewal function $g(t)$ for the following Weibull distribution function with the scale parameter α and shape parameter β. These are: (1) $\alpha = 1$, $\beta = 2$; (2) $\alpha = 1$, $\beta = 4$; and (3) $\alpha = 1$, $\beta = 7$. Find the corresponding renewal function and then plot each Weibull distribution in a chart.

Problem 6.22   Find the transient renewal function $g(t)$ for the following Weibull distribution function with the scale parameter α and shape parameter β. These are: (1) $\alpha = 0.1$, $\beta = 2$; (2) $\alpha = 0.01$, $\beta = 4$; and (3) $\alpha = 0.001$, $\beta = 7$. Find the corresponding renewal function and then plot each Weibull distribution in a chart.

Problem 6.23   Find the transient renewal function $g(t)$ for the following Weibull distribution function with the scale parameter α and shape parameter β. These are: (1) $\alpha = 10$, $\beta = 2$; (2) $\alpha = 100$, $\beta = 4$; and (3) $\alpha = 1000$, $\beta = 7$. Find the corresponding renewal function and then plot each Weibull distribution in a chart.

Problem 6.24   There were 17 new products released and installed over the past one year. The sales data shows that these systems are installed in the following days of the year: 0, 20, 35, 50, 67, 85, 101, 125, 142, 157, 180, 210, 230, 260, 289, 320, and 350. Assume an exponential lifetime of the product with $\lambda = 1$ failures/year. Estimate the number of failures generated by the fleet: (1) at the end of 6 months (183 days) and (2) at the end of 2 years.

Problem 6.25   Using the same installation data as in the previous problem, estimate the lead-time renewal during the replenishment of the inventory by the end of 6 months given that the lead time is (1) 10 days; (2) 30 days; and (3) 60 days. What is the lead-time renewal by the end of one year?

7

Performance-Based Maintenance

7.1 Introduction

Maintenance, repair, and overhaul plays an important role in sustaining the operation of engineering systems in the private and public sectors. These systems usually possess the following features: capital intensive, long service life, high downtime cost, and mission-critical. The goal of maintenance is to proactively replace aging components to ensure high system reliability and availability with minimum cost. In Sections 7.2–7.6, we introduce three maintenance strategies: corrective maintenance (CM), usage-based preventive maintenance (PM), and condition-based maintenance (CBM). Their operational principle and the way to determine the optimal replacement time is discussed. For CBM, we focus on the gamma process, inverse-Gaussian process, and non-stationary Gaussian process (NSGP) models. In Section 7.7, we introduce the performance-based maintenance (PBM) concept in which the maintenance cost is minimized based on the system availability and the payment scheme, not on the materials, labor, or time transacted. A unified system availability model incorporating various performance drivers is derived under uncertain usage. In Section 7.8, we develop a principal-agent game model for the design and planning of PBM contracts. Challenges arising from the actual implementation are discussed, such as adverse selection, moral hazard, and information asymmetry. The chapter concludes with a case study in which NSGP is employed to predict the remaining useful life of electronics equipment.

7.2 Corrective Maintenance

7.2.1 Classification of Maintenance Policy

At its core, maintenance is to manage, control, and execute the activities that will reasonably ensure that the desired availability and performance of equipment are achieved subject to resource or budgetary constraints. Since maintenance expenditure is often viewed as the necessary premium to be paid for availability insurance, ideally all maintenance decisions need to be directed towards the maximum returns on that investment, such as improved reliability and availability or a reduced safety risk to human life. Thus maintenance is a risk control activity, and the expenditure of maintenance on risk management (e.g. condition monitoring, process control) should be directly related to the reduction of probability and consequences of failures. For capital

Reliability Engineering and Services, First Edition. Tongdan Jin.
© 2019 John Wiley & Sons Ltd. Published 2019 by John Wiley & Sons Ltd.
Companion website: www.wiley.com/go/jin/serviceengineering

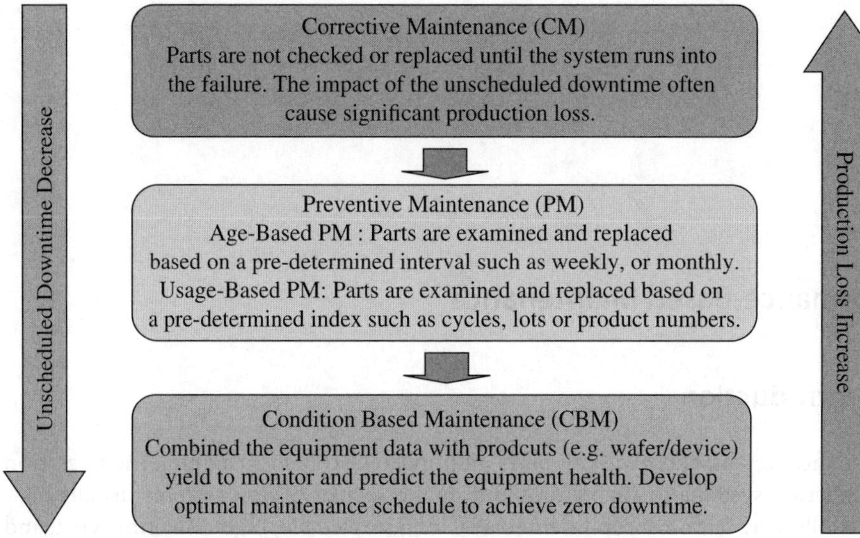

Figure 7.1 The evolution of maintenance technologies.

equipment, core maintenance activities are usually defined by the design and process. Additional maintenance activity often results from premature or unexpected failure. The costs incurred from premature or unexpected failures, such as lost production, loss of reputation, and penalties for late delivery, are usually much larger than the actual repair costs of the failure.

Figure 7.1 classifies the maintenance methods into three categories based on whether the action is triggered by the time or the system condition: (i) corrective maintenance (CM), (ii) preventive maintenanec (PM), and (iii) condition-based maintenance (CBM). Generally speaking, PM is usually triggered by the time (i.e. age) or usage while CM and CBM are system health-driven maintenance strategies. CBM differs from CM in that the replacement actions in CBM are proactively taken prior to the incipient failure.

CM is also known as run-to-failure maintenance (RTFM). Maintenance and repair actions are applied only if the system or the component enters the faulty state (Kenne and Boukas 1997). It is best suited for non-critical systems where the consequence of unexpected downtime is small or negligible, or the failures can be easily identified and removed (Mao and Balasubramaniyan 2009). Typical applications of CM policy include automobiles, personal computers, residential water systems, roof-top solar photovoltaic (PV) panels, and home appliances, among others. For instance, Ding et al. (2009) proposed a method for determining an optimal series of CA contracts for multistate air-conditioning equipment.

PM is maintenance that is regularly performed on a piece of equipment in order to lessen the likelihood of failure. The maintenance is scheduled based on a time or usage trigger. A typical example of an asset with a time-based PM schedule is an escalator that is serviced every year. PM has certain limitations when applied to the equipment that involves usage uncertainty (Wang and Jin 2010). For instance, the lifetime of a safety-critical component is often measured by the engine's actual operating hours. Each engine may have different daily fly hours and the future fly hours also vary day-by-day. In addition, the field failure data collected from the same engine and component type may come from different airlines coupled with different usage conditions. Because of customer and usage variation, the reliability estimate exhibits uncertainty, which deserves careful considerations during the maintenance decision.

CBM is a predictive maintenance triggered by certain predefined metrics or signals, such as vibration or voltage, that indicate the deteriorated system "health" condition (Gebraeel et al. 2009; Bai et al. 2014). The state of the component or system is constantly monitored through sensory data in order to track their level of deterioration. This can be done through the embedded sensors, built-in self-test (BIST) programs, or more recently the Internet of Things. For instance, automated test equipment (ATE) is an electronic instrument used for testing semiconductor wafers and devices. ATE has a built-in self-diagnostic program that routinely checks the functionality and performance of itself to make sure the signals generated to test the devices meet the fidelity criteria. In a power grid system, phasor measure units (PMUs) form a real-time sensor network that measures the voltage and phase angles of transmission lines using a common time source for synchronization (Singh et al. 2013). As such, the grid reliability and heath condition can be monitored and predicted through the aggregation and analysis of these electric parameters in the central control room.

Variations of maintenance strategies are also developed, including opportunistic maintenance (Pham and Wang 2000; Tian et al. 2011), selective maintenance (Cassady et al. 2001), and replacement last (Zhao and Nakagawa 2012). Opportunistic maintenance (OM) gives the maintenance technicians an opportunity to replace those items that are likely to fail and need replacement in the immediate future, during the maintenance of a machine or component. OM gains popularity in the wind power industry because access to offshore wind turbines are rather difficult and costly. Cassady et al. (2001) investigated a selective maintenance policy in which the decision-maker is given multiple maintenance options: minimal repair on failed units, replacement of failed units, or PM. Zhao and Nakagawa (2012) combine both age and periodic PM policies, where the unit could be replaced once it reaches a total operating time T or at a random working cycle Y, whichever occurs last, which is called replacement last. It shows that under a replacement last policy, the unit can be operating for a longer time and avoid premature replacements. This section discusses the working principle and management of CM, and PM and CBM will be elaborated in Sections 7.3 and 7.4.

7.2.2 Corrective Maintenance Management

CM is considered as the simplest maintenance strategy. Assets are deliberately allowed to operate until they break down, at which point reactive maintenance is performed. Since the time-to-failure of systems is random, it is difficult to precisely predict the occurrence time of each failure during the service time. To reduce the unexpected system downtime under CM policy, asset management can estimate the expected number of failures and proactively store sufficient amounts of spare parts for replacement. This is often done by estimating the system mean time to failure (MTTF) in conjunction with the fleet size and usage rate. The following example is provided to assist the estimation of required spare parts under CM policy.

Table 7.1 Fleet operating hours and expected amount of spare inverters from years 1–12.

Year	Operating hours/year	Cumulative operating hours	Cumulative Failures	Failures/year
1	140 400	140 400	1.4	1
2	361 800	502 200	5.0	4
3	388 800	891 000	8.9	4
4–9	388 800	1 279 800	12.8	4
10	388 800	3 612 600	36.1	4
11	361 800	3 974 400	39.7	4
12	140 400	4 114 800	41.1	1

Example 7.1   A DC–AC inverter is a critical component in a roof-top solar PV system. The lifetime of the inverter is exponentially distributed with a constant failure rate $\alpha = 10^{-5}$ failures/hour. Sun Energy is contracted to install multiple roof-top PV systems in a local community. The installation rate is 5 systems/month and the total expected installation is 90 systems in 18 months. The company is also responsible for maintaining these PV systems in the next 10 years after the installation. Assuming CM policy is adopted, calculate how many spare inverters need to be stored annually for Sun Energy to perform failure replacements during the contract period?

Solution:
First we estimate the annual operating hours of the residential solar PV fleet over the 10-year service period. Since solar energy is only available in the daytime, the monthly operating hour of a PV system is 12 (hours/day) × 30 (days/month) = 360 hours/month.

It takes 18 months for Sun Energy to complete the installation of 90 systems. Therefore, the cumulative fleet operating hours increase 1800 hours/month (i.e. 360 × 5) from January of Year 1 to June of Year 2. Once the installation is completed, the monthly PV operating time reaches the maximum with 90 × 360 = 3240 hours/month. This number remains constant until the end of Year 10, then the cumulative operating hours decline from January of Years 11 to June of Year 12 due to the phase-out of the maintenance contract. The yearly demand for replacing failed inverters is obtained by dividing the annual operating hours by the inverter's MTB of 10^5 hours. Table 7.1 summarizes the annual operating hours of the installed PV fleet and the spare inverters required per year for failure replacement.

7.3   Preventive Maintenance

7.3.1   Block Replacement

PM is a scheduled maintenance policy conducted at predefined time intervals or system usages (Chitra 2003). PM is perhaps the most widely used method for equipment maintenance in private and public sectors, including manufacturing, transportation,

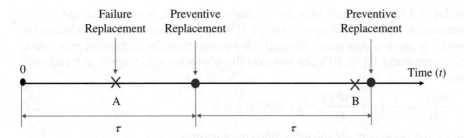

Figure 7.2 Block replacement policy.

energy, healthcare, and defense industries. PM is best suited for those systems that have a clear wear-out characteristic, such as degradation signatures or failure trends (Starr 1997). PM actions include inspection, replacement, and overhaul, and these are made at a predetermined time point to prevent the component from unexpected failure. To that end, it is necessary to understand the component lifetime characteristics, such as mean-time-to-failure, and the cost of sudden failures. The latter happens when the component fails before reaching the predetermined schedule.

Block replacement policy, also known as periodic maintenance policy, is one of the types of PM policy. The replacement is performed periodically at a fixed length of period τ, which is also called the replacement interval. Figure 7.2 shows a typical block replacement scenario. A system or component starts to operate at $t = 0$ and a planned replacement is performed at the end of τ regardless of the number of failures that occurred within that period. If an unexpected failure happens during τ, the replacement is performed immediately, as shown at point A. However, even if a replacement is just made (like B) right before reaching τ, the planned replacement is still performed at the end of τ. Since failures are random, the actual number of failures during τ is uncertain and varies between $[0, \infty)$.

We assume that the maintenance plan is developed over a relatively long period. The intervals between two consecutive replacements are relatively short compared to the planning horizon. When this is the case, we need to consider only one cycle of the operation and estimate a total cost in that cycle. We want to determine the optimal interval τ such that the expected cost per unit time is minimized. The following notation is defined:

c_p = cost of a preventive replacement
c_f = cost of a failure replacement
$R(t)$ = reliability of the component
$F(t)$ = cumulative distribution function (CDF)
$f(t)$ = probability density function (PDF)
$M(t)$ = expected number of failures during τ

The total cost incurred in the interval of $[0, \tau]$ consists of a one-time preventive replacement cost c_p at time τ and the cumulative cost of unexpected failure replacements in that period. Let $M(\tau)$ be the expected number of failures in $[0, \tau]$. Then the goal is to find τ such that the following objective is minimized:

$$\text{min}: \quad C(\tau) = \frac{c_p + c_f M(\tau)}{\tau}, \qquad \text{for } \tau > 0 \tag{7.3.1}$$

where Eq. (7.3.1) aims to minimize the cost per unit time (i.e. the cost rate), and $M(\tau)$ is the number of expected replacements in $[0, \tau]$ and each failure replacement brings the system to as good as new state. Thus $M(\tau)$ is equivalent to the number of renewals in $[0, \tau]$. To minimize Eq. (7.3.1), one can take the derivative with respect to τ and set it to zero:

$$\frac{dc(\tau)}{d\tau} = \frac{1}{\tau^2}\left(c_f\tau\frac{dM(\tau)}{d\tau} - c_p - c_f M(\tau)\right) = 0 \tag{7.3.2}$$

Since $\tau > 0$, we essentially solve the following equation:

$$c_f\tau\frac{dM(\tau)}{d\tau} - c_p - c_f M(\tau) = 0 \tag{7.3.3}$$

The generic renewal equation is given as follows (also see Section 6.2 of Chapter 6):

$$M(t) = F(t) + \int_0^t M(t-x)f(x)dx \tag{7.3.4}$$

A closed-form solution to $M(t)$ is generally not available for most empirical lifetime distributions like Weibull, normal, and lognormal distributions. Therefore, simulation or approximation methods need to be adopted to estimate $M(t)$, and readers are referred to Chapter 6.

Example 7.2   A system follows the Erlang lifetime distribution with $\lambda = 0.1$ failures/month and $k = 2$. The cost of a failure replacement is \$200 and the cost of a preventive replacement is \$50. Find the optimal value of τ to minimize the maintenance cost rate assuming that a periodic maintenance policy is adopted.

Solution:
For an Erlang distribution with $k = 2$ and $\lambda = 1$, the PDF is

$$f(t) = \frac{\lambda^k t^{k-1}\exp(-\lambda t)}{(k-1)!} = 0.01te^{-0.1t}, \qquad \text{for } t \geq 0 \tag{7.3.5}$$

and its Laplace transform is given as follows:

$$L\{f(t)\} = f_L(s) = \frac{0.01}{(s+0.1)^2} \tag{7.3.6}$$

Substituting Eq. (7.3.6) into (6.2.19), the renewal function in the s-domain is obtained as

$$M(s) = \frac{0.01}{s^3 + 0.2s^2} = \frac{0.05}{s^2} - \frac{0.25}{s} + \frac{0.25}{s+0.2} \tag{7.3.7}$$

Hence the time-domain renewal function is

$$M(t) = 0.05t + 0.25(e^{-0.2t} - 1) \tag{7.3.8}$$

and the derivative of $M(t)$ is

$$\frac{dM(t)}{dt} = 0.05 - 0.05e^{-0.2t} \tag{7.3.9}$$

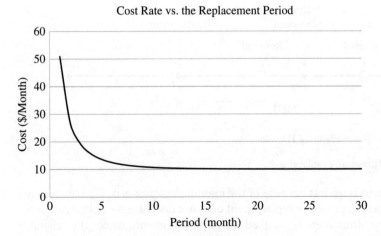

Figure 7.3 Cost rate of block replacement policy.

Substituting Eqs. (7.3.8) and (7.3.9) into (7.3.3) along with $c_f = 200$ and $c_p = 50$, we have

$$(50 + 10t)e^{-0.2\tau} = 0 \qquad (7.3.10)$$

Solving the above equation yields $\tau^* = \infty$ months. This implies that there is no need for block replacement, as the most economical maintenance policy for this system is CM. The cost per unit time versus the length of τ is also plotted in Figure 7.3. It shows that the cost rate reduces monotonically with the period length. This is consistent with the analytical solution obtained in Eq. (7.3.10).

7.3.2 Age-Based Replacement

In age-based PM, the replacement time is determined by the component or system age, not necessarily on the calendar time. Figure 7.4 graphically shows an age-based replacement scenario that involves random failures and the proactive replacements. Note that τ is the scheduled replacement interval depending upon the system age. Assume that the system starts to operate at $t = 0$ and survives to the end of τ. Then a proactive replacement is performed accordingly. Note that MDT stands for the mean downtime (MDT) for performing the replacement. The system returns to the operation at t_2, but fails again at t_3 with $t_3 - t_2 < \tau$. Since the failure occurs prior to reaching τ, a failure replacement is performed at t_3. The system is then restored at t_4 and a new maintenance cycle starts thereafter. Therefore, age-based replacement differs from the block replacement in that τ in the former is always reset after a replacement, regardless of whether it is a failure or planned replacement. In addition, a fundamental assumption for an age-based PM policy is that the component failure rate is monotonically increasing (Jardine and Buzacott 1985). This implies that if the underlying failure rate is constant or monotonically decreasing, any proactive replacement is considered as a waste of resources. For a component with a constant failure rate, the lifetime is exponential and

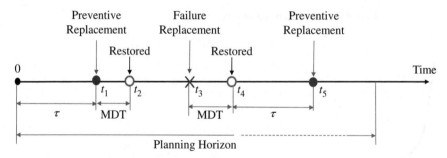

Figure 7.4 Age-based preventive replacement.

proactive replacement is useless because of the memoryless property. If a component's failure rate is decreasing, proactive replacement does not decrease the failure rate at the time of replacement. Thus there is no need for proactive maintenance. The reliability under-age-based PM is often characterized by the mean-time-between-replacements (MTBRs) as opposed to the mean-time-between-failures (MTBFs) in CM. Let T_{MTBR} be the system MTBR; then

$$T_{MTBR} = \int_0^\tau tf(t)dt + \tau R(\tau) = \int_0^\tau R(t)dt \qquad (7.3.11)$$

where $f(t)$ is the PDF of the product life and $R(t)$ is the reliability function. Figure 7.4 shows that replacements under age-based PM consist of two streams: failure replacements and planned replacements. Assume that MDT is small and negligible compared to MTBR. Our goal is to minimize the maintenance cost per unit time or the cost rate, which is given as

$$\min: \quad C(\tau) = \frac{c_p R(\tau) + c_f F(\tau)}{\int_0^\tau R(t)dt}, \qquad \text{for } \tau > 0 \qquad (7.3.12)$$

where $F(t)$ is the cumulative lifetime distribution and $F(t) = 1 - R(t)$. The numerator essentially captures the statistical average cost resulting from both preventive and failure replacements during an MTBR cycle. To find the optimal τ that minimizes $C(\tau)$, we take the derivative with respect to τ in Eq. (7.3.12) and set it to zero:

$$\frac{dC(\tau)}{d\tau} = \frac{(c_f - c_p)f(\tau)\int_0^\tau R(t)dt - [c_p R(\tau) + c_f F(\tau)]R(\tau)}{\left(\int_0^\tau R(t)dt\right)^2} = 0 \qquad (7.3.13)$$

Now the optimal τ, denoted as τ^*, is determined by solving the following equation:

$$(c_f - c_p)f(\tau^*) \int_0^{\tau^*} R(t)dt - (c_p R(\tau^*) + c_f F(\tau^*))R(\tau^*) = 0 \qquad (7.3.14)$$

In general an explicit solution is hardly derived from Eq. (7.3.14) as it involves integration of $R(t)$. Below we present an example to illustrate how to find an optimal τ via the numerical method.

Example 7.3 When a product exhibits a deteriorating lifetime trend, its reliability can be modeled by a Weibull distribution with the following form:

$$R(t) = \exp(-(\alpha t)^\beta), \qquad \text{for } t \geq 0 \qquad (7.3.15)$$

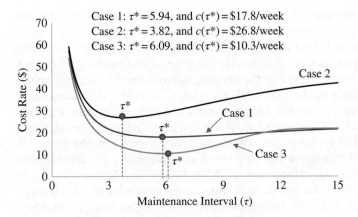

Figure 7.5 The optimal maintenance intervals for three cases.

where α and β are the scale and shape parameters, respectively. It is decided to use the age-based PM policy to maintain the operation of the product. Find the optimal τ to minimize the cost rate in the following three cases:

Case 1: For $\alpha = 0.1$ failures/week and $\beta = 2$ with $c_f = \$200$ and $c_p = \$50$.
Case 2: For $\alpha = 0.1$ failures/week and $\beta = 2$ with $c_f = \$400$ and $c_p = \$50$.
Case 3: For $\alpha = 0.1$ failures/week and $\beta = 4$ with $c_f = \$200$ and $c_p = \$50$.

Solution:
We develop a numerical method to search for τ^* in Eq. (7.3.14). The algorithm is implemented in Matlab and the code is shown in Appendix 7.A. The code starts with "clear," and four input parameters α, β, c_f, and c_p shown at the beginning. Users can copy and paste the codes to run the algorithm in their own Matlab environment. The optimal τ^* and $c(\tau^*)$ represented by "tao_op" and "cost_op" will be displayed upon execution. We solve Cases 1 to 3 and the results are depicted in Figure 7.5. By comparing Cases 1 and 2, it was found that the maintenance interval reduces with the increased failure cost of failure replacements.

7.4 Condition-Based Maintenance

7.4.1 Principle of Condition-Based Maintenance

Traditional maintenance policies, such as PM and CM, have been widely used in public and private sectors. Albeit their popularity, these methods suffer from two main issues: under-maintenance and over-maintenance. The under-maintenance occurs when the system health is not appropriately monitored, resulting in an unexpected failure. The over-maintenance is caused due to the lack of precise or sufficient information about the system health. Hence, we tend to be conservative and excessively replace aging units, especially in risk-averse and safety-critical systems, such as airplanes or medical devices. This results in material waste, increased labor cost, and low utilization of equipment.

Condition-based maintenance (CBM) emerges as an enabling technology with the goal of achieving just-in-time maintenance with a zero-failure target (Lopez 2007; Vichare and Pecht 2006). In CBM the equipment health is monitored by in-situ sensors and the real-time reliability is tracked through a prognostics and health management system (PHMS). CBM is able to predict the pending failure caused by a single failure mode or by multiple failure modes (Niu et al. 2011; Liu et al. 2013). The maintenance job is initiated just before the incipient failure so that the system can remain in uptime as long as possible. Thus CBM is sometimes also referred to as a predictive maintenance technology. Though CBM technology is still in its early development phase, the US Joint Strike Fighter (JSF) Program has developed and integrated a comprehensive PHMS in the J-35 aircraft fleet (Brown et al. 2007). PHMS has also been shown to be successful in monitoring the health of a lithium-ion battery based on advanced regression, classification, and state estimation algorithms (Goebel et al. 2008). More discussions on modeling and applications of PHMS and CBM are available in the papers by Sun et al. (2007) and Tsui et al. (2015) and the book by Pecht (2009).

Figure 7.6 graphically shows the degradation path of three systems over time. Note that $S(t)$ is a random variable representing the degradation level at time t and d is the failure threshold which can be pre-defined. For mechanical units, $S(t)$ can represent the wear-out of a tire or the crack size of ceramic materials. For an electronics instrument, voltage, power and frequency are often designated as the health indicator. Starting at $t = 0$, the value of $S(t)$ increases randomly with time. This stochastic aging process is caused by multiple factors, such as operating environment, load variation, customer usage, and product quality variations, among others. Since the degradation trajectory is often stochastic intermingled with observation uncertainties (Pan and Wang 2014), the system may fail at t_1, t_2, or t_3 depending on the progression speed of $S(t)$. Now the reliability function under CBM can be expressed as

$$R(t) = P\{T \ge t\} = P\{S(t) \le d\} \tag{7.4.1}$$

Equation (7.4.1) states that the system is in a good state provided that $S(t)$ does not exceed the threshold value d. The time when $S(t)$ first surpasses d is called the first hitting time, or first passage time. Therefore, the main task of CBM is to predict the first

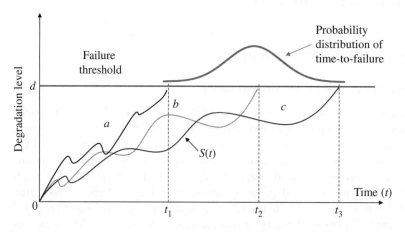

Figure 7.6 Degradations under different trajectory or paths.

passage time given the current health condition of the system. A main hurdle against the adoption of CBM is its relatively high expenses in installing sensors and deploying prognostics tools in order to obtain high-fidelity health information. At present CBM is often implemented in the most "critical" systems or components – those affecting the safety, the capital loss, and the productivity (Starr 1997), such as aircraft engines, wind turbines, and battery systems. With the advent of big data analytics, machine learning, and the Internet of Things, the effective applications of CBM in complex systems will become increasingly mature in electronics, aerospace, transportation, and wind generation systems.

Example 7.4   Assume that the degradation path $S(t)$ in Figure 7.6 is deterministic and exhibits the following form: (i) $S(t) = a + bt$, (ii) $S(t) = a + bt^2$, and (iii) $S(t) = a(e^{bt} - 1)$. Estimate the first passage time t when $S(t)$ exactly exceeds the failure threshold d.

Solution:
(1) Find τ such that we simply solve the following equation:

$$S(t) = a + bt = d$$

which yields

$$t = \frac{d - a}{b}$$

(2) Using the same concept, we can find t by solving the following equation:

$$S(t) = a + bt^2 = d$$

which yields

$$t = \sqrt{\frac{d - a}{b}}$$

(3) We can find t by solving the following equation:

$$S(t) = a(e^{bt} - 1) = d$$

which yields

$$t = \frac{1}{b} \ln\left(\frac{b}{a} + 1\right)$$

7.4.2   Proportional Hazard Model

Statistical models relating to managing CBM policy have been frequently reported in the literature (Gebraeel et al. 2009; Tian 2009; Jin and Mechehoul 2010). A valuable statistical procedure for estimating the risk of equipment failure when it is under condition monitoring is the proportional hazards model (PHM). Originally proposed by Cox (1972), a PHM can take various forms, but all combine a baseline hazard function with an exponential term that takes into account the key factors affecting the progress of the degradation as follows:

$$h(t, \mathbf{z}(t)) = h_0(t)e^{\mathbf{b} \times \mathbf{z}(t)} \tag{7.4.2}$$

where $h_0(t)$ is the baseline hazard function. Items in $\mathbf{z}(t) = [z_1(t), z_2(t), ..., z_m(t)]$ are referred to as the covariates that are time-dependent in general. Each $z_i(t)$ for $i = 1, 2,$ ..., m corresponds to a monitored performance variable at the time of inspection t, such as the vibration, temperature, and humidity, power, and voltage. These condition data are collectively called covariates or stresses. Items in $\mathbf{b} = [b_1, b_2, ..., b_m]$ are the covariate coefficients. Together with $z_i(t)$, the product of $\mathbf{b} \times \mathbf{z}(t)$ indicates the degree of influence each covariate has on the baseline hazard function.

If the Weibull hazard function is adopted in Eq. (7.4.2), this particular form used in this section is known as a Weibull PHM model as follows:

$$h(t, \mathbf{z}(t)) = \frac{\beta}{\eta}\left(\frac{t}{\eta}\right)^{\beta-1} \exp\left(\sum_{i=1}^{m} b_i z_i(t)\right) \tag{7.4.3}$$

with

$$h_0(t) = \frac{\beta}{\eta}\left(\frac{t}{\eta}\right)^{\beta-1} \tag{7.4.4}$$

$$g(\mathbf{z}(t)) = \exp\left(\sum_{i=1}^{m} b_i z_i(t)\right) \tag{7.4.5}$$

where η and β are the Weibull scale and shape parameters, respectively. The Weibull PHM in Eq. (7.4.3) consists of two parts: $h_0(t)$ is the Weibull baseline hazard function that captures the normal aging process and $g(\mathbf{z}(t))$ contains the variables that represent the primary factors and their corresponding weights for tracking the system health condition.

Example 7.5 In the study by Anderson et al. (1982), the form of the hazard model for the aircraft engine was $\eta = 24\,100$ hours, $\beta = 4.47$, $b_1 = 0.41$, $b_2 = 0.98$, and $z_1(t)$ is the ion concentration and $z_2(t)$ is the chromium concentration in parts per million. Its Weibull PHM is in the form

$$h(t, \mathbf{z}(t)) = \frac{4.47}{24\,100}\left(\frac{t}{24\,100}\right)^{4.47-1} \exp(0.41 z_1(t) + 0.98 z_2(t)) \tag{7.4.6}$$

Assuming that the ion and chromium concentrations are time-independent, do the following: (i) obtain the cumulative hazard rate function; (ii) obtain the system reliability function; and (iii) plot the PHM model at different combinations of $z_1(t)$ and $z_2(t)$.

Solution:
1) Since the covariates are time-independent, it implies $z_1(t) = z_1$ and $z_2(t) = z_2$. The cumulative hazard rate function is

$$H(t, \mathbf{z}) = \int_0^t h(x, \mathbf{z})dx = \int_0^t \frac{4.47}{24\,100}\left(\frac{x}{24\,100}\right)^{4.47-1} e^{0.41 z_1 + 0.98 z_2} dx$$

$$= e^{0.41 z_1 + 0.98 z_2}\left(\frac{t}{24\,100}\right)^{4.47}$$

2) The reliability function of the system considering the effects of covariates is

$$R(t, \mathbf{z}) = \exp(-H(t; \mathbf{z})) = \exp\left(-e^{0.41 z_1 + 0.98 z_2}\left(\frac{t}{24\,100}\right)^{4.47}\right)$$

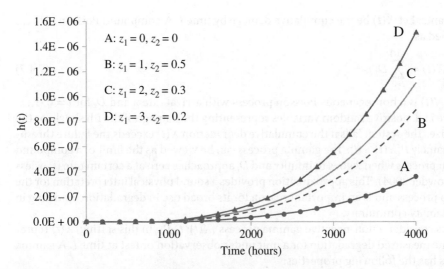

Figure 7.7 Hazard rate of aircraft engines at different covariate levels.

For instance, at $t = 10\,000$ hours,

$$R(t, z_1 = 1, z_2 = 1) = \exp\left(-e^{0.41\times1+0.98\times1}\left(\frac{10\,000}{24\,100}\right)^{4.47}\right) = 0.93$$

$$R(t, z_1 = 2, z_2 = 2) = \exp\left(-e^{0.41\times2+0.98\times2}\left(\frac{10\,000}{24\,100}\right)^{4.47}\right) = 0.764$$

The results show that doubling the iron and chromium concentrations reduces the reliability by $(0.93-0.764)/0.93 = 18\%$ at $t = 10\,000$ hours.

3) The PHMs under different covariate or stress levels are shown in Figure 7.7. Note that A represents the baseline hazard rate $h_0(t)$ with $z_1 = z_2 = 0$. The actual hazard rate $h(t)$ increases with z_1, z_2, or both, as shown in B, C, and D.

7.4.3 Gamma Degradation Process

Many degradation phenomena, such as wear and tear, crack growth, chemical corrosion, and resistance build-up, can be treated as accumulations of additive and irreversible damage to the system caused by a series of random shocks. The arrival sequence of the shocks may be approximated by a Poisson counting process, in which each shock causes a small yet random amount of damage to the system (Desmond 1985; Singpurwalla 1995; van Noortwijk 2009). For example, in machining an engine cylinder, a boring tool is used to enlarge a hole that has already been drilled (or cast). A boring cycle typically consists of five steps: (1) rough-cut on the entry stroke, (2) chamfer, (3) step-off, (4) compensation movement, and (5) exit stroke. Each cycle increases the wear-out of the tool with a small and random amount, while the number of cycles may be modeled as a Poisson counting process. As another example, the capacity of a lithium-ion battery in electric vehicles usually decreases with charge–discharge cycling. Each cycle robs the battery capacity of a small amount, while the number of cycles can be appropriated by a Poisson process (Singpurwalla and Wilson 1998; Lawless and Crowder 2010). The aging physics of these examples suggests that degradation can be modeled by a compound Poisson process or its variants. Let $X(t)$ be the cumulative damage by time t. A compound Poisson process is defined as

$$X(t) = \sum_{i=0}^{N(t)} D_i \tag{7.4.7}$$

where $N(t)$ is a homogeneous Poisson process with arrival rate λ and D_i for $i = 0, 1,...,$ $N(t)$ are i.i.d. positive random variables representing the damage of the ith shock, or the jump size. The system fails if the cumulative degradation $X(t)$ exceeds the failure threshold s, namely $P\{X(t) > d\}$. The gamma process can be viewed as the limit of a compound Poisson process when λ goes to infinity and D_i approaches zero at a certain rate (Lawless and Crowder 2004). This approximation provides a sound physical interpretation for the gamma process and is an important reason for its broad use in degradation modeling in the reliability community.

Let us consider a non-negative gamma process $\{X(t); t \geq 0\}$. In this setting, $X(t)$ represents the measured degradation for a unit under observation or test at time t. A gamma process has the following properties:

(1) $X(0) = 0$ with probability one.
(2) The increment $\Delta X(t) = X(t+\tau) - X(t)$ are independent of t.
(3) $\Delta X(t)$ has a gamma distribution $\Gamma(u, v(t+\tau) - v(t))$, where $v(\cdot)$ is a monotone increasing function.

Thus, with the convention that $X(0) = 0$ and $v(0) = 0$, the degradation $X(t)$ has distribution $\Gamma(u, v(t))$ associated with the following PDF:

$$g(x; v(t), u) = \frac{u^{v(t)} x^{v(t)-1} e^{-ux}}{\Gamma(v(t))}, \quad \text{for } x \geq 0 \tag{7.4.8}$$

with

$$\Gamma(v(t)) = \int_0^\infty z^{v(t)-1} e^{-z} dz \tag{7.4.9}$$

where Eq. (7.4.9) is the gamma function. If $v(t)$ is an integer, then $\Gamma(v(t)) = (v(t)-1)!$ Otherwise, it has to be evaluated numerically. The mean and the variance of $X(t)$ are given as follows:

$$E[X(t)] = \frac{v(t)}{u}, \qquad Var(X(t)) = \frac{v(t)}{u^2} \tag{7.4.10}$$

Suppose that $X(t)$ is observed at time $t_0 < t_1 < \cdots < t_n$, yielding $x_0, x_1, ..., x_n$. Define the x-increment $\Delta x_j = x_j - x_{j-1}$ for $j = 1, 2, ..., n$. Then by properties of (1) to (3), their joint distribution has the following PDF:

$$f(\Delta x) = \prod_{j=1}^n g(\Delta x_j; u, \Delta v_j) \tag{7.4.11}$$

where

$$\Delta v_j = v(t_j) - v(t_{j-1}) = v_j - v_{j-1}, \qquad \text{for } j = 1, 2, ..., n \tag{7.4.12}$$

Note that $\Delta \mathbf{x} = [\Delta x_1, \Delta x_2, ..., \Delta x_n]$. The joint PDF is the multiplication of n gamma distribution with parameter $\{\mu, \Delta v_j\}$. This result is obtained simply because of an independent increment of Δx_j for $j = 1, 2, ..., n$.

Recall the degradation failure time T in Eq. (7.4.1) in which a unit fails when its degradation path first crosses the threshold d. For the gamma process model considered here, the degradation path is monotonic and the lifetime function is obtained as follows:

$$F(t) = P\{T < t\} = P\{X(t) > d\} = \int_d^{\infty} g(x; u, v(t))dx$$

$$= \frac{1}{\Gamma(v(t))} \int_d^{\infty} u^{v(t)} x^{v(t)-1} e^{-ut} dx = \frac{\Gamma(v(t), du)}{\Gamma(v(t))} \tag{7.4.13}$$

with

$$\Gamma(v(t), du) = \int_{du}^{\infty} z^{v(t)-1} e^{-z} dz \tag{7.4.14}$$

Equation (7.4.13) seamlessly connects the time-domain T with the physics-domain $X(t)$ of a unit's degradation path, thus making the gamma process amenable to quantitative analysis. Note that Eq. (7.4.14) is the incomplete gamma function for $du > 0$ and $v(t) > 0$. In general there is no closed-form solution for $\Gamma(v(t), du)$, which needs to be evaluated using the numerical method.

Under the assumption that the shape parameter $v(t)$ is continuous and differential in t, the PDF of T can be obtained by applying the chain rule to $F(t)$ in Eq. (7.4.12). That is,

$$f(t) = \frac{v'(t)}{\Gamma(v(t))} \int_{du}^{\infty} [\ln(z) - \psi(v(t))] z^{v(t)-1} e^{-z} dz \tag{7.4.15}$$

where $\psi(x)$ is called the digamma function, which is obtained by computing the ratio of the first derivative of $\Gamma(v(t))$ over itself as follows:

$$\psi(v(t)) = d \ln(\Gamma(v(t))) = \frac{1}{\Gamma(v(t))} \times \frac{\partial \Gamma(v(t))}{\partial v(t)} \tag{7.4.16}$$

In mathematics, the digamma function is defined as the logarithmic derivative of the gamma function. It is worth mentioning that the derivative is taken with respect to $v(t)$ only in the digamma function, not to t. Both maximum likelihood estimate and semi-parametreic estimation methods can be used to estimate the distribution parameters of the gamma process (Ye et al. 2014).

Example 7.6 Provide the step-by-step deviation of the PDF in Eq. (7.4.15).

Solution:
First substituting Eq. (7.4.8) into Eq. (7.4.13), we have

$$F(t) = P\{X(t) > d\} = \frac{1}{\Gamma(v(t))} \int_d^{\infty} u^{v(t)} x^{v(t)-1} e^{-ux} dx = \frac{1}{\Gamma(v(t))} \int_{du}^{\infty} z^{v(t)-1} e^{-z} dz \tag{7.4.17}$$

The last term is obtained by redefining $z = ux$. Taking the derivative with respect to t in Eq. (7.4.17) yields

$$\frac{\partial}{\partial t} F(t) = \frac{1}{\Gamma^2(v(t))} \left(\Gamma(v(t)) v'(t) \int_{du}^{\infty} \ln(z) z^{v(t)-1} e^{-z} dz \right.$$

$$\left. - \frac{\partial \Gamma(v(t))}{\partial v(t)} v'(t) \int_{du}^{\infty} z^{v(t)-1} e^{-z} dz \right)$$

$$= \frac{v'(t)}{\Gamma^2(v(t))} \left(\Gamma(v(t)) \int_{du}^{\infty} \ln(z) z^{v(t)-1} e^{-z} dz - \frac{\partial \Gamma(v(t))}{\partial v(t)} \int_{du}^{\infty} z^{v(t)-1} e^{-z} dz \right)$$

$$= \frac{v'(t)}{\Gamma(v(t))} \int_{du}^{\infty} \left(\ln(z) - \frac{1}{\Gamma(v(t))} \times \frac{\partial \Gamma(v(t))}{\partial v(t)} \right) z^{v(t)-1} e^{-z} dz$$

$$= \frac{v'(t)}{\Gamma(v(t))} \int_{du}^{\infty} (\ln(z) - \psi(v(t)) z^{v(t)-1} e^{-z} dz \qquad (7.4.18)$$

The final result is obtained by referring to the definition of the digamma function in Eq. (7.4.16). This complete the proof.

7.4.4 Stationary Gamma Degradation Process

Empirical studies show that the shape parameter of the gamma deterioration process at time t is often proportional to a power law as follows:

$$v(t) = ct^b, \quad \text{for } t > 0 \qquad (7.4.19)$$

with $c > 0$ and $b > 0$. For instance, Ellingwood and Mori (1993) investigated the aging process of concrete materials according to a power law, and found that the expected degradation of concrete materials exhibits three patterns: linear with $b = 1$ due to corrosion of reinforcement, quadratic with $b = 2$ under sulfate attack, and square root with $b = 0.5$ under diffusion-controlled aging. The gamma process is called a stationary process if $b = 1$ and non-stationary otherwise. Let t be the current time and τ be the incremental time. For $b = 1$, we have

$$v(t + \tau) - v(t) = c \times (t + \tau) - ct = c\tau \qquad (7.4.20)$$

which is independent of the time t and only depends on τ. Below we discuss the properties of the stationary gamma process, while the non-stationary process is available in Van Noortwijk (2009). When the expected deterioration is linear over time, it is convenient to rewrite Eq. (7.4.8), the PDF for $X(t)$, by using the following reparameterization:

$$f_{X(t)}(t) = \Gamma(x; \mu^2 t/\sigma^2, \mu/\sigma^2), \quad \text{for } x \geq 0 \qquad (7.4.21)$$

If one compares Eq. (7.4.21) with Eq. (7.4.8), it is found that $v(t) = ct = \mu^2 t/\sigma^2$ and $u = \mu/\sigma^2$. Now the mean and variance for the stationary gamma process are

$$E[X(t)] = \mu t, \qquad Var(X(t)) = \sigma^2 t \qquad (7.4.22)$$

Due to the stationarity or the linearity of the stochastic deterioration process in Eq. (7.4.21), both the mean and the variance of the deterioration are linear in time.

A gamma process is called a standardized gamma process if $\mu = 1$ and $\sigma = 1$. By using Eq. (7.4.13), the lifetime distribution can be rewritten as (Van Noortwijk 2009)

$$F(t) = P\{T \leq t\} = \int_d^{\infty} f_{X(t)}(x) dx = \frac{\Gamma(\mu^2 t/\sigma^2, d\mu/\sigma^2)}{\Gamma(\mu^2 t/\sigma^2)} \qquad (7.4.23)$$

The stationarity feature of the gamma process basically results from the property that the increments are independent and have the same type of distribution as their sum, as shown in Eq. (7.4.11). In mathematical terms, this property belongs to the so-called infinite divisibility.

The exact analysis of the gamma distribution and density functions are quite involved, as shown in Eqs. (7.4.23) and (7.4.14). To facilitate the computation, Park and

Padgett (2005) proposed to approximate the cumulative gamma process distribution function (7.4.23) using the Birnbaum–Saunders distribution (1969) as follows:

$$F(t) = P\{X(t) \geq d\} \cong \Phi\left(\frac{\mu t - d}{\sigma\sqrt{t}}\right) = \Phi\left(\sqrt{\frac{\mu d}{\sigma^2}}\left[\sqrt{\frac{\mu t}{d}} - \sqrt{\frac{d}{\mu t}}\right]\right) \tag{7.4.24}$$

where $\Phi(\bullet)$ is the CDF of the standard normal distribution. This approximation is applicable provided $\mu \gg \sigma$. The Birnbaum–Saunders PDF is given as

$$f(t) = \frac{\partial}{\partial t}\Phi\left(\frac{\mu t - d}{\sigma\sqrt{t}}\right) = \frac{1}{\sqrt{2\pi}}\left(\frac{\mu t + d}{2\sigma t^{3/2}}\right)\exp\left(-\frac{(\mu t - d)^2}{2\sigma^2 t}\right) \tag{7.4.25}$$

Example 7.7 Prove the reparameterization in Eq. (7.4.21), namely show that

$$v(t) = ct = \frac{\mu^2 t}{\sigma^2} \text{ and } u = \frac{\mu}{\sigma^2} \tag{7.4.26}$$

Proof:
Based on the mean and variance in Eqs. (7.4.10) and (7.4.22), we have the following connections between the stationary gamma process and the general gamma process:

$$E[X(t)] = \frac{v(t)}{u} = \frac{ct}{u} = \mu t \tag{7.4.27}$$

$$Var(X(t)) = \frac{v(t)}{u^2} = \frac{ct}{u^2} = \sigma^2 t \tag{7.4.28}$$

These equations are obtained by realizing $v(t) = ct$ for the stationary gamma process. Both Eqs. (7.4.27) and (7.4.28) are equivalent to the following system of equations:

$$\begin{cases} v(t) = u\mu t \\ v(t) = u^2\sigma^2 t \end{cases} \tag{7.4.29}$$

Solving this equation system yields $v(t) = \mu^2 t/\sigma^2$ and $u = \mu/\sigma^2$. This completes the proof.

Sometimes a component is subject to two degradation mechanisms, like wear and shock. Then deterioration can be modeled as the sum of normal wear and cumulative shocks. For instance, Peng et al. (2011) designed a PM policy for complex systems that experienced multiple dependent competing failure processes: soft failures due to degradation and catastrophic failures caused by random shocks. Zhu et al. (2015) dealt with a similar problem in which the component wear damage is modeled by a gamma process and the random shock damage is modeled by a generalized counting process. The joint effect on the mean deterioration rate is modeled by covariates via the Cox proportional hazards model.

7.5 Inverse Gaussian Degradation Process

7.5.1 Distribution of Inverse Gaussian Process

Another attractive degradation model with monotone path is the inverse Gaussian (IG) process. Originally proposed by Wasan (1968), the characteristics of IG distribution has been well explored because of its close relation with the Wiener process with drift. However, its application in modeling system degradation and remaining life started quite recently. One possible reason could be ascribed to its unclear physical interpretation to reliability engineers as opposed to the Wiener and gamma processes. Recently Wang and Xu (2010) and Ye and Chen (2014) used IG processes to model the degradation trajectory of GaAs laser devices, and the results show that the IG process with a random volatility model fits the data reasonably well. Yet neither the Wiener nor the gamma process is able to provide such a good fit (Feng et al. 2012).

It has been shown that the IG process is also a limiting compound Poisson process. However, unlike the gamma process, the IG process has different jump size distributions (Wang and Xu 2010). This statistical justification shows that the IG process is qualified as a degradation model. Ye and Chen (2014) have proved that when the shock arrival rate goes to infinity and the jump size approaches zero in a certain way, the limiting compound Poisson process is indeed an IG process. The IG process $\{X(t), t \geq 0\}$ is defined as the stochastic process that satisfies the following properties:

(1) $X(t)$ has independent increments, namely, $X(t_2) - X(t_1) > 0$ and $X(s_2) - X(s_1) > 0$ for $t_2 > t_1 \geq s_2 > s_1$.
(2) $X(t) - X(s)$ follows the IG distribution with $IG((\Lambda(t) - \Lambda(s), \eta[(\Lambda(t) - \Lambda(s))^2])$ for $t > s \geq 0$.

where $\Lambda(t)$ and $\Lambda(s)$ are a monotone increasing function and $IG(a, b)$ denotes its distribution. According to Ye and Chen (2014), the PDF of Inverse Gaussian process is

$$f(x; a, b) = \sqrt{\frac{b}{2\pi x^3}} \exp\left(-\frac{b(x - a)^2}{2a^2 x}\right), \quad \text{for } x > 0 \tag{7.5.1}$$

where a and b are the parameters of the IG distribution with $a > 0$ and $b > 0$. According to Property (2), we can immediately infer the values of both parameters, that is,

$$a = \Lambda(t) - \Lambda(s) \tag{7.5.2}$$
$$b = \eta[\Lambda(t) - \Lambda(s)]^2 \tag{7.5.3}$$

and the corresponding CDF is

$$F(x; a, b) = \Phi\left[\sqrt{\frac{b}{x}}\left(\frac{x}{a} - 1\right)\right] + e^{\frac{2b}{a}}\Phi\left(-\sqrt{\frac{b}{x}}\left(\frac{x}{a} + 1\right)\right), \quad \text{for } x > 0 \tag{7.5.4}$$

where $\Phi(\bullet)$ is the standard normal CDF. The mean and variance of the IG distribution are (Wang and Xu 2010)

$$E[X(t)] = a \tag{7.5.5}$$

$$Var(X(t)) = \frac{a^3}{b} \tag{7.5.6}$$

Let $\Lambda(s = 0) = 0$ and $Y(s = 0) = 0$. According to property (2), $X(t)$ follows $IG(\Lambda(t), \eta\Lambda(t)^2)$ with the mean $\Lambda(t)$ and variance $\Lambda(t)/\eta$. Both the mean and variance are obtained by substituting $a = \Lambda(t)$ and $b = \eta\Lambda(t)^2$ into Eqs. (7.5.5) and (7.5.6), respectively.

Suppose that $X(t)$ is observed at time $t_0 < t_1 < \cdots < t_n$, and the measurements are x_0, $x_1, \ldots, x_n$. Defining the x-increment, $\Delta x_j = x_j - x_{j-1}$ for $j = 1, 2, \ldots, n$. The property of independent increments of the IG process (see Properties (1) and (2)) makes the degradation prediction fairly straightforward. That is,

$$\{X(t) \mid X(t_1), \ldots, X(t_n)\} \sim IG\{\Lambda(t) - \Lambda(t_n), \eta[\Lambda(t) - \Lambda(t_n)]^2\},$$
$$\text{for } t > t_n > \cdots > t_1 \geq 0 \tag{7.5.7}$$

7.5.2 Probability Density Function of First Passage Time

In condition-based monitoring, the time-to-failure of an item, denoted as T, is defined as the first passage time at which the degradation path crosses a predetermined threshold d for the first time. Hence, knowing the distribution of the first passage time, we are able to predict the remaining useful life (RUL) of the item (Van Noortwijk 2009). Consider the IG process with $IG(\Lambda(t), \eta\Lambda(t)^2)$. The CDF for T can be readily obtained from the following relation:

$$
\begin{aligned}
F(t) &= P\{T \le t\} = P\{X(t) \ge d\} \\
&= 1 - F(d; \Lambda(t), \eta\Lambda^2(t)) \\
&= 1 - \Phi\left[\sqrt{\frac{\eta}{d}}(d - \Lambda(t))\right] - e^{2\eta\Lambda(t)}\Phi\left(-\sqrt{\frac{\eta}{d}}(\Lambda(t) + d)\right) \\
&= \Phi\left[\sqrt{\frac{\eta}{d}}(\Lambda(t) - d)\right] - e^{2\eta\Lambda(t)}\Phi\left(-\sqrt{\frac{\eta}{d}}(\Lambda(t) + d)\right)
\end{aligned}
\tag{7.5.8}
$$

As time t progresses and increases, $X(t)$ can be approximated as a normal distribution with mean $\Lambda(t)$ and variance $\Lambda(t)/\eta$ (Chhikara and Folks 1989). Then the corresponding CDF for T in Eq. (7.5.8) can be approximated as

$$
F(t) \cong 1 - \Phi\left(\frac{d - \Lambda(t)}{\sqrt{\Lambda(t)/\eta}}\right) = \Phi\left(\sqrt{\eta\Lambda(t)} - d\sqrt{\frac{\eta}{\Lambda(t)}}\right)
\tag{7.5.9}
$$

which belongs to the Birnbaum–Saunders-type distribution (Tang and Chang 1995). In addition, closed-form expressions for PDF for T are readily available by taking the derivatives of Eqs. (7.5.8) and (7.5.9) with respect to t. This will be elaborated in the next section. In contrast to Eq. (7.4.13), the first passage time of the gamma process does not possess an explicit PDF form, making it more difficult to perform an exact statistical inference.

Example 7.8   Plot the IG distribution function given the following parameters: $\{a = 1, b = 1\}$, $\{a = 1, b = 4\}$, $\{a = 1, b = 0.5\}$, $\{a = 4, b = 1\}$, and $\{a = 0.5, b = 1\}$.

Solution:
According to the PDF in Eq. (7.5.1), we plot the distribution function by varying $x \in (0, 2]$ and the results are plotted in Figure 7.8.

For the IG degradation process, the PDF of time-to-failure of the degradation is obtained by differentiating $F(t)$ in Eq. (7.5.8). The result is

$$
f_T(t) = \Lambda'(t)\left\{\sqrt{\frac{\eta}{d}}\phi\left[\frac{\sqrt{\eta}(\Lambda(t) - d)}{\sqrt{d}}\right] - 2\eta e^{2\eta\Lambda(t)}\Phi\left[-\frac{\sqrt{\eta}(\Lambda(t) + d)}{\sqrt{d}}\right] + e^{2\eta\Lambda(t)}\sqrt{\frac{\eta}{d}}\phi\left[-\frac{\sqrt{\eta}(\Lambda(t) + d)}{\sqrt{d}}\right]\right\}
\tag{7.5.10}
$$

where $\phi(\cdot)$ and $\Phi(\cdot)$ represent the PDF and CDF of the standard normal distribution, respectively. Correspondingly, when the Birnbaum–Saunders-type normal approximation in Eq. (7.5.9) is used, the PDF for T becomes

$$
f_T(t) \cong \frac{\Lambda'(t)\sqrt{\eta}}{2}\left(\frac{1}{\sqrt{\Lambda(t)}} + \frac{d}{(\Lambda(t))^{3/2}}\right)\phi\left(\sqrt{\eta\Lambda(t)} - d\sqrt{\frac{\eta}{\Lambda(t)}}\right)
\tag{7.5.11}
$$

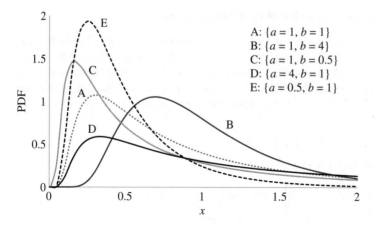

A: $\{a = 1, b = 1\}$
B: $\{a = 1, b = 4\}$
C: $\{a = 1, b = 0.5\}$
D: $\{a = 4, b = 1\}$
E: $\{a = 0.5, b = 1\}$

Figure 7.8 Probability density function of inverse Gaussian distribution.

where d is the failure threshold of the degrading unit. Degradation phenomena such as corrosion, fatigue crack growth, and wear-and-tear can be considered as accumulations of additive and irreversible damage caused by external random shocks. If each shock arrives randomly and causes a small amount of damage to the system, the shock arrival process can be approximated by a Poisson counting process (Desmond 1985; Singpurwalla 1995; van Noortwijk 2009; Ye et al. 2014).

Example 7.9 Provide several degradation phenomena that can be modeled as an IG process.

Solution and Discussion:
For example, the wear of the air bearing slider in a hard disk drive is caused by intermittent contacts between the trailing edge of the slider and the glide avalanche of the disk. The number of contacts over time may be approximated as a Poisson process, while the wear caused by each contact is tiny yet random. Sometimes an external shock may cause systems or infrastructure to fail immediately, such as earthquakes, hurricanes, and power surges. Rafiee et al. (2015) proposed a CBM framework in which the external shocks are classified into fatal and non-fatal types, though both arrive according to the Poisson process. The model is tested on microelectromechanical systems (MEMS) to evaluate the efficiency of a mixed shock-based CBM model. When the shock arrival rate becomes large and the jump size approaches zero, the compound Poisson process could be approximated by computationally more tractable processes like the IG process or the gamma process.

7.6 Non-Stationary Gaussian Degradation Process

7.6.1 The Degradation Model

The degradation process of electronics is usually different from that of mechanical units. In general the degradation phenomenon of a mechanical unit is visible and can be measured by physical parameters such as material removal rate, the density of oil debris,

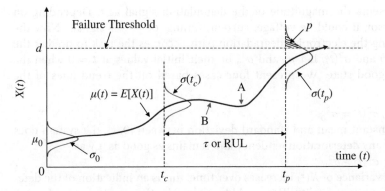

Figure 7.9 System degradation with increasing $\mu(t)$ and $\sigma(t)$. Source: Reproduced with permission of IEEE.

or the length of the crack. Another important feature of mechanical degradation is additivity and non-reversibility, implying that the expected RUL reduces monotonically as time evolves. RUL is the residual life left on an asset at a particular time of operation. Because of the non-reversibility, both the gamma process and IG process are sound statistical methods to model the degradation path of mechanical items. For electronic systems, a monotone degradation path may not exist as degradation signals such as voltage or power may exhibit a sporadic pattern, which is manifested as intermittent failures or no-fault-found issues (Jin et al. 2006). It is not uncommon that electronics systems may fail at one time, but it resumes the normal state after power-off and rebooting or temperature cooling.

Let $X(t)$ be the degradation signal of electronics units; typical signals include voltage, power, current, and frequency (i.e. timing). Figure 7.9 shows how $X(t)$ trends up towards the threshold over time. Obviously maintenance is triggered just before $X(t)$ crosses the limit for the first time. The RUL, denoted as τ, is the period between the current time t_c and the first passage time t_p. It plays an important role in allocating the materials, spare parts, and human resources for the incipient maintenance service. The values of τ are often predicted based on the statistical information of $\mu(t)$ and $\sigma^2(t)$. An interesting observation is that the degradation path is not strictly monotone, and in fact $E[X(t)]$ at point A is smaller than that at B. Figure 7.9 only shows the one-sided failure threshold, but the concept is equally applicable to two-sided degradation failures, which is often the case for voltage and timing parameters that involve plus and minus thresholds.

In electrical engineering, Gaussian distribution is widely used to model the distribution of voltage, current, and timing signals due to the law of large numbers. When the system is in a good state, the amplitude of the signals follows the stationary Gaussian process with constant mean and variance. When the system starts to deteriorate, either the mean of signal drifts, the variance changes (usually increases) or both alter over time. This suggests that the degradation path be modeled as a non-stationary Gaussian process (NSGP) with time-dependent mean and variance (Urmanov 2007; Park and Padgett 2006; Padgett and Tomlinson 2004). Let $\mu(t)$ and $\sigma^2(t)$ be the mean and standard deviation of the NSGP at time t. Then it is defined as

$$X(t) \sim NSG\{\mu(t), \ \sigma^2(t)\} \tag{7.6.1}$$

where $X(t)$ represents the magnitude of the degradation signal at t. Depending on the health precursor, it could be voltage, current, timing, or power signals. Now the issue of monitoring the electronics degradation path becomes the task to predict the trend lines of $\mu(t)$ and $\sigma^2(t)$. Let μ_0 and σ^2_0 be their initial values at $t = 0$ when the system is in the good state. We present four cases based on the trend lines of the $\mu(t)$ and $\sigma(t)$.

Case 1: $\mu(t) = \mu_0, \sigma^2(t) = \sigma^2_0$

Given constant mean and standard deviation between $[0, t]$, the system does not show any deterioration tendency and remains as good as it was.

Case 2: $\mu(t) = \mu_0, \sigma^2(t) > \sigma^2_0$

Since the variance of $X(t)$ increases over time, this is an indication of the deterioration of signal fidelity. This could be caused by the interferences of noise, ambient temperature, and other environmental factors. Nevertheless, a consistent growing variance usually indicates the degradation of the system.

Case 3: $\mu(t) > \mu_0, \sigma^2(t) = \sigma^2_0$

If the mean of $X(t)$ increases over time while the variance remains unchanged, similar to Case 2, this indicates that the systematic deterioration is due to the upward trend of the degradation path.

Case 4: $\mu(t) > \mu_0, \sigma^2(t) > \sigma^2_0$

This represents a more realistic situation where the mean and variance of $X(t)$ increase as time evolves, and is the generalization of Cases 2 and 3. It strongly indicates that the system has entered the degradation process.

Example 7.10   The measurements of the voltage signal at $t = 0$ and the initial eight months exhibit the following distribution characteristics (see Table 7.2). Assuming the failure threshold $d = 4\,\text{V}$, estimate the RUL based on the measurement at the end of month 8.

Solution:
The mean of voltage degradation shows the steady-state upward trend, while the variance remains relatively stable in eight months. Therefore, this degradation pattern belongs to Case 3. We use the regression model to extrapolate the underlying mean value of $\mu(t)$, and the following quadric model yields the best fit in terms of minimum means square error. That is,

$$\mu(t) = 0.015t^2 + 0.03t + 2$$

To predict the first passage time t_p, we need to solve the equation

$$0.015t^2 + 0.03t + 2 = 4$$

Table 7.2 The mean and variance of the voltage degradation in each month.

t (month)	0	1	2	3	4	5	7	8
Mean (V)	2.095	2.020	2.319	2.202	2.290	3.058	3.308	2.095
Variance (V^2)	0.5	0.52	0.51	0.52	0.49	0.50	0.51	0.48

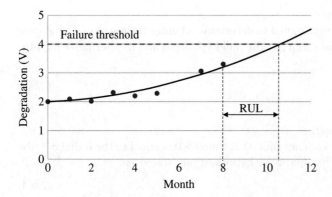

Figure 7.10 The expected trajectory path of voltage degradation.

which yields $t_p = 10.6$ months. Namely, the RUL $= 10.6 - 8 = 2.6$ months. Figure 7.10 graphically depicts the evolution of the voltage degradation path from months 1 to 8 and the projected degradation path over the RUL period. Since we use $\mu(t)$ to estimate RUL, this implies that the probability of reaching the failure threshold is 50% at $t_p = 10.6$ months.

7.6.2 Hypothesis Testing

In this section we discuss how the data sampled from two different time epochs can be analyzed using hypothesis testing and the confidence interval procedure for comparing their mean and variance. Let $x_{01}, x_{02}, \ldots, x_{0n_0}$ represent the n_0 observations of the initial condition of the system state at $t = 0$ and $x_{21}, x_{22}, \ldots, x_{2n_t}$ represent the n_t observations of the system condition at $t > 0$. We assume that these samples are drawn from two normal distributions. Figure 7.11 illustrates the sampling sequences at $t = 0$ and $t > 0$, respectively.

Under the NSGP framework, tracking the system degradation is equivalent to judging whether $\mu(t)$ or $\sigma(t)$ differs significantly from μ_0 or σ_0 for $t > 0$. Below we present the statistical method to test the mean and the variance.

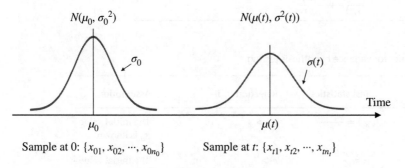

Figure 7.11 Two normal samples at different time epochs.

Case 1: Hypothesis test for $\mu(t)$

The hypothesis test is designed to determine whether the mean of $X(t)$ at time $t > 0$ equals the initial mean μ_0. The null hypothesis and its alternative hypothesis are stated as

$$H_0 : \mu(t) = \mu_0 \tag{7.6.2}$$
$$H_1 : \mu(t) \neq \mu_0 \tag{7.6.3}$$

Case 2: Hypothesis test for $\sigma(t)$

To test whether the variance of $X(t)$ at time $t > 0$ is equal to the initial σ_0^2, the null hypothesis and its alternative hypothesis are stated as

$$H_0 : \sigma^2(t) = \sigma_0^2 \tag{7.6.4}$$
$$H_1 : \sigma^2(t) \neq \sigma_0^2 \tag{7.6.5}$$

The statistics for performing the mean and variance hypothesis tests are summarized in Tables 7.3 and 7.4, respectively. Notice that α is the conference level or the probability of a Type I error. If a hull hypothesis is rejected when it is true, a Type I error has occurred. Statistically it can be stated as

$$P\{\text{type I error}\} = P\{\text{reject } H_0 \mid H_0 \text{ is true}\} = \alpha \tag{7.6.6}$$

Table 7.3 Statistics for mean hypothesis testing.

Hypothesis	Test statistics	Rejection H_0 if	Assumption
$H_0 : \mu(t) = \mu_0$ $H_1 : \mu(t) \neq \mu_0$	$Z = \dfrac{\mu(t) - \mu_0}{\sqrt{\dfrac{\sigma^2(t)}{n_t} + \dfrac{\sigma_0^2}{n_0}}}$	$\lvert Z \rvert > Z_{\alpha/2}$	True value of σ_0 and $\sigma(t)$ are known
$H_0 : \mu(t) = \mu_0$ $H_1 : \mu(t) < \mu_0$	$t = \dfrac{\mu(t) - \mu_0}{\sqrt{\dfrac{s_t^2}{n_t} + \dfrac{s_0^2}{n_0}}}$	$t < t_{\alpha/2, v}$, $t > t_{1-\alpha/2, v}$	True value of σ_0 and $\sigma(t)$ are unknown, and v is the degree of freedom of t-distribution
$H_0 : \mu(t) = \mu_0$ $H_1 : \mu(t) > \mu_0$	$v = \dfrac{\left(\dfrac{s_t^2}{n_t} + \dfrac{s_0^2}{n_0}\right)^2}{\dfrac{(s_t^2/n_t)^2}{n_t-1} + \dfrac{(s_0^2/n_0)^2}{n_0-1}}$		

Table 7.4 Statistics for variance hypothesis testing.

Hypothesis	Test statistics	Rejection H_0 if	Assumption
$H_0 : \sigma^2(t) = \sigma_0^2$	$\chi^2 = \dfrac{(n_t-1)s_t^2}{\sigma_0^2}$	$\chi^2 > \chi^2_{\alpha/2, n_t-1}$ or $\chi^2 < \chi^2_{1-\alpha/2, n_t-1}$	The initial value σ_0 is known
$H_1 : \sigma^2(t) \neq \sigma_0^2$	$F = \dfrac{s_t^2}{s_0^2}$	$F > F_{\alpha/2, n_t-1, n_0-1}$ or $F > F_{1-\alpha/2, n_t-1, n_0-1}$	The initial value σ_0 and $\sigma(t)$ are unknown

Ideally, a Type I error should be as small as possible to minimize the incorrect judgment. This can be achieved by increasing the sample size or reducing the measurement errors through routine calibration of in-situ sensors.

If the true population mean and variance are known, then Z-test and the chi-squared test statistics can be used for identifying the changes in the mean and the variance, respectively. In reality, the true population mean and variance are often difficult to obtain. However, they can be substituted by the statistical mean and variance, which are estimated as follows:

$$\widehat{\mu}_0 = \frac{1}{n_0} \sum_{i=1}^{n_0} x_{0i} \quad \text{and} \quad \widehat{\mu}(t) = \frac{1}{n_t} \sum_{i=1}^{n_t} x_{ti} \tag{7.6.7}$$

$$s_0^2 = \frac{1}{n_0 - 1} \sum_{i=1}^{n_0} (x_{0i} - \widehat{\mu}_0)^2 \quad \text{and} \quad s_t^2 = \frac{1}{n_t - 1} \sum_{i=1}^{n_t} (x_{0i} - \widehat{\mu}_t)^2 \tag{7.6.8}$$

In that case, the t-test and F-test should be used for conducting the hypothesis testing on the mean and the variance accordingly. Readers are referred to Montgomery (2009) for additional information about hypothesis testing of the mean and variance of normal random variables.

Example 7.11 In what situations can the Z-test and chi-squared test be chosen to test the mean and the variance?

Solution:
If the sample size n_0 is large enough, both $\widehat{\mu}_0$ and s_0^2, although estimated, can be treated as the true mean and variance of the initial system state. The similar idea can be applied to $\widehat{\mu}(t)$ and $s^2(t)$ for large n_t. In these situations, the Z-test and the chi-squared test can be used for the mean and the variance tests, respectively.

7.6.3 Estimation of Remaining Useful Life

The system of the remaining useful life (RUL) is defined as the time elapsed between the current time t_c and the time t_p when the degradation path $X(t)$ crosses the failure threshold d for the first time (see Figure 7.9). Then the RUL, denoted as τ, can be determined via the criterion below:

$$P\{X(t_c + \tau) < d\} \geq 1 - p \tag{7.6.9}$$

where p is the probability that $X(t)$ will cross d for the first time at time $t_c + \tau$. In other words, the probability that the system is still in a good state at $t_c + \tau$ is $1 - p$. For a risk-averse system user, a small p is often preferred because the downtime cost due to an unexpected failure is much higher than a scheduled maintenance. If $X(t)$ is a NSGP with projected mean $\mu(t_c + \tau)$ and variance $\sigma^2(t_c + \tau)$, the value of τ can be estimated by solving the following equation:

$$\frac{d - \mu(t_c + \tau)}{\sigma(t_c + \tau)} = Z_{1-p} \tag{7.6.10}$$

where Z_{1-p} is the standard normal value at $(1 - p) \times 100\%$. Equation (7.6.10) shows that τ can be calculated if the explicit form of the projected mean and variance are known. Below we estimate the value of τ based on three different forms of $u(t)$ and $\sigma^2(t)$, but the concept is also applicable to other forms of $u(t)$ and $\sigma^2(t)$.

Case 1: $\mu(t) = b_0 + b_1 t$ and $\sigma^2(t) = \sigma^2_0$

When the trend line for $\mu(t)$ is linear and $\sigma^2(t)$ is time-invariant, τ can be calculated analytically by substituting these formulas into (7.6.10) and the result is given as

$$\tau = \frac{1}{b_1}(d - b_0 - \sigma_0 Z_{1-p}) - t_c \tag{7.6.11}$$

Case 2: $\mu(t) = b_0 + b_1 t$ and $\sigma^2(t) = \sigma^2_0 + at$

Note that both $\mu(t)$ and $\sigma^2(t)$ are linearly increasing over time. After substituting $\mu(t)$ and $\sigma^2(t)$ into Eq. (7.6.10), τ can be obtained by solving the following quadratic equation:

$$b_1^2(t_c + \tau)^2 + k_1(t_c + \tau) + k_2 = 0 \tag{7.6.12}$$

where

$$k_1 = 2b_0 b_1 - 2b_1 d - aZ^2_{1-p} \tag{7.6.13}$$

$$k_2 = d^2 + b_0^2 - 2b_0 d - \sigma_0^2 Z^2_{1-p} \tag{7.6.14}$$

Case 3: $\mu(t) = b_0 + b_1 t + b_2 t^2$ and $\sigma^2(t) = \sigma^2_0 + at$

Note that $\mu(t)$ is a quadratic function and $\sigma^2(t)$ is a linear function. After substituting $\mu(t)$ and $\sigma^2(t)$ into Eq. (7.6.10), τ can be estimated from the following equation:

$$d - (b_0 + b_1(t_c + \tau) + b_2(t_c + \tau)^2) - Z_{1-p}(\sigma_0^2 + a(t_c + \tau))^{1/2} = 0 \tag{7.6.15}$$

It is difficult to obtain the closed-form solution to Eq. (7.6.15). Thus numerical methods need to be employed to solve for τ. Besides linear or quadratic forms, the trend lines of $\mu(t)$ and $\sigma^2(t)$ in reality may exhibit other forms such as exponential or high-degree polynomial functions. In these cases, τ can still be solved based on Eq. (7.6.10) via numerical methods.

The information of RUL is very useful as it can be used to make a sound decision on maintenance scheduling and spare parts provisioning. For instance, to ensure the spare parts availability, the order can be placed just in time when the RUL is slightly larger than the delivery lead time. Thus CBM potentially saves or reduces the inventory holding cost, while in age-based PM policy excessive parts stocking is often required to prevent any unknown failures. CBM also prevents the occurrence of backorders that may happen when using the CM policy.

Example 7.12 Suppose the mean and variance of $X(t)$ trend lines exhibit the forms as $\mu(t) = 2 + 0.1\,t$ and $\sigma^2(t) = 0.3 + 0.02\,t$. For the failure threshold $d = 4$, given $t_c = 2.5$ days, estimate the RUL for $p = 0.05, 0.2$, and 0.5, respectively.

Solution:
Expanding Eq. (7.6.12) yields

$$b_1^2 \tau^2 + (2t_c b_1^2 + k_1)\tau + b_1^2 t_c^2 + k_1 t_c + k_2 = 0 \tag{7.6.16}$$

Thus

$$\tau = \frac{-(2t_c b_1^2 + k_1) \pm \sqrt{(2t_c b_1^2 + k_1)^2 - 4b_1^2(b_1^2 t_c^2 + k_1 t_c + k_2)}}{2b_1^2} \qquad (7.6.17)$$

where k_1 and k_2 are given in Eqs. (7.6.13) and (7.6.14). Going back to the original problem, the Z-values under different $p = 0.05$, 0.1, and 0.5 are given as follows:

$$Z_{1-0.05} = 1.64, \quad Z_{1-0.2} = 0.84, \quad \text{and} \quad Z_{1-0.5} = 0.5 \qquad (7.6.18)$$

Let us compute the RUL under $p = 0.05$ because the solution procedure for the other two is similar. After substituting $Z_{1-0.05} = 1.64$ into Eqs. (7.6.13) and (7.6.14) along with $b_0 = 2$, $b_1 = 0.1$, $a = 0.02$, and $\sigma_0^2 = 0.3$, we have

$$k_1 = 2 \times 2 \times 0.1 - 2 \times 0.1 \times 4 - 0.02 \times 1.64^2 = -0.454 \qquad (7.6.19)$$

$$k_2 = 4^2 + 2^2 - 2 \times 4 \times 2 - 0.3 \times 1.64^2 = 3.188 \qquad (7.6.20)$$

Now substituting k_1 and k_2 along with $t_c = 2.5$ days and other known parameters into Eq. (7.6.17) yields

$$\tau = \frac{-(2 \times 2.5 \times 0.1^2 - 0.454) - \sqrt{\begin{array}{c}(2 \times 2.5 \times 0.1^2 - 0.454)^2 \\ -4 \times 0.1^2(0.1^2 \times 2.5^2 - 0.454 \times 2.5 + 3.188)\end{array}}}{2 \times 0.1^2}$$

$$= 8.68 \ \text{days}$$

Similarly, we can estimate the RUL at $p = 0.2$ and 0.5, which are $\tau = 11.13$ days and $\tau = 17.50$ days, respectively. These results show that the value of RUL depends on the degree of acceptable failure risks. If the system user is risk-averse, a smaller p value is preferred. Otherwise, $p = 0.5$ can be adopted for estimating the RUL under the risk-neutral condition.

Most distribution-based methods predicting a remaining lifetime assume that the degradation path is known or the degradation states could be predetermined. Zhang et al. (2014) propose an adaptive discrete-state model to estimate the system remaining lifetime based on Bayesian belief network theory. Their approach does not require an explicit distribution function to characterize the degradation process and hence it avoids the state identification errors under limited feature data.

7.7 Performance-Based Maintenance

7.7.1 The Rise of Performance-Driven Service

Performance-based maintenance (PBM) emerged as a new business paradigm for the acquisition, operation, and support of capital equipment in the last decade. PBM is also referred to as "power by the hour" (PBH) in commercial airlines or called "performance-based logistics" (PBL) in the defense industry (Kim et al. 2007). PBM differs from time- and material-based service contracts in that the maintenance decision is driven by the financial incentives, not necessarily the system age or the health condition. In a time- and material-based contract, the customer usually bears the costs of spare parts and labor each time the service is accomplished. Hence the original equipment manufacturer (OEM) has less motivation to improve the product reliability as maintenance service can bring lucrative revenue to the supplier in the after-sales market. Studies show that the after-sales service represents as much as 40–50% of the firm's profit, though it only accounts for 20–25% of total revenue (Cohen et al. 2006). The goal of PBM is to design and implement a maintenance service program to incentivize the supplier or the OEM to attain the system performance goal while lowering customer's ownership cost.

Figure 7.12 graphically describes the evolution of asset management strategy driven by the continuous cost reduction of system ownership. It is worth mentioning that maintenance evolution is driven primarily by the lifecycle cost savings, but it does not guarantee the reliability growth. If the costs associated with spare parts, repair labor, and transportation are far below the reliability growth cost, the service provider may opt to achieve the system performance goal by deploying responsive yet cost-effective repair facilities, instead of reliability growth. Later on we also show that a long-term PBM contract motivates the OEM to improve the product reliability in conjunction with the deployment of a responsive service logistics network.

Figure 7.13 depicts an integrated product-service system where the OEM designs and manufactures capital goods and also provides the repair and maintenance services.

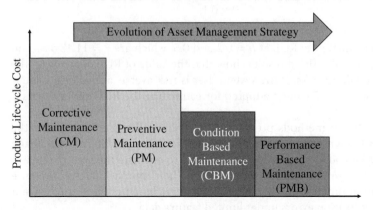

Figure 7.12 The evolution of asset management strategy.

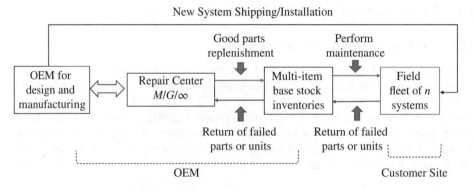

Figure 7.13 An integrated product-service system.

The central repair center is approximated as an $M/G/\infty$ queue model with ample capacities. To expedite the replacement service, a local spare parts inventory is placed near the customer site. A system failure corresponds to a part failure. Since the defective part is repairable, it is returned to the repair center for root-cause analysis and renewal. When fixed, it is put back to the inventory for future replacement. The total costs of the service logistics include the spare parts inventory, material holding cost, transportation, repair facilities, and repair technicians. Since a complex system comprises multiple component types and each component has a different failure rate, the local stock actually holds different part types with different safe-stock quantities.

7.7.2 Procedures for PBM Implementation

According to Richardson and Jacopino (2006), the development and implementation of the PBM can be viewed as a four-step process. Table 7.5 presents the detailed objectives and tasks in each step that needs to be taken in order to meet the performance goal, such as system operational availability (OA), cost per unit usage (CUU), and average system downtime.

It is worth mentioning that defining appropriate performance measures in Step 2 is the key to the effective planning and implementation of PBM contracts. This is because performance measures are used as quantitative means to determine the contractual objectives, decision variables, or decision constraints. With this in mind, it is critical to define the measures that are quantifiable, commonly used, and mutually acceptable between the customer and the supplier. In the PBL domain, logistics footprints, operational availability (OA), equipment reliability, and service response times are often used to measure the performance outcomes.

Mean downtime (MDT) is the average time that a system is in a non-operational state. This includes all downtime associated with repair, maintenance, self-imposed downtime, and any administrative or logistics delays. In particular, administrative delay time (ADT) accounts for the response time from the moment when the system is down to when the repair decision is taken. Mean logistics delay time (MLD) represents the

Table 7.5 A four-step process of performance-based maintenance.

Step	Objectives	Detailed tasks
1	Identifying key performance outcomes	Fleet readiness, mission completion, assurance of spare parts supply
2	Defining performance measures	Operational availability, spare parts availability, stocking-out probability, part failure rate, stock fill rate, expected backorders, MTBF, MTTR, cost per unit usage, logistics response time, logistics footprint
3	Determining performance goals	Minimum system or parts availability, minimum backorders, maximum failure rate, percentage of fill rate, maximum repair waiting time
4	Designing incentive mechanisms	Fixed cost payment, cost plus payment, cost sharing or cost reimbursement

waiting time for the arrival of the repair crew, spare parts, and necessary tooling equipment. Finally, we use mean-time-to-replacement (MTTR) to capture all the hands-on time associated with repair, maintenance, and self-imposed downtime. Let T_{ADT}, T_{MLD}, and T_{MTTR} stand for ADT, MLD, and MTTR, respectively; then

$$MDT = T_{ADT} + T_{MLD} + T_{MTTR} \qquad (7.7.1)$$

For repairable inventory systems, MTTR is highly dependent upon the availability of spare parts and repair turn-around time. A large inventory level usually guarantees the supply of spare parts, hence reducing the MTTR. When the spares stock is not physically close to the customer site, extended transportation time may delay the recovery process of the broken system. A short repair turn-around time can speed up the material circulation between the spares inventory and repair center, hence reducing the backorder probability (Sherbrooke 1992). Below we introduce five overarching performance measures that are frequently used in designing PBM service contracts in the US defense industry.

7.7.3 Five Overarching Performance Measures

The US DoD (2005) proposed five overarching performance measures to assess the effectiveness of a PBM program. These are OA, mission reliability (MR), CUU, logistics response time (LRT), and logistics footprint. The definition of these performance measures are elaborated below.

According to the US DoD (2009), OA is defined as "the probability that a system, when in an actual operating environment, will be ready for commitment to system mission operations at any point in time." FAA (FAA-HDBK-006A, 2008) states that "OA include all sources of downtime, both scheduled and unscheduled. It is an operational measure for deployed systems." Thus a steady-state OA can be estimated by

$$A = \frac{MTBR}{MTBR + MDT} = \frac{T_{MTBR}}{T_{MTBR} + T_{ADT} + T_{MLD} + T_{MTTR}} \qquad (7.7.2)$$

In Eq. (7.7.2), T_{MTBR} stands for the mean-time-between-replacements (MTTRs), which includes scheduled and failure replacements. MDT is given in Eq. (7.7.1), which captures the time elapsed due to ADT, MLD, and MTTR.

MR measures the capability of a system or item to accomplish the operational goal for a specified time period. Obviously, metrics such as MTBF, MTBR, and failure rate can be used to characterize the MR. Let T_i represent the system uptime in the maintenance cycle i for $i = 1, 2, \ldots, n$. Then MTBR can be estimated by

$$MTBR = \lim_{n \to \infty} \frac{1}{n} \sum_{i=1}^{n} T_i \qquad (7.7.3)$$

Logistics response time (LRT) is the duration of calendar time from when a failure occurred to the time that the failure has been resolved. Since LRT are highly correlated with MTTR and MLDT, it can be approximated by the following estimate:

$$LRT = MLDT + MTTR \qquad (7.7.4)$$

CUU is defined as the total operations and support (O&S) costs of a system divided by the usage factor, such as hours, miles, rounds, or launches. CUU aims to quantify the

Table 7.6 Summary of five overarching performance measures.

No.	Performance measures	Quantitative metrics
1	Operational availability	System availability associated with MTBF, MTTR, and MLDT
2	Mission Reliability	MTBF, failure rate, and rate of occurance of failures
3	Cost per unit usage	O&S costs, usage factors (e.g. hours, miles, rounds, and launches)
4	Logistic response time	MTTR, MLDT, and mean downtime
5	Logistics footprint	Spare parts stocking, manpower, training cost, facility size, transportation fleet

rate of the cost after the system is purchased and deployed in the field:

$$CUU = \frac{\text{Operations and support cost}}{\text{Usage factor}} \tag{7.7.5}$$

Logistics footprint quantifies the size of logistics support needed to sustain the operation of a fleet of systems. Measurable elements include but are not limited to: spare parts inventory, personnel, facilities, transportation, and supply. Table 7.6 summarizes the five overarching performance measures and related assessment metrics.

7.7.4 Reliability and MTBF Considering Usage Rate

Given two systems having the same inherent failure rate, the system with a higher utilization rate will fail more quickly than the underutilized one. Hence it is important to incorporate the usage rate in the reliability estimation model. Let $\overline{T}_o$ and $\overline{T}_s$ be the cumulative operating time and the standby or idle time between two consecutive failures, respectively. Then the system utilization rate, denoted as ρ, is defined as

$$\rho = \frac{\overline{T}_o}{\overline{T}_o + \overline{T}_s} \tag{7.7.6}$$

Obviously the value of ρ falls in $[0, 1]$, and a larger ρ implies that the system is more utilized during the course of its lifetime. The unitization rate of transformers in a power system is close to unity because electricity is supplied in 24/7 mode. Most tools in the semiconductor manufacturing industry also operate in three shifts with $\rho \approx 1$. However, the utilization rate of solar PV systems typically is less than 0.5 because there is no power generation in the night. The inherent failure rate λ is the measured failure rate assuming the system is fully utilized with zero idle time ($\rho = 1$) between two consecutive failures. Now the actual MTBF after considering the system utilization rate can be expressed by

$$MTBF = \overline{T}_o + \overline{T}_s = \frac{1}{\rho \lambda} \tag{7.7.7}$$

For instance, $\lambda = 0.001$ failures/hour. The mission MTBF is 1000 hours if the usage rate $\rho = 1$, but it increases to 2000 hours if ρ is down by 50%. This example shows that the actual MTBF is jointly determined by the inherent failure rate and the usage rate.

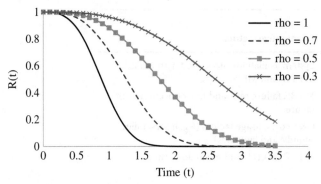

Weibull Reliability vs. Utilization Rate

Figure 7.14 Weibull reliability ($\alpha = 1$, $\beta = 3$) under different utilization rates.

If the lifetime of a system or components can be modeled by a two-parameter Weibull distribution, its failure rate and reliability considering the utilization rate becomes

$$\lambda(t) = \alpha\rho\beta(a\rho t)^{\beta-1} \tag{7.7.8}$$

$$R(t) = \exp(-(\alpha\rho t)^{\beta}) \tag{7.7.9}$$

where α and β are the inherent scale and the shape parameters of the Weibull distribution, respectively. Equations (7.7.8) and (7.7.9) are derived assuming that β does not change with the usage rate. Figure 7.14 shows how the actual reliability changes with ρ. The baseline reliability is plotted with $\alpha = 1$, $\beta = 3$, and $\rho = 1$. Reliability functions with $\rho = 0.7$, 0.5, and 0.3 are also plotted for comparison. It is found that ρ has little influence on the reliability when t is small. However, at $t = 2$, the product with $\rho = 1$ and 0.7 has failed with high confidence, while the reliability of the product with $\rho = 0.3$ still remains above 0.73.

The actual MTBF considering the utilization rate is

$$MTBF = (\alpha\rho)^{-1}\Gamma(1 + \beta^{-1}) = \alpha_{\rho}^{-1}\Gamma(1 + \beta^{-1}) \tag{7.7.10}$$

where $\alpha_{\rho} = \alpha\rho$ is the actual scale parameter in the calendar time. It is worth mentioning that Eqs. (7.7.8–7.7.10) can be extended to other lifetime distributions such as lognormal and gamma distributions by incorporating ρ into their scale parameters.

7.7.5 Operational Availability under Corrective Maintenance

Operational availability (OA) is considered as the core performance measure that governs and influences the other four performance measures, either directly or indirectly. To ensure high OA, the OEM could improve MR through design for reliability or the LRT could shrink via a fast repair process. On the one hand, a higher level of MR implies a longer MTBF, resulting in fewer failures, less repairs, and a lower spares stocking size. On the other hand, to reduce the LRT, the OEM may have to increase the spare parts stocking level, expedite the shipping process, or deploy a larger repair crew. These decisions further influence the costs of a logistics footprint and CUU. Figure 7.15 graphically shows the correlation between OA and the other four performance measures.

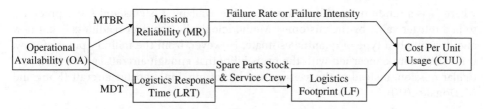

Figure 7.15 The core performance measure in PBM.

Studies (Oner et al. 2010; Jin and Liao 2009; Mirzahosseinian and Piplani 2011; Jin and Wang 2012) have shown that a system under a corrective maintenance policy, such as OA, is determined by six factors, namely, the intrinsic failure rate, system usage, spare parts stocking, the fleet size, repair turn-around time, and hands-on replacement time. Repair turn-around time is the cycle time of repairing a failed item in the repair center. Hands-on time, also known as touch labor, is the time duration of replacing a failed part with a good item. Assuming ample repair capacoty, Jin and Wang (2012) derived a unified availability model by synthesizing six factors into one formula, that is,

$$A(\lambda, s, \rho, n, t_r, t_s) = \frac{1}{1 + \rho \lambda t_s + \rho \lambda t_r \left(1 - \sum_{k=0}^{s} \frac{(n\rho\lambda t_r)^k e^{-n\rho\lambda t_r}}{k!}\right)} \qquad (7.7.11)$$

where
λ = system failure rate
ρ = system usage, where $\rho \in [0, 1]$
s = spare parts stocking level
n = number of operating systems in the field or fleet size
t_r = repair turn-around time between the local inventory and the repair center
t_s = hands-on time of replacing the failed item

Equation (7.7.11) synthesizes reliability theory and the logistics model into a unified metric to assess their interactions. It guides the OEM to achieve the system performance goal by wisely investing the resources in reliability design, product manufacturing, and after-sales maintenance. For a system comprised of different types of components, each having its own OA, the system-level OA is given by

$$A_{sys} = \prod_{k=1}^{m} A_k(\lambda_k, s_k, \rho_k, n_k, t_{kr}, t_{ks}) \qquad (7.7.12)$$

where m is the component types, and $A_k(\lambda_k, s_k, \beta_k, n_k, t_{kr}, t_{ks})$ is given in Eq. (7.7.11) for component type k. In the next section, we formulate a profit-centric model to aid the OEM to maximize the service reward while satisfying the reliability criteria mandated by the customer.

Assume that the fleet consists of K identical systems for which their annual usage rate is similar. Then the fleet readiness is defined as the probability that at least k systems are available at any random point in time. This can be expressed as

$$P\{N \geq k\} = \sum_{i=k}^{K} \binom{K}{i} A_{sys}^i (1 - A_{sys})^{K-i}, \quad \text{for } k = 0, 1, \ldots, K \qquad (7.7.13)$$

where N is a random variable representing the available systems and k is a predetermined integer value by the customer. Methodologically, the fleet readiness is not new and it is simply a type of quantile estimate. However, from the military point of view, it is often more concerned with the probability that enough aircraft can fly for a particular mission, instead of the average number of missions capable aircraft (Kang and McDonald 2010).

7.7.6 Operational Availability under Preventive Maintenance

In this section, we extend the unified availability model in Eq. (7.7.11) from CM to age-based PM applications. PM differs from CM in that the replacement interval τ is predefined. Thus an item may survive through τ or may fail before reaching τ. The former is called scheduled replacement and the latter is called failure replacement. Technically an item surviving at τ is a still a good unit, though it has deteriorated. Therefore, less time and effort are required to recondition a degraded unit than repairing a failed unit in the repair center. We use t_p to denote the mean turn-around time for reconditioning a deteriorated unit. Similarly, we use t_r to denote the mean turn-around time for repairing a failed unit, with $t_r > t_p$. Figure 7.16 shows a typical product-service system where the OEM markets the new product as well as offering PM services in the after-sales market.

For repairable parts, the unit cost typically is high and their failure time is sporadic, precluding the mass storage of spare items. Therefore, parts backorders may occur either in a failure or in a planned replacement. Let T_{MDT} denote the system MDT under an age-based PM policy. Then T_{MDT} can be estimated by (Jin et al. 2015)

$$
\begin{aligned}
T_{MDT} &= (t_s + t_p P\{O > s\})R(\tau) + (t_s + t_r P\{O > s\})F(\tau) \\
&= t_s + (t_p R(\tau) + t_r F(\tau))P\{O > s\}
\end{aligned}
\tag{7.7.14}
$$

where

O = the demand to the inventory, a random variable
s = base stock level
t_s = MTTR or the hands-on time
$R(\tau), F(\tau)$ = reliability or failure probability at τ

Note that $t_p R(\tau) + t_r F(\tau)$ is the mean turn-around time for renewing an item. For a repairable inventory with infinite repair capacity, according to Palm's theorem, O is a random variable that follows the Poisson distribution with the mean demand

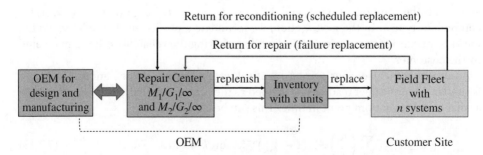

Figure 7.16 Integrated product-service under PM policy for single-item systems.

$\mu = n\lambda(\tau)(t_p R(\tau) + t_r F(\tau))$ provided the fleet size $n \geq 10$ (Wang 2012). Now the parts backorder probability can be estimated as

$$P\{O > s\} = 1 - \sum_{j=0}^{s} \frac{\mu^j e^{-\mu}}{j!}, \qquad \text{for } s = 0, 1, 2, \ldots \qquad (7.7.15)$$

If $\tau \to \infty$, the age-based PM policy simply reduced to a CM policy, which has been discussed in a previous section. By substituting Eqs. (7.3.11) and (7.7.14) into (7.7.2), the OA for a single-item system under an age-based PM policy is obtained as

$$A = \frac{\int_0^\tau R(t; \rho) dt}{\int_0^\tau R(t; \rho) dt + t_s + (t_p R(\tau; \rho) + t_r F(\tau; \rho)) P\{O > s\}} \qquad (7.7.16)$$

where $R(t; \rho)$ is the reliability function considering the usage rate. Equation (7.7.16) comprehends eight performance drivers, namely inherent reliability $R(t)$, usage rate ρ, maintenance interval τ, spare part stock level s, fleet size n, MTTR t_s, parts reconditioning turn-around time t_p, and parts repair turn-around time t_r.

Recently, Xiang et al. (2017) derived a system availability model based on CBM policy; they further proposed a PBM contract to attain three objectives: minimizing the cost rate, maximizing the availability, or maximizing the service profit. The advantage of CBM over age- or time-based PM in that it allows for unit-to-unit degradation heterogeneity.

7.8 Contracting for Performance-Based Logistics

7.8.1 Incentive Payment Schemes

Three types of cost-plus oriented payment schemes are frequently used to reward the supplier or the contractor: cost plus fixed fee (CPFF), cost plus award fee (CPAF), and cost plus incentive fee (CPIF).

In CPFF, the contractor receives a predetermined fee that was agreed upon at the time of contract formation with the customer. It basically transfers a substantial amount of risk to the supplier. However, if a product is well designed and tested and the majority of operational uncertainties and failure mechanisms are known to the supplier, it can charge the customer a fair amount of money to compensate the risks.

In CPAF, the customer pays a fee to the contractor based on the contractor's work performance. Sometimes the fee is determined subjectively by an award fee board whereas in others the fee is based upon the quantitative performance metrics.

CPIF offers incentives to reward the supplier for achieving a high level of performance. It is quite natural for the supplier to shift the concentration from purely cost reduction to profit maximization. In CPIF, the customer pays a larger fee as incentives to the contractor who meets or exceeds the performance targets based on the mutually agreed service criteria. Nowicki et al. (2008) proposed a linear CPIF function that consists of a fixed fee and a reward fee, and the latter is proportional to the achieved availability. That is,

$$S(A) = \begin{cases} a + b_1(A - A_{\min}), & A \geq A_{\min} \\ a + b_2(A - A_{\min}), & A < A_{\min} \end{cases} \qquad (7.8.1)$$

where $A_{\min}$ be the target system availability stipulated in the contract. Here a is the fixed payment regardless of the system availability. Parameters b_1 and b_2 are the reward or penalty rate depending on whether $A \geq A_{min}$ or $A < A_{min}$. A larger value of b_1 (or b_2) implies that the supplier receives more compensation (or penalty) given the same level of A.

Realizing that it becomes more difficult to further increase the system availability if the value is already high, Nowicki et al. (2008) also propose the exponential CPIF model to compensate the efforts of the service provider, and the model is given as

$$S(A) = \begin{cases} \exp(c + d_1(A - A_{min})), & A \geq A_{min} \\ \exp(c + d_2(A - A_{min})), & A < A_{min} \end{cases} \tag{7.8.2}$$

Note that c, d_1, and d_2 are model parameters. Compared with the linear CPIF model, the exponential model has a better controllability on the cost because it becomes more difficult or requires much more effort for the supplier to improve the availability as it approaches. Similarly, the penalty also increases exponentially if A drops below A_{min}.

7.8.2 Game-Theoretic Contracting Model

Several game-theoretic models have been proposed to design and manage performance based contracting (PBC) services. For instance, Lin et al. (2017) designed a PBC program where the spare parts and maintenance service are carried by third party logistics. We apply the principal-agent game theory to construct a PBC framework where the integrated firm provides the product and offers the after-sales services. Two players involved in the game are the supplier and the customer. The supplier behaves as an agent whose goal is to coordinate maintenance, spares inventory, and repair capacities to maximize the service profit. The customer behaves as the principal who minimizes the CUU by stipulating the reward function, reliability, and system availability criteria. The following generic model, denoted as Model 7.1, is formulated to maximize the service profit and to minimize the CUU for a fleet of systems.

Model 7.1   Objective functions:

Max: Service profit	(7.8.3)
Min: Cost per unit usage (CUU)	(7.8.4)

Subject to:

Maintenance interval $\geq \tau_{min}$	(7.8.5)
System operational availability $> A_{min}$	(7.8.6)
System reliability $> MTBR_{min}$	(7.8.7)
Logistics response ime $< T_{max}$	(7.8.8)
Logistics footprint $< C_{max}$	(7.8.9)

where A_{min} is the target availability, $MTBT_{min}$ is the required reliability, T_{max} is the maximum allowed response time, and C_{max} is the maximum logistics footprint including inventory, repair facilities, and transportation. The supplier's objective is to maximize the profit in Eq. (7.8.3), while the customer intends to minimize the CUU in general. Constraints (7.8.5–7.8.8) represent the performance criteria stipulated by the customer and Constraint (7.8.9) is the logistics budget limit of the supplier. As such, Model 7.1

effectively institutionalizes five overarching performance measures of DoD (2005) under the game theoretic framework.

Three main challenges in deploying the principal-agent models are given as follows:

- *Adverse selection* refers to a situation where the supplier has the information that users do not, or vice versa, about aspects of product reliability. In the case of a PBM contract, adverse selection is the tendency of over-use the equipment to get a free maintenance service.
- *Moral hazard* is a situation where one player is more likely to take risks because the costs or negative consequence that could result will not be borne by that player.
- *Information asymmetry* arises from the transactions where one party has more or better information than the other. This asymmetry creates problems like adverse selection, moral hazard, and information monopoly.

In summary, the literature pertaining to maintenance and service contracting is huge. This chapter only provides conceptual discussions on three types of maintenance policies, i.e. CB, PM, and CBM, as well as the emerging PBM practice. The works by Wang (2002) and Nakagawa (2008) summarize the early achievement in maintenance planning and optimization. For recent advancements and trends, readers are referred to the surveys by Sharma et al. (2011), Alaswada and Xiang (2017), and Olde Keizera et al. (2017).

7.9 Case Study – RUL Prediction of Electronics Equipment

7.9.1 Built-in Self-Test Program

Failure symptoms of electronics equipment are often manifested as an anomaly in voltage, current, and timing (i.e. frequency). To ensure functionality, automatic diagnostic tools, known as built-in self-test (BIST) program, are routinely executed to monitor the health condition of instruments made of printed circuit boards (PCBs). Reliability degradation of a PCB is typically characterized by the upper and/or lower tolerance limits of electrical or timing parameters. For instance, voltage thresholds can be defined in a range around the mean value; if it is a timing or frequency parameter, the reliability threshold can be defined by seconds, microseconds, or picoseconds. The instrument, or more specifically, the PCB fails if the readings of electrical or time parameters either fall above the upper limit or below the lower limit.

Originally from Jin et al. (2011), we take a set of BIST data from a voltage test of a PCB to demonstrate the application of the NSGP prediction model in Section 7.6. As shown in Figure 7.17, each data point generated by the BIST program represents the voltage level at a particular time instance. The PCB fails if the voltage readings exceed $V_U = 3.1984$ V or drops below $V_L = 3.1675$ V.

7.9.2 A Four-Step Process for RUL Prediction

In the following, the NSGP prognostic model will be applied to track and estimate the mean and variance of voltage signals, and the RUL is predicted accordingly based on the degradation trend lines. The NSGP prediction consists of four steps as follows:

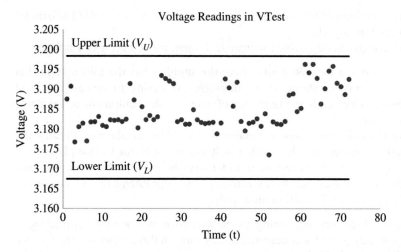

Figure 7.17 Chronological readings of voltage test. Source: Reproduced with permission of IEEE.

Step 1: Computing the initial mean and variance

Figure 7.17 shows the system is healthy and no degradation is observed in its early lifetime. We use the first 20 samples (i.e. $1 \le t \le 20$) to characterize the initial reliability state. We perform the normality test based on the first 20 readings to justify whether the voltage fits the normal distribution, and the result is summarized in Figure 7.18. With 95% conference, the data fits the normal distribution with $\mu_0 = 3.183$ V and $\sigma_0 = 0.003863$ V.

Step 2: Estimating the mean and variance $t > 20$

As time progresses, the sampling window is moving forward while the window size, n, is fixed. Without loss of generality, $n = 20$ is adopted for the moving window. As new voltage data are sampled by the BIST program, the mean and variance are

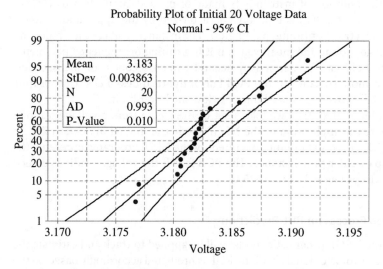

Figure 7.18 Normality test of initial voltage data.

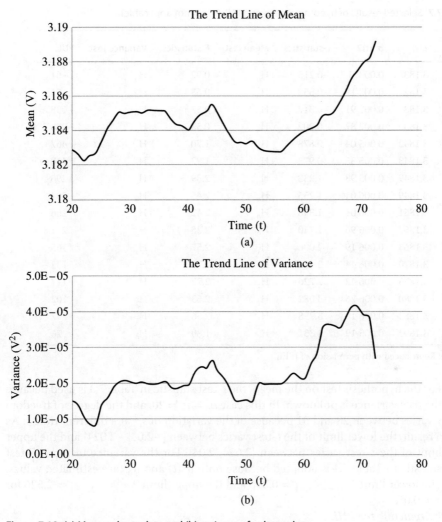

Figure 7.19 (a) Mean voltage data and (b) variance of voltage data.

recomputed and updated. Figure 7.19a and b depicts the mean and the variance, respectively. It is observed that $\mu(t)$ remains relatively stable between $0 \leq t \leq 60$, and then it trends up. The same observation can be made on $\sigma^2(t)$.

Step 3: Performing hypothesis testing

Based on the current $\mu(t)$ and $\sigma^2(t)$, hypothesis testing is performed to determine whether the mean and variance differ appreciably from their initial values $\mu_0 = 3.183$ V and $\sigma_0 = 0.003\ 863$ V. A 95% confidence level is chosen with $\alpha = 0.05$ and the testing results are summarized in Table 7.7 for each time instance. No significant changes are observed for the mean and variance for $t \in [21, 66]$. From $t = 67$, both the mean and the variance began to trend up as manifested by acceptance of the alternative hypothesis (H_1). The testing results match the visual observations on $\mu(t)$ and $\sigma^2(t)$ in Figure 7.19a and b. Due to the space limitation, Table 7.7 only shows the data for $t \in [21, 24]$ and $t \in [61, 72]$.

Table 7.7 Selected results of hypothesis tests and RUL (n/a = not applicable).

Time	$\mu(t)$	$\sigma^2(t)2$	t-statistics	Mean test	F-statistics	Variance test	RUL
21	3.183	0.00371	−0.215	H_0	0.92	H_0	−61
22	3.182	0.003 19	−0.558	H_0	0.68	H_0	−51
23	3.183	0.002 91	−0.317	H_0	0.57	H_0	−138
24	3.183	0.002 87	−0.202	H_0	0.55	H_0	−288
61	3.1842	0.005 04	0.878	H_0	1.70	H_0	467
62	3.1843	0.005 35	0.973	H_0	1.92	H_0	410
63	3.1849	0.005 98	1.233	H_0	2.39	H_0	296
64	3.1849	0.006 04	1.255	H_0	2.45	H_0	294
65	3.1851	0.006 01	1.398	H_0	2.42	H_0	266
66	3.1857	0.005 96	1.740	H_0	2.38	H_0	211
67	3.1863	0.006 19	2.096	H_1	2.57	H_1	165
68	3.1870	0.006 44	2.456	H_1	2.78	H_1	132
69	3.1875	0.006 42	2.726	H_1	2.77	H_1	117
70	3.1880	0.006 26	3.082	H_1	2.63	H_1	102
71	3.1882	0.006 18	3.258	H_1	2.56	H_1	97
72	3.1892	0.005 19	4.331	H_1	1.80	H_0	76

Source: Reproduced with permission of IEEE.

For the hypothesis test on the mean, the t-test statistic in Table 7.3 is used because the true variance is unknown. In this case $n_0 = n_t = 20$, and the degree of freedom v varies between 28 and 31 because of the variation of s_t^2 at a different time t. As a result, the lower limit of the t-test varies between $[-2.04, -2.024]$ and the upper limit of the t-test varies between $[2.024, 2.04]$. For the variance test, the F-test statistic in Table 7.4 is adopted because both $\sigma^2(t)$ and σ_0^2 are estimated values. The lower limit $F_{\alpha/2,n_t-1,n_0-1} = 0.396$ and the upper limit $F_{1-\alpha/2,n_t-1,n_0-1} = 2.526$ for $\alpha = 0.05$.

Step 4: Predicting the RUL

The mean and the variance data are fed in the regression model, and it turns out that $\mu(t)$ exhibits a quadratic form. The variance remains unchanged for $t \in [0, 72]$ based on the hypothesis testing at the 95% significance level. Thus $\mu(t) = b_0 + b_1 t + b_2 t^2$ and $\sigma^2(t) = \sigma_0^2$ are the appropriate models to estimate RUL. The RUL at each sampling epoch can be predicted using Eq. (7.6.9), and RUL predictions are listed in the last column of Table 7.7. Notice that $p = 15\%$ is adopted to estimate the RUL.

Appendix 7.A. Finding the Optimal "Tao" under Age-Based PM

```
% date is 9/27/2017
% this program is design to estimate the best "tao" for systems under
% age-based maintenance
% this codes are not completed
```

```
Clear
% input data
alpha=0.1     % Weibull scale parameter
beta=2        % Weibull shape parameter

cf=80  %failure replacement cost
cp=50  %planned replacement cost

%estimate the MTTF
MTTF=(1/alpha)*gamma(1+1/beta);
t_step=MTTF/10000;
n=ceil(2*MTTF/(t_step*10));   % define the number of increment
 from 0.5MTTF to 1.5 MTTF
% if the optimal solution is within the 0.5 MTTF and 1.5MTTF do the
% following
tao=0.1*MTTF ;  % this is the initial "tao"
for j=1:n;
tao=tao+t_step*10;
m=ceil(tao/t_step);     % estimate the number of integration steps
result(j)=0;
tao_data(j)=0;
cost(j)=0;
temp1=0;
t=-t_step;
for i=1:m;
    t=t+t_step;
    temp1=temp1+exp(-(alpha*t)^beta)*t_step ;
end  % end of m
    temp2=(cf-cp)*alpha*beta*(alpha*tao)^(beta-1)*exp(-(alpha*tao)^beta);
    temp3=(cf+(cp-cf)*exp(-(alpha*tao)^beta))*exp(-(alpha*tao)^beta);
    result(j)=temp2*temp1-temp3;
    cost(j)=(cp*exp(-(alpha*tao)^beta)+cf*(1-exp(-(alpha*tao)^beta)))/temp1;
    tao_data(j)=tao;
end % end of n

% sort the data and identify the solution
for j=1:(n-1);

    if (result(j)*result(j+1))<0;
      tao_op=(tao_data(j)+tao_data(j+1))/2;
      cost_op=(cost(j)+cost(j+1))/2;
    end;
end

% output the optimal results
tao_op
cost_op
```

References

Alaswada, S. and Xiang, Y. (2017). A review on condition-based maintenance optimization models for stochastically deteriorating system. *Reliability Engineering and System Safety* 157 (1): 54–63.

Anderson, M., Jardine, A.K.S., and Higgins, R.T. (1982). The use of concomitant variables in reliability estimation. *Modelling and Simulation* 13: 73–81.

Bai, G., Wang, P., Hu, C., and Pecht, M. (2014). A generic model-free approach for lithium-ion battery health management. *Applied Energy* 135: 247–260.

Birnbaum, Z.W. and Saunders, S.C. (1969). A new family of life distributions. *Journal of Applied Probability* 6 (2): 319–227.

Brown, E.R., McCollom, N.N., Moore, E.-E., Hess, A., (2007), "Prognostics and health management: A data-driven approach to supporting the F-35 lightning II," in *Proceedings of IEEE Aerospace Conference*, March 3–10, 2007, Big Sky, MT, 12 pages, DOI: 10.1109/AERO.2007.352833.

Cassady, C.R., Murdock, W.P. Jr., and Pohl, E.A. (2001). Selective maintenance for support equipment involving multiple maintenance actions. *European Journal of Operational Research* 129 (2): 252–258.

Chhikara, R.S. and Folks, J.L. (1989). *The Inverse Gaussian Distribution: Theory, Methodology, and Applications*. New York, NY: Marcel Dekker, Inc. 0-8247-7997-5.

Chitra, T., "Life based maintenance policy for minimum cost," in *Annual Reliability and Maintainability Symposium*, 27–30 January 2003, pp. 470–474.

Cohen, M., Agrawal, N., Agrawal, V., (2006), "Winning in the aftermarket," *Harvard Business Review*, May, 2006, pp. 129–138.

Cox, D.R. (1972). Regression model and life table with discussion. *Journal of the Royal Statistical Society, Series B, Statistical Methodology* 34: 187–220.

Desmond, A. (1985). Stochastic models of failure in random environments. *Canadian Journal of Statistics* 13: 171–183.

Ding, Y., Lisnianski, A., Frenkel, I., and Khvatskin, L. (2009). Optimal corrective maintenance contract planning for aging multi-state system. *Applied Stochastic Models in Business and Industry* 25 (5): 612–631.

DoD, (2005), DoD Guide for Achieving Reliability, Availability, and Maintainability, The US Department of Defense, Washington, DC, 2005, http://www.dote.osd.mil/reports/RAMGuide.pdf (last accessed on October 6, 2017).

DoD, (2009), Department of Defense Reliability, Availability, Maintainability, and Cost Rationale Report Manual, Prepared by the Office of the Secretary of Defense in Collaboration with The Joint Staff, June 1, 2009. Available at: http://www.acq.osd.mil/se/docs/DoD-RAM-C-Manual.pdf (accessed on October 4, 2017).

Ellingwood, B.R. and Mori, Y. (1993). Probabilistic methods for condition assessment and life prediction of concrete structures in nuclear power plants. *Nuclear Engineering Design* 142 (2/3): 155–166.

FAA-HBDK-006A, (2008), Reliability, Maintainability, and Availability (RMA) Handbook, January 2008, https://www.quanterion.com/wp-content/uploads/2014/09/FAA-HDBK-006A.pdf (accessed on October 3, 2017).

Feng, J., Sun, Q., and Jin, T. (2012). Storage life prediction for high performance capacitor using multi-phase Wiener degradation model. *Communication in Statistics-B: Simulation and Computation* 41 (8): 1317–1335.

Gebraeel, N., Elwany, A., and Pan, J. (2009). Residual life predictions in the absence of prior degradation knowledge. *IEEE Transactions on Reliability* 58 (1): 106–117.

Goebel, K., Saha, B., Saxena, A. et al. (2008). Prognostics in battery health management. *IEEE Instrumentation and Measurement Magazine* 2008, 11: 33–40.

Jardine, A.K.S. and Buzacott, J.A. (1985). Equipment reliability and maintenance. *European Journal of Operational Research* 19 (3): 285–296.

Jin, T. and Liao, H. (2009). Spare parts inventory control considering stochastic growth of an installed base. *Computers and Industrial Engineering* 56 (1): 452–460.

Jin, T. and Mechehoul, M. (2010). Minimize production loss in device testing via condition based equipment maintenance. *IEEE Transactions on Automation Science and Engineering* 7 (3): 1–6.

Jin, T., Tian, Z., and Xie, M. (2015). A game-theoretical approach for optimizing maintenance, spares and service capacity in performance contracting. *International Journal of Production Economics* 31 (3): 31–43.

Jin, T. and Wang, P. (2012). Planning performance based contracts considering reliability and uncertain system usage. *Journal of the Operational Research Society* 63 (2): 1467–1478.

Jin, T., Wang, P., Huang, Q., (2006), "A practical MTBF estimate for PCB design considering component and non-component failures", in *Proceedings of Annual Reliability and Maintainability Symposium*, pp. 604–610.

Jin, T., Wang, P., Sun, Q., (2011), "Reliability prognostics based on equipment built-in diagnostic tools," in *Proceedings of Annual Reliability and Maintainability Symposium*, pp. 1–7.

Kang, K., McDonald, M., (2010), "Impact of logistics on readiness and life cycle cost: a design of experiments approach," in *Proceedings of 2010 Winter Simulation Conference*, pp. 1336–1346.

Kenne, J.P., Boukas, E.K., (1997), "Production and corrective maintenance planning problem of a failure prone manufacturing system," in *Proceedings of the American Control Conference*, vol. 2, 4–6 June 1997, pp. 1013–1014.

Kim, S.H., Cohen, M.A., and Netessine, S. (2007). Performance contracting in after-sales service supply chains. *Management Science* 53 (12): 1843–1858.

Lawless, J. and Crowder, M. (2004). Covariates and random effects in a gamma process model with application to degradation and failure. *Lifetime Data Analysis* 10: 213–227.

Lawless, J. and Crowder, M. (2010). Models and estimation for systems with recurrent events and usage processes. *Lifetime Data Analysis* 16: 547–570.

Lin, Y.-K., Lin, J.-J., and Yeh, Y.-H. (2017). Coordinating a service supply chain under arms offset program's intervention by performance-based contracting. *Mathematical Problems in Engineering* 2017: 1, 8590371–10.

Liu, X., Li, J., Al-Khalifa, K.N. et al. (2013). Condition-based maintenance for continuously monitored degrading systems with multiple failure modes. *IIE Transactions* 45 (4): 422–435.

Lopez, L. (2007). Advanced electronic prognostics through system telemetry and pattern recognition methods. *Microelectronics Reliability* 47 (12): 1865–1873.

Mao, J., Balasubramaniyan, A., "Successful application of reliability centered maintenance to semiconductor equipment," In *Proceedings of The Applied Reliability Symposium*, June 9–12, 2009, San Diego, CA.

Mirzahosseinian, H. and Piplani, R. (2011). A study of repairable parts inventory system operating under performance-based contract. *European Journal of Operational Research*. 214 (2): 256–261.

Montgomery, D. (2009). *Design and Analysis of Experiment, Chapter 2*, 7e. Wiley.

Niu, G., Singh, S., Holland, S.W., and Pecht, M. (2011). Health monitoring of electronic products based on Mahalanobis distance and Weibull decision metrics. *Microelectronics Reliability* 51 (2): 279–284.

Nakagawa, T. (2008). *Advanced Reliability Models and Maintenance Policies*. London: Springer-Verlag.

Nowicki, D., Kumar, U.D., Steudel, H.J., and Verma, D. (2008). Spares provisioning under performance-based logistics contract: profit-centric approach. *Journal of the Operational Research Society* 59 (3): 342–352.

Olde Keizera, M.C.A., Flapper, S.D.P., and Teunter, R.H. (2017). Condition-based maintenance policies for systems with multiple dependent components: a review. *European Journal of Operational Research* 261 (2): 405–420.

Öner, K.B., Kiesmüller, G.P., and van Houtum, G.J. (2010). Optimization of component reliability in the design phase of capital goods. *European Journal of Operational Research* 205 (3): 615–624.

Padgett, W.J. and Tomlinson, M.A. (2004). Inference from accelerated degradation and failure data based on Gaussian process models. *Lifetime Data Analysis* 10: 191–206.

Pan, R., W. Wang, (2014), "Analysis of degradation process with measurement errors," in *Proceedings of Reliability and Maintainability Symposium*, pp. 1–5.

Park, C. and Padgett, W.J. (2005). Accelerated degradation models for failure based on geometric Brownian motion and gamma processes. *Lifetime Data Analysis* 11 (4): 511–527.

Park, C. and Padgett, W.J. (2006). Stochastic degradation models with several accelerating variables. *IEEE Transactions on Reliability* 55 (2): 379–390.

Pecht, M. (2009). *Prognostics and Health Management of Electronics*, vol. 2009. Wiley.

Peng, H., Feng, Q., and Coit, D.W. (2011). Reliability and maintenance modeling for systems subject to multiple dependent competing failure processes. *IIE Transactions* 43 (1): 12–22.

Pham, H. and Wang, H. (2000). Optimal (τ, T) opportunistic maintenance of a k-out-of-n:G system with imperfect PM and partial failure. *Naval Research Logistics* 47 (3): 223–239.

Rafiee, K., Feng, Q., and Coit, D.W. (2015). Condition-based maintenance for repairable deteriorating systems subject to generalized mixed shock model. *IEEE Transactions on Reliability* 64 (4): 1164–1170.

Richardson, D., Jacopino, A., (2006), "Use of r&m measures in Australian defense aerospace performance based contracts," in *Proceedings of Annual Reliability and Maintainability Symposium*, pp. 331–336.

Sharma, A., Yadava, G.S., and Deshmukh, S.G. (2011). A literature review and future perspectives on maintenance optimization. *Journal of Quality in Maintenance Engineering* 17 (1): 5–25.

Sherbrooke, C.C. (1992). Multiechelon inventory systems with lateral supply. *Naval Research Logistics* 39 (1): 29–40.

Singh, A., Bapat, J., Das, D., (2013), "Distributed health monitoring system for control in Smart Grid network," in *Proceedings of 2013 IEEE Innovative Smart Grid Technologies – Asia (ISGT Asia)*, pp. 1–6.

Singpurwalla, N.D. (1995). Survival in dynamic environments. *Statistical Science* 10: 86–103.

Singpurwalla, N.D. and Wilson, S. (1998). Failure models indexed by two scales. *Advances in Applied Probability* 30: 1058–1072.

Starr, A.G., (1997), "A structured approach to the selection of condition based maintenance," in *Proceedings of 5th International Conference on FACTORY 2000*, 2–4 April 1997, IEE Conference Publication No. 435, pp. 131–138.

Sun, B., Kang, R., and Xie, J. (2007). Research and application of the prognostic and health management system. *Systems Engineering and Electronics* 29 (10): 1762–1767.

Tang, L.C. and Chang, D.S. (1995). Reliability prediction using nondestructive accelerated-degradation data: case study on power supplies. *IEEE Transactions on Reliability* 44: 562–566.

Tian, Z., (2009), "An artificial neural network approach for remaining useful life prediction of equipments subject to condition monitoring," in *Proceedings of The International Conference on Reliability, Maintainability and Safety*, pp. 143–148.

Tian, Z., Jin, T., Wu, B., and Ding, F. (2011). Condition-based maintenance optimization for wind power generation systems under continuous monitoring. *Renewable Energy* 36 (5): 1502–1509.

Tsui, K.L., Chen, N., Zhou, Q. et al. (2015, 793161). Prognostics and health management: a review on data driven approaches. *Mathematical Problems in Engineering* 2015: 1–17. doi: 10.1155/2015/793161.

Urmanov, A., (2007), "Electronic prognostics for computer servers," in *Proceedings in Reliability and Maintainability Symposium*, pp. 65–70.

Van Noortwijk, J.M. (2009). A survey of the application of gamma processes in maintenance. *Reliability Engineering and System Safety* 94: 2–21.

Vichare, N. and Pecht, M. (2006). Prognostics and health management of electronics. *IEEE Transactions on Components and Packaging Technologies* 29 (1): 222–229.

Wang P., Jin, T., (2010), "Optimal replacement of safety-critical aircraft parts with utilization uncertainties," In *Proceedings of 2010 Annual Reliability and Maintainability Symposium (RAMS)*, pp. 1–6.

Wang, H.Z. (2002). A survey of maintenance policies of deteriorating systems. *European Journal of Operational Research* 139: 469–489.

Wang, W. (2012). A stochastic model for joint spare parts inventory and planned maintenance optimization. *European Journal of Operational Research* 216 (1): 127–139.

Wang, X. and Xu, D. (2010). An inverse Gaussian process model for degradation data. *Technometrics* 52: 188–197.

Wasan, M.T. (1968). On an inverse Gaussian process. *Skandinavisk Aktuarietidskrift* 51: 69–96.

Xiang, Y., Coit, D.W., Feng, Q., and Zhu, Z. (2017). Condition-based maintenance under performance-based contracting. *Computers and Industrial Engineering* (in print).

Ye, Z.S. and Chen, N. (2014). The inverse Gaussian process as a degradation model. *Technometrics* 56 (3): 302–311.

Ye, Z.S., Xie, M., Tang, L.C., and Chen, N. (2014). Semiparametric estimation of gamma processes for deteriorating products. *Technometrics* 56 (4): 504–513.

Zhao, X. and Nakagawa, T. (2012). Optimization problems of replacement first or last in reliability theory. *European Journal of Operational Research* 223 (1): 141–149.

Zhang, X., Kang, J., and Jin, T. (2014). Degradation modeling and maintenance decision using Bayesian brief network. *IEEE Transactions on Reliability* 63 (2): 620–633.

Zhu, W., Fouladirad, M., and Bérenguer, C. (2015). Condition-based maintenance policies for a combined wear and shock deterioration model with covariates. *Computers and Industrial Engineering* 85 (July): 268–283.

Problems

Problem 7.1   State the general criteria or conditions in which corrective maintenance are appropriate to be applied for system maintenance in private and public industries.

Problem 7.2   The system MTBF is 1000 hours and there are 10 systems currently used in the field. Under corrective maintenance policy, estimate the number of failures that will be expected in the next five years, if the system lifetime follows: (1) exponential distribution; (2) the Weibull distribution with shape parameter $\beta = 0.7$; and (3) the Weibull distribution with $\beta = 2.6$.

Problem 7.3   The lifetime of the product follows the gamma distribution as follows:

$$f(t) = \frac{\lambda^k t^{k-1} \exp(-\lambda t)}{\Gamma(k)}, \qquad \text{for } t \geq 0$$

The product will be sold from years 1 to 3 with a monthly installation rate of 20 systems/month. The expected service life of each product is seven years. Since the manufacturer is responsible for supplying the spare parts during the product lifetime, estimate the monthly spare parts stock level to meet the failure replacement demand. Assuming (1) $\lambda = 0.5$ failures/year and $k = 2$; (2) $\lambda = 0.5$ failures/year and $k = 4$; and (3) $\lambda = 0.5$ failures/year and $k = 3.5$.

Problem 7.4   State the general criteria or conditions in which block replacement is preferred over corrective maintenance and age-based maintenance in industrial applications.

Problem 7.5   A system lifetime follows the lognormal distribution with parameter $\{\mu, \sigma\}$. The cost of a failure replacement is $250 and the cost of a scheduled replacement is $70. Under the periodic maintenance, determine the optimal τ to minimize the maintenance cost rate. The following cases are considered: Case 1: $\mu = 1$ and $\sigma = 1$; Case 2: $\mu = 1$ and $\sigma = 5$; and Case 3: $\mu = 2$ and $\sigma = 1$.

Problem 7.6   For a product whose lifetime follows the Weibull distribution, the cost for a failure replacement is $200 and the cost for a scheduled replacement is $100. Under periodic maintenance policy, determine the optimal τ that minimizes the maintenance cost rate. The following cases are considered: Case 1: $\alpha = 1$ and $\beta = 2$; Case 2: $\alpha = 1$ and $\beta = 4.5$; and Case 3: $\alpha = 0.5$ and $\beta = 2$.

Problem 7.7   State the general criteria or conditions in which age-based replacement is preferred over corrective maintenance and block replacement for capital equipment in both private and public industries.

Problem 7.8   Redo Problem 7.5 and find the optimal replacement age if an age-based maintenance policy is adopted.

Problem 7.9   Dedo Problem 7.6 and find the optimal replacement age if an age-based maintenance policy is adopted.

Problem 7.10   A customer has purchased a new machine and is not quite sure which maintenance policy should be adopted. According to the reliability data from the OEM, the machine's life follows the Weibull distribution with MTBF = 1500 hours and the standard deviation is 500 hours. The customer realizes that an unexpected downtime cost would be $c_f = \$2000$ and a scheduled downtime cost is $c_p = \$700$. If block or age-based maintenance is adopted, the customer needs to stock spare parts in advance and the inventory holding cost is \$1000/part/week. Determine which one is the most cost-effective: CM, block replacement, or age-based maintenance?

Problem 7.11   Assume that the degradation path $S(t)$ in Figure 7.6 is deterministic with the following trend line: (1) $S(t) = 10t$; (2) $S(t) = 2 + 0.3t^3$; and (3) $S(t) = 1.5 + 2te^{0.1t}$. Estimate the first passage time τ when $S(\tau)$ exactly exceeds the failure threshold $d = 50$.

Problem 7.12   Assume that the mean trend lines exhibit one of the following forms: (1) $E[S(t)] = 10t$; (2) $E[S(t)] = 2 + 0.3t^3$; and (3) $E[S(t)] = 1.5 + 2te^{0.1t}$. The standard deviation of the degradation is 3, which is time-invariant. At a 90% confidence level, determine the first passage time of $S(t)$ in Figure 7.6 assuming that the failure threshold $d = 50$.
1) Assume that the $S(t)$ is uniformly distributed for given t.
2) Assume that the $S(t)$ is normally distributed for given t.

Problem 7.13   According to Example 7.5, the hazard rate of the aircraft engine given in Eq. (7.4.6) fits the Weibull PHM. The customer requires 5500 hours MTBF under $z_1 = 2.5$ and $z_2 = 0.4$. Age-based maintenance is applied to the engine with $c_f = 90\,000$ and $c_p = 35\,000$. Do the following:
1) Estimate the required scale parameter η given $\beta = 4.47$.
2) Determine the optimal τ that minimizes the maintenance cost rate.

Problem 7.14   For a degradation process approximated by a compound Poisson process with the form of $X(t) = \sum_{i=0}^{N(t)} D_i$, where $N(t)$ is the number of random shocks that follow the Poisson process with rate λ and D_i is the damage size under the ith shock, estimate:
1) The mean and the variance of $X(t)$ when D_i is uniformly distributed with $[a, b]$.
2) The mean and the variance of $X(t)$ when D_i is normally distributed with $N \sim (\mu, \sigma^2)$.
3) Use a simulation program to validate the results in (1) and (2) above.

Problem 7.15 Use numerical methods to compute the gamma function in Eq. (7.4.9) under different $v(t)$, which is given as follows:
1) Constant $v(t) = 4$, 4.5, or 8.7.
2) Linear function $v(t) = 1 + t$ or $v(t) = 1 + 2t$.
3) Polynomial function $v(t) = 1 + 0.2t + 0.3t^2$.
4) Power law model $v(t) = 0.4t^3$.

Problem 7.16 Plot the PDF of Eq. (7.4.15) with the forms of u and $v(t)$ given as follows:
1) For $v(t) = 2t$, assume $u = 1$ and 3, respectively.
2) For $v(t) = 1 + t$, assume $u = 1$ and 3, respectively.
3) For $v(t) = 0.4t^3$, assume $u = 1$ and 3, respectively.

Problem 7.17 The CDF function for a gamma degradation process is shown in Eq. (7.4.13). Assume that $d = 30$ and estimate RUL with u and $v(t)$ being given as follows:
1) For $v(t) = 2t$, assume $u = 1$ and 3, respectively.
2) For $v(t) = 1 + t$, assume $u = 1$ and 3, respectively.
3) For $v(t) = 0.4t^3$, assume $u = 1$ and 3, respectively.

Problem 7.18 Use the Birnbaum–Saunders distribution in Eqs. (7.4.24) or (7.4.25) to estimate the RUL in Problem 7.17 using the same forms of u and $v(t)$.

Problem 7.19 For an IG process with IG(a, b), provide the detailed derivation to show the mean and variance, namely $E[X(t)] = a$ and $Var(X(t)) = a^3/b$.

Problem 7.20 Given the PDF of IG in Eq. (7.5.1), provide the step-by-step process to show the CDF in Eq. (7.5.4).

Problem 7.21 A degradation process follows the IG process, obtain the reliability function $R(t)$ given the following conditions:
(1) $d = 30$ and $\Lambda(t) = 1$.
(2) $d = 30$ and $\Lambda(t) = 1 + 0.2t$.
(3) $d = 40$ and $\Lambda(t) = 0.1t^2$.
(4) $d = 40$ and $\Lambda(t) = 2 + 0.1t^2$.
(5) $d = 25$ and $\Lambda(t) = 0.1t^3$.
(6) $d = 50$ and $\Lambda(t) = 0.1t^3$.

Problem 7.22 The degradation of resistors is often monitored based on the dissipated power. Samples are regularly collected during the period of a week, and the mean and standard deviation of the power dissipated are estimated and summarized in the table below. Perform the hypothesis test by comparing the mean and variance at $t = 0$. The same size n is assumed to be sufficiently large.

t (week)	0	1	2	3	4	5	7	8	9	10
μ (W)	2.00	1.97	2.25	2.02	2.44	2.82	3.32	3.57	3.88	4.18
σ (W)	0.20	0.24	0.22	0.21	0.16	0.22	0.24	0.22	0.20	0.22

The mean and standard deviation of the power dissipation of resistor

Problem 7.23 Assuming the failure threshold of the resistor $P = 5.5$W, estimate the RUL based on the measurements across the 10 weeks in Problem 7.22.

Problem 7.24 State the interactions between the five overarching performance measures.

Problem 7.25 Assuming the repair center has unlimited capacity and is modeled as an M/G/∞ queue, provide the detailed proof of the unified OA model in Eq. (7.7.11).

Problem 7.26 Assume that the repair center's capacity is unlimited and can process two parallel queues: those that deal with failure returns and those that deal with the returns for reconditioning. Provides the detailed proof of the unified OA model in Eq. (7.7.16).

Problem 7.27 For the principal-agent contract model in Model 7.1, what type of moral hazard issues would the OEM meet if the PBM service is offered? What types of moral hazard issues might the customer face?

Problem 7.28 The OEM decide to adopt the logistics network in Figure 7.16 to implement the principal-agent contract model in Model 7.1. That is, the OEM is responsible for the implementation of age-based replacement, spare parts supply, and the parts repair and reconditioning services. Do you think that the customer has full observability on the equipment reliability, the stock level (OEM owns the stock), and the repair center activities?

8

Warranty Models and Services

8.1 Introduction

A product warranty is an after-sales service contract between the customer and the supplier under which the supplier (i.e. manufacturer or retailer) is obliged to repair, replace, or make compensation to the buyer in the case of product failure for a pre-specified time period, referred to as the "warranty period." Based on the rebate policy, a warranty program can be classified into a free replacement warranty, a pro-rata warranty (PRW), or a combination of both. From the perspective of the coverage mechanism, a warranty can be divided into a renewing warranty (RW) and a non-renewing warranty (NRW). This chapter aims to introduce different types of warranty rebate policy for non-repairable and repairable products. Particularly, Section 8.2 provides an overview on warranty service models that are adopted by industries or studied in academia. Section 8.3 discusses the warranty rebate models of non-repairable products and Section 8.4 extends them to repairable products. Section 8.5 discusses how to design and implement a warranty servicing contract during the new product introduction by considering the effect of the product's installed base. Section 8.6 investigates the effects of reliability growth on the cost saving of a product warranty in the aftermarket. In Section 8.7, we discuss the challenges of implementing an extended warranty (EW) in dual distribution channels, and also present the framework of a modeling two-dimensional warranty policy. The case study in Section 8.8 demonstrates how to achieve a design-for-warranty by choosing a component and a non-component from multiple suppliers with the goal of minimizing the warranty cost during a new product introduction.

8.2 Warranty Concept and Its Roles

8.2.1 Overview of Warranty Services

To maintain the competitiveness in the global market, many products nowadays are sold under a warranty contract. During the warranty period, consumers can claim for repair, replacement, or compensation for a failed or underperforming product as stipulated in the initial purchase contract. Warranty services play three important roles. First, they protect consumers against defective items. Second, they protect the producers against consumers' excessive claims. Third, warranty services promote and expand sales, especially for new products.

Reliability Engineering and Services, First Edition. Tongdan Jin.
© 2019 John Wiley & Sons Ltd. Published 2019 by John Wiley & Sons Ltd.
Companion website: www.wiley.com/go/jin/serviceengineering

The last three decades witnessed a trend of offering longer periods of warranty services to customers in automobiles, consumer electrics, and capital equipment. This tendency is driven primarily by improved quality control and fierce competition in the global market. For example, Nissan has been offering a 10-year unlimited mileage warranty for cars (Nissan 2011). In 2016 Samsung announced lifetime guarantee against long-term image retention for SUHD Quantum Dot TV in the US market (Samsung Newsroom 2016). In a survey paper by Shafiee and Chukova (2013), it was found that in the 1990s, the warranty period for airplanes was typically 2.5 years or 2500 flight-hours, whichever came first. Today it has been extended to 5–10 years or 5000–10 000 flight-hours. With a long-term warranty commitment to product repair and replacement, manufacturers or suppliers need to develop a lifetime service strategy to handle the infant mortality failures, random failures during useful life, and degradation failures in the product wear-out phase. Recently, Shang et al. (2018) proposed a condition-based renewable replacement warranty policy based on the inverse Gaussian degradation process. The goal is to maximize the manufacturer's after-sales profit by optimizing the sale price and replacement threshold in the warranty period, as well as the preventive maintenance policy in post-warranty service.

Though a longer warranty program usually attracts more customers to buy the product, it may escalate the financial burden on the manufacturers or sellers. The warranty servicing cost is the expense of rectifying a faulty item during the warranty period, which includes labor, materials, and logistics. Depending on the nature of the product and manufacturing processes, this cost can vary between 2% and 10% of an item's sale price. According to Warranty Week (2016a), the US-based automotive companies paid $15.6 billion on worldwide warranty claims in 2014, reaching a new record that was 20% higher than 2013. This corresponds to 2% of the product revenue. Similarly, the US-based auto part suppliers spent $2.1 billion annually on warranty claims in 2014. Since the warranty servicing costs directly affect the manufacturer's profit margin, finding an effective approach to reducing the servicing costs is critical to the manufacturers or original equipment manufacturers (OEMs). To alleviate the warranty servicing costs, it is often preferred and indeed more cost-effective to detect the reliability problems and make corrective actions in the early design stage.

Recently both manufacturers and retailers have started to offer and sell an extended warranty on top of the standard or base warranty. An extended warranty, also called a service contract or a maintenance agreement, is a prolonged warranty offered to consumers after the expiration of the base warranty on new items. Under the customer's expenses, an extended warranty may or may not cover peripheral items, wear and tear, damage by computer viruses, normal maintenance, accidental damage, or any consequential loss. In the consumer electronics market, extended warranties cost 20–30% of the original price, and also give sales associates up to 15% commission at some retailers (Warranty Week 2006b). Therefore, extended warranty services should be designed from the viewpoint of the sellers as well as the consumers. It is important to guide the consumers to decide whether or not to buy an extended warranty service or how to choose the starting date and coverage length to lower the total ownership cost.

Traditionally, products are sold with one-dimensional warranties that expire after a specific time period. However, the reliability of a car, for instance, is often influenced by the age and the mileage. Thus a two-dimensional (2-D) warranty is jointly defined by the age and usage that forms a region in a 2-D plane. The motivation of implementing the

2-D warranties is the heterogeneity or diversity in customer usage of the product. Most products sold with a 2-D warranty have a rectangular region framed by two parameters W and U, representing the warranty age and the usage coverage, respectively. For example, if the drivetrain of new car is offered with a 3-year or 30 000 miles warranty, it implies $W = 3$ years and $U = 30\,000$ miles, whichever comes first. Products sold with a 2-D warranty include aircraft landing gear, jet engines, electric motors, and copy machines, among others.

8.2.2 Classification of Warranty Policy

Overall, a product warranty is an agreement offered to a consumer to repair or replace a faulty item, or to partially or fully reimburse the consumer in the case of failure. The form of reimbursement or compensation to the customer on failure of the product is one of the most important characteristics of a warranty policy. In general, there are four commonly used warranty rebate policies: 1) the ordinary free replacement warranty (OFRW), or simply FRW; 2) the unlimited free replacement warranty (UFRW); 3) the pro-rata warranty (PRW), and 4) the mixed warranty (Reliasoft 2009).

Ordinary free replacement warranty (OFRW): If a product fails prior to the expiration of the warranty period because of quality or reliability issues, it will be replaced or repaired at no cost to the customer. The repaired product is then covered by an OFRW. The length of the warranty is equal to the remaining length of the original warranty. Many vehicle and home appliance manufacturers adopt the ordinary free replacement/repair warranty policy.

Unlimited free replacement warranty (UFRW): If a product fails before the end of the warranty period, it will be replaced or repaired at no cost to the customer. The replaced or repaired product is then covered by a new UFRW. Namely, the duration of the new warranty is equal to the length of the original warranty. The UFRW is used for consumer electronics such as cell phones that have a relatively short product lifecycle.

Pro-rata warranty (PRW): If an item fails before the end of the warranty, it is replaced at a cost to the customer that depends on the age of the item at the time of failure. The replacement item is then covered by an identical new warranty. This policy is often used for non-repairable items such as tires and batteries, and typically the reimbursement decreases with the age of the item.

Mixed warranty: This is a combination of the preceding warranty polies. Usually a mixed warranty starts with a free replacement up to a specified time, say w_1, and switches to the pro-rated cost procedure for the remainder of the warranty period, say $w_2 - w_1$, where w_2 is the total warranty coverage length. This is called FRW/PRW.

In the literature, FRW usually means an OFRW. In fact, the replacement can be treated as a particular type of repair that brings the item to an as-good-as-new state. Regarding the mechanism of the warranty coverage, there are two types of warranty policy used in the industry and discussed in the literature.

Renewing warranty (RW): The warrantor (e.g. manufacturer) repairs or replaces any faulty item from the time of the purchase up to the end of the warranty period. At the time of each repair within the current warranty period, the item is warranted anew for a period of original length w. The warranty coverage expires when the lifetime of the item (the original one or its repaired version) exceeds w. For example, light bulbs are often covered by a renewing free replacement warranty (RFRW). Assume a light bulb has an initial warranty period of $w = 500$ hours. If it fails during this period and is replaced by a new bulb, the warranty time starts anew as if purchasing a new product.

Non-renewing warranty (NRW): Under this policy, the warrantor assumes all or a portion of the expenses associated with the failure of the product over a fixed period of calendar time. For example, an item is purchased on January 1, 2017 with a one-year warranty coverage. The item may fail and be replaced or repaired at any time in 2017, but the warranty associated with this item will expire on December 31, 2017. Most household appliances, such as vacuum cleaners, refrigerators, washing machines, dryers, and televisions, are covered by an NRW.

A large body of warranty models have been developed and published in litearture. Books focusing on warranty models include the ones by Murthy and Blischke (2006), Thomas (2006), Rai and Singh (2009), and Blischke et al. (2011). The recent book by Jiang (2015) also reports new development between the interface of product reliability, warranty, and quality control. Several book chapters are also available including those by Murthy and Djamaludin (2001), Wang and Pham (2006), and Iskandar and Jack (2011). Finally, review papers by Karim and Suzuki (2005), Wu (2012), and Shafiee and Chukova (2013) provide systematic discussions on the development of warranty models and its interactions with the maintenance management. For this chapter, the main reference sources are the books by Elsayed (2012) and Murthy and Blischke (2006).

8.3 Warranty Policy for Non-repairable Product

8.3.1 Warranty Reserve under Pro-Rata Rebate

For PRW policy, the customer will receive the cost compensation that proportionally decreases with the age of the item being replaced. Let w be the warranty period. Based on the product reliability, the manufacturer estimates the expected number of products that will fail in a small time interval $[t, t + dt]$ for $0 \leq t < w$. Then multiply the expected number of failures by the replacement cost at time t to obtain the warranty reserve that must be allocated for failure replacements during $[t, t + dt]$. Under the assumption that all failures in the warranty period are claimed, the total amount of warranty reserve is obtained by aggregating the incremental warranty cost for period $[0, w]$. Before we present the cost model of Menke (1969), the following notation is defined:

c = unit product price including the warranty cost
w = length of the warranty period
$f(t)$ = probability density function (PDF) of the product lifetime
$F(t)$ = cumulative lifetime distribution of the product
$N(t)$ = expected number of failures in $[0, t]$
$b(t)$ = pro-rata customer rebate at time t for $0 \leq t \leq w$
c_r = the amount of warranty reserve in $[0, w]$ per unit product
α and β = Weibull scale and shape parameters, respectively
r = warranty reserve cost per unit of product

Let $f(t)$ be the PDF of the lifetime of a product. The expected number of failures in $[t, t + dt]$, denoted as $dN(t)$, can be estimated as

$$dN(t) = f(t)dt \tag{8.3.1}$$

Figure 8.1 Linear and exponential pro-rata rebate functions.

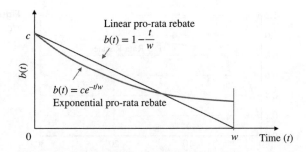

Multiplying Eq. (8.3.1) with the pro-rata rebate function $b(t)$ yields the incremental warranty cost incurred to the manufacturer during $[t, t+dt]$. That is,

$$dc_r = b(t) \times dN(t) = b(t)f(t)dt \tag{8.3.2}$$

Hence, the total amount of warranty reserve for all failure replacements occurring in $[0, w]$ is obtained by integrating t from 0 to w. That is,

$$r = \int_0^w b(t)f(t)dt \tag{8.3.3}$$

Next, we investigate four cases by considering different forms of $b(t)$ and $f(t)$. In particular, we consider linear and exponential rebate function $b(t)$ that are shown in Figure 8.1. For modeling the product lifetime, we focus on exponential and Weibull distributions, but the analysis can be extended to other lifetime distributions.

Case 1: $b(t) = c(1 - t/w)$ and $f(t) = \lambda e^{-\lambda t}$

In this case, the manufacturer chooses to implement a linear pro-rata rebate policy and the product lifetime is exponential with failure rate λ. The total warranty reserve now becomes

$$r = \int_0^w c\left(1 - \frac{t}{w}\right)\lambda e^{-\lambda t}dt = c\left(1 - \frac{1}{\lambda w}(1 - e^{-\lambda w})\right) \tag{8.3.4}$$

Thus the ratio between the warranty reserve and the unit price is

$$\frac{r}{c} = 1 - \frac{1}{\lambda w}(1 - e^{-\lambda w}) = 1 - \frac{1}{w/\tau_m}(1 - e^{-w/\tau_m}) \tag{8.3.5}$$

where $\tau_m = 1/\lambda$ is the mean-time-to-failure (MTTF) of the product. The ratio between the warranty reserve and the product cost goes up as the ratio between w and τ_m increases, as shown in Figure 8.2.

Case 2: $b(t) = ce^{-t/w}$ and $f(t) = \lambda e^{-\lambda t}$

In this case, the manufacturer opts to implement an exponential cost rebate policy. At $t = 0$, a full rebate is offered to the customer and at $t = w$, an amount of $0.367c$ is rebated to the customer upon failure. When the product lifetime is exponential with failure rate λ, the warranty reserve is

$$r = \int_0^w ce^{-t/w}\lambda e^{-\lambda t}dt = \frac{cw}{1 + \lambda w}\left(1 - e^{-(1+\lambda w)}\right) \tag{8.3.6}$$

Thus the ratio between the warranty reserve and the unit price of the product is

$$\frac{r}{c} = \frac{w}{1 + \lambda w}(1 - e^{-(1+\lambda w)}) \tag{8.3.7}$$

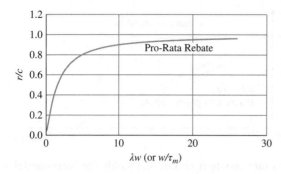

Figure 8.2 Plot the relation with λw and r/c.

Case 3: $b(t) = c(1 - w/t)$ and $f(t) = \alpha\beta(\alpha t)^{\beta}\exp(-(\alpha t)^{\beta})$

In this case, the manufacturer chooses to implement a linear pro-rata rebate policy, and the product lifetime follows the Weibull distribution. The total warranty reserve now becomes

$$r = \int_0^w c\left(1 - \frac{t}{w}\right)\alpha\beta(\alpha t)^{\beta-1}e^{-(\alpha t)^{\beta}}\,dt$$

$$= c(1 - e^{-(\alpha w)^{\beta}}) - \frac{c}{w}\int_0^w t\alpha\beta(\alpha t)^{\beta-1}e^{-(\alpha t)^{\beta}}\,dt \tag{8.3.8}$$

Case 4: $b(t) = ce^{-t/w}$ and $f(t) = \alpha\beta(\alpha t)^{\beta}\exp(-(\alpha t)^{\beta})$

In this case, the manufacturer adopts the exponential cost rebate function, and the product lifetime follows the Weibull distribution. The warranty reserve can be estimated as

$$r = \int_0^w ce^{-t/w}\alpha\beta(\alpha t)^{\beta-1}e^{-(\alpha t)^{\beta}}\,dt \tag{8.3.9}$$

In Cases 3 and 4, an explicit analytical solution is difficult to obtain, so numerical methods need to be applied to compute r.

8.3.2 Warranty Reserve under Fixed Rebate

From the customer perspective, the rebate under the PRW decreases over time. This makes sense for products like consumer electronics for which the useful lifetime are relatively short due to rapid technological progress. For products like washing machines and air-conditioning equipment, their average lifetime is much longer, typically 7–10 years. Then the manufacturer may wish to consider an alternative approach by paying a lump-sum or fixed rebate to the customer for any failure occurring before the expiration of the warranty. Again, manufacturers are required to adjust the price of the product and allocate the warranty reserve to meet customer claims. Figure 8.3 shows a fixed rebate cost model $b_2(t)$ in comparison with a linear pro-rata rebate model $b_1(t)$.

Let ρ for $0 < \rho < 1$ be the proportion of the unit cost c to be refunded as a lump-sum rebate and s be the unit lump-sum rebate. Then $s = \rho c$ or $b(t) = \rho c$ for the fixed rebate policy. We consider two cases for exponential and Weibull lifetimes, respectively.

Case 1: $b(t) = \rho c$ and $f(t) = \lambda\exp(-\lambda t)$

We substitute $b(t) = \rho c$ and $f(t) = \lambda\exp(-\lambda t)$ into Eq. (8.3.3), which yields

$$r = \rho c(1 - e^{-\lambda w}) \tag{8.3.10}$$

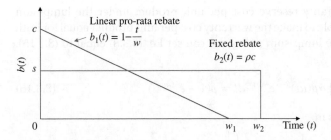

Figure 8.3 Fixed rebate versus pro-rata rebate.

where r is the warranty reserve cost per unit product under the lump-sum rebate. If it is desirable to make the warranty cost per unit of product equal for both the pro-rata and the lump-sum policies, we can set Eq. (8.3.4) equal to (8.3.10), which yields

$$1 - \frac{1}{\lambda w}(1 - e^{-\lambda w}) = \rho(1 - e^{-\lambda w}) \tag{8.3.11}$$

Solving for ρ results in

$$\rho = \frac{1}{1 - e^{\lambda w}} - \frac{1}{\lambda w} \tag{8.3.12}$$

Under the lump-sum rebate, the proportion of the unit cost to be refunded as a function λw is shown in Figure 8.4.

Hence the lump-sum rebate per unit product becomes

$$s = \rho c = c\left(\frac{1}{1 - e^{-\lambda w}} - \frac{1}{\lambda w}\right) \tag{8.3.13}$$

For a production batch with m units of sold items, the total amount of warranty reserve, C_r, should be allocated for the customer claims in the period of $[0, w]$, which is

$$C_r = ms(1 - e^{-\lambda w}) \tag{8.3.14}$$

Case 2: $b(t) = \rho c$ and $f(t) = \alpha\beta(\alpha t)^\beta \exp(-(\alpha t)^\beta)$

We substitute $b(t) = \rho c$ and $f(t) = \alpha\beta(\alpha t)^\beta \exp(-(\alpha t)^\beta)$ into Eq. (8.3.3), which yields

$$r = \rho c(1 - e^{-(\alpha w)^\beta}) \tag{8.3.15}$$

Figure 8.4 Plot the relation between ρ and λw.

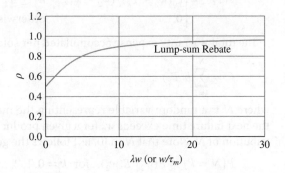

where r is the warranty reserve cost per unit product under the lump-sum rebate. If it is desirable to make the warranty cost per unit product equal for both the pro-rata and the lump-sum policy, we can set Eq. (8.3.8) equal to (8.3.15). That gives

$$\int_0^w \left(1 - \frac{t}{w}\right) \alpha\beta(\alpha t)^{\beta-1} e^{-(\alpha t)^\beta} dt = \rho(1 - e^{-(\alpha w)^\beta}) \tag{8.3.16}$$

Solving for ρ results in

$$\rho = 1 - \frac{1}{w\left[1 - \exp(-(\alpha w)^\beta)\right]} \int_0^w t \times \alpha\beta(\alpha t)^{\beta-1} e^{-(\alpha t)^\beta} dt \tag{8.3.17}$$

More discussions about warranty reserve under a pro-rata and fixed rebate are also available in the book by Elsayed (2012).

8.3.3 Mixed Warranty Policy

In this section, we present a mixed warranty policy comprised of two different rebate mechanisms that depend on when the claim takes place. In particular, the customer receives a fixed amount of rebate for any failure occurring in $0 \le t \le w_1$ and a pro-rata rebate compensation is applied if a failure occurs in $w_1 \le t \le w_2$. In order to simplify the analysis, we define the following notation (Ritchken 1985):

w_1 = length of the fixed rebate period for the mixed policy
$w_2 - w_1$ = length of the pro-rated period for the mixed policy and $w_2 > w_1$
b = cost of replacement under the fixed rebate
T_k = interarrival time of failure k and $k-1$ and all T_k are i.i.d.
$c(T_k)$ = cost of a failure to the manufacturer under the mixed policy
C = total warranty cost accrued for each sold product
$F(t), R(t)$ = cumulative distribution function (CDF) and reliability function of T_k
$f(t)$ = PDF of T_k
$F^{(k)}(t) = \int_0^x F^{(k-1)}(w)dw$, for $k = 1, 2, \ldots$ with $F^{(0)}(t) = F(t)$
N = number of failures that occurred until the next failure time exceeds the warranty period

The mixed rebate policy is depicted in Figure 8.5. The mathematical expression associated with the mixed policy is

$$b(T_k) = \begin{cases} c_0, & 0 \le T_k \le w_1 \\ c_0(w_2 - T_k)/(w_2 - w_1), & w_1 \le T_k \le w_2 \\ 0, & \text{otherwise} \end{cases} \tag{8.3.18}$$

The total warranty expense accumulated per sold product is

$$C = \sum_{k=1}^N b(T_k) \tag{8.3.19}$$

where N is a random variable representing the number of failures that occurred until the next failure time exceeds w_2 for a given product item. Next we characterize the distribution of N. Note that N in $[0, w_2]$ follows the geometric distribution as

$$P\{N = k\} = F(w_2)^k R(w_2), \quad \text{for } k = 0, 1, 2, \ldots \tag{8.3.20}$$

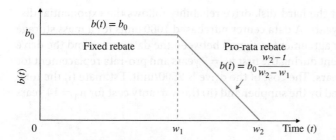

Figure 8.5 Mixed rebate function.

The rationality is that the product keeps failing for k times consecutively before the next replaced item survives beyond w_2. Thus the expected number of failures is

$$E[N] = \sum_{k=0}^{\infty} kR(w_2)F(w_2)^k = \frac{F(w_2)}{R(w_2)} \tag{8.3.21}$$

Now we derive the expected total warranty cost for a sold item under the mixed rebate by taking the expectation to Eq. (8.3.19). That is,

$$E[C] = E\left[\sum_{k=1}^{N} b(T_k)\right] = E\left[\sum_{k=1}^{N} E[b(T_k \mid N)]\right] \tag{8.3.22}$$

Note that two random variables N and T_k are involved in $E[C]$. Since the consecutive failures are mutually independent, Eq. (8.3.22) can be expressed as

$$E[C] = E[N]E[b(T_k)] \tag{8.3.23}$$

Moreover, from Ritchken (1985), Thomas (2006), and Elsayed (2012), we have

$$E[b(T_k)] = b_0 \int_0^{w_1} f(t)dx + \frac{b_0}{w_2 - w_1} \int_{w_1}^{w_2} (w_2 - t)f(t)dt = \frac{b_0}{w_2 - w_1} \int_{w_1}^{w_2} F(t)dt \tag{8.3.24}$$

where $f(t)$ is the PDF of the product time-to-failure distribution or the interarrival time between two consecutive failures. Substituting Eqs. (8.3.21) and (8.3.24) into Eq. (8.3.23), we obtain the expected warranty cost of the product under the mixed rebate policy as

$$E[C] = \frac{b_0 F(w_2)}{(w_2 - w_1)R(w_2)} \int_{w_1}^{w_2} F(t)dt \tag{8.3.25}$$

Two special cases can be derived from the mixed warranty policy. If $w_1 = w_2$, then the mixed policy becomes the full-rebate policy, and we take the following limit to estimate $E[C]$:

$$\lim_{w_1 \to w_2} E[C] = \frac{b_0 F(w_2)}{R(w_2)} \lim_{w_1 \to w_2} \frac{\int_{w_1}^{w_2} F(t)dt}{(w_2 - w_1)} = \frac{b_0 F^2(w_2)}{R(w_2)} \tag{8.3.26}$$

On the other hand, if $w_1 = 0$, the mixed warranty policy becomes the pro-rata rebate model. That is,

$$\lim_{w_1 \to 0} E[C] = \frac{b_0 F(w_2)}{w_2 R(w_2)} \int_0^{w_2} F(t)dt \tag{8.3.27}$$

Example 8.1   Assume that the hard-disk drive reliability follows the exponential distribution with MTTF = 50 years. A data center purchased 1000 units for a mass storage purpose. A mixed warranty agreement is signed between the data center and the drive supplier with free replacement during the first $w_1 = 7$ years and pro-rata replacement for an additional $w_2 - w_1 = 3$ years. The cost of the drive is \$500/unit. Estimate (i) the total warranty cost that is incurred by the supplier and (ii) the warranty cost for $w_1 = 14$ years and $w_2 - w_1 = 6$ years.

Solution:

1) Since a full rebate is adopted in $[0, w_1]$, then $b_0 = \$500$/drive. Given MTTF = 50 years, the drive's failure rate is $\lambda = 1/50 = 0.02$ failures/year. The corresponding reliability and CDF functions are given respectively as

$$R(t) = e^{-0.02t} \text{ and } F(t) = 1 - e^{-0.02t} \tag{8.3.28}$$

Substituting all related known conditions into (8.3.25) along with 1000 items of drives, the total warranty cost would be

$$E[C_1] = 1000 \times \frac{500[1 - \exp(-0.02 \times 10)]}{(10 - 7) \times \exp(-0.02 \times 10)} \int_7^{10} (1 - \exp(-0.02t))dt = \$17\,293$$

The total purchasing cost is $1000 \times 500 = \$500\,000$. Thus the warranty cost accounts for $17\,293/500\,000 = 3.5\%$ of the total cost. In other words, among the sale price of \$500/drive, the base price (without warranty overheads) should be set at $\$500 \times (1–3.5\%) = \$482.71$/drive.

2) When the mixed warranty period doubles from $w_1 = 7$ to 14 years and $w_2 - w_1 = 3$ to 6 years, the total warranty cost becomes

$$E[C_2] = 1000 \times \frac{500[1 - \exp(-0.02 \times 20)]}{(20 - 14) \times \exp(-0.02 \times 20)} \int_{14}^{20} (1 - \exp(-0.02t))dt = \$70\,774$$

The warranty cost under the 20-year policy is 4.1 times that of $E[C_1]$ in the original 10-year policy.

8.3.4   Optimal Preventive Maintenance under Warranty

One potential method to reduce the expected warranty service cost is to jointly plan the warranty service and maintenance program. This is because warranties (base warranty and extended warranty) and maintenance (corrective and preventive) are strongly correlated and of great interest to both manufacturers and customers. For instance, a typical age-based replacement requires a spare part upon failure or at a fixed time, whichever occurs first. In this section we present a joint model due to Ritchken and Fuh (1986) who determined the optimal age replacement policy for items under warranty, such that the average cost per unit time is minimized. Related notation is listed below:

t = calendar time or operating time of the product
w = length of the warranty period
T_i = time to the ith failure, a random variable
$F(t), f(t)$ = CDF and PDF of the time-to-failure
$R(t)$ = reliability up to time t

$h(t)$ = hazard rate at time t

$F_Z(z), f_Z(z)$ = CDF and PDF of residual life beyond w

$N(t)$ = number of failures an installed item has in [0, t], a random number

$M(t), m(t)$ = renewal function of $N(t)$ and the renewal density function

c_f = cost for replacing a failed item

c_a = cost for replacing a functioning, yet aged, item

X_i = time between two failures i and $i - 1$, a random variable

Y = mean time between two replacements of an item

C_w = warranty cost accrued for each installed product, a random variable

$C(\tau)$ = sum of the warranty cost and the replacement cost post w

The cost of replacing the jth failure that occurs before the expiration of the warranty period w is

$$c_j = b(T_j), \quad \text{for} \quad T_j \in [0, \ w] \tag{8.3.29}$$

where $b(T_j)$ is the rebate policy. Let $N(w)$ be the number of warranty replacements in [0, w]. The expected cumulative warranty replacement cost over the period of [0, w] is

$$E[C_w] = E\left[\sum_{j=1}^{N(w)} b(T_j)\right] \tag{8.3.30}$$

If the item survives beyond w, the residual life of the item, Z is a random variable with CDF given by Ross (1970):

$$F_Z(z) = F(w + z) - \int_0^w R(w + z - x)dM(x)$$

$$= F(w + z) - \int_0^w R(w + z - x)m(x)dx, \quad \text{for} \ z \ge 0 \tag{8.3.31}$$

where $m(x)$ is the renewable density function with $m(x) = dM(x)/dx$ and $R(x) = 1 - F(x)$. Taking the derivative with respect to z yields the PDF of the residual life as follows:

$$f_Z(z) = \frac{\partial F_Z(z)}{\partial z} = f(w + z) + \int_0^w f(w + z - x)dM(x) \tag{8.3.32}$$

The cost incurred over the full replacement cycle is the sum of the warranty expenses over [0, w] and the replacement cost of either a failed item before reaching $w + \tau$ or an aged item at time $\tau + w$, where τ is the scheduled replacement time.

Let $C(\tau)$ be the expected cost over the replacement cycle including w and τ. Then

$$C(\tau) = E[C_w] + c_f F_Z(\tau) + c_a(1 - F_Z(\tau)) \tag{8.3.33}$$

Two scenarios will be observed on the item survived beyond w. As shown in Figure 8.6, in Scenario 1 the item continues to operate until it reaches time $w + \tau$ for a planned replacement. In Scenario 2, the item failed between w and $w + \tau$ and is replaced immediately. Let Y be the mean time between two replacement cycles; then

$$E[Y] = w + \tau R_Z(\tau) + \int_0^\tau zf_Z(z)dz$$

$$= w + \int_0^\tau R_Z(z)dz \tag{8.3.34}$$

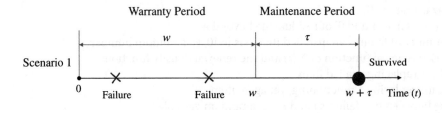

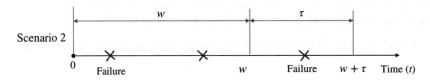

Figure 8.6 Warranty and post-warranty replacement.

Therefore, the steady-state replacement cost rate is obtained by dividing Eq. (8.3.33) by Eq. (8.3.24) as follows:

$$r(\tau) = \frac{E[C_w] + c_f F_Z(\tau) + c_a(1 - F_Z(\tau))}{w + \int_0^\tau R_Z(z)dz} \tag{8.3.35}$$

The objective is to find the optimal value τ^* to minimize $r(\tau)$ in Eq. (8.3.35). If an item's hazard rate $h(t)$ is continuous and monotonically non-decreasing, Ritchken and Fuh (1986) proved that a unique solution exists that minimizes Eq. (8.3.35).

Example 8.2 Let c be the unit product price. Consider a linear pro-rata rebate policy. The cost of replacing an item at the jth failure within the warranty period w is

$$b(T_j) = T_j \frac{c}{w}, \quad \text{for } j = 0, 1, 2, \dots, N(w) \tag{8.3.36}$$

where $T_j \in [0, w]$. The mean replacement cost over the warranty period is

$$E[C_w] = E\left[\sum_{j=1}^{N(w)} b(T_j)\right] = \frac{c}{w} E\left[\sum_{j=1}^{N(w)} T_j\right] \tag{8.3.37}$$

where T_j and $N(w)$ are random variables. Suppose the product's lifetime follows the Erlang distribution with the PDF as follows:

$$f(t) = te^{-t}, \quad \text{for } t \geq 0 \tag{8.3.38}$$

Determine the optimal replacement time τ given the warranty period $w = 1$.

Solution:

Step 1: Since the rebate cost depends on the number of failures in $[0, w]$, let us find the Erlang renewal function $M(t)$ and renewal intensity $m(t)$. The Laplace transform of $f(t)$ is given as follows:

$$f_L(s) = \int_0^\infty te^{-t}e^{-st}dt = \frac{1}{(s+1)^2}e^{-(s+1)t}\Big|_\infty^0 = \frac{1}{(s+1)^2} \tag{8.3.39}$$

Hence the renewal function in the s-domain can be avoided as

$$M_L(s) = \frac{f_L(s)}{s(1 - f_L(s))} = \frac{1}{s^3 + 2s^2} = \frac{0.5}{s^2} - \frac{0.25}{s} + \frac{0.25}{s+2} \qquad (8.3.40)$$

Perform the inverse Laplace transform. We obtain $M(t)$ and $m(t)$ as follows:

$$M(t) = 0.5t + 0.25(e^{-2t} - 1) \qquad (8.3.41)$$

$$m(t) = 0.5 - 0.5e^{-2t} \qquad (8.3.42)$$

Step 2: Find the CDF of the residual lifetime. First we find the CDF of the product life based on the Erlang PDF:

$$F(t) = \int_0^t xe^{-x}dx = 1 - e^{-t} - te^{-t} \qquad (8.3.43)$$

From Eq. (8.3.31), the CDF of the residual lifetime is

$$F_Z(z) = [1 - e^{-(w+z)} - (w+z)e^{-(w+z)}] - \int_0^w [e^{-(w+z-x)} + (w+z-x)e^{-(w+z-x)}]$$

$$\times [0.5 - 0.5e^{-2x}]dx$$

$$= 1 - e^{-(w+z)} - (w+z)e^{-(1+z)}$$

$$- \left(0.5 \int_0^w e^{-(w+z-x)}dx - 0.5 \int_0^w e^{-(w+z+x)}dx \right.$$

$$\left. + 0.5 \int_0^w (w+z-x)e^{-(w+z-x)}dx - 0.5 \int_0^w (w+z-x)e^{-(w+z+x)}dx \right)$$

$$= 1 - e^{-z} - ze^{-z} \qquad (8.3.44)$$

The PDF of Z is obtained by taking the derivative with respect to z in Eq. (8.3.43) as follows:

$$f_Z(z) = ze^{-z} \qquad (8.3.45)$$

The hazard rate function of the residual lifetime is

$$h_Z(z) = \frac{f_Z(z)}{1 - F_Z(z)} = \frac{ze^{-z}}{e^{-z} + ze^{-z}} \qquad (8.3.46)$$

Given $E[C_w]$ in Eq. (8.3.37) and the failure replacement cost c_f and planned replacement cost c_a, the optimal maintenance interval τ is determined by minimizing $r(\tau)$ in Eq. (8.3.35), which yields τ^*.

8.4 Warranty Models for Repairable Products

8.4.1 Warranty Cost under as-Good-as-New Repair

A warranty for repairable systems may have a fixed duration in terms of calendar time or other measures of usage. Such products may also have a lifetime warranty, which means the manufacturer commits the repair or replacement of failed parts during the whole warranty period when consumers own and use the product. The lifetime of the product may terminate due to technology obsolescence, changes in design, changes in ownership of the product, or failure of a critical component/part, which is not under warranty. In this section, we present several warranty models for repairable products.

Products like the wind turbine and jet engine are repairable and if failed they are restored to the good condition by repair. There are three types of repair effects on a failed product. The repair can bring the system to as-good-as-new state, to as-good-as-old state, or to somewhere in between (i.e. imperfect repair). Examples of bringing to an as-good-as-new state include an overhaul service that is applied to a system. For instance, an aircraft engine is removed from the wing and goes through a complete renovation or renewal process that can be treated as an as-good-as-new repair. Small or regular repairs often bring the system to an as-good-as-old state; for example, changing the broken tire of the vehicle or fixing a power line after it was broken by a falling tree. An imperfect repair often brings the system to a state between the new and old. For instance, in opportunistic maintenance, the broken part is replacement along with the change of several aging or deteriorating components that will fail but have not failed yet.

In this section, we consider the case when the repair brings the system to an as-good-as-new state. To make the model applicable in various applications, a general failure time distribution is assumed. We also assume that the repair downtime is negligible compared to the lifetime. The following notation is presented to facilitate the model presentation:

T_i = time to failure between the ith and $(i-1)$th failures (i.e. repairs), a random variable
S_n = time to the nth failure, with $S_n = T_1 + T_2 + \cdots + T_n$
$f^{(n)}(t)$ = PDF of S_n
$F^{(n)}(t)$ = CDF of S_n
$N(t)$ = number of failures in $[0, t]$, a random variable
$M(t)$ = expected number of failures in $[0, t]$, where $M(t) = E[N(t)]$
w = warranty service period or length
L = product lifecycle from the first shipment to the last sale
q = new sales rate or the number of products sold per unit time
Q = cumulative amount of products sold in $[0, L]$
r_i = expected repair cost of ith repair for $i = 0, 1, 2, \ldots, n$; note that $r_0 = 0$
C_n = cumulative warranty cost given that exactly n repairs were performed in $[0, w]$
C_w = warranty cost per unit of product in $[0, w]$, a random variable
C_{fleet} = warranty cost for a product fleet in $[0, w]$, a random variable

Let $P\{N = n\}$ be the probability of having n failures (i.e. repairs) in the interval $[0, w]$. Then the expected cost per unit of product during the warranty period, $E[C_w]$, can be estimated by

$$E[C_w] = \sum_{n=0}^{\infty} C_n P\{N = n\} \tag{8.4.1}$$

$$\text{with } C_n = \sum_{i=0}^{n} r_i \tag{8.4.2}$$

Since each repair brings the system to the as-good-as-new state, the failure-and-restoration sequence is equivalent to a renewal process. Therefore, we have

$$P\{N = n\} = P\{S_n < w < S_{n+1}\} = F^{(n)}(w) - F^{(n+1)}(w) \tag{8.4.3}$$

where

$$F^{(n)}(w) = \int_0^\infty F^{(n-1)}(w - t)f(t)dt \tag{8.4.4}$$

with $F^{(1)}(w) = F(w)$ and $F^{(0)}(w) = \int_0^\infty f(t)dt = 1$, with $f(t)$ being the PDF of the lifetime. Substituting Eq. (8.4.3) into Eq. (8.4.1), we obtain the expected warranty cost per unit product as

$$E[C_w] = \sum_{n=0}^\infty C_n[F^{(n)}(w) - F^{(n+1)}(w)] = \sum_{n=0}^\infty r_n F^{(n)}(w) \tag{8.4.5}$$

Assume that the product's sales rate is $q(t)$ and L is the product life cycle that is defined as the time interval from when the first new product is sold to when the last product was sold. The expected total warranty cost of all sold units, $E[C_{fleet}]$, can be estimated by

$$E[C_{fleet}] = \left(\sum_{n=0}^\infty r_n F^{(n)}(w) \right) \int_0^L q(t)dt \tag{8.4.6}$$

Similarly, the variance of C_w is given by

$$Var(C_W) = \sum_{n=0}^\infty C_n^2 P\{N = n\} - (E[C_w])^2 \tag{8.4.7}$$

By substituting Eq. (8.4.3) into Eq. (8.4.7), the variance can also be expressed as

$$Var(C_w) = \sum_{n=1}^\infty (C_n^2 - C_{n-1}^2)F^{(n)}(w) - (E[C_w])^2 \tag{8.4.8}$$

Under the special case, if the repair cost is time-invariant with $r_i = r$ for all $i = 1, 2, \ldots, n$, then the expected warranty cost and the variance of the cost per unit product can be further simplified. The results are given below:

$$E[C_w] = \sum_{n=0}^\infty rF^{(n)}(w) = r \sum_{n=0}^\infty F^{(n)}(w) \tag{8.4.9}$$

$$Var(C_w) = r^2 \sum_{n=1}^\infty (2n - 1)F^{(n)}(w) - (E[C_w])^2 \tag{8.4.10}$$

If a total Q units of items are sold to customers in the product life cycle, the expected total warranty cost of the entire fleet is given by

$$E[C_{fleet}] = QE[C_w] = rQ \sum_{n=0}^\infty F^{(n)}(w) \tag{8.4.11}$$

$$Var(C_{fleet}) = QVar(C_w) \tag{8.4.12}$$

The result of Eq. (8.4.12) is obtained because we assume products purchased and used by customers fail independently. Next we consider the minimum repair policy in which the system or product is brought to as-good-as-old state upon repair.

8.4.2 Warranty Cost under Minimum Repair

When an item is under minimum repair, it is restored to the same failure rate at the time of failure. The rationality is that repairing or replacing one or several components of a large and complex system have a limited effect on the overall failure rate. This is because the aging process of the majority components remain unchanged. We use the same notations in Section 8.4.1 to develop the warranty cost model for a single item system and a group of items, respectively.

The failure number of a single repairable item can be characterized by a counting process $\{N(t), t \geq 0\}$ and the probability of having exactly one failure in $[t, t + dt]$ is $h(t)dt$ where $h(t)$ is the failure rate. Ross (1970) shows that the process is a non-homogeneous Poisson process (NHPP) because $h(t)$ changes with time. Let $N(w)$ be the number of failures (i.e. repairs) in the warranty period $[0, w]$. Then

$$P\{N(w) = n\} = \frac{M^n(w)e^{-M(w)}}{n!}, \quad \text{for } n = 0, 1, 2, \dots \tag{8.4.13}$$

where

$$M(w) = \int_0^w h(t)dt = -\ln(1 - F(w)) \tag{8.4.14}$$

The last term in Eq. (8.4.14) is obtained by realizing that

$$F(w) = 1 - R(w) = 1 - e^{-\int_0^w h(t)dt} = 1 - e^{-M(w)} \tag{8.4.15}$$

For instance, if the failure rate is a linear function $h(t) = 2t$, then $M(w) = w^2$. According to Eq. (8.4.13), the probability of having no failure in $[0, w]$ is equal to $P\{N(w) = 0\} = \exp(-w^2)$. Probabilities for $n \geq 1$ can also be estimated accordingly given $M(w)$. From Eq. (8.4.3), the nth convolution of $F(w)$ is

$$F^{(n)}(w) = F^{(n-1)}(w) - P\{N(w) = n - 1\} \tag{8.4.16}$$

Similarly we have

$$F^{(n-1)}(w) = F^{(n-2)}(w) - P\{N(w) = n - 2\} \tag{8.4.17}$$

$$F^{(2)}(w) = F^{(1)}(w) - P\{N(w) = 1\} \tag{8.4.18}$$

$$F^{(1)}(w) = F^{(0)}(w) - P\{N(w) = 0\} \tag{8.4.19}$$

If we sum all the equations from (8.4.16) to (8.4.19), it leads to the following result:

$$F^{(n)}(w) = F^{(0)}(w) - \sum_{k=0}^{n-1} P\{N(w) = k\}$$

$$= 1 - \sum_{k=0}^{n-1} \frac{M^k(w)e^{-M(w)}}{k!} \tag{8.4.20}$$

The result is obtained by noticing that $F^{(0)}(w) = 1$ for $w > 0$. Equation (8.4.20) represents an important result as it shows that the CDF of the time-to-nth repair (i.e. S_n) can be expressed by NHPP provided $M(w)$ is known or given.

Let us consider a product that is subject to failure and the minimum repair is performed to return the product to the condition prior to its failure. This product is warranted for a period of w and after repair no warranty extension is provided beyond w. We now present a model originally from Park (1979) and Park and Yee (1984) to determine the present worth of the repair in $[0, w]$. The following notation is defined:

$R(t)$ = reliability at time t
α, β = Weibull scale and shape parameters, respectively
$f_n(t)$ = PDF of nth time-to-failure
b = average cost of a repair
C_w = present worth of a repair during $[0, w]$
$C\infty$ = present worth of a repair with a lifetime warranty

Since the minimum repair is performed upon failure, the hazard rate resumes at $h(t)$ instead of $h(0)$. The system failure times are not renewal points, but can be described by an NHPP. The probability density of the time to the nth failure is (Park 1979; Elsayed 2012)

$$f_n(t) = \frac{h(t)(\alpha t)^{(n-1)\beta} \exp(-M(t))}{(n-1)!} = \frac{\alpha\beta(\alpha t)^{\beta-1}(\alpha t)^{(n-1)\beta} \exp(-(\alpha t)^{\beta})}{(n-1)!} \tag{8.4.21}$$

where $M(t)$ is the cumulative hazard and $M(t) = (\alpha t)^{\beta}$ for the Weibull distribution. The present worth of repairs during the warranty period is

$$C_w = \sum_{n=1}^{\infty} \int_0^w be^{-it} f_n(t) dt$$

$$= b\beta \int_0^{\alpha w} e^{-iz/\alpha} z^{\beta-1} dz$$

$$= b\beta(\alpha w)^{\beta} e^{-iw} \sum_{k=0}^{\infty} \frac{(iw)^k}{\beta(\beta+1)\cdots(\beta+k)} \tag{8.4.22}$$

where i is the interest rate. For a lifetime warranty, the cost is obtained as

$$C_\infty = b \int_0^{\infty} h(t)e^{-it} dt = b\left(\frac{\alpha}{i}\right)^{\beta} \Gamma(\beta+1) \tag{8.4.23}$$

In theory a system can be brought to as-good-as-new state or as-good-as-old state depending on the available resources. When both types of repair are possible, an important problem is to determine the repair policy. Cui et al. (2004) study an optimal resource allocation problem in which both perfect and minimum repair can be applied with the goal of maximizing the system lifetime.

8.5 Warranty Service for Variable Installed Base

8.5.1 Pro-Rata Rebate Policy

For repairable products sold with a free-replacement or PRW policy, planning of repair facilities involves the forecasting of the demand for repair during the warranty period. The repair capacity including technicians, tools, and logistics footprint depends on the types of repair actions and on the anticipated sales over the product lifecycle. In this section, we present a warranty reserve model to cover the customer claims during the product lifecycle under a variable or dynamic sales rate. The following notation is due to Blischke and Murthy (2000):

t = calendar time
l = product lifecycle
w = warranty period
$h(t)$ = hazard rate function
$s(t)$ = product sales rate
S = cumulative sales or the installed base between $[0, l]$
$f(t)$ = PDF of the product lifetime
b = base price of the item or the maximum refund at age of zero
$v(t)$ = expected refund rate (i.e. the monetary quantity per unit time) at time t
V = cumulative refund of the install base in $[0, l + w]$

It is assumed that the product's lifecycle l is longer than the warranty period w, and that new products are put into use immediately upon purchase. The manufacturing must provide necessary repair or replacements for items that fail before reaching age w. Since the last sales stops at the end of l, the manufacturer must continuously deliver the warranty services from $t = 0$ to $t = l + w$. Let S be the total sales over the product lifecycle in the period of $[0, l]$. Then we have

$$S = \int_0^l s(t)dt \tag{8.5.1}$$

where $s(t)$ for $0 \le t \le l$ denotes the sales rate (i.e. sales per unit time). This includes both the first and repeated purchase for the products by different customers. Figure 8.7 graphically describes the relation between $s(t)$ and $v(t)$ as time evolves.

Suppose that a product is sold under a linear pro-rata rebate policy with a non-renewing contract. The rebate over the time interval $[t, t + dt]$ is due to failures of items sold in the interval $[t - u, t)$, where

$$u = \max\{0, t - w\} \tag{8.5.2}$$

and fail in the interval $[t, t + dt]$. Let $v(t)$ be the expected refund rate (i.e. the monetary quantity per unit time) at time t. Then Blischke and Murthy (1994) show that

$$v(t) = b \int_u^t s(x)\frac{t - x}{w}f(t - x)dx, \quad \text{for } 0 \le t \le l + w \tag{8.5.3}$$

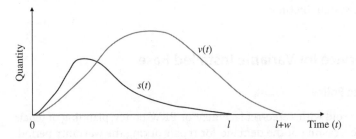

Figure 8.7 Warranty planning for variable sales.

where $f(t)$ is the PDF of the product lifetime. The cumulative warranty reserve required to meet the replacements of the installed base over $[0, l + w]$ is given by

$$V = \int_0^{l+w} v(t)dt \tag{8.5.4}$$

Example 8.3   Consider a smart phone base price $b = \$200$/item and $w = 2$ years. The product life cycle is expected to be $l = 5$ years and the sales rate is constant with $s(t) = 10^6$ units/year over l. The product lifetime is exponentially distributed with a constant failure rate $\lambda = 0.1$ failure/year. Under PRW policy, estimate the expected refund rate and the total warranty reserve needed for all sold items.

Solution:
We consider two cases for estimating the refund rate based on Eq. (8.5.3).

Case 1: For $0 \leq t \leq 1$ (i.e. $0 \leq t \leq w$),

$$v(t) = b \int_0^t s(t) \frac{t - x}{w} \lambda e^{-\lambda(t-x)} dx = \frac{bs}{w} \int_0^t (t - x) \lambda e^{-\lambda(t-x)} dx$$

$$= \frac{bs}{w} \left(\frac{1}{\lambda} - \frac{1}{\lambda} e^{-\lambda t} - t e^{-\lambda t} \right) \tag{8.5.5}$$

Note that $s = s(t)$ because of the constant sales rate. Substituting $b = \$200$, $w = 2$, and $s = 10^6$ into Eq. (8.5.5) yields

$$v(t) = (10 - 10e^{-0.1t} - te^{-0.1t}) \times 10^8$$

Case 2: For $1 < t \leq 6$ (i.e. $1 < t \leq l + w$),

$$v(t) = b \int_{t-w}^t s(t) \frac{t - x}{w} \lambda e^{-\lambda(t-x)} dx = \frac{bs}{w} \int_{t-w}^t (t - x) \lambda e^{-\lambda(t-x)} dx$$

$$= \frac{bs}{w} \left(\frac{1}{\lambda} - \frac{1}{\lambda} e^{-\lambda w} - w e^{-\lambda w} \right) \tag{8.5.6}$$

Substituting $b = \$200$, $w = 2$, and $s = 10^6$ into Eq. (8.5.6) yields

$$v(t) = (10 - 9e^{-0.1}) \times 10^8 = 1.8565 \times 10^8$$

8.5.2   Warranty Service Demands with Failures Not Claimed

The warranty models discussed thus far assume that customers always claim the repair or replacement on each failure during the warranty period. If a customer feels that the product is close to the lifecycle or the rebate is trivial at the end of the warranty period, the failure may not be reported to the manufacturer. For instance, electronics products like cell phones and computers are updated so fast that the current generation becomes outdated in one or two years, largely due to the rapid technology advancement. As a result, the customers may opt to buy a new generation instead of requesting a warranty repair for the old items.

Figure 8.8 shows that Item 3 is not claimed for warranty as it approaches the end of w. Wu (2011) called such behavior a failure-not-claimed (FNC) or failed-but-not-reported (FBNR) phenomenon. Wang et al. (2018) analyzed the field returns and found that FBNR

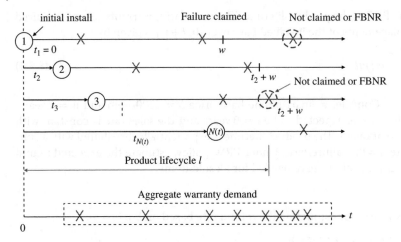

Figure 8.8 Unclaimed failures in warranty and post-warranty periods.

phenomenon is quite common for low and medium priced products. Ignorance of the FBNR may lead to an overestimate of reliability or warranty cost. Similar situations are observed in the post-warranty period in which products like Items 1 and 3 in Figure 8.8 are not reported for repair or replacement. Therefore, it is imperative to incorporate the FBNR phenomena into the warranty and post-warranty demand forecasting.

In this section we consider the probability of customers not claiming warranty repairs even if the item is still under the warranty coverage. We focus on non-renewable free minimal-repair warranty policy because it is widely used for consumer electronics (e.g. microwave ovens, refrigerators, and computers) and industrial equipment (e.g. motors, pumps, and power generators). The same notations in Section 8.5.1 are used unless indicated. According to Xie et al. (2014), the following assumptions are made:

1) Both warranty repair and post-warranty repair are minimal, meaning a failed item is restored to a functioning state with the same failure intensity at the time of failure.
2) The new product is sold only during a certain period l. Let $s(t)$ be the sales intensity. The cumulative sales $N(t)$, or the installed base, up to time t can be characterized by NHPP truncated at time l with

$$N(t) = \begin{cases} \int_0^t s(x)dx, & 0 \le t \le l \\ \int_0^l s(x)dx, & t \ge l \end{cases} \tag{8.5.7}$$

3) If a failed unit is not reported to the manufacturer, it is removed from service permanently.

Under minimal repair, the hazard function of the product is equal to its failure intensity $h(t)$. Suppose an item is sold at time $t = t_i$. If every failure is reported and then fixed during the warranty period, the expected number of warranty demands, which is the same as the number of failures, can be easily expressed as

$$H_i(w) = \int_{t_i}^{t_i+w} h(x - t_i)dx \tag{8.5.8}$$

However, due to the FBNR phenomena, $E[D_i(w)] \le H_i(w)$. Let $0 \le \rho_1 \le 1$ be the probability that an installed item reports the warranty claim each time it fails during w. In a special case where the failure process is governed by an homogeneous Poisson process (HPP) with $h(t) = \theta$ (i.e. interarrival failure times follow the exponential distribution), the repair demand generated by each installed item can be modeled as a continuous-time Markov chain. Xie et al. (2014) obtained a closed-form expression for the expected repair demand as

$$E[D_i(w)] = \frac{\rho_1}{1 - \rho_1}(1 - e^{-(1-\rho_1)\theta w}), \quad \text{for } 0 \le \rho_1 \le 1 \tag{8.5.9}$$

Clearly, $E[D_i(w)] = 0$ when $\rho_1 = 0$, which means that all failed items do not report the repair claims. On the other hand, using the L'Hospital rule, we have

$$\lim_{\rho_1 \to 1} E[D_i(w)] = \lim_{\rho_1 \to 1} \frac{\rho_1}{1 - \rho_1}(1 - e^{-(1-\rho_1)\theta w}) = \theta w \tag{8.5.10}$$

which is exactly the expected number of renewals of an HPP with parameter θ during the period of w when all failures are reported and fixed.

The aggregate warranty repair demand of an installed base depends on the cumulative sales $N(t)$ and the number of units being taken out of service with probability $1 - \rho_1$. Let T_i for $i = 1, 2, \ldots, N(t)$ be the random time of purchasing the ith item. Let us define $\tau_i = \min\{t - T_i, w\}$. The aggregate warranty repair demands Z can be obtained by

$$E[Z] = \begin{cases} \displaystyle\sum_{i=1}^{N(t)} E[D_i(\tau_i)], & t \le l \\ \displaystyle\sum_{i=1}^{N(l)} E[D_i(\tau_i)], & t > l \end{cases} \tag{8.5.11}$$

Another aspect that is also crucial to the after-sales services is the estimation of post-warranty repair demand. As shown in Figure 8.8, some customers will continue to rely on the manufacturer's repair service after the warranty period. Obviously, the FBNR phenomenon becomes more common for post-warranty repairs because the repairs are no longer free. Let $0 \le \rho_2 \le 1$ be the probability that a failed item will be reported to the manufacturer for post-warranty repairs. The value of ρ_2 depends on various factors, such as the length of the base warranty, the product's expected life, and the post-warranty repair expense. For the case of the exponential failure time distribution, the expected repair demand in post-warranty is obtained as follows:

$$E[Z_{post}(w)] = \Lambda(l)\frac{\rho_2}{1 - \rho_2}e^{-(1-\rho_1)\theta w}(1 - e^{-(1-\rho_2)\theta(u-w)}) \tag{8.5.12}$$

where u is the expected product design life and $\Lambda(l)$ is the expected installed base across the product lifecycle with $\Lambda(l) = E[N(l)]$.

8.6 Warranty Service under Reliability Growth

8.6.1 Warranty Decision from Customer Perspective

Spare parts play a critical role in warranty and post-warranty repairs. This section investigates a spare parts stock allocation problem in which the reliability of the product grows during the warranty period. This situation often arises when a new product is introduced to the market, and the manufacturer continues to improve the reliability through corrective actions, such as engineering change orders or retrofit activities. In industries, onsite spare parts stocking and on-call repair are the two common approaches to replacing the faulty items at a customer site. For onsite spares stocking, customers will purchase and store certain amounts of spare units locally in case of failures. A down machine can be quickly restored upon the replacement of good parts because the lead time of delivering spare parts from a central warehouse is avoided. During the warranty period, all defective parts are returned to the manufacturer or OEM for the exchange of good units under no cost to the customer, but the customer has to invest in the initial inventory.

Under the on-call repair mode, the OEM or the vendor provides the spare parts and repair service, and the customer simply generates a repair request when the machine failed. Generally there are two levels of spare parts services offered by the OEM: regular service and emergency service. Under both services the spare part is delivered to the customer site within a fixed time, the lead time of emergency delivery being much shorter than the regular delivery mode. For instance, in the semiconductor equipment industry, the lead time of emergency delivery is 2–4 hours in continental USA, while it takes 12–24 hours under the regular delivery. If an emergency service is chosen, the customer needs to pay extra handling fees, though the spare part is still free in the warranty period. Figure 8.9 depicts the three different modes for spare parts supply during the warranty period, including the onsite stocking, emergency delivery, and regular delivery.

The customer needs to decide whether an onsite spare parts inventory or the on-call service should be adopted in order to lower the warranty cost while ensuring system availability. Hence the warranty cost model is formulated from the customer perspective subject to the reliability growth of the new product. Factors that are essential to the decision-making include the fleet size (or installed base), the reliability growth rate, and the inventory cost. Therefore the warranty service decision problem is investigated from four different cases in Table 8.1.

In the following sections, we will show that an onsite spare facility is favorable to the customer if the system reliability is low or the installed base is large. The on-call repair becomes more attractive only if reliability reaches the maturity phase. For capital-intensive units with sporadic failures, a one-for-one replenishment policy turns out to be more cost-effective than a batch orders inventory policy (Sherbrooke 1992). Hence we focus on a one-for-one spare parts replenishment policy under a reliability growth and free replacement condition.

Zhang et al. (2010) found that parametric distribution models conventionally used in a hard disk drive (HDD) reliability analysis do not characterize the field lifetime well.

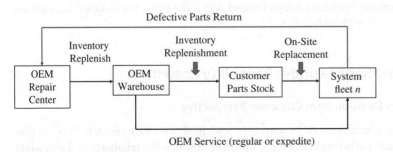

Figure 8.9 Repair and spare parts supply during the warranty period.

Table 8.1 Different customers during a new production introduction.

Case	Fleet size	Reliability	Comments
1	Small	Low	Early new introduction
2	Large	Low	Strategic customers in the early introduction phase
3	Small	High	Late joined customers
4	Large	High	Strategic customers at the end of the introduction phase

Rather they use a biological growth model, i.e. a Bertalanfy–Richards growth model, to predict HDD survivals within the warranty period. The tests using actual HDD field data suggest the biological growth model performs better in predicting warranty returns.

8.6.2 System Availability Under Warranty Services

System availability is a key criterion for the customer to decide which service mode shall be adopted. A high system availability is always preferred as the machines can reach full utilization with less downtime losses. The steady-state system operational availability is often measured by computing the ratio of MTBF versus the sum of MTBF and MDT (mean downtime). Before presenting the availability model, the following notation is defined:

$f(t)$, $F(t)$, and $R(t)$ = PDF, CDF, and reliability of the product lifetime, respectively
T_1, T_2 = regular and expedited spare parts delivery lead time, respectively
T_3 = replacement time given a spare part is available on site
s = base stock level of spare parts at customer site
O = steady-state spare parts on order, a random variable
n = fleet size or the number of field operating units
λ = failure rate of the product
T_{MTTF} = mean-time-to-failure

We first estimate the system downtime when the customer chooses to allocate his or her own spare parts inventory. The customer will experience two replacement scenarios upon the system failure. If the number of failures is less than the inventory level, the replacement can be performed immediately. On the other hand, if the spares demands exceed the in-stock units, the system remains in a down state until the arrival of the spare parts. Since the inventory is out-of-stock, the customer must place orders to obtain spare units from OEM's warehouse (see Figure 8.9). By taking both scenarios into consideration, the system MDT can be estimated as

$$T_r^{(c)} = T_3 P\{O \leq s\} + (T_1 + T_3) P\{O > s\} \tag{8.6.1}$$

where

$$P\{O \leq s\} = \sum_{k=0}^{s} \frac{(n\lambda T_1)^k e^{-n\lambda T_1}}{k!} \tag{8.6.2}$$

where $T_r^{(c)}$ represents the MDT using the regular inventory replenishment mode given that the customer's stock is empty. O is the steady-state spare parts demand of the system fleet. Note that s is the base stock level, λ is the product failure rate, and n is the fleet size. According to Cox and Smith (1954), when there are n identical items operating at the customer site, the aggregate spare parts demand O follows the HPP provided that n approaches infinity. Wang (2012) further showed that as long as $n \geq 10$, the occurrence times between two successive failures can be approximated by the exponential distribution. Therefore Eq. (8.6.2) is derived based on the rationality that the distribution of O follows the HPP for sufficient large n. If the inventory is replenished through expedited delivery mode, the system MDT becomes

$$T_e^{(c)} = T_3 P\{O \leq s\} + (T_2 + T_3)P\{O > s\} \tag{8.6.3}$$

where

$$P\{O \leq s\} = \sum_{k=0}^{s} \frac{(n\lambda T_2)^k e^{-n\lambda T_2}}{k!} \tag{8.6.4}$$

Equation (8.6.3) differs from Eq. (8.6.1) in that if the customer's stock is empty, the former relies on the regular repair while the latter seeks an expedited repair service. Therefore, the inventory replenishment cycle T_2 is shorter than T_1.

When the customer opts to rely on the on-call repair service, the OEM provides two types of repair modes: regular repair and expedited repair. In a regular repair, the spare part is delivered to the customer within T_1 hours. If the expedited repair mode is adopted, the customer receives the spare unit in a much shorter time $T_2 \ll T_1$. Upon receipt of the spare part, it takes an additional T_3 hours to exchange the faulty unit with a good one. Hence T_3 is also referred to as the mean-time-to-replacement (MTTR). Thus the MDT under an on-call repair service consists of the time of shipping the part from the central warehouse to the customer and the hands-on replacement time. That is,

$$T_r^{(o)} = T_1 + T_3 \tag{8.6.5}$$

$$T_e^{(o)} = T_2 + T_3 \tag{8.6.6}$$

where $T_r^{(o)}$ represents the MDT using the regular delivery mode and $T_e^{(o)}$ represents the MDT under the expedited delivery mode, with $T_e^{(o)} < T_r^{(o)}$. Since the on-call repair does not require the setup of local stockroom, T_1 and T_2 are relatively stable because OEM's central warehouse usually holds sufficient amounts of spare units.

Let T_{MTTF} be the MTTF of individual working items in the field. The system availability under the customer's own stocking option and OEM's on-call repair is given in the following:

Option 1: Customer's own stocking with a regular parts delivery service:

$$A_1 = \frac{T_{MTTF}}{T_{MTTF} + T_3 + T_1(1 - P\{O \leq s\})} \tag{8.6.7}$$

Option 2: Customer's own stocking with an expedited parts delivery service:

$$A_2 = \frac{T_{MTTF}}{T_{MTTF} + T_3 + T_2(1 - P\{O \leq s\})} \tag{8.6.8}$$

Option 3: On-call repair with a regular delivery mode:

$$A_3 = \frac{T_{MTTF}}{T_{MTTF} + T_1 + T_3} \tag{8.6.9}$$

Option 4: On-call repair with an expedited delivery mode:

$$A_4 = \frac{T_{MTTF}}{T_{MTTF} + T_2 + T_3} \tag{8.6.10}$$

In the special case where the reliability of field items follows the Weibull distribution,

$$R(t) = \exp\left(-\left(\frac{t}{\theta}\right)^k\right), \qquad \text{for } t \geq 0 \tag{8.6.11}$$

where θ and k are the scale and the shape parameter, respectively. Then T_{MTTF} can be estimated by

$$T_{MTTF} = \theta\Gamma(1 + 1/k) \tag{8.6.12}$$

where $\Gamma(\bullet)$ is the gamma function. Since we are interested in how the item reliability and fleet size influence the spares stocking policy, variables like T_{MTTF}, n, and s are identified as the key drivers behind the warranty servicing operation.

8.6.3   Fleet Downtime Cost to Customer

Assume that the customer deploys n machines, with each being a single-item system. A system failure corresponds to an item failure and the machine is recovered after the faulty item is replaced with a spare item. The following notation is defined for the analysis and estimation of the system downtime cost from the customer's perspective:

T_d = MDT per failure event
C_{DT} = total downtime cost per unit system per failure event
c_{eq} = initial system purchase cost (\$/system)
n_d = depreciation time (e.g. n_d = 5 years)
t_a = annual production time
c_{pl} = production loss per unit time
c_{lb} = labor cost per unit time
c_{fc} = facility cost per unit time
c_{sp} = spare part unit cost.
s = base stock level of the customer-owned inventory
α, β = parameters of the Crow/AMSAA model
w = warranty period or duration
L = product lifecycle
t_0 = initial purchasing time of product, with $0 \leq t_0 \leq L$

The downtime cost incurred by the customer usually consists of three terms: (i) machine depreciation C^{dp}; (ii) profit loss due to the interrupted production C^{pl}; (iii) idle labor and underutilized facility C^{lf}. Thus the total downtime cost, C_{DT}, per system per failure is the sum of all three cost terms as follows:

$$C_{DT} = C_{eq} + C_{pl} + C_{lf}$$
$$= T_d\left(\frac{c_{eq}}{n_d t_a} + c_{pl} + c_{lb} + c_{fc}\right) \tag{8.6.13}$$

where $C^{dp} = T_d c_{eq}/n_d t_a$, $C^{pl} = T_d c_{pl}$, and $C^{lf} = T_d(c_{lb} + c_{fc})$. For a detailed derivation of Eq. (8.6.12), interested readers can refer to Granado and Jin (2010). In practice, C_{DT}

can also be broken into two categories: hard cost and soft cost. The hard cost is the non-recoverable loss associated with equipment depreciation, idle labor, and wasted facility. The soft cost could be fully or partially recovered depending on the nature of products. For instance, testing of semiconductor wafers due to machine failure can be resumed once it is restored; hence the production loss can be partially recovered. If the product quality is time sensitive, the losses due to the equipment failures cannot be recovered. Typical examples include food industry, energy production, and data center.

The Crow/AMSAA model (Crow 2004) is widely used to predict the reliability growth trend of a product. The underlying assumption is that the power law model is appropriate to represent the failure intensity rate. That is,

$$\lambda(t) = \alpha\beta t^{\beta-1}, \qquad \text{for } t \geq 0 \tag{8.6.14}$$

where α and β are model parameters. For $\beta < 1$, the product exhibits a growth trend with a decreasing failure intensity over time. Let t_0 be the initial purchasing time and $0 \leq t_0 \leq L$ with L being the product lifecycle. Since warranty service is offered to cover a period of time w, the expected number of failures for the purchased item during $[t_0, t_0 + w]$ can be predicted based on Eq. (8.6.14). That is,

$$M(w) = \int_{t_0}^{t_0+w} \alpha\beta t^{\beta-1}dt = \alpha\left((t_0 + w)^\beta - t_0^\beta\right), \quad \text{for } 0 \leq t_0 \leq L \tag{8.6.15}$$

By combining Eqs. (8.6.13) and (8.6.14), the total downtime cost for a system fleet during $[t_0, t_0 + w]$ can be estimated as

$$
\begin{aligned}
C_{ft} &= nM(w)C_{DT} + sc_{sp} \\
&= \alpha nT_d\left((t_0 + w)^\beta - t_0^\beta\right)\left(\frac{c_{eq}}{n_d t_a} + c_{pl} + c_{lb} + c_{fc}\right) + sc_{sp}
\end{aligned}
\tag{8.6.16}
$$

where n is the number of identical items operating at the customer site. Note that the spare parts inventory cost is also included if the customer adopts a local sparing policy, where s is the base stock level and c_{sp} is the part unit cost including the holding cost.

8.6.4 Minimizing the Downtime Cost

The goal of the customer is to minimize the total cost incurred during the system downtime. This can be translated into the following optimization model, denoted as Model 8.1.

Model 8.1 Minimize:

$$C_{ft}(T_1, T_2, s) = \alpha nT_d((t_0 + w)^\beta - t_0^\beta)\left(\frac{c_{eq}}{n_d t_a} + c_{pl} + c_{lb} + c_{fc}\right) + sc_{sp} \tag{8.6.17}$$

Subject to

$$A(T_1, T_2, s) \geq A_{\min} \tag{8.6.18}$$

$$x_1 + x_2 = 1 \tag{8.6.19}$$

$$s \in \{0, 1, 2, \ldots\} \tag{8.6.20}$$

In this formulation, the decision variables are the choice of the repair service (i.e. T_1 or T_2) and the spares inventory level s. Constraint (8.6.18) defines the system operational availability requirement in which A_{min} is the minimum target. Since the system availability may differ appreciably under different warranty servicing modes, Eq. (8.6.18) depends on the decisions on T_1, T_2, and s. The actual availability formulas are available in Eqs. (8.6.7)–(8.6.10).

Sometimes, the fleet readiness is preferred over the availability criterion of individual systems. For instance, in a manufacturing setting, a minimum number of machines, say m, is required to be available in order to meet the production needs. In the air force sector, a squadron is an operational unit consisting of two or more flights of aircraft and the personnel required to fly them. Fleet readiness means two or more planes are ready to execute the mission. The fleet readiness, A_{fleet}, can be estimated as

$$A_{fleet} = P\{N \geq m\} = \sum_{i=m}^{n} \binom{n}{i} A^k (1 - A)^{n-i} \tag{8.6.21}$$

where n is the fleet size and A is the availability of a single system. Similarly, A can be estimated from Eqs. (8.6.7)–(8.6.10) once T_1, T_2, and s are determined.

This section addresses a warranty servicing decision problem confronted by customers who purchased new equipment with low reliability. Two service mechanisms, i.e. self-stocking and vendor repair, are discussed and the corresponding downtime costs are compared. Self-stocking has the merits of shortening the downtime and sharing the inventory cost across multiple systems. A vendor-based on-call repair service might be delayed due to the uncertainties in parts shipment. However, a trade-off must be made between the spares inventory, the fleet size, and the reliability growth trend. In general, customer self-stocking is preferred if the system reliability is low and the fleet size is large. On-call repair becomes more favorable to the customer if the production losses during the downtime can be recovered.

8.7 Other Warranty Services

8.7.1 Extended Warranty Contracts

Extended warranty (EW) is a prolonged service program in which the seller provides repair, replacement, or maintenance for an additional period of time beyond the manufacturer's base warranty period. While a base warranty is an integral part of the product sale, extended warranty is an optional service and customers can purchase it voluntarily. Some studies (Berner 2004) show that EW has profit margins of 50–60%, which is nearly 5–10 times the profit margin on product sales. Because of the lucrative after-sales business, today extended warranties have been offered on almost all consumer electronics and domestic appliances, ranging from laptop computers to washing machines and refrigerators.

Extended warranty can be offered by the manufacturer, the retailers, or third-party providers. This differs from the base warranty as it is usually offered by the manufacturers or OEM. Given the variation in service provider's practices, it is important to understand the strategic implications of different distribution channels of extended warranty. Meanwhile, consumers in the market differ in terms of their ability of risk-taking. Some opt to buy the product without the EW while risk-averse consumers prefer to buy the EW in addition to the base warranty. Should the manufacturer sell the EW directly or through the retailer or third-party provider? Do different service distribution channels influence the product's unit prices and consequently the sales volume? And how do the different service distribution channels affect the profit margin of the manufacturer and the retailers? These questions will be taken into account in designing and deploying extended warranty services.

Figure 8.10 shows a dual warranty distribution mechanism in a vertical supply chain network comprised of manufacturer, retailer, and consumers. The manufacturer markets a new product through the retailer at a wholesale price p_x. The retailer resells the product to end consumers at price p_r. In the meantime, the manufacturer offers the base warranty with length w which is included in p_x. However, the extended warranty service can be sold through the retailer or directly by the manufacturer, as shown by Figure 8.10a and b, respectively. The price of the extended warranty is p_e with length η. When consumers buy products, they need to decide whether the extended warranty should be purchased or simply take the base warranty package. Since both the manufacturer and the retailer are risk neutral, consumers believe that the product is subject to a certain non-trivial probability of failure, $r \in (0, 1)$, and the number of product failures occurs with an intensity rate $h(t)$.

When the extended warranty is sold through the retailer, there exists double marginalization problem for both the product and the service at the retailer. Double marginalization is the phenomenon in which two or more firms in the same industry having their own market powers but at different vertical levels in the supply chain, for example, upstream and downstream, can apply their own markups in prices. When the manufacturer sells the extended warranty directly, the double marginalization problem disappears. However, the retailer does not consider the complementary good effect of price of the product on the demand for extended warranty. In economics, a complementary good is a product with a negative cross-elasticity of demand, in contrast to a substitute product. This means a product's demand increases when the price of another product is decreased (e.g. gas vs. car). Thus, the optimal sales channel for the manufacturer depends on the trade-off between the double marginalization effect and the complementary good effect. Recently, Chen et al. (2017) showed that the manufacturer is better off by offering extended warranty indirectly when the warranty costs are sufficiently small

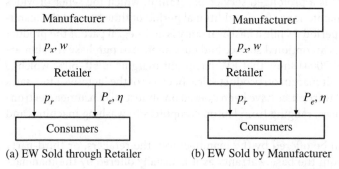

(a) EW Sold through Retailer   (b) EW Sold by Manufacturer

Figure 8.10 Extended warranty service models: (a) through retailer; (b) from manufacturer.

and is attractive to the consumer. Such attractions are manifested by a small co-payment or deductible policy commonly used in consumer electronics.

Consumers who decline the extended warranty and just purchase the product with the base warranty will obtain a monetary transfer, wv, if the product fails prior to w (Desai and Padmanabhan 2004), where v is monetary benefit if the product is operational. Define θ as the consumer's risk aversion degree. The utility function of the consumer is U_b, which is given by

$$U_b = v(1 - r) + wvr - \theta v^2 r(1 - r)(1 - w)^2 - p_r \tag{8.7.1}$$

Consumers who purchase both the product and the extended warranty will obtain a total monetary transfer, ηv, if the product fails prior to η. The utility function of the consumer, denoted as U_e, is given by

$$U_e = v(1 - r) + \eta vr - \theta v^2 r(1 - r)(1 - \eta)^2 - p_r - p_e \tag{8.7.2}$$

Figure 8.11 shows three types of consumers in terms of their degree of risk-averseness. Consumers with $\theta \in [0, \theta_b]$ only purchase the product bundled with a base warranty. Consumers with $\theta \in (\theta_b, \theta_e]$ will purchase the product and the extended warranty. Consumers with $\theta > \theta_e$ are zero-risk taken and they buy nothing at all. To make the discussion meaningful, the relevant parameters must meet the criterion $0 < \theta_b < \theta_e$, which is the necessary condition for providing the extended warranty program. Here

$$\theta_b = \frac{p_e - vr(\eta - w)}{v^2 r(1 - r)[(1 - w)^2 - (1 - \eta)^2]} \tag{8.7.3}$$

$$\theta_e = \frac{v(1 - r) + \eta vr - p_r - p_e}{v^2 r(1 - r)(1 - \eta)^2} \tag{8.7.4}$$

Thus the total demand for the product is $Q = \theta_e/q$, the demand for the product sold without the extend warranty $Q_b = \theta_b/q$, and the demand for the products sold with extended warranties is $Q_e = (\theta_e - \theta_b)/q$. Note that $Q = Q_b + Q_c$.

Chen et al. (2017) show that the double marginalization problem and the complementary goods both affect the selection of the EW distribution channel. If the manufacturer sells the extended warranty directly, the double marginalization issue can be avoided. If the manufacturer is able to control the retail price, it can incorporate the complementary goods effect. If he has less control of the retail price, dual service distribution

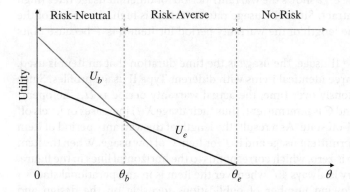

Figure 8.11 Illustration of warranty demand at different risk levels.

turns out to be the best option for the manufacturer. The study further indicates that the distribution channel affects the EW length and the profit margin of the manufacturer. The profitability of the manufacturer is highest when the manufacturer sells the product and the EW program directly and the profit is least when the product and EW are offered by the retailer.

8.7.2 Two-Dimensional Warranty

Multidimensional warranties usually involve a time-axis similar to that of one-dimensional warranties and one or more axes that represent other key characteristics of the product, such as usage intensity and output energy (like a wind turbine). We confine our discussion to two-dimensional warranties that are characterized by a rectangular region $\Omega \equiv W \times U$ where the warranty expires when the item reaches an age W or the usage reaches a level U, whichever comes first. According to Blischke et al. (2011) and Manna et al. (2007), usage may be defined in one of the following types:

Type I usage: Many products like machines are bought to produce something and usage is the amount of the output. We use the term "output" in a very general sense, as illustrated by the following examples, where $X(t)$ represents the distance traveled over the interval $[0, t]$ by an automobile or $X(t)$ represents the amount of components polished over the interval $[0, t]$ by a machine tool.

Type II usage: In this case, $X(t)$ represents the total duration of usage over the interval $[0, t]$. Here $0 \leq X(t) \leq t$. For an aircraft engine, $X(t)$ represents the number of flight hours by the engine over the interval $[0, t]$.

Type III usage: Here $X(t)$ represents the number of times the product is used in the interval $[0, t]$. Examples include the landing gear of an airplane, where $X(t)$ represents the number of landings over the interval $[0, t]$ and for a lithium-ion battery, $X(t)$ represents the number of cycles the battery has been discharged and recharged over the interval $[0, t]$.

The two-dimensional warranty is characterized by two parameters, namely age W and usage U. Depending on a number of factors, the warranty period for which an item is covered under warranty could be less than or greater than W. Such factors include the usage type, the usage intensity rate and the warranty renewing agreement. For a product under Type I usage, Figure 8.12 shows the warranty period for different usage rates (high and low) in an NRW contract. Since the usage rate of Item A is high, the length of the warranty period is U. The length of the warranty period for Item B is T because of its low usage rate.

For products with Type II usage, the usage is the time duration that an item is used. Figure 8.13 depicts the three identical items with different Type II usage profiles. Since Item A operates continuously over time, the actual warranty period is U. The operational profile of Items B and C is intermittent; thus their usage $X(t)$ increases or levels off depending on the on-and-off state. As a result, the length of the warranty period of Item B is U because of high intermittent usage and is T for Item C of low usage. When the item is not used, the usage rate is zero, which corresponds to the horizontal lines in the figure. The slope of the trajectory is always 45° whenever the item is in an operational state.

Recently there is a growing number of publications focusing on the design and optimization of 2-D warranty policies. For instance, Ye and Murthy (2016) investigate

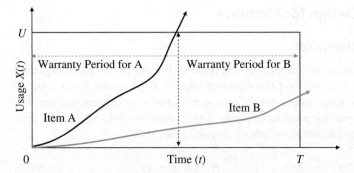

Figure 8.12 A 2-D warranty period for a non-renewing warranty of Type I usage.

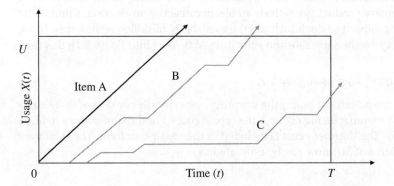

Figure 8.13 A 2-D warranty period for a non-renewing warranty of Type II usage.

the design of a flexible 2-D warranty policy that contains a number of rectangular regions that allows the consumer to choose a favorable one according to their usage behavior. Park et al. (2016) deal with a renewable 2-D warranty policy using failure time and warranty servicing time as two main factors for minimizing the warranty service cost rate from the manufacturer's perspective. Xie et al. (2017) develop a 2-D aggregate warranty demand forecasting model with sales uncertainty. Wang et al. (2017) further extend and generalize the forecasting model to accommodate the situation where the FBNR events are observed. Wang et al. (2017) study the consumer's and the manufacturer's optimal strategies for products sold with periodic PM under a 2-D warranty policy. Profit maximization and cost minimization for both sides are analyzed by a game-theoretic approach.

Studies were also carried out to design and optimize 2-D extended warranty contracts. Under a 2-D extended warranty contract, Tong et al. (2014) proposed a pricing model from the perspective of automobile manufacturers by comparing two service purchasing times, one being the sales time and the other the expiration of the base warranty. Park et al. (2018) developed an optimal post-warranty periodic preventive maintenance strategy by minimizing the expected cost rate incurred during the lifecycle of the system. The goal is to determine an optimal PM cycle and thus to minimize the post-warranty cost rate for the user. For reviews and discussions on the 2-D warranty models, refer to Karim and Suzuki (2005) and Wang and Xie (2017).

8.8 Case Study: Design for Warranty

8.8.1 Minimizing Warranty Cost

For capital-intensive systems, warranty planning requires the evaluation of the expected repair and replacement demand over the warranty period. The forecast is made based on the component and non-component failures during the warranty period and the anticipated new sales over the product lifecycle. The cumulative field failures by time t, denoted as $S(t)$, can be estimated as follows (Jin and Liao, 2009):

$$S(t) = \sum_{n=0}^{N(t)} \widehat{\lambda}_s(t - t_n) = \widehat{\lambda}_s t(N(t) + 1) + \frac{1}{2}\widehat{\lambda}_s N(t)(N(t) + 1) \tag{8.8.1}$$

where $\widehat{\lambda}_s$ is the product failure rate and $N(t)$ is the installed base by t. The sales rate of a good capital-intensive product is relatively stable, because the production is limited by the manufacturing capacity, which includes the availability of skilled technicians, tools, and materials. Let r be the sales rate and substitute $N(t) = rt$ into Eq. (8.8.1); this then becomes

$$S(t) = \widehat{\lambda}_s((0.5r^2 + r)t^2 + (0.5r + 1)t) \tag{8.8.2}$$

Under the free misplacement policy the warranty cost typically consists of three components: the item manufacturing cost c_m, the repair cost c_r, and the inventory holding cost c_h. Notice that the shipment cost is included in the repair cost item. The total warranty cost for failures of $S(t)$ now can be estimated as

$$E[C(t)] = S(t)(c_m + c_r + c_h)$$
$$= \widehat{\lambda}_s(c_m + c_r + c_h)((0.5r^2 + r)t^2 + (0.5r + 1)t) \tag{8.8.3}$$

Eq. (8.8.3) establishes the explicit relation between the failure rate λ_s, the sales rate r, and the total warranty cost $C(t)$. Minimizing the warranty cost is equivalent to reducing the spare parts inventory and logistics overhead in the after-sales market; thus it increases the revenue and profit margin of the manufacturer. The following notation further defines the formulation of the optimization model:

x_{ij} = decision variable for component type i, $x_{ij} = \{0, 1\}$ for $j = 1, 2, \dots, d_i$
y_{ij} = decision variable of the ith non-component, $y_{ij} = \{0, 1\}$ for $j = 1, 2, \dots, h_i$
c_{ij} = cost of the ith type of component from supplier j, for $j = 1, 2, \dots, d_i$
b_{ij} = cost of the ith non-component of the jth choice, for $j = 1, 2, \dots, h_i$
C_1 = total design budget for components
C_2 = total design budget for non-components

Jin and Ozalp (2009) formulated the following optimization model, denoted as Model 8.2, to minimize the product warranty cost.

Model 8.2

$$\text{Min:} \quad f(\mathbf{x}) = E[C(t)] \tag{8.8.4}$$

Subject to

$$\sum_{i=1}^{k} \sum_{j=1}^{d_i} n_i c_{ij} x_{ij} \le C_1 \tag{8.8.5}$$

$$\sum_{i=1}^{5} \sum_{j=1}^{h_i} b_{ij} y_{ij} \leq C_2 \tag{8.8.6}$$

$$\sum_{j=1}^{d_i} x_{ij} = 1 \quad \text{for} \quad i = 1, 2, \ldots, k. \tag{8.8.7}$$

$$\sum_{j=1}^{h_i} y_{ij} = 1 \quad \text{for} \quad i = 1, 2, \ldots, k. \tag{8.8.8}$$

Note that k is the number of component types in the product. The model was formulated as a nonlinear integer optimization program. Decision variables are component and non-component choices such that the mean of the warranty cost is minimized at time t. For the ith component type used in the product, only one selection can be made among a total of d_i choices, each of which has a different unit cost c_{ij} and failure rate λ_{ij}. The ith non-component also has h_i options, each of which has a different failure rate and cost. For a non-component index, $i = 1$ for design, 2 for manufacturing, 3 for software, 4 for process, and 5 for no-fault found (NFF) (see Section 4.2 of Chapter 4). For example, two teams could be available to develop the software program for the new product: an in-house team and a subcontractor. In that case, $h_3 = 2$. The non-component failure rate is modeled by triangular distributions (see Section 3.5 of Chapter 3). The software program developed by the in-house team may yield fewer bugs than the subcontractor developer, but the cost for the in-house design could be higher than the subcontractor.

8.8.2 Application to Electronics Product Design

Model 8.2 is applied for designing a high-performance electronic instrument. The instrument consists of a single printed circuit board (PCB), which is capable of generating high-speed digital singles. The hardware of the instrument is comprised of 156 different component types with a total of 2900 parts. Design engineers have identified 21 component types with a total of 634 parts, which are crucial to the instrument reliability. Other types of components can be treated as failure-free during the product useful lifetime based on historical data.

Table 8.2 lists the component choices as well as the unit cost and the associated failure rate. Operating temperatures for each component type are also listed in the table, where T_l and T_u stand for the lower and upper temperature limits, respectively. The low and upper limits of electrical derating are specified by p_l and p_u. The activation energy E_a and the fitting parameter β were found from Bellcore (1995) and are also listed in the table. The total design budget for the hardware components is $C_1 = 1500$.

Table 8.3 provides the triangular distribution parameters for non-component failure rates. For each non-component choice, two options are available, and each is associated with different distribution parameters and costs. Parameters a, b, and h are estimated from predecessor products that have a similar complexity to the new product. The total design budget for non-components is $C_2 = 2560$.

Other parameters for the warranty cost model include $c_m = 150$, $c_r = 30$, $c_h = 3$, and $r = 1.2$ product/hour. The optimal solution is searched by a genetic algorithm (GA). The optimization codes are implemented using Matlab and the detailed computational algorithm is available in Jin and Ozalp (2009). Table 8.4 lists the Pareto optimal solution

Table 8.2 Parameters for component choices.

i	Component type	n_i	Supplier 1 $(j=1)$		Supplier 2 $(j=2)$		E_a	Temperature (°C)		Derating (%)		β
			λ_{ij}	c_{ij}	λ_{ij}	c_{ij}		T_l	T_u	p_l	p_u	
1	5 V dry relay	60	2.50E-8	3.5	3.87E-8	2.8	0.7	40	55	65	70	0.013
2	Mini SMD relay	48	3.00E-8	2.5	6.96E-8	2	0.7	40	55	65	70	0.013
3	10 μF 16 V capacitor	45	1.25E-9	1.5	1.80E-9	1.2	0.7	45	70	50	65	0.035
4	2.2 μF 20 V capacitor	21	2.50E-9	1.5	4.50E-9	1.2	0.7	45	70	50	65	0.035
5	22 μF 20 V capacitor	19	2.50E-9	1.5	3.90E-9	1.2	0.7	45	70	50	65	0.035
6	ASIC	12	2.50E-8	16	4.14E-8	12.8	1.6	50	60	55	65	0.024
7	EEPROM	2	1.48E-7	4.7	2.37E-7	3.76	1.4	40	60	55	65	0.024
8	100 pF 250 V capacitor	40	5.00E-9	1.4	7.80E-9	1.12	0.7	45	70	50	60	0.035
9	0.5 A diode	121	1.35E-9	2.2	2.45E-9	1.76	0.9	60	75	60	80	0.029
10	Switch diode	28	5.63E-9	2.6	9.83E-9	2.08	0.8	40	75	60	80	0.029
11	0.09 W resistor	36	1.67E-9	0.8	3.49E-9	0.64	0.5	55	70	65	80	0.015
12	Zener diode	40	1.35E-9	3.1	2.56E-9	2.48	1	45	60	60	80	0.029
13	DAC	5	1.05E-8	6.3	2.62E-8	5.04	1.2	45	65	60	70	0.024
14	Active loader	6	7.00E-9	4.6	1.25E-8	3.68	0.8	50	75	55	75	0.024
15	PPMU	12	3.00E-9	10.3	5.46E-9	8.24	1.3	45	55	60	75	0.024
16	1 A diode	24	1.35E-9	2.8	3.11E-9	2.24	0.9	55	65	60	80	0.029
17	Transistor	9	3.25E-9	3.3	7.02E-9	2.64	1.2	45	60	60	75	0.024
18	0.1 μF 50 V capacitor	82	3.50E-10	1.3	5.42E-10	1.04	0.7	45	70	60	80	0.041
19	Logic gates	8	3.25E-9	2.4	6.19E-9	1.92	1.1	40	60	65	80	0.024
20	IC translator	8	3.25E-9	3.2	6.93E-9	2.56	1.2	45	60	65	75	0.024
21	Differential driver	8	3.25E-9	4.5	6.84E-9	3.6	1.3	40	55	60	70	0.024

Source: Reproduced with the permission of Inderscience Enterprises Limited. (Inderscience retains copyright of the original table).

Table 8.3 Parameters for non-component failure rate γ.

i	Issue	In-house Team $(j=1)$				Subcontractor $(j=2)$			
		a	b	h	Cost ($)	a	b	h	Cost ($)
1	Design	3.14E-6	1.43E-5	5.23E-6	630	4.05E-6	1.48E-5	6.77E-6	480
2	Software	9.50E-7	3.77E-6	2.18E-6	690	1.12E-6	3.86E-6	2.82E-6	660
3	Mfg	1.26E-6	2.85E-6	2.09E-6	480	1.40E-6	3.60E-6	2.28E-6	555
4	Process	3.03E-6	1.01E-5	5.94E-6	435	3.69E-6	1.01E-5	7.64E-6	420
5	NFF	1.88E-6	8.08E-6	7.20E-6	510	2.35E-6	8.46E-6	8.32E-6	315

Source: Reproduced with the permission of Inderscience Enterprises Limited. (Inderscience retains copyright of the original table).

Table 8.4 Optimal design of the digital instrument.

Component Type	1	2	3	4	5	6	7	8	9	10	11	12	13	14
Choice	1	2	1	2	2	2	1	1	2	2	1	1	1	1

Component Type	15	16	17	18	19	20	21	D	S	M	P	NFF	$E[C(t)]$
Choice	2	2	1	1	1	1	2	1	1	1	2	2	1 248 204

Note: D = design, S = software, M = manufacturing, P = process.

after an extensive GA search. The anticipated warranty cost is \$1 248 204 for the installed base during one year period. Though the solution from GA does not guarantee global optimality, it is usually sufficient to guide the reliability practitioners or management to make a sound decision.

As the complexity and integration of new products continues to increase, product reliability must be addressed when it is still in the early development phase. Both the component and non-component failures need to be considered and appropriately mitigated by proactively identifying and eliminating high-risk items. This objective can be achieved through various design for Six Sigma techniques. We have applied one such technique, the design for reliability (DFR), to minimize the warranty cost through the optimal selection of component suppliers, design teams, and manufacturing process in the early development phase. The benefit of DFR is substantial when reliable products are shipped to customers with fewer warranty returns and repairs. Meanwhile, it also enhances the quality image and market competitiveness of new products.

References

Bellcore, (1995), Reliability Prediction Procedure for Electronic Equipment, TR-332, Issue 5, published by Bell Communication Research-Bellore.

Berner, R. (2004). The warranty windfall: service contracts are cash cows—but retailers are mum about their importance. *Business Week* 12 (20): 84–86.

Blischke, W.R. and Murthy, D.N.P. (1994). *Warranty Cost Analysis*, Chapter 9. New York: Marcel Dekker.

Blischke, W.R. and Murthy, D.N.P. (2000). *Reliability: Modelling, Prediction and Optimization*, Chapter 17, 1e. Wiley.

Blischke, W.R., Karim, M.R., and Murthy, D.N.P. (2011). *Warranty Data Collection and Analysis*. Springer.

Chen, Y., J. Qin, Y. Chen, T. Jin, (2017), "The impacts of failure rate characteristic on extended warranty service supply chain coordination," Working paper submitted for review on August 18, 2017.

Cox, D.R. and Smith, W.L. (1954). On the superposition of renewal processes. *Biometrika* 41: 91–99.

Crow, L.H. (2004). An extended reliability growth model for managing and assessing corrective actions. In: *Proceedings of Annual Reliability and Maintainability Symposium*, 73–80.

Cui, L., Kuo, W., Loh, H.T., and Xie, M. (2004). Optimal allocation of minimal & perfect repairs under resource constraints. *IEEE Transactions on Reliability* 53 (2): 193–199.

Desai, P.S. and Padmanabhan, V. (2004). Durable good, extended warranty and channel coordination. *Review of Marketing Science* 2 (2): 1–23.

Elsayed, E. (2012). *Reliability Engineering*, Chapter 9, 2e. Hoboken, NJ: Wiley.

Granado, J.A. and Jin, T. (2010). Spares provisioning for system maintenance under reliability growth: a case study. *Reliability Review* 30 (3): 4–16.

Iskandar, B.P. and Jack, N. (2011). Warranty servicing with imperfect repair for products sold with a two-dimensional warranty. In: *Replacement Models with Minimal Repair* (ed. L. Tadj, M.-S. Ouali, S. Yacout and D. Ait-Kadi), 163–175. Springer.

Jiang, R. (2015). *Introduction to Quality and Reliability Engineering*, vol. 2015. Berlin, Heidelberg: Science Press, Beijing and Springer-Verlag.

Jin, T. and Liao, H. (2009). Spare parts inventory control considering stochastic growth of an installed base. *Computers and Industrial Engineering* 56 (1): 452–460.

Jin, T. and Ozalp, Y. (2009). Minimizing warranty cost by design for reliability in product development phase. *International Journal of Six Sigma and Competitive Advantage* 5 (1): 42–58.

Karim, M.R. and Suzuki, K. (2005). Analysis of warranty claim data: a literature review. *International Journal of Quality and Reliability Management* 22: 667–686.

Manna, D.K., Pal, S., and Sinha, S. (2007). A use-rate based failure model for two dimensional warranty. *Computers and Industrial Engineering* 52: 229–240.

Menke, W.W. (1969). Determination of warranty reserves. *Management Sciences* 15 (10): B542–B549.

Murthy, D.N.P. and Blischke, W.R. (2006). *Warranty Management and Product Manufacture*. Springer.

Murthy, D.N.P. and Djamaludin, I. (2001). Warranty and quality. In: *Integrated Models in Production Planning, Inventory, Quality and Maintenance* (ed. M.A. Rahim and M. Ben-Daya), 323–358. New York: Kluwer Academic Publishers.

Nissan (2011), Warranty Information Booklet available at: https://owners.nissanusa.com/content/techpub/warranty/2011_N_WIB.pdf (accessed on November 2, 2017).

Park, K.S. (1979). Optimal number of minimum repairs before replacement. *IEEE Transactions on Reliability* R-28: 137–140.

Park, K.S. and Yee, S.R. (1984). Present worth of service cost for consumer product warranty. *IEEE Transactions on Reliability* R-33 (5): 424–426.

Park, M., Jung, K.M., and Park, D.H. (2016). Optimal warranty policies considering repair service and replacement service under the manufacturer's perspective. *Annals of Operations Research* 244 (1): 117–132.

Park, M., Jung, K.M., and Park, D.H. (2018). Optimization of periodic preventive maintenance policy following the expiration of two-dimensional warranty. *Reliability Engineering and Systems Safety* 170: 1–9.

Rai, B.K. and Singh, N. (2009). *Reliability Analysis and Prediction with Warranty Data: Issues, Strategies, and Methods*. Boca Raton, FL: CRC Press.

Reliasoft, (2009), "A pro-rata warranty model for non-repairable products," The Reliability Hotwire, No. 100, available at: http://www.weibull.com/hotwire/issue100/relbasics100.htm (accessed on November 7, 2017).

Ritchken, P.H. (1985). Warranty policy for non-repairable items under risk aversion. *IEEE Transactions on Reliability* R-34 (2): 147–150.

Ritchken, P.H. and Fuh, D. (1986). Optimal replacement policies for irreparable warrantied items. *IEEE Transactions on Reliability* R-34 (2): 147–150.

Ross, S.M. (1970). *Applied Probability Models with Optimization Applications*, 44–55. San Francisco, CA.: Holden-Day Publisher.

Samsung Newsroom, (2016), "Samsung announces lifetime guarantee against long-term image retention for 2016 SUHD quantum dot TVs in the U.S.", available at: https://news .samsung.com/us/samsung-announces-lifetime-guarantee-long-term-image-retention-2016-suhd-quantum-dot-tvs-u-s (accessed on October 30, 2017).

Shafiee, M. and Chukova, S. (2013). Maintenance models in warranty: a literature review. *European Journal of Operational Research* 229 (3): 561–572.

Shang, L., Si, S., Sun, S., and Jin, T. (2018). Optimal warranty design and post-warranty maintenance for products under inverse Gaussian degradation. *IISE Transactions* available at: https://www.tandfonline.com/doi/abs/10.1080/24725854.2018.1448490.

Sherbrooke, C.C. (1992). Multiechelon inventory systems with lateral supply. *Naval Research Logistics* 39 (1): 29–40.

Thomas, M.U. (2006). *Reliability and Warranties: Methods for Product Development and Quality*. Boca Raton, FL: CRC Press.

Tong, P., Liu, Z., Men, F., and Cao, L. (2014). Designing and pricing of two-dimensional extended warranty contracts based on usage rate. *International Journal of Production Research* 52 (21): 6362–6380.

Wang, W. (2012). A stochastic model for joint spare parts inventory and planned maintenance optimization. *European Journal of Operational Research* 216 (1): 127–139.

Wang, H.Z. and Pham, H. (2006). Warranty cost models with independence and imperfect repair. In: *Reliability and Optimal Maintenance* (ed. H.Z. Wang and H. Pham), 203–257. Springer.

Wang, X. and Xie, W. (2017). Two-dimensional warranty: a literature review. *Journal of Risk and Reliability* 2017.

Wang, J., Zhou, Z., and Peng, H. (2017). Flexible decision models for a two-dimensional warranty policy with periodic preventive maintenance. *Reliability Engineering and Systems Safety* 162: 14–27.

Wang, X., Ye, Z.S., Hong, Y.L., and Tang, L.C. (2018). Analysis of field return data with failed-but-not-reported events. *Technometrics* 60 (1): 90–100.

Warranty Week (2006), "Extended warranty pricing," available at: http://www .warrantyweek.com/archive/ww20061024.html (accessed on October 31, 2017).

Warranty Week, (2016), "Automotive warranty report," available at: http://www .warrantyweek.com/archive/ww20150326.html (accessed on February 12, 2017).

Wu, S. (2011). Warranty claim analysis considering human factors. *Reliability Engineering and Systems Safety* 96 (1): 131–138.

Wu, S. (2012). Warranty data analysis: a review. *Quality and Reliability Engineering International* 28 (8): 795–805.

Xie, W., Liao, H., and Zhu, X. (2014). Estimation of gross profit for a new durable product considering warranty and post-warranty repairs. *IIE Transactions* 46 (2): 87–106.

Xie, W., Shen, L., and Zhong, Y. (2017). Two-dimensional aggregate warranty demand forecasting under sales uncertainty. *IISE Transactions* 49 (5): 553–565.

Ye, Z.-S. and Murthy, D.N.P. (2016). Warranty menu design for a two-dimensional warranty. *Reliability Engineering and Systems Safety* 155: 21–29.

Zhang, S., Sun, F.B., and Gough, R. (2010). Application of an empirical growth model and multiple imputation in hard disk drive field return prediction. *International Journal of Reliability, Quality and Safety Engineering* 17 (6): 565–578.

Problems

Problem 8.1   There are three non-repairable products: cell phones, vehicle tires, and hair dyer. You are required to develop the base warranty service model by adopting one of the following cost rebate models: free replacement, pro-rata rebate, and mixed model. Which rebate model is more appropriate to these repairable products? Justify your statement.

Problem 8.2   There are four different repairable products: electric vehicle battery, personal computer, and airplane engine. You are required to develop the base warranty service model by adopting one of the following cost rebate models: free replacement, free repair, PRW. Which rebate model is more appropriate to these repairable products? Justify your statement.

Problem 8.3   A two-dimensional warranty is applied to products that are subject to time and usage factors. For example, the reliability of an automobile depends on the age of the vehicle and the mileage driven. Could you give several additional products or systems for which the multidimensional warranty model is more appropriate than the one-dimensional warranty? Explain your reason.

Problem 8.4   The difference between a renewing warranty (RW) and an NRW is that the product upon replacement or repair, the warranty period, is restarted from 0 to w in the RW case, while in an NRW the warranty time is simply the remaining time prior to w. Obviously the RW service incurs a higher cost to the manufacturer. Explain and justify why RW policy is preferred over NRW for certain products like light bulbs and cell phones.

Problem 8.5   Suppose that item is non-repairable and the failure time follows the Weibull distribution. The item is covered by a free replacement warranty with $w = 1$ year. The product MTTF is three years and the shape parameter is 0.5, 1, 1.5, and 2. Compute the expected warranty cost for different scale and shape parameters.

Problem 8.6   Solve the same problem in Problem 8.5 assuming that the warranty policy is pro-rata with $w = 1$ year.

Problem 8.7   Solve the same problem in Problem 8.5 assuming that mixed warranty policy is adopted. From time 0 to $w_1 = 6$ months is free replacement and from w_1 to $w_2 = 1$ year is a pro-rata rebate.

Problem 8.8   Consider a free replacement warranty policy. The cost of replacing an item at the jth failure time T_j within the warranty period w is $b(T_j) = c$, for $j = 0, 1, 2, \ldots, N(w)$. Note that $N(w)$ is the total failure in $[0, w]$. Suppose the product's lifetime follows Erlang distribution with the PDF

$f(t) = te^{-t}$ for $t \geq 0$. Given $w = 1$, 2, and 3, determine (i) the expected number of replacements in $[0, w]$ and (ii) the optimal replacement time τ for different w.

Problem 8.9 The product lifetime follows the exponential distribution with a failure rate $\lambda = 2$ failures/year. When it fails, the product is restored to as-good-as-new state. The cost per repair is \$100 and the size of the installed base is 20 units. Estimate the mean and variance of the warranty cost given $w = 1$, 2, and 3 years, respectively.

Problem 8.10 The product lifetime follows the Weibull distribution with a scale parameter of two years and a shape parameter of 2.5. When it fails, the product is resorted to "as-good-as-old" state. The cost per repair is \$100 and the size of the installed base is 20 units. Estimate the mean and variance of the warranty cost given $w = 1$, 2, and 3 years, respectively.

Problem 8.11 For a free replacement policy, the expected number of replacements N during the warranty period $[0, w]$ is $E[N] = F(w)/R(w)$, where $R(t)$ and $F(t)$ are the reliability and CDF of the product lifetime.

Problem 8.12 Let S_n be the time to the nth failure and $S_n = T_1 + T_2 + \cdots + T_n$, where T_i for $i = 1$, 2, ..., n are identically and independently distributed with CDF $F(t)$. Show that

$$P\{N = n\} = P\{S_n < t < S_{n+1}\} = F^{(n)}(t) - F^{(n+1)}(t)$$

where $F^{(n)}(t)$ and $F^{(n+1)}(t)$ are the nth and $(n+1)$th convolution of $F(t)$.

Problem 8.13 The Bass model or Bass diffusion model consists of a simple differential equation that describes the process of how new products get adopted in a population. Let $N(t)$ be the quantity of adopted new products in the market in $[0, t]$; then the diffusion model is

$$\frac{dN(t)}{dt} = (m - N(t)) \left(a_1 + \frac{a_2}{m} N(t) \right), \quad \text{for } t \geq 0$$

where a_1 is the innovator factor and a_2 is the imitator factor. The sum of a_1 and a_2 usually falls in the range of $[0.3, 0.7]$. Also, m is the size of the potential market of the new product, which depends on the product's price and warranty. Solve for $N(t)$.

Problem 8.14 Based on the Bass model from the previous problem, plot $N(t)$ for the following cases: (i) $a_1 = 0.3$, $a_2 = 0.3$; (ii) $a_1 = 0.3$, $a_2 = 0.1$; (iii) $a_1 = 0.1$, $a_2 = 0.3$.

Problem 8.15 The product lifecycle of a new cell phone is expected to be $l = 5$ years and the sales rate over l is given by $s(t) = kte^{-t}$, where k is the model parameter. Estimate the value of k given the total sales during the lifecycle would be 1, 2, and 4 million, respectively. Also draw $s(t)$ and $S(t)$ for three cases for $0 \le t \le l$.

Problem 8.16 Use the sales rate from the previous problem to estimate the warranty cost assuming that the warranty period $w = 1$ year and the base item price is \$150/item in two cases: (i) free replacement, (ii) linear pro-rata rebate. In both cases, the product lifetime is exponentially distributed with a constant failure rate $\lambda = 0.1$ failure/year.

Problem 8.17 In Eq. (8.5.10), show that

$$\lim_{\rho_1 \to 1} E[D_i(w)] = \lim_{\rho_1 \to 1} \frac{\rho_1}{1 - \rho_1}(1 - e^{-(1-\rho_1)\theta w}) = \theta w$$

Problem 8.18 Assume that the MTTF of a single-item system is 5000 hours and there are 20 units operating in a customer site. To ensure the system availability, the customer can either keep up two spare parts locally for replacement or resort to on-call repairs. The lead-time for an emergency on-call repair is four hours and the lead-time for a regular on-call repair is 12 hours. The hands-on replacement time is one hour. Estimate the system availability under different options.

Problem 8.19 In a production floor, a minimum number of machines are required to be operational in order to meet the production demands. From the previous problem, the probability that at least 18 systems must be available in any random point in time should not be less than 95%. Show whether onsite stocking or the on-call repair meets this criterion? If not, how can the goal be attained?

Problem 8.20 Explain what is the double marginalization problem? Why may it occur if the extended warranty service is distributed through the retailer? How can this phenomenon be avoided?

Problem 8.21 Show the detailed steps that lead to Eq. (8.3.44):

$$F_Z(z) = [1 - e^{-(w+z)} - (w + z)e^{-(w+z)}]$$

$$- \int_0^w [e^{-(w+z-x)} + (w + z - x)e^{-(w+z-x)}][0.5 - 0.5e^{-2x}]dx$$

$$= 1 - e^{-z} - ze^{-z}$$

given that $F(t) = 1 - e^{-t} - te^{-t}$ and $m(t) = 0.5 - 0.5e^{-2t}$.

9

Basic Spare Parts Inventory Models

9.1 Introduction

An inventory system is like a battery energy storage system that deals with charging and discharging of electricity. When items are taken out of the inventory to meet customer demand, it behaves like discharging energy from the battery. When the inventory reaches a low level and is replenished with new items, it is similar to recharging energy into a battery. Therefore, inventory planners address two fundamental questions: when to order and how much is needed for replenishment? Inventory costs include purchasing, ordering, holding, and backorders. Lead time plays a critical role in stochastic inventory management because it is directly correlated with inventory performance, such as the probability of stockout, fill rate, and expected backorders. In this chapter, we introduce the basic inventory planning and optimization models under both continuous and periodic review policy. Section 9.2 overviews the inventory models based on review policy and demand patterns. Sections 9.3 and 9.4 present the economic order quantity (EOQ) and newsvendor models as both form the foundation of inventory theory. Section 9.5 introduces the (q, r) continuous review model with stochastic demand. Section 9.6 presents an approximation solution to the (s, S, T) periodic review model. Section 9.7 introduces the concept of an installed base driven inventory planning model. Section 9.8 discusses spare parts demand forecasting with the focus on the installed base concept.

9.2 Overview of Inventory Model

9.2.1 Inventory Cost and Review Policy

Inventory systems prevail in process manufacturing, discrete/repetitive production, service parts logistics, product and goods distributors, and repair and maintenance services. Regardless of the application, the inventory problem typically involves the placement of order, replenishment of stock, and supply and delivery of an item to meet the customer demand. There are two fundamental questions in inventory management: when to order and how much? In theory a large inventory level can guarantee the customer demands, but the cost becomes high as well. Let C_{inv} be the cost of the inventory; then

$$C_{inv} = C_p + C_o + C_h + C_b \tag{9.2.1}$$

Reliability Engineering and Services, First Edition. Tongdan Jin.
© 2019 John Wiley & Sons Ltd. Published 2019 by John Wiley & Sons Ltd.
Companion website: www.wiley.com/go/jin/serviceengineering

where

C_p = the purchasing cost of ordered items. It is the multiplication of the number of orders or batch size with the price per unit item. Sometimes, unit price under a larger order quantity is given a discount or price break in the name of economy of scale.

C_o = the ordering cost that represents the setup cost whenever an order is placed, independent of the order quantity. Transportation and logistics costs can be treated as part of the ordering cost.

C_h = the holding cost that represents the cost of maintaining inventory in stock. It typically consists of the interest on capital and the cost of storage, maintenance, and handling.

C_b = the backorder cost incurred by the supplier when the demand is not fulfilled, resulting in either lost sales or loss of customer goodwill because of the delayed supply.

Periodic and continuous inventory reviews are two common policies used to track the inventory status for accounting and ordering purposes. Periodic review monitors and documents the inventory counts at specified calendar times or in a fixed time interval. For example, a car dealer operating under a periodic review policy might count the number of new vehicles available at the beginning of each week. The advantages of a periodic inventory review include the savings of the time a business owner or inventory planner spends collecting the inventory data, which allows more time available for focusing on core business operations. However, a periodic review may not provide accurate or up-to-date inventory status for a high-volume sales business, which creates potential challenges to decide when and how much reordering items are necessary in the next replenishment cycle. It may also cause an inventory obsolesce issue if the item or product is seasonal or perishable, like fresh products or Christmas gifts.

A continuous inventory review, also known as an instantaneous review, requires a computer-based information system to track individual items and update inventory counts each time an item is removed from the stock. For example, retailers like Walmart often use bar code scanners or radio-frequency identification (RFID) technology to record customer purchases and incoming packages such as stock-keeping units to update inventory counts in a real-time manner. As an advantage, a continuous inventory review allows for real-time updates of inventory counts, making it easier for the planner to decide when to place a reorder and how much is needed for inventory replenishment. Due to the advancement of information technology, equipment required for performing real-time counting, such as bar code scanners, inventory software, and computers, become accessible to and cost-effective for general commodities. Thus, a continuous inventory review policy is becoming a prevalent approach to managing commodities and products in manufacturing, distribution, and retail stores. The following terminologies are defined to facilitate the modeling and planning of inventory systems throughout this chapter:

- Lead time = time from placing an order to receiving it
- Lead time variability = the uncertainty or variance of the lead time
- Replenishment orders = orders being placed but not arrived yet
- On-hand inventory = physical inventory available in the stockroom
- Backorder level = quantity of demand units being backordered
- Stockout = the event that a demand unit is backordered

- Net inventory level = on-hand inventory minus backorder level
- Inventory position = net inventory level plus replenishment orders
- Fill rate = fraction of demands that are filled from stock immediately
- Safety stock = amount by which the reorder point exceeds the average lead time demand
- Bullwhip effect = a larger swing or amplification of inventory in response to changes in downstream customer demand, as one looks at upstream firms in the supply chain for a product.

9.2.2 Inventory Demand Patterns

Inventory demand patterns can be broadly classified into deterministic demand and stochastic demand. Deterministic demand includes static demand and dynamic demand. Static demand means that the amount of demand per order is constant or fixed over time. Dynamic demand means that the demand varies from one period to the other, but the quantity is known in advance. Examples of dynamic demand include advance demand information from the customers before the order is placed by the customer.

For stochastic demand, the quantities and/or the lead times are unknown with ambiguity. Typically demands involving a stochastic nature include the demands for Christmas trees and the new generation of cellular phones, as well as the demand for electricity and fuel in the upcoming winter season. Stochastic demands can be further classified into stationary demand and non-stationary demand. For stationary demand, the mean and variance of the demand remain constant while the actual demand at a random point in time is uncertain. For non-stationary demand, the mean and/or the variance of the demand is time-dependent and varying. Hence a non-stationary inventory demand is also referred to as a time-varying demand process. Figure 9.1 depicts four types of demand patterns corresponding to constant, dynamic, stationary, and non-stationary processes. The majority of the inventory systems are confronted with stochastic demands due to the random nature of the market size and consumer behavior.

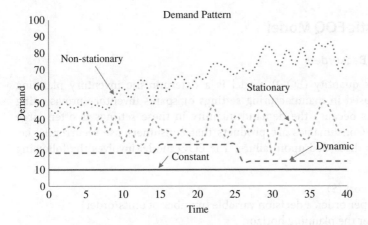

Figure 9.1 Various types of demand patterns.

Table 9.1 Parameters and features of inventory system.

No.	Parameters/factors	Features
1	Demand	Constant, variable/dynamic but deterministic, stationary stochastic, non-stationary stochastic
2	Lead time	Zero, deterministic, and variable or uncertain
3	Horizon	Single period, multiperiod, infinite horizon
4	Item quantity	Single item, multi-item
5	Supply capacity	Unlimited, limited
6	Echelon	Single, multiple
7	Service criteria	Allow backorders (or backlog), no backorders or lost sales
8	Product type	Perishable, unperishable, seasonal, unseasonal

Based on the demand characteristics, inventory models can be classified into deterministic and stochastic (or probabilistic) planning models. Deterministic inventory models are used to determine the optimal inventory of a single item to meet the demand when the amount of the demand is most likely to be fixed. Since it conceives the system to be deterministic, it automatically means that one has full information about the system. Stochastic inventory models can be further classified into stationary and non-stationary inventory models. Stochastic inventory systems consisting of probabilistic demand and supply are more suitable in many real-world circumstances. However, such models also create larger challenges in analysis and implementation, which may become uncontrollable.

Table 9.1 lists eight parameters or factors that need to be taken into account in modeling, planning, and operating inventory systems: demand pattern, lead time, planning horizon, items, supply capacity, echelon level, service criteria, and product type. Together they jointly influence the way to choose the inventory control method and replenishment policy.

9.3 Deterministic EOQ Model

9.3.1 EOQ with No Backorders

The economic order quantity (EOQ) model is a deterministic inventory planning method. It is often used in manufacturing settings or spares inventory provisioning for consumable parts because the demand quantity in these settings is often large yet relatively stable. Consumables are products that consumers use recurrently, i.e. items are used up or discarded upon failure. The EOQ model involves the following parameters:

C_T = total inventory cost ($)
q = quantity of units per order, a decision variable (number of units/order)
D = total demand over the planning horizon
K = cost for placing an order ($/order)

p = unit price ($/unit)
h = holding cost per unit time ($/product/unit time)
$\bar{I}$ = average inventory level (number of units)
T = inventory planning horizon (such as a year)

If no backorders are allowed, the EOQ model consists of three cost items: (i) ordering cost C_o in $[0, T]$; (ii) total purchasing cost C_p; and (iii) total holding cost C_h. These cost items can be further estimated as follows:

$$C_o = (\text{ordering cost/order}) \times (\text{number of orders}) = KD/q \qquad (9.3.1)$$
$$C_p = (\text{unit price}) \times (\text{total demand}) = pD \qquad (9.3.2)$$
$$C_h = (\text{holding cost per unit time}) \times (\text{planning horizon})$$
$$\times(\text{average inventory}) = hT\bar{I} \qquad (9.3.3)$$

In Eq. (9.3.3), $\bar{I}$ represents the average inventory level over $[0, T]$. The average inventory is used because the inventory counts change over time due to the incoming demands and new replenishments. Figure 9.2 depicts how the instantaneous inventory level changes over time assuming that the inventory level decreases linearly. The initial stock level is q at $t = 0$ and tapers off as more demands are coming. The inventory is depleted at $t = q/D$ and is replenished with the same amount of quantity q. Then a new cycle begins and the process is repeated thereafter.

To estimate the holding cost C_h, we need to estimate the average inventory level $\bar{I}$. Let τ be the length of the inventory cycle, which is the time interval between two consecutive replenishments. From Figure 9.2, the instantaneous inventory level at t in period $[0, q/D]$, denoted as $I(t)$, is given as

$$I(t) = -\frac{q}{\tau}(t - \tau) = -D\left(t - \frac{q}{D}\right) \qquad (9.3.4)$$

Hence the holding cost per cycle, denoted as C_{h_cycle} can be estimated as follows:

$$C_{h\_cycle} = \int_0^\tau hI(t)dt = \int_0^{q/D} h\left(-D\left(t - \frac{q}{D}\right)\right) dt = \frac{hq^2}{2D} \qquad (9.3.5)$$

The result is obtained by realizing that $\tau = q/D$. Now the total holding cost in $[0, T]$ is

$$C_h = C_{h\_cycle} \times (\text{number of cycles}) = \frac{hq^2}{2D}\left(\frac{D}{q}\right) = \frac{hq}{2} \qquad (9.3.6)$$

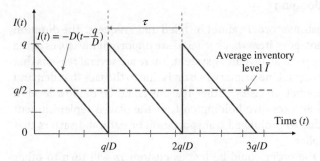

Figure 9.2 EOQ inventory control with no backorders.

By summing Eqs. (9.3.1), (9.3.2), and (9.3.6), the total inventory cost is obtained as

$$C_T = \frac{KD}{q} + pD + \frac{hq}{2} \tag{9.3.7}$$

To find the optimal value of q that minimizes C_T, we take the derivative with respect to q in Eq. (9.3.7) and set to zero:

$$\frac{dC_T}{dq} = -\frac{KD}{q^2} + \frac{h}{2} = 0 \tag{9.3.8}$$

Solving Eq. (9.3.8) yields

$$q^* = \sqrt{\frac{2kD}{h}} \tag{9.3.9}$$

which minimizes the total inventory cost in Eq. (9.3.7). Notice that q^* is independent of the unit purchasing cost p. However, if the discount rate or economy of scale is applicable to the inventory system, q^* will be influenced by p. By substituting Eq. (9.3.9) into (9.3.7), the minimum inventory cost is

$$C_T = \sqrt{2hkD} + pD \tag{9.3.10}$$

Example 9.1   Consider a retail store that sells 5000 units of iPad per year. Suppose further that the fixed cost of an order is $1000 and the yearly holding cost is $20/unit. The price per unit is $400. Backorders are not allowed. Find the best order quality such that the annual inventory cost is minimized

Solution:
Note that $K = \$1000$/order and the annual demand is $D = 5000$ items, with the unit holding cost $h = \$20$/unit/year. Based on Eq. (9.3.9), the optimal order quantity is

$$q^* = \sqrt{\frac{2KD}{h}} = \sqrt{\frac{2 \times 1000 \times 5000}{20}} = 707 \text{ units/order} \tag{9.3.11}$$

Now the total inventory cost is obtained based on Eq. (9.3.7):

$$C_T = \frac{(1000)(5000)}{707} + (400)(5000) + \frac{(20)(707)}{2} = \$2\,014\,142$$

9.3.2   EOQ Model with Backlogging

A backorder occurs when a customer order cannot be filled upon request. The duration of the backorder and the percentage of items backordered are important measures of the service quality and responsiveness of an inventory system. There are several reasons that cause the occurrence of backorders. One primary reason is due to the fact that demand is uncertain and difficult to predict precisely. Other reasons could be the low level of safety stock in order to avoid an excessive holding cost, or the placed replenishment order has not been received due to prolonged transportation or natural disasters that interrupted the upstream supplier.

When a backorder occurs, the order could be lost as customers will turn to other suppliers (i.e. lost sales) or the customer is willing to wait for a period of time until

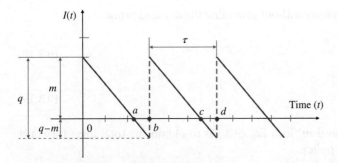

Figure 9.3 An inventory system with backlogs.

the products or items are available. In our analysis, we consider the second case, which is also called *backlogged*. The cost terms below are defined in a period of $[0, T]$. The replenishment lead time is assume to be zero. The following new notation is defined while the others are the same as the basic EOQ model:

c_s = shortage or backlog cost ($/unit)

m = the reorder point

$q - m$ = maximum shortage that was allowed in one cycle

Figure 9.3 depicts the cycles of an inventory system with backlogs and zero replenishment lead time. Under the assumption that backlogs are allowed, the inventory system consists of ordering cost, purchasing cost, holding cost, and shortage cost. That is,

$$C_T(q, m) = C_o + C_p + C_h + C_b \tag{9.3.12}$$

Both C_o and C_p are the same as in the basic EOQ model without backorders. Therefore, our efforts are focused on the analytical expressions of C_h and C_s. The total inventory holding cost is

$$C_h = (\text{holding } \cos t \text{ per cycle}) \times (\text{cycles}) = h\left(\frac{m}{2}\right)\left(\frac{m}{D}\right)\left(\frac{D}{q}\right) = \frac{m^2 h}{2q} \tag{9.3.13}$$

Note that D/q is the number of inventory cycles in $[0, T]$ and the annual shortage cost is

$$C_s = (\text{shortage } \cos t \text{ per cycle}) \times (\text{cycles})$$

$$= \frac{1}{2}c_s(q - m)\left(\frac{q - m}{D}\right)\left(\frac{D}{q}\right) = \frac{c_s(q - m)^2}{2} \tag{9.3.14}$$

Finally, the total inventory cost is obtained by substituting Eqs. (9.3.1), (9.3.2), (9.3.13), and (9.3.14) into Eq. (9.3.12). That is,

$$C_T(q, m) = \frac{KD}{q} + \frac{m^2 h}{2q} + pD + \frac{(q - m)^2 c_s}{2q} \tag{9.3.15}$$

To find the optimal policy, we take the partial derivative with respect to q and m in Eq. (9.3.15) and set them to zero, respectively:

$$\frac{\partial C_T(q, m)}{\partial q} = -KDq^{-2} - \frac{m^2 h q^{-2}}{2} - \frac{(q - m)^2 c_s q^{-2}}{2} = 0 \tag{9.3.16}$$

$$\frac{\partial C_T(q, m)}{\partial m} = \frac{mh}{q} - \frac{(q - m)c_s}{q} = 0 \tag{9.3.17}$$

The results are given as follows without providing the detailed proof:

$$q^* = \sqrt{\frac{2KD(h + c_s)}{hc_s}} \qquad (9.3.18)$$

$$m^* = \sqrt{\frac{2KDc_s}{h(h + c_s)}} \qquad (9.3.19)$$

Finally, we substitute q^* and m^* into Eq. (9.3.15) to obtain the total inventory cost under the optimal ordering policy.

Example 9.2 Consider a computer retail outlet that sells 2000 laptops per year of a popular model. Suppose further that the fixed cost of an order is $600 and the yearly holding cost per laptop is $70. Also assume the purchase price is $300/laptop and the shortage cost is 40$/laptop. Do the following: (i) compute the optimal order quantity and re-order point, (ii) estimate the inventory cost, and (iii) plot the inventory cycle diagram.

Solution:
(1) $D = 2000$ units/year, $K = \$600$/order, $h = \$70$/unit/year, $p = \$300$/unit, and $c_s = 40\$$/unit. Based on Eqs. (9.3.18) and (9.3.19), we obtain the optimal inventory policy as

$$q^* = \sqrt{\frac{2KD(h + s)}{hs}} = \sqrt{\frac{2 \times 600 \times 2000(50 + 80)}{50 \times 80}} = 279 \qquad (9.3.20)$$

$$m^* = \sqrt{\frac{2KDs}{h(h + s)}} = \sqrt{\frac{2 \times 600 \times 2000 \times 80}{50(50 + 80)}} = 172 \qquad (9.3.21)$$

(2) Substituting Eqs. (9.3.20) and (9.3.21) into (9.3.15), the total cost would be

$$C_T(q^*, m^*) = \frac{600 \times 2000}{279} + \frac{172^2 \times 50}{2 \times 279} + 300 \times 2000$$
$$+ \frac{(279 - 172)^2 \times 80}{2 \times 279} = \$608\,593 \qquad (9.3.22)$$

(3) The inventory cycle diagram is shown in Figure 9.4. Under this control policy, the cycle length is 51 days and there are $2000/279 = 7.2$ replenishments across a year.

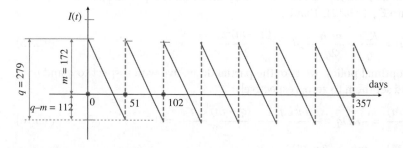

Figure 9.4 The cycle times of the laptop inventory system.

9.4   The Newsvendor Model

9.4.1   The Single-Period Inventory Model

The newsvendor model is a stochastic inventory policy used to determine optimal stock levels for uncertain demand products with seasonality characteristics. It is a single-period inventory decision model that handles perishable products for maximizing the expected profit or minimizing the cost at the end of the period. The inventory model takes into account two uncertain scenarios. First, given the initial inventory level q, any demand above q is lost for potential sales at the end period. Second, if the total demand at the end of the period is less than q, the overall cost is incurred due to the inventory leftovers. This model is also known as the newsvendor problem by analogy with the situation faced by a newspaper vendor who must decide the quantity of the daily papers to stock at the beginning of a day. Facing the uncertain demand, the newsvendor realizes that any leftover copies will be worthless at the end of the day. Many seasonable goods such as Christmas trees and Halloween pumpkins are similar to this lost sales situation. The following notation is defined to facilitate the model formulation:

D = random daily demand of the newspapers with probability density function (PDF) $f_D(x)$
q = the amount of newspapers to be stored at the beginning of a day, decision variable
c = purchasing cost (wholesale price) paid for each newspaper
r = revenue for selling each copy of newspaper
a = salvage value of a newspaper at the end of the day or cost to dispose of a paper ($a > 0$ for salvage value and $a < 0$ for disposal cost)
d = shortage cost for being unable to provide the newspaper when it is needed
h = cost to hold any extra newspaper at the end of the day
Let $f_D(x)$ be the PDF of D; we approach the problem from two scenarios:

Scenario 1: $D \leq q$.
In this case, the profit is the sum of the revenue and the salvage minus the purchase cost of the newspaper, namely

$$P_1 = rD - cq + a(q - D), \ \text{ for } D \leq q \tag{9.4.1}$$

The expected profit is

$$E[P_1] = \int_0^q [rx - cq + a(q - x)]f_D(x)dx \tag{9.4.2}$$

Scenario 2: $D > q$.
The profit equals the revenue minus the sum of the purchase cost and shortage cost (i.e. penalty):

$$P_2 = rq - cq - d(D - q), \ \text{ for } D > q \tag{9.4.3}$$

Thus the expected profit is

$$E[P_2] = \int_q^\infty [rq - cq - d(x - q)]f_D(x)dx \tag{9.4.4}$$

The goal of the newsvendor is to maximize the expected profit over a single period. This can be achieved by maximizing the sum of $E[P_1]$ and $E[P_2]$ as follows:

$$\text{Max: } E[P(q)] = \int_0^q [rx - cq + a(q - x)]f_D(x)dx + \int_q^\infty [rq - cq - d(x - q)]f_D(x)dx$$

$$(9.4.5)$$

After rearranging, it becomes

$$E[P(q)] = rE[D] - cq + a \int_0^q (q - x)f_D(x)dx - (d + r) \int_q^\infty (x - q)f_D(x)dx \quad (9.4.6)$$

and the total profit consists of the revenue, purchase cost, salvage value, and the shortage cost corresponding to the four terms in Eq. (9.4.6). The optimal stock is obtained by solving q for the following equation. Namely,

$$\frac{dE[P(q)]}{dq} = -c + a \int_0^q f_D(x)dx - (d + r) \int_q^\infty f_D(x)dx = 0 \quad (9.4.7)$$

The detailed derivations are omitted here. The optimal value of q is governed by the following critical fractile formula:

$$q^* = F_D^{-1}\left(\frac{r + d - c}{r + d - a}\right) \quad (9.4.8)$$

where $F_D^{-1}(\cdot)$ is the inverse cumulative distribution function (CDF) of D and $F_D(x) = \int_0^x f_D(y)dy$.

If the goal of the newsvendor is to minimize the total inventory cost, instead of maximizing the profit, the cost objective becomes

$$\text{Min: } E[C(q)] = cq + h \int_0^q (q - x)f_D(x)dx + p \int_q^\infty (x - q)f_D(x)dx \quad (9.4.9)$$

where c is the unit purchase cost, h is the holding cost per item, and p is the penalty cost per item in shortage. The optimal order quantity is obtained as

$$q^* = F_D^{-1}\left(\frac{p - c}{p + h}\right) \quad (9.4.10)$$

The solution in Eq. (9.4.10) is the same as the one in Eq. (9.4.8) by exchanging $h = -a$ and $p = d + r$.

Example 9.3   An aircraft carrier has a very critical component that has a reliability problem. The aircraft carrier is beginning a six-month cruise mission and the logistics department must determine how many spare components to stock in the ship. The component's time-to-failure follows the exponential distribution with a mean of three months. A failed component cannot be repaired on board, yet must be replaced with a spare unit. If the onboard stock is exhausted, every additional failure requires an expensive resupply with a cost of $80 000/item. The unit cost of the component is $15 000 if purchased prior to the cruise. The holding cost is $30 000/item/mission at the end of the trip. There is essentially no value left by the spares at the end of the trip because of technical obsolescence. Determine the spare parts stock.

Solution:

This is a single-period inventory decision problem only for the current mission. According to the problem statement, $c = \$15\,000$/unit, $h = \$30\,000$/unit/period, and $d = \$80\,000$/unit. Since the goal is to minimize the cost, Eq. (9.4.10) is used to determine the optimal stock level

$$F_D(q^*) = \frac{p - c}{p + h} = \frac{80\,000 - 15\,000}{80\,000 + 30\,000} = 0.591$$

Given that the mean time-to-failure is three months, the expected number of failures during the six-month trip is 2. The number of failures, denoted X, has a Poisson distribution with a mean of 2, namely

$$F_D(x) = \sum_{k=1}^{x} \frac{2^x e^{-2}}{x!}, \quad \text{for } x = 0, 1, 2, \ldots \tag{9.4.11}$$

Thus we have

$$F_D(0) = 0.135, \quad F_D(1) = 0.406, \quad F_D(2) = 0.677, \quad \text{and } F_D(3) = 0.875$$

Since it is a discrete distribution, the smallest value of q is selected such that $F_D(q)$ exceeds 0.591. This occurs for $q^* = 2$, which means only two spare parts are sufficient. They are in addition to the component initially installed in the ship.

9.4.2 Inventory Performance Measures

This section provides some discussions on the service level metrics that are commonly used to assess the inventory performance. Typical metrics include stockout probability, fill rate, and expected backorders.

Stockout probability, also known as stockout rate, is the chance that the demand is not satisfied immediately from the on-hand inventory. Take the newsvendor problem as an example. Among the 365 days, there are 50 days in which the demand for newspaper is larger than the stock. The probability of not meeting the needs is $50/365 = 13.7\%$. Note that the stockout rate does not count how many units of demand are not met. It merely records if there is a stockout event of items or not. It does not differentiate between whether we are unable to meet the demand of one customer or 10 customers. It also does not indicate whether it is a shortage of one item or a shortage of 100 items. Let D be the demand and s be the on-hand stock level. The stockout probability is estimated by

$$\text{Stockout probability} = P\{D > s\} \tag{9.4.12}$$

Service level is defined as the probability of not incurring a stockout during a lead-time period. For instance, a service level of 90% means there is a 0.9 chance that demand will be met during the lead time and a 0.1 chance of a stockout. Let D be the lead-time demand and s is the reorder point; then

$$\text{Service level} = P\{D \leq s\} \tag{9.4.13}$$

Fill rate is the fraction of customer demand that is met through an on-hand or available inventory, without backorders or lost sales. Unlike the stockout probability, the fill rate is able to track the percentage of customer demand units that are fulfilled during the lead time. The fill rate can be estimated as

$$\text{Fill rate} = \frac{\min\{s, D\}}{D} \tag{9.4.14}$$

If the demand D is less than the on-hand inventory level s, the fill rate is 100%. Otherwise, the fill rate is the ratio of s to D.

Backorders is the number of customer orders that are not fulfilled by the supplier. A backorder generally indicates that customer demand for a product or service exceeds a supplier's capacity or on-hand inventory. Total backorders to an inventory, also known as backlog, may be expressed in terms of the dollar amount or units. Let B be the random variable representing the number of backorders. For an inventory facing random and continuous demand, the expected value of B can be expressed by

$$E[B] = \int_s^\infty (x - s) f_D(x) dx = \mu_X - s - \int_0^s (x - s) f_D(x) dx \tag{9.4.15}$$

where $f_D(x)$ is the PDF of the demand and μ_X is the mean value of X. If the demand is discrete, then

$$E[B] = \sum_{x=s+1}^\infty (x - s) P\{D = x\} = \mu_X - s - \sum_{x=0}^s (x - s) P\{D = x\} \tag{9.4.16}$$

where $P\{D = x\}$ is the probability that the demand equals x for $x = 0, 1, \ldots$.

Example 9.4   A spare parts inventory is responsible for providing capital-intensive units to local customers. Since the demand is sporadic and intermittent, the demand stream is treated as a Poisson process with rate $\lambda = 2$ items/month. The inventory is managed under a one-for-one replenishment policy, and the replenishment lead time is $L = 1.5$ months. The current reorder point is $s = 5$. Estimate the stockout probability, the fill rate, and the expected backorders.

Solution:
(1) The stockout happens when the demand during the lead time exceeds the safety stock level. Let X be the lead-time demand. According to the Poisson process, the stockout probability is

$$P\{X > s\} = \sum_{k=s+1}^\infty \frac{(\lambda L)^k e^{-\lambda L}}{k!} = 1 - \sum_{k=0}^s \frac{(\lambda L)^k e^{-\lambda L}}{k!} = 1 - \sum_{k=0}^5 \frac{(2 \times 1.5)^k e^{-2 \times 1.5}}{k!} = 0.084$$

(2) Since the expected demand in the lead time is $E[X] = \lambda L = 2 \times 1.5 = 3$, the expected fill rate is

$$\text{Fill rate} = \frac{\min\{s, E[X]\}}{E[X]} = \frac{\min\{5, 3\}}{3} = 1$$

However, if the lead-time demand increases and becomes 10, then the fill rate is $\min\{5, 10\}/10 = 0.5$, implying that the inventory is short of $10 - 5 = 5$ items.

(3) Since the demand is discrete, the expected number of backorders is

$$E[B] = \sum_{k=s+1}^\infty (k - s) \frac{(\lambda L)^k e^{-\lambda L}}{k!} = \sum_{k=6}^\infty (k - 5) \frac{(2 \times 1.5)^k e^{-2 \times 1.5}}{k!} \cong 0.13$$

The result is obtained by truncating the probability up to $k = 30$ by ignoring the higher demands. This is reasonable because $P\{X = 31\} = 3.7 \times 10^{-21} \approx 0$.

9.5   The (q, r) Inventory System under Continuous Review

9.5.1   Optimal Policy with Fixed Lead Time

For the (q, r) inventory policy under continuous review, q is the batch size per order and r represents the re-order point. Figure 9.5 graphically describes the working principle of this continuously reviewed inventory system. The demands arrive at a random point in time, driving down the inventory level. If the inventory level reaches or drops below r, an order with a quality of q units is placed for replenishment. The inventory is filled after a certain length of lead time L. Backorders may occur during the lead time and all the backorders are fulfilled upon the receipt of q. Note that the cycle time varies over the planning horizon under the continuous review procedure. Since the demand during L is uncertain, upon replenishment the total on-hand inventory varies from cycle to cycle.

The performance of the (q, r) inventory system largely depends on the lead-time demand profile. As shown in Figure 9.5, there are three possible inventory operating scenarios: (i) in Cycle 1 the inventory exactly meets the demand right before the replenishment; (ii) in Cycle 2 the inventory experiences backorder situations; and (iii) in Cycle 3 surplus inventory is left at the time of the replenishment. Let D be the random demand to the inventory during the planning horizon. To further analyze these scenarios, let us define a new random variable X representing the lead-time demand. According to (Winston 2004), the mean and variance of X are estimated by

$$E[X] = L \times E[D] \tag{9.5.1}$$

$$Var(X) = L \times Var(D) \tag{9.5.2}$$

where $E[D]$ and $Var(D)$ are the mean and variance of D. The formula in Eq. (9.5.2) is due to the fact that the variance for a sum of independent random variables is equal to the sum of the variance of individual random variables.

The objective is to determine the optimal settings of q and r such that the inventory cost over the planning horizon (e.g. annual) is minimized. Recall from Section 9.2 that

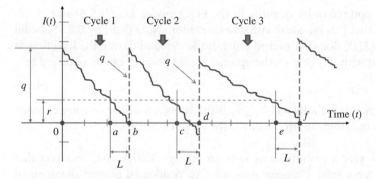

Figure 9.5 Operation of (q, r) inventory under continuous review.

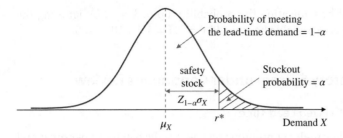

Figure 9.6 The relation between reorder point and safety stock.

the inventory cost is comprised of: (i) ordering cost; (ii) purchase cost; (iii) holding cost; and (iv) shortage or backorder cost. Under the (q, r) control policy, the expected inventory cost can be expressed as

$$C_T(q, r) = \frac{KE[D]}{q} + h\left(\frac{q}{2} + r - E[X]\right) + \frac{c_s E[B]E[D]}{q} + pE[D] \tag{9.5.3}$$

Note that the first term is the ordering cost. The second term represents the holding cost. The third term is the shortage cost (or cost due to backlogs) with c_s being the shortage cost per item, and the last term is the purchase cost of all items. Note that $E[D]$ is the expected total demand. $E[X]$ is the expected demand during the lead time and $E[B]$ is the expected backorders. All other notations are the same as in the EOQ model. Taking the derivatives with respect to q and r, and setting them to zero, we have

$$\frac{\partial C_T(q, r)}{\partial q} = -q^{-2}KE[D] + 0.5h - q^{-2}c_s E[B]E[D] = 0 \tag{9.5.4}$$

$$\frac{\partial C_T(q, r)}{\partial r} = h + q^{-1}c_s E[D]\frac{\partial E[B]}{\partial r} = 0 \tag{9.5.5}$$

The optimal value of q and r are obtained by solving the above equation system, which yields

$$q^* = \sqrt{\frac{2KE[D]}{h}} \tag{9.5.6}$$

$$P\{X > r^*\} = \frac{hq^*}{c_s E[D]} \tag{9.5.7}$$

If one recalls the optimal order quantity in the EOQ model, Eqs. (9.5.6) and (9.3.9) are identical except that D is replaced with its expectation. Note that Eq. (9.5.7) is valid provided that $hq^*/c_s E[D]$ does not exceed unity. If the demand during the lead time is normally distributed with $N(\mu_X, \sigma^2_X)$, the optimal record point r^* can be obtained by

$$r^* = \mu_X + Z_{1-\alpha}\sigma_X \tag{9.5.8}$$

where α is the stockout probability and $Z_{1-\alpha}\sigma_X$ is the safety stock. The relation between the safety stock and the stockout probability is shown in Figure 9.6.

Example 9.5 Each year a manufacturer sells on average 500 DC–AC inverters that convert DC power from solar PV generators into AC residential power. The demand is normally distributed with a standard deviation of 60. The lead time for ordering batch inverters is $L = 15$ days, and $K = \$100$/order and the annual holding cost $h = \$50$/unit/year. The cost for stockout is $c_s = \$30$/unit. The purchase cost is $200 per item. Assuming that the manufacturer adopts the (q, r) inventory system with a continuous review, determine the optimal value of q and r, the expected backorders and the total inventory cost.

Solution:
$D \sim N(u, \sigma^2) = N(500, 60^2); L = 15/365 = 0.0411$ year; $K = \$100$/order; $h = \$50$/item/year; $c_s = \$30$/item; and $p = \$200$/item.

(1) The q^* value can be obtained by directly substituting the known parameters into Eq. (9.5.6):

$$q^* = \left(\frac{2KE[D]}{h}\right)^{1/2} = \left(\frac{2 \times 100 \times 500}{50}\right)^{1/2} = 45 \text{ items}$$

(2) To find the optimal reorder point, we need to estimate the mean and variance of the lead-time demand X, namely

$$\mu_X = E[X] = L \times E[D] = 0.0411 \times 500 = 20.5$$

$$\sigma_X^2 = Var(X) = L \times Var(D) = 0.0411 \times 3600 = 147.9$$

From Eq. (9.5.7), we have

$$P\{X \geq r^*\} = \frac{(50)(44.7)}{(30)(500)} = 0.149 \tag{9.5.9}$$

Given that the stockout probability $\alpha = 0.149$, we obtain $Z_{(1-0.149)} = 1.04$ from the standard normality table. Thus the optimal reorder point and the safety stock are obtained, respectively, as follows:

$$r^* = \mu_X + Z_{(1-0.149)}\sigma_X = 20.5 + (1.04)(12.2) = 34 \tag{9.5.10}$$

$$\text{The safety stock} = r^* - \mu_X = 34 - 20.5 = 13.5 \tag{9.5.11}$$

(3) In theory, the expected backorders can be obtained from Eq. (9.4.15), that is,

$$E[B] = \int_{r*}^{\infty} (x - r^*)f_X(x)dx = \int_{r*}^{\infty} (x - r^*)\frac{1}{\sqrt{2\pi}\sigma_X}e^{-\frac{(x-\mu_X)^2}{2\sigma_X^2}} dx$$

where $\mu_X = 20.5$ and $\sigma_X = 147.9$. Since the closed-form solution is unavailable for $E[B]$ under the normal distribution, the following approximation is proposed to estimate $E[B]$:

$$E[B] \approx \sum_{x=r*+1}^{\infty} (x - r^*)\frac{1}{\sqrt{2\pi}\sigma_X}e^{-\frac{(x-\mu_X)^2}{2\sigma_X^2}} \cong \sum_{x=36}^{78} (x - 35)\frac{1}{12.2\sqrt{2\pi}}e^{-\frac{(x-20.5)^2}{2\times147.9}} = 0.823$$

The summation is truncated at $x = 2r^* = 78$. This is because $(x - r^*)f_x(x)$ for $x > 78$ is close to zero and makes little contribution to $E[B]$.

(4) The total inventory cost in Eq. (9.5.3) is obtained as follows:

$$C_T(q^*, r^*) = \frac{100 \times 500}{45} + 50\left(\frac{45}{2} + 34 - 20.5\right) + \frac{30 \times 0.823 \times 500}{45} + 200 \times 500 = 103,183.17$$

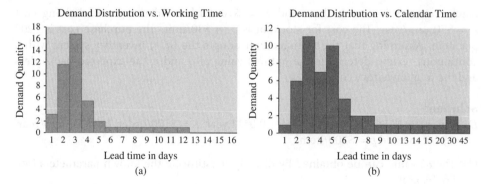

Figure 9.7 Lead time distribution: (a) single mode; (b) dual mode.

9.5.2 Record Point with Variable Lead Time

The inventory models discussed thus far assume that the lead time is known and fixed. There are several factors contributing to the variation of the lead time, such as shipping and transport, supplier-side shortages, and administrative delay. Vermorel (2009) studied the inventory system at Lokad and found that the root-cause of lead-time variations is oftentimes a supplier-side shortage, while shipping and transport tends to be fairly deterministic. Figure 9.7 shows the lead-time distribution for two products. Obviously, the distribution of both lead times is not necessarily normal; rather they are actually skewed with left tails. Indeed the lead-time distribution in Figure 9.7b exhibits dual modes with a mean time of three days and five days, respectively.

If the lead time L varies randomly with the mean of $E[L]$ and variance of $Var(L)$, the mean and variance of the lead-time demand X become (Ravindran and Warsing 2013)

$$E[X] = E[L]E[D] \tag{9.5.12}$$

$$Var(X) = E[L]Var(D) + (E[D])^2 Var(L) \tag{9.5.13}$$

If one compares the above equations with Eqs. (9.5.1) and (9.5.2), the mean remains the same while the variance with a stochastic lead time becomes larger. More discussions about the effect of lead-time uncertainty on an inventory decision are available in Song (1994) and Simchi-Levi and Zhao (2005).

Example 9.6   Assume that the lead time L in Example 9.5 is random with a mean of 15 days and a standard deviation of 10 days. Recalculate the recorder point and the safety stock assuming that the stockout probability given $\alpha = 0.149$.

Solution:
We convert the lead time into units of years such that $E[L] = 15/365 = 0.0411$ year and $Var(L) = (10/365)^2 = 0.000\,751$ (year)^2. Based on Eqs. (9.5.12) and (9.5.13), we have the mean and the variance of X as follows:

$$\mu_X = E[L]E[D] = 0.0411 \times 500 = 20.5$$

$$\sigma_X^2 = E[L]Var(D) + (E[D])^2 Var(L) = 0.0411 \times 3600 + (500)^2 \times 0.000\,751 = 335.6$$

The record point with $\alpha = 0.149$ is

$$r^* = \mu_X + Z_{(1-0.149)}\sigma_X = 20.5 + (1.04)\sqrt{335.6} = 40$$

The safety stock $= r^* - \mu_X = 40 - 20.5 = 19.5$.

Compared with the results in Eqs. (9.5.10) and (9.5.11), this example shows that both the reorder point and the safety stock need to increase in order to safeguard the same service level if the lead-time uncertainty increases.

9.5.3 The Base Stock Policy

The continuously reviewed (q, r) inventory system becomes the so-called base stock inventory system if $q = 1$. This implies that a replenishment order is immediately placed whenever the inventory position drops by one unit. In that case $r + q = r + 1$ is referred to as the order-up-to level. To simplify the model notation, we use s to denote the order-up-to level and the re-order point is $r = s - 1$. The base stock system starts with an initial amount of inventory s. Each time a new demand arrives, a replenishment order is placed to the external supplier. After a period of lead time L, the replenishment order arrives to the inventory. Since the demand is stochastic, multiple orders (or inventory on-order) can be placed and pending for arrival over a time interval. Before presenting the optimal policy, the following assumptions are made on the base stock system: (i) no ordering cost; (ii) replenishment lead time is known and fixed; (iii) demands occur randomly, independently, and continuously over time; and (iv) backorders are allowed and will be satisfied.

The performance of a base stock inventory system, denoted as $(s, s-1)$, is largely dependent on the amount of demand occurring during L. It is interesting to know that under a base stock policy, the lead-time demand and the inventory on-order are always the same. This is because the inventory is provisioned under a one-for-one replenishment policy. Backorders take place if the lead-time demand exceeds s. Let us define the key parameters:

$s =$ the base stock level, also known as the order-up-to level

$I_h =$ on-hand inventory or the number of units physically in the inventory

$B =$ backorders or the number of demands that are not met or fulfilled

$I =$ inventory level, which is $I = I_h - B$

$I_o =$ on-order inventory, also known as the pipeline inventory, is the number of units that have been ordered but not received

$I_p =$ inventory position and $I_p =$ on-hand inventory + inventory on-order backorders

$X =$ lead-time demand, a random variable

Under a base stock policy with *base stock level r*, the inventory position is always kept at s. That is,

$$I_p = I + I_o = I_h - B + I_o = s \tag{9.5.14}$$

or, expressed as the expected values,

$$E[I_p] = E[I_h] - E[B] + E[I_o] = s \tag{9.5.15}$$

Figure 9.8 graphically describes the relation between these inventory parameters at a random point in time.

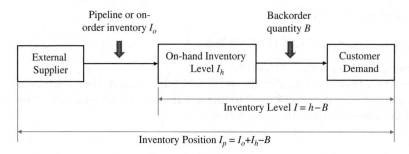

Figure 9.8 The relation of base stock inventory parameters.

The lead-time demand X depends on characteristics of L and D. The mean and variance of X, denoted as $E[X]$ and $Var(X)$, are specified in expressions (9.5.1) and (9.5.2), respectively. If L and D are uncertain and varying, then Eqs. (9.5.12) and (9.5.13) should be used to compute $E[X]$ and $Var(X)$. The performance of the base stock system can be measured by the following performance measures.

1) The stockout probability:

$$P\{\text{stocking out}\} = P\{X \geq s\} \text{ or } P\{X > s\} \tag{9.5.16}$$

2) The non-stockout probability (no backorders in lead time):

$$P\{\text{not stocking out}\} = P\{X \leq s - 1\} \tag{9.5.17}$$

3) The average inventory fill rate:

$$\text{Fill rate} = \frac{E[X]P\{X \leq s-1\}}{E[X]} = P\{X \leq s-1\} \tag{9.5.18}$$

For Eq. (9.5.16), the replenishment order will not be placed until the inventory position drops to $s - 1$. Hence stockout occurs as long as $X > s - 1$. If one replaces I_o with the lead-time demand X, Eq. (9.5.15) can be rewritten as follows:

$$E[I_h] = s - E[X] + E[B] \tag{9.5.19}$$

Equation (9.5.19) represents the physical inventory level prior to the replenishment. For instance, if $s = 20$, $E[X] = 10$, and $E[B] = 3$, then there are $E[I_h] = 20 - 10 + 3 = 13$ units in the stockroom.

Next we determine a value for r that minimizes the base stock inventory cost. Ordering cost in base stock inventory, if have, can be treated as part of the purchasing cost. Since the purchase cost does not influence the decision on s, the base inventory cost only consists of the holding cost and the backorder cost. Let h be the unit holding cost per unit time and b the backorder cost per item per unit time. The inventory cost function is

$$\begin{aligned} C(s) &= hE[I_h] + bE[B] \\ &= h(s - E[X] + E[B]) + bE[B] \\ &= h(s - E[X]) + (h + b)E[B] \\ &= h(s - E[X]) + (h + b)\sum_{x=s}^{\infty}(x - s)P\{X = x\} \end{aligned} \tag{9.5.20}$$

The objective function $C(s)$ is a convex function; hence the optimal value of s is the smallest integer that satisfies the following condition:

$$C(s+1) > c(s) \text{ and } C(s-1) > C(s) \tag{9.5.21}$$

Let us compute the difference between $C(s+1)$ and $C(s)$ as follows:

$C(s+1) - C(s)$

$$= h(s+1 - E[X]) + (h+b) \sum_{x=s+1}^{\infty} (x-s-1)P\{X=x\} - h(s - E[X])$$

$$-(h+b) \sum_{x=s}^{\infty} (x-s)P\{X=x\}$$

$$= -b + (h+b)P\{X \le s\} \tag{9.5.22}$$

The condition in expression (9.5.22) implies that

$$(h+b)P\{X \le s\} - b > 0 \tag{9.5.23}$$

or equivalently

$$P\{X \le s\} > \frac{b}{b+h} \tag{9.5.24}$$

Now choosing the smallest integer s that satisfies inequality (9.5.24) is equivalent to choosing the smallest integer s that satisfies inequality (9.5.23).

Example 9.7 The demand for a critical part of a radar system can be approximated as a Poisson counting process with rate $\lambda = 0.1$ part/day. The lead time of replenishing the inventory is $L = 20$ days. The holding cost is $h = \$200$/part/day and the backorder cost is $b = \$800$/part/day. Since the demand is intermittent and the unit cost is high, the base stock inventory system is used to manage this type of spare part. Determine the optimal base stock level s such that the holding and backorder cost is minimized in the long run.

Solution:
Since the demand for the critical parts is Poisson, the lead-time demand X also observes the Poisson process of which the probability mass function is given as follows:

$$P\{X=k\} = \frac{(\lambda L)^k e^{-\lambda L}}{k!} = \frac{(0.1 \times 20)^k e^{-0.1 \times 20}}{k!} = \frac{2^k e^{-2}}{k!}, \text{ for } k = 0, 1, 2 \dots \tag{9.5.25}$$

By substituting Eqs. (9.5.25) into (9.5.24) along with $h = 200$ and $b = 800$, we have

$$\sum_{k=0}^{s} \frac{2^k e^{-2}}{k!} > \frac{800}{200 + 800} = 0.8 \tag{9.5.26}$$

Since there is no explicit solution for s^* in the above inequality, we find s^* using trial-and-error in the Excel spreadsheet tabulated in Table 9.2. Finally, we choose $s^* = 3$ because this is the smallest integer that makes the cumulative Poisson probability larger than 0.8.

Table 9.2 Cumulative distribution of Poisson process.

k	0	1	2	3*	4
$P\{X = k\}$	0.135	0.271	0.271	0.18	0.09
$P\{X \leq k\}$	0.135	0.406	0.677	0.857	0.947

9.6 The (s, S, T) Policy under Periodic Review

9.6.1 The Inventory Control Mechanism

The inventory systems discussed so far are managed in a continuous review fashion. An alternative approach is to check the inventory level periodically, say every T length of period, where T is the review interval. In cases where physical inspection or manual counting of any given item is necessary, it is desirable to have these activities done only periodically. It is quite likely that the amount of labor required to monitor the physical inventory is drastically reduced with the adoption of the enterprise resource planning (ERP) system, RFID technology and Internet of Things (IoT). Perhaps the most important reason for periodic review is to allow groups of items to be recorded simultaneously over a common interval. This potentially leads to cost savings in logistics and transportation, and possibly to enjoy a price discount due to the economy of scale.

Among periodic-review systems, the (s, S, T) periodic-review system is the most directly parallel version of the (q, r) continuous-review inventory system. Note that s stands for the reorder point, S is for order-up-to-level, and T is the length of the review period. The operating mechanism of the (s, S, T) system proceeds as follows. Every T time units, the inventory position I_p is reviewed. If $I_p > s$, there is no need for placing a replenishment order. If $I_p \leq s$, order $Q = S - I_p$ items to bring the inventory position back to the level S. The main difference between the (q, r) system and the (s, S, T) system is that the order quantity Q in each period is not the same under the periodic review. Figure 9.9 shows how the inventory level of the (s, S, T) system varies over the planning horizon. Two observations are made. First, the inventory position is capped by S. Second, the re-order quantity Q_i for $i = 1$, 2, and 3 varies from cycle to cycle depending on the current inventory position $I(t)$ when the order is placed.

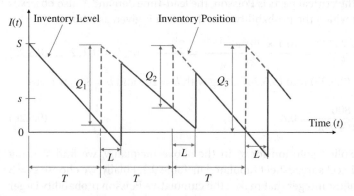

Figure 9.9 Using inventory position to control of the (s, S, T) system.

9.6.2 Approximation Solution to Optimal Policy

Scarf (1960) established the optimality of the (s, S, T) policies under general conditions incorporating random demand and shortage cost. Ravindran and Warsing (2013) point out that the practical use of (s, S, T) policy is often inhibited by computational difficulties that involve non-linear integer decision or stochastic dynamic programming, especially when the delivery lead time is non-zero. A viable approach would be to present an approximation method that relieves the computational burden, yet is proven to be near-optimal under various conditions. We introduce a two-stage approximation solution that first computes s^* and then determines S^* in a sequential manner.

To compute the s^*, we shall characterize and estimate the lead time demand. The (s, S, T) inventory system is at risk of stockout over the replenishment lead time L, but we must also keep in mind that the inventory position will not be reviewed again until a period of T interval has passed. Therefore, s must be set to a level that protects the inventory against the stockout over a duration of $T + L$. Thus in order to determine the optimal value of s, we must specify the demand distribution over the review and lead-time period. Let X be the random variable representing the demand over $T + L$. If T and L are known and constant, the mean and variance of X are given as follows:

$$E[X] = (T + L)E[D] \tag{9.6.1}$$

$$Var(X) = (T + L)Var(D) \tag{9.6.2}$$

where D is the total demand during the inventory planning horizon. When the demand during the period of $T + L$ follows the normal distribution, the optimal re-order point, denoted as s^*, can be explicitly computed as

$$s^* = E[X] + Z_{1-\alpha}\sqrt{Var(X)} \tag{9.6.3}$$

where, $Z_{1-\alpha}$ is the Z-value of a standard normal distribution at the critical fractile of in-stock probability $1 - \alpha$ for $0 \le \alpha \le 1$. Obviously a smaller α implies a higher level of service quality. If X does not follow the normal distribution, we can still determine $s^* = F_X^{-1}(1 - \alpha)$, where $F_X(x)$ is the CDF of the demand during $T + L$.

Next we focus our attention on specifying the order-up-to level S such that the inventory cost is minimized. Assume that the cost of the (s, S, T) inventory system consists of ordering, purchasing, and holding expenses. One straightforward way to do this is to treat S^* as the sum of s^* and Q^*, and the latter can be estimated by utilizing the EOQ model discussed in Section 9.3. Accordingly, we compute

$$S^* = s^* + Q^* = s^* + \sqrt{\frac{2KD}{h}} \tag{9.6.4}$$

The rationality behind this approximation is justified as follows. At each review period, the inventory position I_p usually fluctuates up and down around s, yet the value of I_p never exceeds S^*. Hence each order with an amount of $S^* - I_p$ units should be close to EOQ. In a steady-state condition, the order quantity in each period should reasonably balance the ordering cost and the holding expense.

Like the (q, r) inventory system, the (s, S, T) system can also accommodate backorders. Let c_s be the unit shortage cost per period. The approximated optimal value of S^*

considering backorders can be estimated as

$$S^* = s^* + Q^* = s^* + \sqrt{\frac{2KD(h + c_s)}{hc_s}} \tag{9.6.5}$$

It is worth mentioning that other approximation schemes are available for computing (s, S, T) inventory parameters. Porteus (1985) made an overview and numerical comparisons of a number of such approximations for computing s and S. For a more recent review on this subject, readers are referred to the study by Silver et al. (1998), where different methods for computing s are compared under various shortage penalties, and several extensions to EOQ-like formulas for computing S are provided.

9.6.3 Variable Lead Time

When the lead time L of the (s, S, T) inventory system is random and uncertain, X consists of a mixed distribution. Particularly a portion of it is comprised of the sum of T random variable demands and a portion is comprised of a randomly sized sum of uncertain demands (the L portion). By taking into account both cases, the variance of X with random L and D is obtained as follows:

$$E[X] = (T + E[L])E[D] \tag{9.6.6}$$

$$\begin{aligned} Var(X) &= (T + E[L])Var(D) + (E[D])^2 Var(L) \\ &= TVar(D) + E[L]Var(D) + (E[D])^2 Var(L) \end{aligned} \tag{9.6.7}$$

The result of $E[X]$ is straightforward. Let use prove the variance in Eq. (9.6.7).

Proof:
Recall the (q, r) continuous-review inventory system. The variance of the lead-time demand X under variable lead time is given in Eq. (9.5.13) and is rewritten below:

$$Var(X) = E[L]Var(D) + (E[D])^2 Var(L) \tag{9.6.8}$$

Now define $L' = T + L$ where T is a constant and L is a random variable. The mean and variance of L' are

$$E[L'] = E[T] + E[L] = T + E[L] \tag{9.6.9}$$

$$Var(L') = Var(T + L) = Var(L) \tag{9.6.10}$$

Now substitute $E[L]$ with $E[L']$ and $Var(L)$ with $Var(L')$ in Eq. (9.6.8), which leads to the result in Eq. (9.6.7). This completes the proof.

Example 9.8 Assume the annual inventory demand is normally distributed with $E[D] = 1200$ and $Var(D) = 200$. We would like to investigate how L and T jointly influence the demand during $T + L$ by considering the following cases: (1) $T = 4$ weeks, $E[L] = 1$ week; (2) $T = 4$ weeks, $E[L] = 2$ weeks; and (3) $T = 2$ weeks, $E[L] = 2$ weeks.

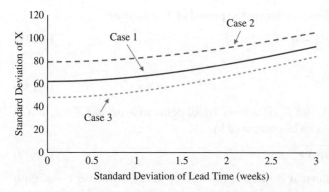

Figure 9.10 The uncertain lead time versus the demand variation.

Solution:

According to Eq. (9.6.6), the mean demand during $T + L$ for three cases are calculated and the results are $E[X] = 115$, 188, and 69, respectively, for Cases 1, 2, and 3. The results make sense because the expected value of $T + L$ in Case 2 is the longest while it is the shortest in Case 3; hence the mean demand is 188 in Case 2 and it is only 69 in Case 3.

Next we compute $Var(X)$ according to Eq. (9.6.7) by increasing the standard deviation of lead time from 0 to 3 weeks for each case. The resulting standard deviation of X is plotted in Figure 9.10 for each case. Obviously, Case 2 always has the largest variation because of larger T and L. By comparing Cases 1 and 3, it is found that reducing L can offset the lead-time demand variation. In other words, if the inventory involves a large $E[L]$ and/or $Var(L)$, then a smaller T is preferred in order to meet the service criteria with a lower inventory level.

9.6.4 The $(S - 1, S, T)$ Periodic Review Policy

As one variation of the periodic review inventory system (s, S, T), let us consider the case where the cost of placing an order is either zero, negligible, or sufficiently small compared with the inventory capital and shortage cost. This results in an order-up-to system where every T length of time an order of $Q = S - I_p$ is placed, where S is the order-up-to level and I_p is the inventory position. Accordingly, this type of periodic review inventory system is referred to as an $(S - 1, S, T)$ system or is simply denoted as an (S, T) system. In an (S, T) inventory system, the order-up-to level S serves the role of protecting the system from stockout events. Thus S behaves more like a re-order point rather than an order quantity. Consider the order placed on time $t = t_1$. Once reviewed at $t = t_1 + L$, this order must keep the system in positive stock through the next review cycle T and the receipt of the order that will be placed at that review, namely $t = t_1 + T + L$. Thus the order-up-to level S needs to be determined by considering the in-stock probability level we would like the system to achieve. This enables us to define the target customer service level (CSL), and thereby set the critical fractile of the distribution of demand the inventory must cover. Like the (s, S, T) system, the demand distribution in the (S, T)

system to be covered is the demand during the period of $T + L$. Thus

$$P\{X \le S^*\} = F_X(S^*) \ge 1 - \alpha \tag{9.6.11}$$

or equivalently

$$S^* = F_X^{-1}(1 - \alpha) \tag{9.6.12}$$

where $1 - \alpha$ is the target CSL and $F_X(x)$ is the CDF of demand X during $T + L$. If X is normally distributed, then S^* can be computed by

$$S^* = E[X] + Z_{1-\alpha}\sqrt{Var(X)} \tag{9.6.13}$$

where $Z_{1-\alpha}$ is the standard normal Z-value at the fractile of the CSL level $1 - \alpha$. Both $E[X]$ and $Var(X)$ can be estimated from Eqs. (9.6.1) and (9.6.2) for a fixed lead time.

9.7 Basic Supply Chain Systems

9.7.1 The Concept of Echelon Inventory

A supply chain is defined as an interconnected system where products or services are moving from upstream suppliers to downstream customers. It often consists of a series of stages that are geographically separated and physically distinct at which the inventory of each stage is either stored or moved in or out. Some of these stages may be owned by a single corporation and others may be jointly owned by one or more firms. For instance, in a vendor managed inventory, the supplier owns the stock items while the customer may be responsible for providing the physical storage room. The key objective of supply chain management (SCM) is to meet customer demands through the best use of resources, including materials, labor, and logistics information. Various aspects of optimizing the supply chain have been developed, including maintaining the right mix and location of factories and warehouses to serve customer needs (Daskin et al. 2002) and using location-allocation models (Mak and Shen 2009), vehicle routing algorithm (Wu et al. 2016), dynamic programming (Yano and Lee 1995), and big data informatics to maximize the distribution efficiency.

From the standpoint of inventory management, a supply chain seeks to match uncertain demand with a variable supply and do so with the minimal cost. Thus all SCM problems are poised to answer two inventory-related questions: when and how much to order? Understanding the interdependencies in computing optimal inventory system parameters is the crux of modeling and operating multi-echelon inventory systems in a supply chain setting. Before we begin to understand what is exactly meant by an inventory echelon, let us first examine and characterize supply chain systems according to the nature of an underlying network configured by various stages.

Figure 9.11 shows a four-stage supply chain example. The network structure is rather simple because it is a series system, in which each stage has only one upstream supplier and one downstream customer. Supply chain stages are numbered from upstream to downstream, with stage 1 being the farthest upstream node, supplying adjacent stage 2, and so forth, continuing to the farthest downstream node, which is 4 in Figure 9.11. Such a notational convention is widely adopted in operations management literature (Zipkin 2000). In addition, the supplier external to the farthest upstream node in a series system

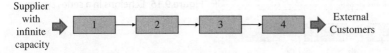

Figure 9.11 A four-stage series system.

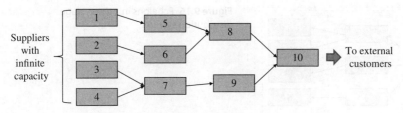

Figure 9.12 Example of an assembly system.

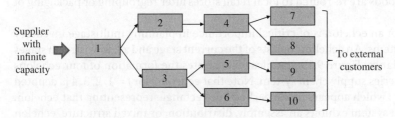

Figure 9.13 Example of a distribution system.

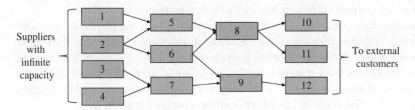

Figure 9.14 Example of a mixed supply chain system.

(i.e. stage 1) is assumed to have infinite capacity. By the same token, the customer external to the farthest downstream stage (i.e. stage 4) is assumed to accept all the output.

In addition to the series system, Figures 9.12–9.14 show three general supply chain networks commonly used in manufacturing, logistics, and service sectors. In particular, Figure 9.12 shows an assembly system in which all nodes have only one downstream customer, but may have multiple upstream suppliers. Again, the nodes that are the farthest upstream are often assumed to have an infinite supply capacity and the node that is the farthest downstream is assumed to deliver the output to the external customers. Figure 9.13 shows a distribution system in which all nodes have only one upstream supplier, but each node may have multiple downstream customers. Finally, Figure 9.14 shows a mixed system in which series, assembly, and distribution functions are mutually embedded. A typical example of a mixed system includes Walmart's transportation and logistics system wherein finished products from the factories are shipped to distribution

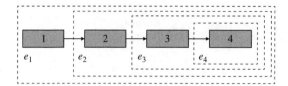

Figure 9.15 Echelons in a series system.

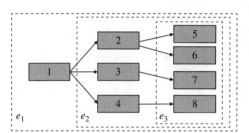

Figure 9.16 Echelons in a three-stage distribution system.

centers where goods are re-routed to local retail stores after regrouping or packaging of goods.

The concept of an echelon is of critical importance in planning multistage inventory supply chain systems. An echelon consists of the current stage and all downstream stages from it. For instance, Figure 9.15 graphically illustrates the formation of four echelons in a four-stage series supply chain system. Note that echelon j for $j = 1, 2, 3, 4$ is denoted by the symbol e_j, which appears in the dashed line rectangle representing that echelon. If a supply chain system exhibits an assembly, distribution, or mixed structure, echelon j consists of all the nodes of the current stage and all the nodes of downstream stages from it. Figure 9.16 is a three-stage distribution system, where, for instance, e_2 contains Nodes 2 to 4 of the current stage and Nodes 5 to 8 downstream.

An echelon inventory consists of two types of inventory. These are physically on-hand and backordered. Together they are called the installation inventory. Those in the transit pipeline between adjacent stages are called in-transit stock. Let $I_j^{(e)} \in \{-\infty, +\infty\}$ denote the installation inventory level at stage j. Furthermore, let W_j be the in-transit inventory to stage j. In addition, define I_i and B_i as the on-hand inventory and backorders of stage i for $i = j, j+1, \ldots, n$. The echelon inventory position at stage j is given by

$$I_j^{(e)} = I_j - B_j + \sum_{i=j+1}^{n} (I_i - B_i + W_i), \qquad \text{for } j = 1, 2, \ldots, \ n \qquad (9.7.1)$$

This implies that the echelon inventory at stage j includes the installation stock at j, and all installation stock and in-transit stock downstream from $j + 1$ to the last stage n. For instance, if $j = n$, the value of $I_n^{(e)}$ equals the sum of on-hand and in-transit subtracted by backorders, which is identical to Eq. (9.5.14). This concept of echelon inventory, established by Clark and Scarf (1960), significantly facilitates the mathematical modeling and analysis related to a multi-echelon inventory decision. Hence, it is important to understand the interaction between adjacent stages in a supply chain regarding their inventory levels and consider using the echelon inventory for calculations and not the local inventory.

For a multi-echelon inventory system, the fundamental issue is to determine how to set the stocking levels at each stage in a way that the inventory decisions and replenishment policies across the supply chain network are well coordinated to achieve the performance target. Due to the complexity of the problem, much attention has been focused on studying series systems, but still by no means are these problems trivially formulated and solved, particularly for non-stationary demand and supply cases. Interested readers can study the seminal work of Graves and Schwarz (1977) on distribution systems or read the widely cited papers by Rosling (1989) on assembly systems. For a comprehensive treatment of inventory models and multi-echelon systems, we refer the readers to the foundational inventory book of Zipkin (2000) and a new book by Snyder and Shen (2011).

9.7.2 Bullwhip Effect and Supply Resilience

A unique characteristic in SCM is the so-called "bullwhip effect," a signature concept that Dr. Hau Lee co-developed in the 1990s, which has become a basic tenet or principle in both academia and industry (Lee et al. 1997). When a person cracks a bullwhip, the small movements at his/her wrist produce huge waves at the other end of the whip, which describes how information on demand is increasingly exaggerated and distorted as it moves from the downstream customers to the upstream manufacturer or supplier, driving up costs and hurting supply chain efficiency. There are many factors causing or contributing to the bullwhip effect in supply chains. The following list names a few: (i) disorganization between each supply chain link; (ii) lack of communication between each link in the supply chain; (iii) free return policies; (iv) order batching; (v) price variations; and (vi) inaccurate forecasting or demand information. Taking order batching as an example, customers may not immediately place an order with their supplier, often accumulating the demand first. This may create a surge in demand at some stages followed by no demand after.

Figure 9.17 shows a simple distribution system facing the demands from two retailers A and B. Let X_A and X_B, respectively, be the weekly demand of A and B. Then the aggregate weekly demand to the supplier C is

$$X_C = X_A + X_B \tag{9.7.2}$$

Assume that X_A and X_B are independent and follow the normal distribution $X_A \sim N(\mu_A, \sigma_A)$ and $X_B \sim N(\mu_B, \sigma_B)$. Then the mean and variance of Xc can be obtained as follows:

$$E[X_C] = E[X_A] + E[X_A] = \mu_A + \mu_B \tag{9.7.3}$$

$$Var(X_C) = Var(X_A) + Var(X_B) = \sigma_A^2 + \sigma_B^2 \tag{9.7.4}$$

For instance, given $\mu_A = 10$ and $\sigma_A = 5$ and $\mu_B = 9$ and $\sigma_B = 3$, then $E[X_C] = 10 + 9 = 19$ and $Var(X_C) = 5^2 + 3^2 = 36$ (or $\sigma_C = 6$). To show how small variations at A and B can

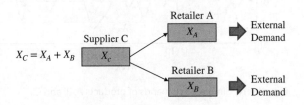

Figure 9.17 A simple distribution system to model the bullwhip effect.

create larger fluctuations at C, we simulate the normal distributions of X_A, X_B, and X_C using the following model:

$$X_k = \mu_k + \sigma_k Z_k, \quad \text{for } k \in \{A, B, C\} \tag{9.7.5}$$

where

$$Z_k = \sin(2\pi \times rand()) \sqrt{-2\ln(rand())} \tag{9.7.6}$$

Note that Z_k is the random variable with $Z \sim N(0, 1)$, which is generated based on the random generator of a standard normal distribution in Eq. (9.7.6). The simulation results are depicted in Figure 9.18 showing that the demand fluctuation perceived by upstream supplier C is much larger than those of A and B. Inventory pooling, lateral supply or transshipment, multisourcing, and keeping a high level of safety stock are the common approaches to addressing the bullwhip effect issue.

Recently supply chain resilience is treated as the next phase in the evolution of traditional, place-centric enterprise structures to highly virtualized, customer-centric structures (Wieland and Wallenburg 2013). For instance, the 2011 flooding disaster in Thailand created a widespread impact on individuals and businesses of all types, including the hard drive industry, disrupting transportation, logistics, power generation, and the availability of labor (Dignan 2011). Hard-drive suppliers Western Digital lost 60% of its drive production capacity in Thailand compared to 40% for the industry overall. Other companies such as Seyyon Semiconductor noted that it will take a year for the production base in Thailand to recover. Nidec, which makes spinning motors for disk drives, is trying to shift production from Thailand to China and the Philippines. Though Seagate's facilities in Thailand were able to be recovered, the company struggles to acquire components for its drives.

Supply chain resilience is defined as the ability to cope with sudden changes or disruptions caused by natural disasters or man-made errors. Therefore, a resilient supply network is capable of aligning its strategy and operations with the changing environment to prevent or mitigate the risks that affect its function. For instance, Medal et al. (2014) study the minimax facility location and hardening problem (MFLHP), which seeks to minimize the maximum distance from a demand point to its closest located facility after facility disruptions. According to Dahlberg (2013), there are four

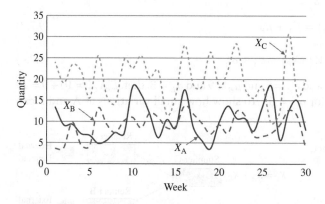

Figure 9.18 Simulated demands of products A, B, and C.

levels of supply chain resilience: (i) reactive SCM, (ii) internal supply chain integration with planned buffers, (iii) collaboration across extended supply chain networks, and (iv) dynamic supply chain adaptation and flexibility. It is worth mentioning that supply chain resilience is not about coping with a one-time crisis or just having a flexible supply structure. Rather it is more concerned with continuously anticipating and handling destructive events that can permanently impair the core business value with the central focus on delivering customer satisfaction. Therefore, supply chain resilience requires continuous innovation with respect to product structures, processes alignment, information sharing, and environmental sustainability.

9.8   Spare Parts Demand Forecasting

9.8.1   Spare Parts Demand Pattern

Capital equipment plays a vital role in sustaining manufacturing processes, energy generation, transportations and logistics, information technology, and military deployment. Modular design is often adopted in these types of systems to facilitate the upgrade, maintenance, and repair services. Upon failure, the defective part, also called the line replaceable unit (LRU), is swapped with a spare unit so that the equipment can be restored to the operational state. The defective part is then sent back to the repair center for root-cause analysis. After it is fixed, the LRU is returned to the spare parts stockroom as a backup unit for future replacement. If the LRU unit is returned from a preventive replacement, the unit is still functional with performance degradation. In that case, the LRU after completing the reconditioning in the repair center will be put back in the spares inventory for future replacements.

The demand for spare parts are often chartered with intermittent, sporadic, and time-varying features. Figure 9.19 shows the demand profiles for four aircraft spare parts A–D over a period of 52 weeks (one year). In Figure 9.19a, the mean demand is relatively stable across the year, but the variance increases over time. In Figure 9.19b, it is opposite in the sense that the mean decreases while the variance remains relatively stable. Figure 9.19c shows the case where both the mean and variance increase over time. Finally, in Figure 9.19d, the mean demand drops sharply from 18 in week 25 to the quantity of 5 in week 26, and then remains at that level thereafter.

Though a variety of inventory demand forecasting models have been developed in the literature, the majority of them revolve around the stationary demands with constant mean and variance. As shown in Figure 9.19, the demand for spare parts often behaves as in a non-stationary process, which is driven by the interactions of several key factors, such as reliability, corrective actions, usage intensity, maintenance policy, operating condition, and installed base. An installed base represents the total number of field systems in operation at a particular time instance. For example, when the manufacturer introduces a new product to the market, the demand for spare parts tends to be increased because of the growth of the installed base. To meet the CSL or mitigate the backorders, one viable approach is to periodically adjust or raise the safety stock level to meet the increasing demand rate.

We focus our attention on slow moving service parts as this is mostly related to the maintenance and repair of capital goods. The area of intermittent demand forecasting

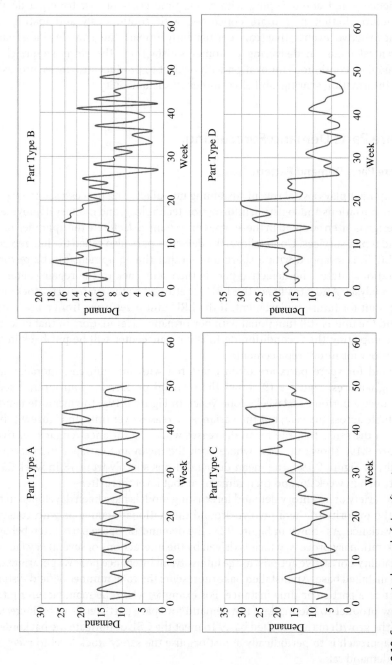

Figure 9.19 Spare parts demand of aircraft parts.

of spare parts has received much attention in the reliability and operations management community, and a significant amount of publications are available in the field which have been elaborated in the review papers of Kennedy et al. (2002), Basten and Van Houtum (2014), and Selviaridis and Wynstrab (2015). Generally speaking, methods to forecast repairable inventory demand can be classified into three categories: (i) data-driven model, (ii) reliability regression model, and (iii) installation-based model. Below, the state-of-the-art of these approaches are reviewed.

9.8.2 Data-Driven Method

Data-driven predicting models include the time series model (i.e. parametric model), distribution-based estimates, and non-parametric models.

Classical parametric approaches, such as the exponential moving average, have been widely used in inventory forecasting, yet they tend to either overestimate or underestimate the mean level of intermittent demand when the demand changes like in Figure 9.19d. To overcome this issue, Croston (1972) suggested a single exponential smoothing method that separates the demand arrival and the demand size, and the ratio of the latter to the former is used to estimate the mean demand. The essence of Croston's method is that the demand for spare parts is treated as a compound Poisson process with a variable batch size. Since then, different variation methods have been proposed. For instance, Syntetos and Boylan (2001) showed that Croston's estimator is biased and suggested an adjustment factor to deflate and smooth Croston's estimate of the mean interarrival time of demands. It is important to resolve the bias issues because the attainment of service level targets is directly linked to the forecast biases. Thus other adjustment strategies to mitigate forecasting bias have also been developed by Boylan and Syntetos (2003) and Shale et al. (2006), and also by the recent paper by Bacchetti and Saccani (2012).

The distribution-based forecasting model turns out to be quite effective in predicting the spare parts demand driven from a static or infinite installed base. The Poisson process is a natural candidate for representing a slow and intermittent demand pattern, and has been widely adopted by service logistics providers and repairable inventory planners. In formulating the well-known multi-echelon technique for recoverable inventory control (METRIC) model, Sherbrooke (1968) proposed use of the compound Poisson process model to estimate the failure intensity generated by the field fleet in multiple bases. Eaves (2002) proposed a negative binomial distribution to represent the intermittent demand patterns when the demand involves a compound distribution with Poisson arrivals and logarithmic size. When extended inventory replenishment lead times are involved, the normality assumption on the demand size becomes more reasonable due to the effects of the central limit theorem.

Bootstrapping is a non-parametric forecasting approach and has received much attention in reliability modeling and estimation. In bootstrapping (Efron 1979), the empirical distribution under study is reconstructed through consecutive sampling with replacements from an available dataset. An important assumption is that the future demand behavior is correlated with or resembles the historical demand profile. Porras and Dekker (2008) develop a bootstrapping approach in which the number of consecutive observations sampled in each replication equals the lead-time demand to address the autocorrelation issues. A large number of replications or resampling is typically required in bootstrapping, yet this technique gains in popularity due to the rapid advances in computing capability and data acquisition. Willemain et al. (2004) proposes a non-parametric inventory forecasting method that synthesizes the Markov process, bootstrapping and "jittering" to simulate the entire distribution of lead-time demand. The Markov process can effectively handle the autocorrelation, whereas jittering ensures that the simulated values differ from those already observed.

9.8.3   Reliability Regression Model

Another research stream predicts the expected number of spare parts based on reliability characteristics of the component, such as usage intensity, failure rate, and operating conditions. Inventory and spare parts demand considering variable product usage have been investigated by Lau et al. (2006) and Bian et al. (2013). For a constant failure rate, queuing theory and Markov models have been widely used to determine the number of spare parts needed to ensure the required availability level of field systems (Mani and Sarma 1984; Jardine and Tsang 2006). The power law model is adopted to predict the number of product failures under time-varying failure intensity (Jung 1993). In fact, the component failure rate depends on the operating conditions or covariates such as the operating time, maintenance, humidity, temperature, and loading stresses. Ghodrati and Kumar (2005) show that ignoring the effect of covariates can lead to a 20% difference in the number of required spares for hydraulic jacks in a case study of Kiruna Mine in Sweden.

The proportional hazard model (PHM) turns out to be quite appealing as it captures the operational effects on the baseline failure rate by coupling the covariate function (Ghodrati 2006; Louit et al. 2011). The main assumption in PHM is that the effect of covariates is time-independent. Recently, Barabadi et al. (2014) proposed a time-dependent covariate function in which the temporal effect of operating stress (e.g. temperature) is incorporated into the baseline failure rate model. They applied this time-dependent covariate PHM model to predict the required number of protective fuses in a power grid under uncertain weather conditions.

9.8.4   Installation-Based Model

As discussed in the previous section, most forecasting methods in literature rarely consider the existence of a relation between explanatory variables and spare parts demand. To apply such "causal" forecasting methods, it is key to understanding the demand generating process (Deshpande et al. 2006). According to Ghobbar and Friend (2002, 2003), causal methods have been shown to have a potentially positive role in a service industry like aircraft parts. It seems intuitive that the size of installed machines and their usage should have a certain impact on the future demand for spare parts (Ritchie and Wilcox 1977). The installed base refers to the whole fleet of systems that are in use at customer sites. Though the idea of forecasting the installed base during the new product introduction was mentioned in the mid 1980s (Farrell and Saloner 1986; Brockhoff and Rao 1993), the development of analytical spare parts forecasting models combining the installed based was rather postponed until the 2000s. Auramo and Ala-Risku (2005) discussed installed base information for service parts logistics management, but their main interest was on how to obtain the installed base information from various sources. The question is how to specially define this installed base information and which information should be chosen and taken into account in forecasting future spare parts demand.

As a very early attempt, Shaunty and Hare Jr. (1960) tried to link spare parts demand to product usage for a static size of airplane fleet. They investigated this connection to the parts demand based on the number of landings of the planes as the base usage, instead of flight hours. The approach of Ritchie and Wilcox (1977) was based on renewal theory for a given product fleet. They assumed random part failure, with a constant rate. Moreover, they introduced the probability that a failed item may not be discarded. Yamashina (1989) related service parts demand to a variable product installed base, the product life characteristics, and the part lifetime. The author, however, noted the difficulties in calculating this demand analytically, and suggested the use of bulk production at time $t = 0$.

Jin et al. (2006) made an early attempt to integrate the installed base information into the Crow/AMSAA model to obtain a quantitative forecasting model for spare parts logistics. Their model was developed to forecast the non-stationary service demand when a new product is introduced in the market coupled with the reliability growth effect. In Hong et al. (2008), the fleet size (i.e. installed base), the product discard rate, the failure rate, and the replacement probability are identified as the key factors needed to construct the spares demand model. Jin and Liao (2009) proposed a superimposed renewal model to estimate the non-stationary spare parts demand under a stochastically expanding fleet that follows the Poisson counting process. Recently, the superimposed renewal model was extended by Jin et al. (2017) to estimate the lead-time inventory demand under an expanding fleet of redundant systems with exponential failures. Jalil et al. (2011) presented several case studies, including IBM, as how the industries leverage the real-time installed base data to manage the spare parts logistics. Xie et al. (2014) designed warranty and post-warranty servicing programs considering the shrinkage of the field fleet size as well as the non-reported repair requests. Kim et al. (2017) leveraged the installed base information to decide the lot size in the final production run to cover the demand of 18 spare parts during the end-of-life phase in a business-to-consumer market. Researchers also used the logistic growth function to add new sales into an existing installed base that are applicable to perishable and high-tech products (Minner 2011; Lapide 2012; Wu and Bian 2015).

In summary, the installed base prediction is a kind of causal forecasting in the sense that the forecast of spare parts is made not only based on the historic demand information but also on data of current installations that triggers the demand for the replacements (Dekker et al. 2013). With the rapid advances of information technology such as the IoT and big data analytics, it becomes more increasingly feasible than ever before to collect historical market data and project the installed base expansion during the new product introduction. For more discussion on this subject, readers are referred to a recent review paper by Van der Auweraer et al. (2017).

9.8.5 Forecasting under Reliability Growth and Fleet Expansion

During the new product introduction, the spare parts inventory system faces a variety of uncertain factors, such as product reliability, market size, and the service capacity. To appropriately allocate the safety stock level against the stockout risk, three key parameters should be appropriately incorporated in the forecasting model. These are the number of operating systems installed over time, the reliability growth rate, and the repair rate of defective parts (Jin et al. 2006). We present a spare parts demand model by taking into account these factors simultaneously. The model is developed for a single-item system fleet, but can be extended to multi-item systems. Let $u(t)$ be the failure intensity function. Then the cumulative failures for a single-item system in $[0, t]$, denoted as $m(t)$, can be estimated by

$$m(t) = \int_0^t u(x)dx \tag{9.8.1}$$

When $u(t)$ observes the Crow/AMSSA estimate (or the power law model), $m(t)$ can be explicitly expressed as

$$m(t) = \int_0^t u(t)d\tau = \int_0^t \lambda\beta x^{\beta-1}dx = \alpha t^\beta \tag{9.8.2}$$

where λ and β are the power law model and both are always non-negative. Given N identical systems operating by time t, the cumulative failures during the period $[0, t]$ can be estimated by

$$M(t) = N \times m(t) = \alpha N t^\beta \tag{9.8.3}$$

Equation (9.8.3) is obtained assuming that all N systems are installed at the same time. For capital equipment like wind turbines, power generators, and aircraft engines, when a new product is introduced to the market, the installed base usually increases over time because of the market expansion. Therefore N is a time-varying parameter, and indeed it is a random number owing to the uncertain market. If the installation follows the Poisson counting process with a rate of λ (i.e. systems per unit time), then by time t, the probability that there are n systems in the field is

$$P\{N(t) = n\} = \frac{(\lambda t)^n e^{\lambda t}}{n!} \quad \text{for } n = 0, 1, 3, \ldots \tag{9.8.4}$$

Now we substitute N with $N(t)$ in Eq. (9.8.4) and the cumulative failures from $N(t)$ systems in $[0, t]$, which can be estimated by

$$M(t) = \sum_{i=1}^{N(t)} m(t - T_i) = \sum_{i=1}^{N(t)} (m(t) - m(T_i)) = \sum_{i=1}^{N(t)} \alpha t^\beta - \sum_{i=1}^{N(t)} \alpha T_i^\beta \tag{9.8.5}$$

where T_i is the installation time of the ith system for $i = 1, 2, \ldots, N(t)$. Note that $0 < T_1 < T_2 < \cdots < T_{N(t)}$. Note that $M(t)$ is a random number because of the stochastic nature of $N(t)$ and T_i. Under a special case when the installation time T_i is known and fixed, then $M(t)$ can be obtained as

$$M(t) = \alpha n_t t^\beta - \sum_{i=1}^{n_t} \alpha t_i^\beta \tag{9.8.6}$$

where n_t is the cumulative installed systems by t, and t_i is the actual installation time of system i, for $i = 1, 2, \ldots, n_t$.

The performance of a spare parts inventory system largely depends on the lead-time demand, i.e. the demand during the replenishment period. Let l be the lead time from when a replenishment order is placed at t till the order arrives to the inventory at $t + l$. Then the lead-time demand, denoted as $D(t)$, is given by

$$D(t) = M(t + l) - M(t) \tag{9.8.7}$$

where $M(t + l)$ represents the aggregate failures of an installed base in $[0, t + l]$. Similarly, $M(t)$ represents the aggregate failures of an installed base in $[0, t]$, respectively. If one substitutes Eq. (9.8.5) into (9.8.7), the following lead-time demand estimate is obtained after some algebraic simplification:

$$D(t) = \left(\sum_{i=1}^{N(t+l)} \alpha(t + l)^\beta - \sum_{i=1}^{N(t+l)} \alpha T_i^{\ \beta} \right) - \left(\sum_{i=1}^{N(t)} \alpha t^\beta - \sum_{i=1}^{N(t)} \alpha T_i^{\ \beta} \right) \tag{9.8.8}$$

It is difficult to obtain the explicit solution because of the stochastic nature of $N(t)$, $N(t + l)$, and T_i for $i = 1, 2, \ldots, N(t + l)$. Therefore, simulation can be used to estimate the value of $D(t)$. Under a special case when the installation time T_i is known and fixed, then $M(t)$ can be obtained as

$$D(t) = \left(\sum_{i=1}^{n_{t+l}} \alpha(t + l)^\beta - \sum_{i=1}^{n_{t+l}} \alpha t_i^\beta \right) - \left(\sum_{i=1}^{n_t} \alpha t^\beta - \sum_{i=1}^{n_t} \alpha t_i^{\ \beta} \right) \tag{9.8.9}$$

where $n_{t + l}$ is the cumulative installed systems by $t + l$ and t_i is the actual installation time of system i, for $i = 1, 2, \ldots, n_{t + l}$.

References

Auramo, J. and Ala-Risku, T. (2005). Challenges for doing downstream. *International Journal of Logistics: Research and Applications* 8 (4): 333–345.

Bacchetti, A. and Saccani, N. (2012). Spare parts classification and demand forecasting for stock control: investigating the gap between research and practice. *Omega* 40 (6): 722–737.

Barabadi, A., Barabady, J., and Markeset, T. (2014). Application of reliability models with covariates in spare part prediction and optimization – a case study. *Reliability Engineering and System Safety* 123 (3): 1–7.

Basten, R.J.I. and Van Houtum, G.J. (2014). System-oriented inventory models for spare parts. *Surveys in Operations Research and Management Science* 19 (1): 34–55.

Bian, J., Guo, L., Yang, Y., and Wang, N. (2013). Optimizing spare parts inventory for time-varying task. *Chemical Engineering Transactions* 33 (2013): 637–642.

Boylan, J.E., A.A. Syntetos, (2003), "Intermittent demand forecasting: size-interval methods based on average and smoothing," In *Proceedings of the International Conference on Quantitative Methods in Industry and Commerce*, Athens, Greece, 2003.

Brockhoff, K.K. and Rao, V.R. (1993). Toward a demand forecasting model for pre-announced new technological products. *Journal of Engineering and Technology Management* 10 (3): 211–228.

Clark, A.J. and Scarf, H.E. (1960). Optimal policies for a multi-echelon inventory problem. *Management Science* 6 (4): 475–490.

Croston, J.D. (1972). Forecasting and stock control for intermittent demands. *Operations Research Quarterly* 23 (3): 289–304.

Dahlberg, G., (2013), "The four levels of supply chain maturity," GT Nexus, available at: http://www.gtnexus.com/resources/blog-posts/4-levels-supply-chain-maturity (accessed on November 20, 2017).

Daskin, M.S., Coullard, C.R., and Shen, Z.-J.M. (2002). An inventory-location model: formulation, solution algorithm and computational results. *Annals of Operations Research* 110 (1–4): 83–106.

Dekker, R., Pince, C., Zuidwijk, R., and Jalil, M.N. (2013). On the use of installed base information for spare parts logistics: a review of idea and industry practice. *International Journal of Production Economics* 143 (2): 536–545.

Deshpande, V., Iyer, A., and Cho, R. (2006). Efficient supply chain management at the US Coast Guard using part-age dependent supply replenishment policies. *Operations Research* 54 (6): 1028–1040.

Dignan, L., (2011), "Thailand floods to lead to hard drive shortages for months," Between the Lines, available at: http://www.zdnet.com/article/thailand-floods-to-lead-to-hard-drive-shortages-for-months (accessed on November 25, 2017).

Eaves, A.H.C., (2002), Forecasting for The Ordering and Stock-Holding of Consumable Spare Parts. PhD Thesis, University of Lancaster, Lancaster, UK, 2002.

Efron, B. (1979). Bootstrap methods: another look at the Jackknife. *The Annals of Statistics* 7 (1): 1–26.

Farrell, J. and Saloner, G. (1986). Installed base and compatibility: innovation, product preannouncement, and predation. *American Economics Review* 76 (5): 940–955.

Ghobbar, A.A. and Friend, C.H. (2003). Evaluation of forecasting methods for intermittent parts demand in the field of aviation: a predictive model. *Computers and Operations Research* 30 (14): 2097–2114.

Ghobbar, A.A. and Friend, C. (2002). Sources of intermittent demand for aircraft spare parts within airline operations. *Journal of Air Transport Management* 8 (4): 221–231.

Ghodrati, B. (2006). Weibull and exponential renewal models in spare parts estimation: a comparison. *International Journal of Performability Engineering* 2 (2): 135–147.

Ghodrati, B. and Kumar, U. (2005). Reliability and operating environment-based spare parts estimation: a cases study from Kiruna Mine, Sweden. *Journal of Quality in Maintenance Engineering* 11 (2): 169–184.

Graves, S.C. and Schwarz, L.B. (1977). Single cycle continuous review policies for arborescent production/inventory systems. *Management Science* 23 (5): 529–540.

Hong, J.S., Koo, H.-Y., Lee, C.-S., and Ahn, J. (2008). Forecasting service parts demand for a discontinued product. *IIE Transactions* 40 (7): 640–649.

Jalil, M.N., Zuidwijk, R.A., Fleischmann, M., and VanNunen, J.A.E.E. (2011). Spare parts logistics and installed base information. *Journal of the Operational Research Society* 62 (3): 442–457.

Jardine, A.K.S. and Tsang, A.H.C. (2006). *Maintenance, Replacement, and Reliability: Theory and Applications.* Boca Raton, FL: CRC/Taylor and Francis.

Jin, T. and Liao, H. (2009). Service parts inventory control considering stochastic growth of an installed base. *Computers and Industrial Engineering* 56 (1): 452–460.

Jin, T., H. Liao, Z. Xiong, C.H. Sung, (2006), "Computerized repairable inventory management with reliability growth and system installations increase," In *Proceedings of Conference on Automation Science Engineering.* October 8–9, 2006, Shanghai, China, pp. 336–341.

Jin, T., Taboada, H., Espiritu, J., and Liao, H. (2017). Allocation of reliability-redundancy and spares inventory under Poisson fleet expansion. *IISE Transactions* 49 (7): 737–751.

Jung, W. (1993). Recoverable inventory systems with time-varying demand. *Production and Inventory Management Journal* 34 (1): 71–81.

Kennedy, W.J., Wayne Patterson, J., and Fredendall, L.D. (2002). An overview of recent literature on spare parts inventories. *International Journal of Production Economics* 76 (2): 201–215.

Kim, T.Y., Dekker, R., and Heij, C. (2017). Spare part demand forecasting for consumer goods using installed base information. *Computers and Industrial Engineering* 103 (1): 201–215.

Lapide, L. (2012). Installed base forecasting. *The Journal of Business Forecasting* 31 (1): 18–21.

Lau, H.C., Song, H.W., See, C.T., and Cheng, S.Y. (2006). Evaluation of time-varying availability in multi-echelon spare parts systems with passivation. *European Journal of Operational Research* 170 (1): 91–105.

Lee, H.L., Padmanabhan, V., and Whang, S. (1997). Information distortion in a supply chain: the bullwhip effect. *Management Science* 43 (4): 545–558.

Louit, D., Pascual, R., Banjevic, D., and Jardine, A.K.S. (2011). Condition-based spares ordering for critical components. *Mechanical Systems and Signal Processing* 25 (5): 1837–1848.

Mak, H.-Y. and Shen, Z.-J.M. (2009). A two-echelon inventory-location problem with service considerations. *Naval Research Logistics* 56 (8): 730–744.

Mani, V. and Sarma, V.V.S. (1984). Queuing network models for aircraft availability and spares management. *IEEE Transactions on Reliability* R-33 (3): 257–262.

Medal, H.R., Pohl, E.A., and Rossettib, M.D. (2014). A multi-objective integrated facility location-hardening model: analyzing the pre- and post-disruption tradeoff. *European Journal of Operational Research* 237 (1): 257–270.

Minner, S. (2011). Forecasting and inventory management for spare parts: an installed base approach. In: *Service Parts Management*, Chapter 8 (ed. N. Altay and L.A. Litteral), 157–169. London: Springer.

Porras, E. and Dekker, R. (2008). An inventory control system for spare parts at a refinery: an empirical comparison of different re-order point methods. *European Journal of Operational Research* 184 (1): 101–132.

Porteus, E. (1985). Numerical comparisons of inventory policies for periodic review systems. *Operations Research* 33 (1): 134–152.

Ravindran, A.R. and Warsing, D.P. Jr., (2013). *Supply Chain Engineering: Models and Applications*, 1e. Boca Raton, FL: CRC Press, Taylor & Francis Group.

Ritchie, E. and Wilcox, P. (1977). Renewal theory forecasting for stock control. *European Journal of Operational Research* 1 (2): 90–93.

Rosling, K. (1989). Optimal inventory policies for assembly systems under random demands. *Operations Research* 37 (4): 565–579.

Scarf, H.E. (1960). The optimality of (S, s) policies in dynamic inventory problems. In: *Mathematical Methods in the Social Sciences* (ed. K.J. Arrow, S. Karlin and P. Suppes), 196–202. Stanford, CA: Stanford University Press.

Selviaridis, K. and Wynstrab, F. (2015). Performance-based contracting: a literature review and future research direction. *International Journal of Production Research* 53 (12): 3505–3540.

Shale, E.A., Boylan, J.E., and Johnston, F.R. (2006). Forecasting for intermittent demand: the estimation of an unbiased average. *Journal of Operational Research Society* 57 (7): 588–592.

Shaunty, J.A., Hare, V.C., Jr., (1960). An airline provisioning problem. *Management Sciences*, vol. MT-1 (2): 66–84.

Sherbrooke, C.C. (1968). Metric: A multi-echelon technique for repairable item control. *Operations Research* 16 (1): 122–141.

Silver, E.A., Pyke, D.F., and Peterson, R. (1998). *Inventory Management and Production Planning and Scheduling*, 3e. New York: Wiley.

Simchi-Levi, D. and Zhao, Y. (2005). Safety stock positioning in supply chains with stochastic lead-times. *Manufacturing and Service Operations Management* 7 (4): 295–318.

Snyder, L.V. and Shen, Z.-J.M. (2011). *Fundamentals of Supply Chain Theory*, 1e. Hoboken, NJ: Wiley.

Song, J.-S. (1994). The effect of lead time uncertainty in a simple stochastic inventory model. *Management Science* 40 (5): 603–613.

Syntetos, A.A. and Boylan, J.E. (2001). On the bias of intermittent demand estimates. *International Journal Production Economics* 71 (1–3): 457–466.

Van der Auweraer, S., Boute, R., Syntetos, A., (2017), Forecasting spare part demand with Installed Base information: a review," FEB Research Report KBI_1721, available at: https://lirias.kuleuven.be/bitstream/123456789/602881/1/KBI_1721.pdf.

Vermorel, J., (2009), "Understanding varying lead time," Published on October 20, 2009, available at: http://blog.lokad.com/journal/2009/10/20/understanding-varying-lead-time.html (accessed on May 24, 2017).

Wieland, A. and Wallenburg, C.M. (2013). The influence of relational competencies on supply chain resilience: a relational view. *International Journal of Physical Distribution and Logistics Management*. 43 (4): 300–320.

Willemain, T.R., Smart, C.N., and Schwarz, H.F. (2004). A new approach to forecasting intermittent demand for service parts inventories. *International Journal of Forecasting* 20 (3): 375–387.

Winston, W.L. (2004). *Operations Research: Applications and Algorithms*, page 856, 4e. Cengage Learning.

Wu, W., Tian, Y., and Jin, T. (2016). A label-based ant colony system for heterogeneous vehicle routing with time window and backhauls. *Applied Soft Computing Journal* 47 (C): 224–234.

Wu, X. and Bian, W., (2015), "Demand analysis and forecast for spare parts of perishable hi-tech products," in *Proceedings of IEEE International Conference on Logistics, Informatics and Service Sciences (LISS)*, pp. 1–6.

Xie, W., Liao, H., and Zhu, X. (2014). Estimation of gross profit for a new durable product considering warranty and post-warranty repairs. *IIE Transactions* 46 (2): 87–106.

Yamashina, H. (1989). The service parts control problem. *Engineering Costs and Production Economics* 16 (3): 195–208.

Yano, C.A. and Lee, H.L. (1995). Lot sizing with random yields: a review. *Operations Research* 43 (2): 311–334.

Zipkin, P.H. (2000). *Foundations of Inventory Management*. New York: McGraw-Hill/Irwin.

Problems

Problem 9.1   Draw the demand profile of spare parts that exhibit one of the following patterns:
1) Constant mean of 100 items/week with standard deviation of 10.
2) Constant mean of 100 items/week. The initial standard deviation in week 1 is 10, yet it increases with 2 items/week.
3) The initial mean demand is 50 items/week and increases with 5 items/week, while the standard deviation is always 10.
4) The initial mean demand is 50 items/week and increases with 5 items/week. The initial standard deviation is 10 and the standard deviation increases with 2 items/week.

Problem 9.2   Simulate the weekly demand based on the four cases in Problem 9.1.

Problem 9.3   Consider an electronics vendor that sells 5000 units of high-speed digital cables per year. Suppose further that the fixed cost of an order is $1000 and the yearly holding cost is $20/unit. The unit price is $400. Backorders are not allowed. Compute the optimal order quality such that the annual inventory cost is minimized?

Problem 9.4   A car dealer sells 500 cars per year of a popular model. Suppose further that the fixed cost of an order is $1000 and the yearly holding cost per car is $500. The average price per car is $20 000. Assuming the inventory does not allow backorders, do the following: (1) compute the optimal order quantity; (2) compute the total annual inventory cost; (3) raw the inventory cycle chart for one year period.

Problem 9.5   Consider an iPad retail outlet that sells 500 units per year of a popular make. Suppose further that the fixed cost of an order is $100 and the yearly holding cost per iPad is $30. The unit cost for each iPad is $300. Assuming the inventory has no backorders, do the following: (1) compute the optimal order quantity; (2) compute the total annual inventory cost; (3) draw the inventory cycle chart for one year period.

Problem 9.6   Each year a university bookstore sells an average of 1000 books of college algebra. The annual demand is normally distributed with a standard deviation of 50 books. The lead time for ordering a batch of books is three weeks and the cost for putting an order is $50. The annual holding cost is $15/book. The cost for stockout is $30/book. Determine: (1) the optimal order quantity; (2) re-order point; (3) the stockout level; (4) draw the distribution for the lead time demand using an Excel file; (5) the fill rate; (6) the expected backorders; (7) the safety stock.

Problem 9.7   Each month a local electronics store sells 200 cameras of a model. The monthly demand is normally distributed with a standard deviation of 10 cameras. The lead time for ordering a batch of cameras is two weeks and the cost for putting an order is $200. The annual holding cost is $20/camera. The purchasing price for each camera is $150. The cost for stockout or shortage is $30/item. Determine: (1) the optimal order quantity; (2) the re-order point; (3) the maximum stockout level; (4) the mean and the standard deviation of the demand during the lead time; (5) the fill rate; (6) the expected backorders; (7) the safety stock.

Problem 9.8   Prove that $q^* = F_D^{-1}((r + d - c)/(r + d - a))$ in Eq. (9.4.8) is the optimal solution for maximizing $E[P(q)]$ in Eq. (9.4.5).

Problem 9.9   It is projected that the demand for a Christmas tree is normally distributed with a mean of 200 and standard deviation of 30. The purchase cost from the supplier is $15/tree and the retail price is $35/tree. There is no salvage value at the end of the season. The shortage cost is $20/tree and the holding cost is $10/tree over the season. Do the following: (1) what is the optimal ordering quantity to maximize the profit; (2) what is the ordering quantity that minimize the cost?

Problem 9.10   Estimate the probability of stockout, the fill rate, and the expected backorders for Problem 9.9.

Problem 9.11   Resolve the two questions in Problem 9.9 if the salvage value is $5/tree at the end of the Christmas season. Besides, what are the probability of stockout, the fill rate, and the expected backorders?

Problem 9.12   For a base-stock inventory system, let X be the lead-time demand and s be the order-up-to level. Show that the expected backorder is

$$E[B] = E[X] - s - \sum_{x=0}^{s}(x - s)P\{D = x\}$$

Problem 9.13   Let L be the lead time with mean $E[L]$ and variance $Var(L)$. Show that the variance of lead-time demand $Var(X)$ is given as

$$Var(X) = (E[L])^2 Var(D) + (E[D])^2 Var(L) + Var(L)Var(D)$$

Problem 9.14   The gearbox is a critical component of the wind turbine (WT) system. A WT repair vendor needs 50 items each year with a standard deviation of 15. The gearbox is capital intensive with $100 000/item. Thus the base stock policy is used to meet the customer needs. The lead time to order a gearbox is $L = 30$ days. The annual holding cost is $15 000/item. The cost for stockout or shortage is $25 000/item. Determine: (1) the mean and the standard deviation of the demand during the lead time; (2) the optimal stock level that minimizes the inventory cost; (3) the risk of stockout; (4) the fill rate; (5) the expected backorders; (6) the safety stock.

Problem 9.15 Assume that the lead time L is uncertain with $E[L] = 30$ days and $Var(L) = 100$ (day)^2, Resolve the questions in Problem 9.14.

Problem 9.16 The demand for a critical part of a missile system can be approximated as a Poisson counting process with rate $\lambda = 0.1$ part/week. The inventory lead time is $L = 25$ days. The holding cost is $h = \$2000$/part/week and the backorder cost is $b = \$8000$/part/week. Since the demand is intermittent and the unit cost is high, the base-stock inventory system is used. Determine the optimal base stock level such that the holding and backorder cost is minimized in a long run. What is the probability of stockout? What is the expected number of backorders?

Problem 9.17 Assume that the inventory of the critical part in Problem 9.15 is managed using the (s, S, T) periodic review policy and the length of the review period $T = 15$ days. Determine the optimal reorder point s and the order-up-to level S such that the holding and backorder cost is minimized in a long run (ordering cost is ignored). What is the probability of stockout? What is the expected number of backorders?

Problem 9.18 Resolve Problem 9.17 assuming a $(S - 1, S, T)$ periodic review policy is adopted. Given $T = 15$ days, what is optimal order-up-to level S such that the holding and backorder cost is minimized in a long run (ordering cost is ignored)? What is the probability of stockout? What is the expected number of backorders?

Problem 9.19 Using your own words to describe what is a supply chain system (20–30 words). Use your own language to state what analytical skills are generally required to solve supply chain design problems (30–40 words).

Problem 9.20 For a distribution network with the following structure, do the following: (1) estimate the mean and the variance of upstream stages; (2) simulate the demand of all stages; (3) if the variance to mean ratio in Stage 1 is limited to 1, how can you achieve it?

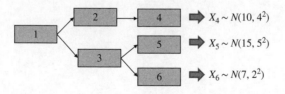

Problem 9.21 For an assembly system with the following structure, do the following: (1) estimate the mean and the variance of upstream stages; (2) simulate the demand of all stages; (3) if the variance to mean ratio in Stage 7 is limited to 1, how can you achieve it?

10

Repairable Inventory System

10.1 Introduction

Repairable inventory theory deals with modeling and design of stocking systems for service or spare parts that can be repaired and reused rather than discarded. It is part of the closed-loop supply chain management that involves product recycling, remanufacturing, and reuse for environmental sustainability. A typical question is to determine the inventory level and/or the stock locations subject to performance criteria or resource constraint. An added complication to the problem is to determine the capacity or location for carrying out the repair job. Performance measures include backorders, inventory cost, repair time, stockout probability, and system availability. Various solution techniques have been developed to solve this type of problems, ranging from deterministic and stochastic programing to Markov chain and queuing networks. In this chapter we introduce basic models for managing repairable inventory systems. Section 10.2 discusses the inventory performance measures at item and system levels. Sections 10.3–10.5 focus on the design and analysis of single echelon systems with uncapacitated and capacitated repair, and infinite and finite fleet size. Section 10.6 investigates a single-echelon system that offers both regular and emergency repairs. Section 10.7 designs multiresolution inventory systems to handle non-stationary demand. Section 10.8 illustrates the principle of the classical multi-echelon model, i.e. the multi-echelon technique for recoverable item control (METRIC) system. The case study in Section 10.9 describes how Teradyne deploys an installed base strategy to manage product reliability and spare parts supply during a new product introduction.

10.2 Characteristics of Repairable Inventory Systems

10.2.1 Spare Parts Supply Mode

The goal of a repairable inventory model is to determine the stock level and location, supply mode, and the repair capacity to meet the uncertain demand in the after-sales market. The history of the spare parts logistics model can date back to the 1960s when Sherbrooke (1968) introduced the METRIC model. METRIC aims to determine the stocking policies at the bases (i.e. customer site) and the depot (i.e. central warehouse) to minimize the backorders at the base subject to cost constraint. The model was derived based on Poisson demand and infinite calling population, and unlimited repair capacity.

Reliability Engineering and Services, First Edition. Tongdan Jin.
© 2019 John Wiley & Sons Ltd. Published 2019 by John Wiley & Sons Ltd.
Companion website: www.wiley.com/go/jin/serviceengineering

Early extensions of the METRIC branched out in three directions. The first direction considers multi-item, multi-indenture (Muckstadt 1973). The second direction derives the inventory policy under the capacitated repair facility (Diaz and Fu 1997). In the third direction, the METRIC has also been extended to incorporating lateral or emergency transshipment (Lee 1987; Alfredsson and Verrijdt 1999). New extensions are continuously reported in the literature, such as accommodating variable usage (Lau et al. 2007), reliability and redundancy (De Smidt-Destombes et al. 2009; Öner et al. 2010), level of repair analysis (Saranga and Kumar 2007), remanufacturing and recycling (Guide et al. 2000), variable installed base (Jin and Liao 2009), and last buy synthesis (Behfard et al. 2015). Several review papers are available on this topic, including the early ones by Guide and Srivastava (1997) and Kennedy et al. (2002), and the recent one by Basten and Van Houtum (2014).

A generic distribution network managing the after-sales spare parts supply is shown in Figure 10.1. Without loss of generality, the distribution network comprises three suppliers, one central stock, and two local stock sites, together serving different customers. Each local stockroom can directly provide spare parts to its adjacent customers within a region or zone. Capital goods like wind turbines and aircraft engines usually consist of a number of critical parts called line replaceable units (LRUs). Hence the local stock actually stores different types of LRU units to meet the demand from the customers.

There are four different service modes for spare parts supply (Axsater 2015): (i) regular supply; (ii) emergency supply; (iii) lateral transshipment; and (iv) direct supply. In regular supply, when a machine fails, the defective part is replaced with a spare item provided by the local stock. If the part is repairable, the faulty item is shipped to the repair center for troubleshooting. When fixed, it is returned to the central warehouse for future use. In the meantime, the local inventory is replenished by placing orders to the central warehouse. In an emergency supply, a spare part is delivered directly from the central warehouse to the customer site, bypassing the local stockroom. This happens when the local stocks are out of spares. Lateral transshipment is adopted when the local stockroom lacks the spare parts, yet the demand is fulfilled by taking a spare item from other local stocks. In the direct supply mode, the customer receives a spare part directly from the repair center, the manufacturing facility, or the subcontractor, without the involvement

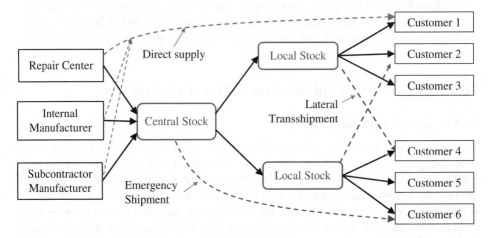

Figure 10.1 A multi-echelon spare parts distribution network.

Table 10.1 Classification of repairable inventory models.

No.	Category	Characteristics	Selected references
1	Supply structure	Single-echelon versus multi-echelon	Sherbrooke (1968), Muckstadt and Thomas (1980)
2	Level of repair	Subassembly versus assembly	Muckstadt (1973), Saranga and Kumar (2007)
3	Service mode	Direct service, emergency repair, transshipment, and local stocking	Jung et al. (2003), Lee (1987), Alfredsson and Verrijdt (1999), Kutanoglu and Mahajan (2009), Patriarca et al. (2016)
4	Demand pattern	Stationary (e.g. Poisson) versus time-varying rate	Jung (1993), Lau and Song (2008), Jin and Liao (2009)
5	Performance measures	Item level versus system level	Kim et al. (2007), Jin et al. (2013), Topan et al. (2017)
6	Installed base	Static fleet versus variable fleet size	Jin and Liao (2009), Jin and Tian (2012), Liu and Tang (2016)
7	System usage	Constant usage versus variable usage	Sherbrooke (1968), Lau et al. (2007), Jin and Wang (2012)
8	Reliability allocation	Constant versus growing	Öner et al. (2010), Louit et al. (2011), Öner et al. (2013), Selçuk and Ağralı (2013), Sleptchenko and Van der Heijden (2017)
9	Redundancy allocation	Series versus parallel system	De Smidt-Destombes et al. (2009), Xie et al. (2014), Van Jaarsveld and Dekker (2011)
10	Maintenance	Corrective versus preventive maintenance	De Smidt-Destombes et al. 2009, Van Horenbeek et al. (2013), Jin et al. (2015).

of central and local stocks. Direct supply is frequently used in the product development time. For instance, when a new product is in the prototyping phase and is assessed by the customers, direct supply is preferred because a failed part can be immediately returned to the manufacturer for root-cause analysis with less time delays. Throughout the chapter, component, part, and item are used interchangeably, representing an LRU.

As shown in Table 10.1, we classify repairable inventory models into 10 categories according to their operational conditions, such as single versus multi-echelon, stationary versus non-stationary demand, and exact versus approximate solutions. Recent literature shows that much attention is paid to incorporate reliability growth and maintenance policy into the repairable inventory planning to achieve the lifecycle cost reduction of the product.

10.2.2 Item-Level Performance Measures

A variety of inventory performance measures have been proposed in the manufacturing and service operations community. We discuss two types of performance measures that are frequently used in repairable inventory management. They are defined either at the component or system level. Performance metrics at the component level measure the service outcome of an individual part type that is independent of other part types. This is a traditional approach to tracking the quality of service from the inventory perspective and is primarily applied to finished goods and spare parts. For slow-moving repairable inventory facing intermittent demands, performance measures at the item level include the following:

1) *Probability of stockout*: This measures the probability that an out-of-stock event will take place during a specific time interval when a customer order arrives to a specific inventory. The generic formula to compute the stockout probability is given in Eq. (9.4.12) in Chapter 9. While the stockout probability is very easy to compute and to understand, it does not reflect the size of the stockout or the number of demand shortages for a given time period.

2) *Fill rate*: This measures the expected percentage of demand that can be fulfilled directly from the on-hand inventory during a specific interval (e.g. replenishment cycle). The formula to compute the fill rate is given in Eq. (9.4.14). The fill rate only considers whether a part is immediately available when the demand arrives; however, it does not indicate how long the customer needs to wait in case of a backlog.

3) *Expected backorders*: This metric measures the expected number of parts that customers are waiting to receive and also answers when the shortage will be fulfilled. From the customer perspective, the delay in receiving a needed part implies a prolonged machine downtime. Depending on whether the demand is continuous or discrete, the expected backorders can be estimated using Eqs. (9.4.15) and (9.4.16), respectively. It is worth mentioning that no backorders occur if unfilled demands are taken over by emergency repair or lateral transshipment.

4) *Part response time (PRT)*: This is the average duration from when the order is placed to when the customer receives the physical unit. If an on-hand part is available in the inventory, the PRT is primarily the transportation time to the customer. If the inventory is out of stock, the demand is either backlogged or transferred to a central warehouse or other locations. Then the PRT consists of the transportation time and any delays caused by waiting or repair.

These item-level performance measures are mutually correlated. For instance, a higher stockout probability often leads to a lower fill rate and results in a longer PRT. Caglar et al. (2004) solve a two-echelon spare parts distribution system where the base stock level in the central warehouse and local inventory are optimized subject to a response time constraint. Recently, Van Jaarsveld et al. (2015) incorporated system operator's requirements as a time-window fill rate criterion for optimizing the repair turn-around times of aircraft components. In fact, the outcomes of these performance measures depend on the inventory level, the replenishment policy, the nature of the demand process, the repair capacity, and the transportation mode.

10.2.3 System-Level Performance Measures

Since the main purpose of the after-sales services is to keep the system operational, the performance measures defined at the system level become more relevant to the customer satisfaction. Below we introduce four types of system-level performance measures.

1) *Mean system uptime.* This quantity measures the cumulative duration of a system that is in good state prior to unexpected failure or scheduled maintenance, whichever comes first. System uptime consists of production time and idle time. The latter indicates that the system is in a standby mode, but ready for production. For a series system, the mean uptime is determined by the reliability of individual components. For a parallel system, the uptime is prolonged because redundant units can take over the load upon the failure of primary units. Hence a low part availability has less impact on the uptime of a parallel system.

2) *Mean system downtime.* This metric measures the expected time involved in repairing and restoring a failed system or measures the duration of performing a maintenance task. For a multi-item system, the downtime is determined by the availability of the spare parts, the PRT, and the repair and maintenance capacity. An unexpected failure often results in a prolonged system downtime because of waiting for the arrival of spare parts and assembling the repair team. This is in contrast to the preventive or scheduled maintenance where both the materials and repair crews are proactively planned.

3) *System availability.* This represents the steady-state probability that the equipment is ready for use during a specific time period. It is defined as the ratio of the system uptime divided by the sum of uptime and downtime. For a multi-item system, its availability is jointly determined by component reliability, redundancy level, part availability, parts response time, maintenance policy, and equipment usage. System availability is perhaps the most widely used metric in private and public sectors because it directly reflects the customer goal of generating value through the use of the equipment.

4) *Fleet readiness.* In mission-critical applications such as airforce and global positioning services, an operation is often carried out by deploying a fleet of systems (e.g. aircraft or satellites), instead of a single system. The success of the mission requires the availability of a number of systems, instead of a single system. Fleet readiness measures the availability of a group of systems as if one single entity. Given a group of n systems, fleet readiness measures the availability of a minimum number of systems, denoted as k, that are available at a random point in time. If all systems have an identical availability, the fleet readiness, denoted as A_{red}, can be estimated by

$$A_{red} = \sum_{i=k}^{n} \binom{n}{i} A_s^i (1 - A_s)^{n-i} \qquad (10.2.1)$$

where A_s is the system availability. To some extent, Eq. (10.2.1) resembles the reliability estimate of k-out-of-n:G system that is operational provided that at least k components are good.

10.2.4 Item Approach Versus System Approach

The item approach can create as much pooling effect as possible by leveraging lateral transshipment, emergency repair, and direct supply modes. Tight parts availability and short downtime constraints require that, in addition to the central warehouse, spare items need to be stocked near the fleet under support. Since the demands to the local stock are intermittent, the use of lateral transshipment and emergency repair can mitigate high spares stocking in the local warehouse. Thus the entire service network behaves virtually as a single inventory aggregated with multiple stock locations. In addition, a pooling effect can also be created between spare part stocks for different machines that use the same components. For example, if machines A and B use the same LRU, the company only needs to create one local stock to support both product fleets. The same idea is also applicable to different customers requiring different service level commitments. To achieve an effective pooling effect, sophisticated pricing strategy and inventory control policy are of importance because the trade-off between the inventory and logistics efficiency must be appropriately addressed.

The system-based approach is also referred to as a multi-item approach, under which the inventory levels of different parts are jointly optimized to meet the service level criteria. Performance measures at item level may not be so relevant to a user of a complex system comprised of multiple parts. Consider a solar photovoltaic (PV) system having an extremely high service level on solar panels. This might not help all that much if failures of the DC–AC inverter cause most of the power outages. To achieve the same system availability or mean downtime goal, the system approach tends to keep less expensive parts while stocking cheaper items. In other words, the system approach may generate the same service performance outcome with less overall inventory investment. For a comprehensive discussion on the relation between system availability and spares inventory, readers are referred to the book by Houtum and Kranenburg (2015).

10.3 Single-Echelon Inventory with Uncapacitated Repair

10.3.1 Operational Structure

Single-echelon repairable inventory is referred to as a logistics supply chain where only one stock location is available to supply the spare parts to all customers. Figure 10.2 depicts a single-echelon repairable inventory system that supports the operation of a group of machines in the field. In this setting, the original equipment manufacturer (OEM) is responsible for producing the machines as well as providing after-sales support, which includes repair and spares supply. Figure 10.2 is also applicable to the situation where the third party logistics (3PL) supplier takes the responsibility for repair and parts supply. In that case, the role of the OEM is mainly to produce new products for the customers. To facilitate the repair-by-replacement tasks, the spares inventory is often placed in a location that is close to all the customers.

The operation of the inventory system is described as follows. Initially there are s spare parts in the inventory and a one-for-one replenishment policy is adopted. This

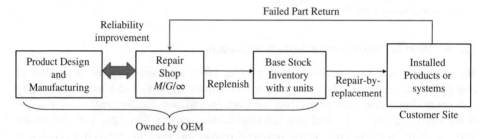

Figure 10.2 A repairable inventory system with ample capacity.

replenishment policy is quite common in capital-intensive spare parts provisioning (Graves 1985). When a demand arrives to the inventory, a good item is shipped to the customer. Meanwhile a replenishment order is placed to the repair shop. Backorders may occur if the stock is out of spares at the time of request. The failed machine is sent to production once the faulty item is replaced with a good part. Since parts are repairable, the faulty item is routed to the repair shop for trouble-shooting and renewal. Assuming that the repair shop has ample capacity, parts repair times are independently and identically distributed with a mean time of $1/\mu$, where μ is the repair rate of a single item. Although this assumption is rather ideal, Sherbrooke (1992) showed that such an approximation is reasonable in circumstances when the demands for repair are slow and intermittent. Let λ be the demand rate of a particular part type. Let $j = 0, 1, 2, \ldots$ be the number of failed parts in the repair shop. Then the repair shop can be modeled as an $M/G/\infty$ queue, and the corresponding transition diagram is shown in Figure 10.3.

The inventory replenishment lead time, also known as the repair turn-around time (TAT), is defined as the time duration from when the part is removed from the down system to when it is repaired and put back in the stockroom. Based on the routing path in Figure 10.4, the TAT consists of the forward-and-backward transportation times and the actual hands-on time (i.e. touch labor) in the repair shop. Since the repair shop is uncapacitated, the faulty part once entering the shop will be repaired immediately without delay.

The repair TAT is indeed a random variable because it is influenced by several uncertain factors, including the severity of failure, the technician's skill level, diagnostic tools, and transportation mode. Let t_f and t_b, respectively, be the average forward and backward transportation times. Let t_r be the average hands-on repair time in the shop. Then the expected TAT is

$$\text{TAT} = t_f + t_b + t_r \tag{10.3.1}$$

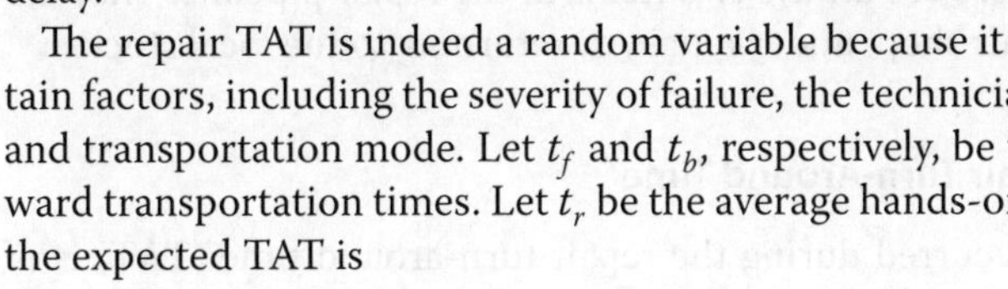

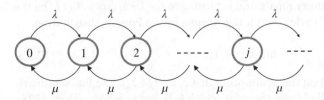

Figure 10.3 *M/G/∞* queue for a repair shop.

Repair Turn-Around Time (TAT)		
To Repair Shop	In Repair Shop	To Stockroom
t_f	t_r	t_b
Forward Transportation	Hands-on Repair in Testbed	Backward Transportation

Figure 10.4 The breakdown of repair turn-around time.

In most applications, if the conditions $t_f \ll t_r$ and $t_b \ll t_r$ hold, then TAT $\cong t_r$. Now the repair rate in the shop can be estimated as

$$\mu = \frac{1}{t_r} \tag{10.3.2}$$

Example 10.1   Assume that the demand rate of a spare part is $\lambda = 1$ failures/month and the average TAT is $t_r \cong 60$ days where both t_f and $t_b \ll t_r$. Estimate: (i) What is the expected number of failures in one year? (ii) What is the average number of parts under repair in the pipeline?

Solution:

(1) Since each year has $m = 12$ months, the expected number of failures in a year is simply $m\lambda = 12 \times 1 = 12$ parts/year.
(2) Let L be the expected number of parts in the repair pipeline that include the units in transit and in the repair shop. According to Little's law (see Section 1.8 of Chapter 1), we have

$$L = \lambda t_r = 1 \times (60/30) = 2 \ \text{(items)}$$

This implies on average there are always two items in the repair pipeline. They are either in transit to the repair shop, under repair, or returning to the local stock.

10.3.2   Demand During Repair Turn-Around Time

The number of failures that occurred during the repair turn-around time (TAT) is of particular importance in determining the service level of the repairable inventory. This is because given a fixed amount of stock, the stockout probability largely depends on the number of demand during TAT. Figure 10.4 shows that the repair pipeline consists of defective parts in transit, under repair, and returning to the local stock. Let D be the number of demands in a TAT cycle. Then it is governed by the Poisson distribution:

$$P\{D = k\} = \frac{(\lambda t_r)^k e^{-\lambda t_r}}{k!}, \qquad \text{for } k = 0, 1, 2 \ldots \tag{10.3.3}$$

Equation (10.3.3) is derived on the assumption that $t_r \gg (t_f + t_b)$. The Poisson distribution results from the so-called Palm theorem, which is stated as follows (Palm 1938):

Palm's Theorem. If demands arrive according to a Poisson process with rate λ and if the times that the faulty parts remain in the repair pipeline are independent and identically distributed according to a given distribution with mean t_r, then the steady-state distribution for the number of parts in the repair pipeline is Poisson with mean λt_r. Since there is no waiting time in the $M/G/\infty$ queue, the times that faulty parts remain in the repair shop become independent and identically distributed.

It is worth mentioning that Eq. (10.3.3) is derived based on the demand call from an infinite system population. If the fleet has a limited size with n systems, we can approximate λ with $\lambda \cong n\lambda_s$ where λ_s is the failure intensity rate of each system. Such an approximation is valid provided that the fraction of down machines at a random point in time is relatively small compared with n. This happens when the machine downtime is short and negligible compared to the uptime or when failures rarely occur during a

TAT cycle. Since $n\lambda_s$ is the upper bound of the demand rate (i.e. $n\lambda_s \geq \lambda$), the inventory policy based on $n\lambda_s$ often yields a better service performance than the one developed based on the actual fleet demand rate.

Example 10.2 There are $n = 70$ machines operating in the field and the failure rate of a critical part in the machine is $\lambda_s = 0.25$ failures/year/machine. The mean repair TAT is 30 days. (1) What is the probability that three machines will fail over $t_r = 30$ days? (2) Calculate the distribution for $k = 0, 1, 2, \ldots$.

Solution:

(1) Since the failure intensity of each machine is $\lambda_s = 0.25$ failures/year/machine, the aggregate spare parts demand rate of the fleet is

$$\lambda = n\lambda_s = 70 \times 0.25 = 17.5 \text{ failures/year} \tag{10.3.4}$$

Now substituting Eq. (10.3.4) into Eq. (10.3.3), we have

$$P\{D = 3\} = \frac{(17.5 \times (30/365)^3 e^{-17.5\times(30/365)}}{3!} = 0.118 \tag{10.3.5}$$

(2) The distribution from $k = 0$ to 7 is computed and the results are given in Table 10.2. It is worth mentioning that Eq. (10.3.5) is valid only if all repairable parts circulate in a closed-loop supply chain with no parts being added to or taken out of the pipeline. In reality, OEM may impose certain restrictions on whether a returned part should be repaired or scrapped. One such restriction in the electronics industry mandates the maximum number of repairs per returned part. Another common restriction is called physical restrictions, stating that the part should be scrapped rather repaired if it has been severely damaged or the version is out of date. If some retuned parts are scrapped, new parts need to be added in the stockroom to balance the shortage. Then TAT is a weighted average of the repair turn-around time for those being repaired and the times of procuring new parts.

In a maintenance program, we are more interested in how many spare parts are needed to avoid the inventory shortage during t_r. Let s be the base stock level. The probability that no more than s failures will occur in a TAT cycle can assist the planner in setting the appropriate base stock level, and its formula is

$$P\{D \leq s\} = \sum_{k=0}^{s} \frac{(\lambda t_r)^k e^{-\lambda t_r}}{k!}, \quad \text{for } s = 0, 1, 2, \ldots \tag{10.3.6}$$

If one wants to estimate the stockout probability, i.e. $P\{D > s\}$, it is simply the commentary of Eq. (10.3.6). That is, $P\{D > s\} = 1 - P\{D \leq s\}$.

Table 10.2 Probability mass function and cumulative distribution.

k	0	1	2	3	4	5	6	7
$P\{N = k\}$	0.237	0.341	0.245	0.118	0.042	0.012	0.003	0.001
$P\{N \geq k\}$	0.237	0.579	0.824	0.942	0.984	0.996	0.999	~1.000

Example 10.3 Using the same data in Example 10.2, the probability that no more than two failures will occur over $t_r = 30$ days (or 0.0822 year) is given as

$$P\{D \le 2\} = e^{-17.5 \times 0.0822}$$

$$\left[\frac{(17.5 \times 0.0822)^0}{0!} + \frac{(17.5 \times 0.0822)^1}{1!} + \frac{(17.5 \times 0.0822)^2}{2!}\right] = 0.824$$

This means that the probability of meeting the demand reaches 82.4% if the local stock has two items.

10.3.3 Performance Characterization

Two scenarios need to be considered when systems fail and a request is made for a spare part. If an on-hand spare part is available in the inventory, the system can be restored quickly by replacing the faulty item with a good item. If the inventory is empty, the replacement job is postponed until the arrival of a spare part. Mean-time-to-replacement (MTTR) is defined as the duration from when the system failed to when it is restored to operation. For a series system, MTTR is highly correlated with the availability of spare parts. The availability of spare parts is further correlated with another two random variables, namely, on-hand inventory Q and backorders B. Assuming the base stock level is set with s units, then Q and s are related through

$$Q = \max\{0, s - O\} = (0, s - O)^+ \tag{10.3.7}$$

where O is a random variable representing the steady-state inventory on-order. For instance, if $s = 5$ and $O = 4$, then $Q = \max\{0, 5 - 4\} = 1$. As another example, if $s = 5$ and $O = 7$, then $Q = \max\{0, 5 - 7\} = 0$. Therefore, Q is a non-negative integer varying between 0 and s.

For a single-echelon inventory system with an ample repair capacity, according to Palm's theorem, O can be modeled as a Poisson distribution with a mean value of λt_r. By combining with Eq. (10.3.3), the expected value of Q is now obtained by

$$E[Q] = \sum_{k=0}^{\infty} (0, s - k)^+ P\{D = k\} = \sum_{k=0}^{s} (s - k)P\{D = k\} = \sum_{k=0}^{s} \left((s - k)\frac{(\lambda t_r)^k e^{-\lambda t_r}}{k!}\right) \tag{10.3.8}$$

where $P\{D = k\}$ is the probability that the demand quantity is k during t_r. Similarly, the backorders B and the base stock s are also co-related through

$$B = \max\{0, O - s\} = (0, O - s)^+ \tag{10.3.9}$$

For instance, if $s = 5$ and $O = 4$, then $B = \max\{0, 4 - 5\} = 0$. As another example, if $s = 5$ and $O = 7$, then $B = \max\{0, 7 - 5\} = 2$. Obviously B is a non-negative integer taking any value between 0 and ∞. The expected backorder is given as

$$E[B] = \sum_{k=0}^{\infty} ((k - s)^+ P\{D = k\})$$

$$= \sum_{k=s+1}^{\infty} ((k - s)P\{D = k\})$$

$$= \lambda t_r - s - \sum_{k=0}^{s} \left((k - s)\frac{(\lambda t_r)^k e^{-\lambda t_r}}{k!}\right) \tag{10.3.10}$$

After substituting Eq. (10.3.8) into Eq. (10.3.10), we can connect $E[B]$ with $E[Q]$ as follows:

$$E[B] = \lambda t_r - s + E[Q] \tag{10.3.11}$$

Since the expected demand during a TAT cycle is $E[D] = \lambda t_r$, the inventory fill rate is

$$\text{Fill rate} = 1 - \frac{E[B]}{E[D]} = \frac{s - E[Q]}{\lambda t_r} \tag{10.3.12}$$

The PRT stands for the part response time and measures the time elapsed from when the order is placed by the customer to when the part is received. If an on-hand spare part is available in the stockroom, the PRT is almost zero (assuming the distance between the local stock and customer is small). If the stock is empty, it takes on average a period of t_r to receive a spare part. By considering both scenarios, the average PRT is

$$PRT = t_r P\{D > s\} = t_r \sum_{k=0}^{s} \frac{(\lambda t_r)^k e^{-\lambda t_r}}{k!} \tag{10.3.13}$$

10.3.4   Variance of Backorders

Though Eq. (10.3.10) allows us to calculate the average part shortage during TAT, it does not capture the variation of the shortage. Understanding the backorder variation is important in terms of setting up the safety inventory level against the volatile demand behavior in the long run. Given the base stock level s, the second moment of the backorders, denoted as $E[B^2]$, is obtained by

$$E[B^2] = \sum_{k=s+1}^{\infty} \left((k-s)^2 P\{D=k\}\right)$$

$$= \sum_{k=0}^{\infty} (k-s)^2 P_k - \sum_{k=0}^{s} (k-s)^2 P_k$$

$$= \sum_{k=0}^{\infty} (k^2 - 2ks + s^2) P_k - \sum_{k=0}^{s} (k^2 - 2ks + s^2) P_k$$

$$= \lambda t_r + (\lambda t_r)^2 - 2s\lambda t_r + s^2 - \sum_{k=0}^{s} k^2 P_k + 2s \sum_{k=0}^{s} k P_k - 2s^2 \sum_{k=0}^{s} P_k \tag{10.3.14}$$

To simplify the notation, we denote $P_k = P\{D = k\}$ in the above equation. By combining Eqs. (10.3.10) with (10.3.14), the variance of B is obtained as follows:

$$Var(B) = E[B^2] - (E[B])^2$$

$$= \lambda t_r + (\lambda t_r)^2 - 2s\lambda t_r + s^2 - \sum_{k=0}^{s} k^2 P_k + 2s \sum_{k=0}^{s} k P_k$$

$$- 2s^2 \sum_{k=0}^{s} P_k - (\lambda t_r - s + E[Q])^2$$

$$= \lambda t_r + \sum_{k=0}^{s} (s^2 - k^2) P_k - (E[Q])^2 - 2\lambda t_r E[Q] \tag{10.3.15}$$

where $E[Q]$ is the expected on-hand inventory given in Eq. (10.3.8).

Example 10.4 A single-echelon repairable inventory is managed under one-for-one replenishment. The part repair lead time $t_r = 15$ days and the demand rate is 0.1 parts/day. The initial base stock level $s = 2$. Assuming that the demand is a Poisson process, estimate: (1) the probability of stockout; (2) the expected on-hand spare parts; (3) the expected backorders; and (4) the variance of the backorders.

Solution:

1) The stockout occurs when the repair pipeline holds more than three parts, that is,

$$P\{D > s\} = 1 - P\{D \le s\} = 1 - \sum_{k=0}^{2} \frac{(0.1 \times 15)^k e^{-0.1 \times 15}}{k!} = 0.19$$

2) The expected on-hand inventory can be directly computed using Eq. (10.3.8) as follows:

$$E[Q] = \sum_{k=0}^{2} \left((2-k) \frac{(0.1 \times 15)^k e^{-0.1 \times 15}}{k!} \right) = 0.78 \ \text{(item)}$$

3) The expected backorders can be directly computed using Eq. (10.3.10) as follows:

$$E[B] = 0.1 \times 15 - 2 + 0.78 = 0.28 \ \text{(item)}$$

4) Finally, the variance of the backorders is obtained using Eq. (10.3.15), and the result is

$$Var(B) = 0.1 \times 15 + \sum_{k=0}^{2} (2^2 - k^2) \frac{(0.1 \times 15)^k e^{-0.1 \times 15}}{k!} - (0.78)^2 - (2)(0.1)(15)$$
$$\times 0.78 = 0.44 \ \text{(item)}^2$$

10.4 Single-Echelon Inventory with Capacitated Repair

10.4.1 Operational Structure

The actual capacity of a repair shop is hardly ample, but rather is often limited by the availability of test beds, the number of trained technicians, and the facility space. Therefore, it is more realistic to treat the repair shop as a capacitated service entity with delay and waiting. Figure 10.5 shows a single echelon inventory system with a limited repair capacity. Compared with the inventory system in Figure 10.2, the main difference is the queuing model in the repair shop. We use an $M/M/m$ queue to model the capacitated repair shop where m is the number of available repair servers. This queueing network is also known as the Erlang-C model in which any incoming parts enter the waiting line if all repair servers are busy.

Figure 10.6 depicts the transition diagram of an $M/M/m$ queue. The following assumptions are made. First, parts fail independently and defective items arrive at the repair shop with Poisson rate λ. Second, the hands-on repair time of a faulty part is exponentially distributed with mean time $1/\mu$. Third, the repair shop has m servers. A faulty part receives the repair service immediately if an idle server is available. If all m servers are busy, the part needs to wait for the next available server based on the first-come, first-served protocol.

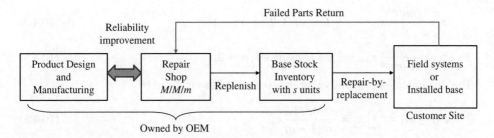

Figure 10.5 A repairable inventory system with limited capacity.

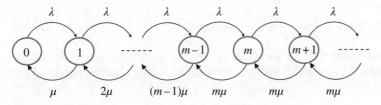

Figure 10.6 *M/M/m* queue for limited repair capacity.

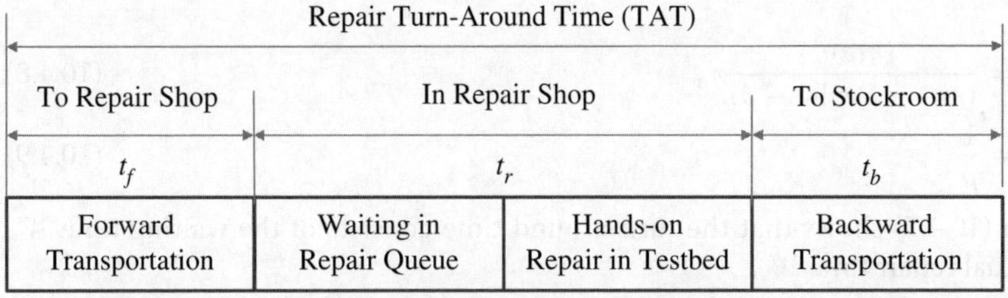

Repair Turn-Around Time (TAT)			
To Repair Shop	In Repair Shop		To Stockroom
t_f	t_r		t_b
Forward Transportation	Waiting in Repair Queue	Hands-on Repair in Testbed	Backward Transportation

Figure 10.7 The breakdown of TAT in a capacitated repair shop.

In a capacitated repair shop, the repair TAT now consists of the forward and backward transportation times, the waiting time in the repair shop, and the actual hands-on repair time (i.e. touch labor). The breakdown of the TAT in a capacitated repair shop is shown in Figure 10.7.

10.4.2 Performance Characterization

Let π_j (for $j = 0, 1, 2, \ldots$) be the probability of having j items in the repair shop including those in transit. We define

$$\rho = \lambda/m\mu \tag{10.4.1}$$

The system is stable as long as $\rho < 1$. Then the steady-state probabilities are given as (Winston 2004)

$$\pi_0 = \frac{1}{\displaystyle\sum_{i=0}^{m-1} \frac{(m\rho)^i}{i!} + \frac{(m\rho)^m}{m!(1-\rho)}} \tag{10.4.2}$$

$$\pi_j = \frac{(m\rho)^m}{j!}\pi_0, \qquad \text{for } j = 1, 2, \ldots, m-1 \tag{10.4.3}$$

$$\pi_j = \frac{(m\rho)^m}{m!m^{j-m}}\pi_0, \qquad j = m, m+1, m+2, \ldots \tag{10.4.4}$$

The steady-state probability that m servers are busy is given by

$$P\{j \geq m\} = \sum_{j=m}^{\infty} \pi_j = \frac{(m\rho)^m}{m!(1-\rho)}\pi_0 \tag{10.4.5}$$

The expected number of parts in the repair shop, denoted as L, consist of the items in waiting and the items under servicing. It can be shown that

$$L = \frac{\rho}{1-\rho}P\{j \geq m\} + \frac{\lambda}{\mu} = \frac{\rho(m\rho)^m}{m!(1-\rho)^2}\pi_0 + m\rho \tag{10.4.6}$$

Let t_r be the total repair time in the shop, W_q be the waiting time, and W_s be the actual hands-on time. Given L in Eq. (10.4.6), according to Little's law the repair TAT is obtained by

$$t_r = \frac{L}{\lambda} = W_q + W_s = \frac{(m\rho)^m}{(m\mu - n\lambda)m!(1-\rho)}\pi_0 + \frac{1}{\mu} \tag{10.4.7}$$

where

$$W_q = \frac{(m\rho)^m}{(m\mu - \lambda)m!(1-\rho)}\pi_0 \tag{10.4.8}$$

$$W_s = \frac{1}{\mu} \tag{10.4.9}$$

Equation (10.4.7) shows that the turn-around time consists of the waiting time W_q and the actual repair time W_s.

10.4.3 Expected Backorders and On-Hand Inventory

The probability is that the local stock is running out of spares only if the demand during a repair TAT cycle exceeds the base stock level s. Since the spares demand D during TAT follows the Poisson distribution, the stockout probability is

$$P\{D > s\} = \sum_{k=s+1}^{s} \frac{(\lambda t_r)^k e^{-\lambda t_r}}{k!} \tag{10.4.10}$$

Here Eq. (10.4.10) is derived assuming $t_r \gg (t_f + t_b)$. For a capacitated repairable inventory system, let L be the number of parts in the repair pipeline (including those in transit) and B is the backorders to the customers. According to Eq. (9.6.1) in Chapter 9, the following condition holds:

$$s = E[D] - E[B] + E[L] \tag{10.4.11}$$

where D is the demand in a TAT cycle and L is the number of parts in the repair pipeline. For a Poisson process with demand rate λ, the mean and variance of the demand during TAT are given by

$$E[D] = Var(D) = \lambda t_r \tag{10.4.12}$$

Now substituting Eq. (10.4.12) into Eq. (10.4.11) and rearranging the equation, we have

$$E[B] = E[D] + E[L] - s$$
$$= \lambda t_r + E[L] - s$$
$$= 2E[L] - s \tag{10.4.13}$$

This implies that in a steady-state condition, backorders will occur if s is less than twice that of $E[L]$.

10.5 Repairable Inventory for a Finite Fleet Size

10.5.1 The Machine-Inventory Model

The inventory models discussed thus far assume that the spares demand is generated from an infinite calling pool. This assumption is valid if one of the following conditions holds: (i) the installed base or the fleet size is quite large or (ii) the mean-time-to-repair is small and can be ignored compared to the uptime. In many applications, the fleet size under the support is not large enough to justify the constant demand rate. For instance, a customer owns a limited number of machines K on the production floor. If all K machines are working, the demand rate of spare parts is $K\lambda$, where λ is the machine failure rate. If k (for $1 < k \le K$) machines failed at the same time, the spare parts demand rate drops to $(K - k)\lambda$. Therefore, the repairable inventory models built upon an infinite calling pool may not be appropriate to characterize the inventory performance under a limited fleet size.

In this section, we address the spare parts provisioning issue under a limited number of machines. Figure 10.8 depicts a single-echelon machine-inventory system to support the operation of K machines. The following assumptions are made:

1) The initial base stock level is s for $s \le K$. When machines fail, a spare part is used to replace the faulty item in the machine. If the inventory is empty, the machine remains in the down state until a spare part is provided by the repair shop.
2) Each machine fails exponentially with constant rate λ. The replenishment lead time of the inventory is also exponentially distributed with a mean duration of $1/\mu$.
3) Given an on-hand spare part, the repair-by-replacement time is much shorter than the inventory replenishment lead time.

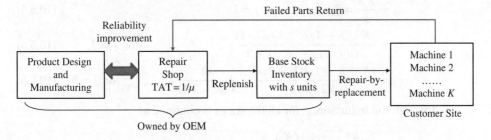

Figure 10.8 The machine-inventory model with a limited fleet size.

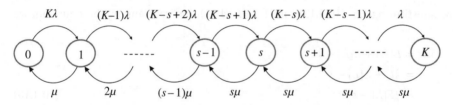

Figure 10.9 The transition diagram of the machine-inventory problem.

Let $j = 0, 1, \ldots, K$ be the state representing the number of spare parts shortages. This also corresponds to the number of down machines at the customer site. For instance, if $j = 0$, there is no part shortage; hence all K machines are in an operational state. For $j = s$, this implies that the inventory is out of stock, and only $K - s$ machines are in an operational state. The transition diagram of the machine-inventory problem is shown in Figure 10.9 below.

10.5.2 Performance Characterization

To characterize the service performance of the machine-inventory model in Figure 10.9, we compute the steady-state probability π_j for $j = 0, 1, 2, \ldots, K$. The generic formulas to calculate π_j are given as follows (Winston 2004):

$$\pi_0 = \frac{1}{1 + \sum_{j=1}^{k} c_j}, \qquad \text{for } j = 0 \tag{10.5.1}$$

$$\pi_j = c_j \pi_0, \qquad \text{for } j = 1, 2, 3, \ldots k \tag{10.5.2}$$

with

$$c_j = \frac{\lambda_0 \lambda_1 \cdots \lambda_{j-1}}{\mu_1 \mu_2 \cdots \mu_j}, \quad \text{for } j = 1, 2, 3, \ldots k \tag{10.5.3}$$

where λ_{j-1} is the transition rate from states $j - 1$ to j, and μ_j is the transition rate from states j to $j - 1$. Based on Eqs. (10.5.1) to (10.5.3), the performance outcome of the machine-inventory model can be estimated, with the detailed steps summarized below:

Step 1: Define $\rho = \lambda/\mu$. Without loss of generality, we examine the first three states for $j = 1, 2,$ and 3. That is,

$$c_1 = \frac{K\lambda}{\mu} = K\rho, \qquad \text{for } j = 1 \tag{10.5.4}$$

$$c_2 = \frac{(K\lambda)(K-1)\lambda}{(\mu)(2\mu)} = \frac{K(K-1)}{1 \times 2}\rho^2, \qquad \text{for } j = 2 \tag{10.5.5}$$

$$c_3 = \frac{(K\lambda)(K-1)\lambda(K-2)\lambda}{(\mu)(2\mu)(3\mu)} = \frac{K!}{3!(K-3)!1}\rho^3, \qquad \text{for } j = 3 \tag{10.5.6}$$

Through the induction, c_j for all the states is obtained as

$$c_j = \frac{K!}{j!(K-j)!}\rho^j = \binom{K}{j}\rho^j, \qquad \text{for } j = 1, 2, \ldots, s \tag{10.5.7}$$

$$c_j = \frac{K!}{j!(K-j)!} \rho^j \frac{j!}{s!s^{j-s}} = \binom{K}{j} \rho^j \frac{j!}{s!s^{j-s}}, \qquad \text{for } j = s+1, \; s+2, \; \ldots, \; K$$

$$(10.5.8)$$

Substituting Eqs. (10.5.7) and (10.5.8) into Eqs. (10.5.1) and (10.5.2), we can obtain the steady-state probability π_j for $j = 0, 1, 2, \ldots, K$.

Step 2: Let L be the expected number of down machines, L_q be the number of machines waiting for repair, and L_s be the number of machines currently under repair. Then

$$L = \sum_{j=0}^{K} j\pi_j \tag{10.5.9}$$

$$L_q = \sum_{j=s}^{K} (j-s)\pi_j \tag{10.5.10}$$

$$L_s = L - L_q \tag{10.5.11}$$

Step 3: Let W be the total expected downtime of the machine, L_q be the machine's waiting time for repair, and L_s be the actual servicing time. Using Little's law, $L = \bar{\lambda} W$, we have

$$W = \frac{L}{\bar{\lambda}} = \frac{L}{\lambda(K-L)} \tag{10.5.12}$$

$$W_q = \frac{L_q}{\bar{\lambda}} = \frac{L_q}{\lambda(K-L)} \tag{10.5.13}$$

$$W_s = W - W_q \tag{10.5.14}$$

where

$$\bar{\lambda} = \sum_{j=0}^{K} \lambda_j \pi_j = \sum_{j=0}^{K} (K-j)\lambda \pi_j = \lambda(K-L) \tag{10.5.15}$$

Example 10.5 Assume a factory has $K = 5$ machines and $s = 2$ spare parts are allocated on-site for failure replacement. The mean life of this critical part is 30 days and the TAT of fixing a faulty part is two days. Calculate π_j, L, L_q, L_s, W, W_q, and W_s.

Solution:
We have $\lambda = 1/30$ failures/day and $\mu = 1/2$ repair/day. Hence $\rho = \lambda/\mu = 0.0777$. Since $s = 2$, c_1 and c_2 are estimated based on Eq. (10.5.7) and c_3, c_4, and c_5 are estimated by using Eq. (10.5.8). The results are given as follows:
For $j = 1$ and 2: $c_1 = 0.333$, $c_2 = 0.044$
For $j = 3$, 4, and 5: $c_2 = 0.0044$, $c_4 = 0.0003$, $c_5 \cong 0$
We then calculate π_j based on Eqs. (10.5.1) and (10.5.2). That is,

$$\pi_0 = \frac{1}{1 + \sum_{j=1}^{K} c_j} = \frac{1}{1 + 0.333 + 0.044 + 0.0044 + 0.0003 + 0} = 0.7233$$

where $\pi_1 = c_1\pi_0 = 0.241$, $\pi_2 = c_2\pi_0 = 0.032$, $\pi_3 = c_3\pi_0 = 0.003$, $\pi_4 = c_4\pi_0 = 0.0002$, and $\pi_5 = c_5\pi_0 \cong 0$.

Using Eq. (10.5.9), the average number of down machines is estimated by

$$L = \sum_{j=0}^{5} j\pi_j = 0 \times 0.7233 + 1 \times 0.241 + 2 \times 0.032 + 3 \times 0.003 + 4 \times 0.0002$$

$$+ 5 \times 0 = 0.317$$

The average number of machines that are waiting for repair is

$$L_q = \sum_{j=3}^{5} (j - s)\pi_j = (3 - 2)\pi_3 + (4 - 2)\pi_4 + (5 - 2)\pi_5 = 0.0037$$

Hence the number of machines that are under servicing is

$$L_s = L - L_q = 0.317 - 0.0037 = 0.312$$

The total downtime W, the waiting time W_q, and actual servicing time W_s are given as follows:

$$\overline{\lambda} = \lambda(K - L) = \frac{1}{30}(5 - 0.316) = 0.157$$

$$W = \frac{L}{\overline{\lambda}} = \frac{0.316}{0.156} = 2.02 \text{ (day)}$$

$$W_q = \frac{L_q}{\overline{\lambda}} = \frac{0.0036}{0.156} = 0.023 \text{ (day)}$$

$$W_s = W - W_q = 2.02 - 0.023 = 2.177 \text{ (day)}$$

10.6 Single-Echelon Inventory with Emergency Repair

10.6.1 Prioritizing Repair and Supply Services

The repairable inventory models studied thus far accommodate backorders. Namely, if the inventory is out of spares, the demand is backlogged and unfulfilled demand is met once the inventory is replenished. There are two implications associated with backorders. First, backorders may damage customer satisfaction because extended machine downtime results in a higher production loss. Second, customers may seek an alternative supplier due to a prolonged PRT. Emergency repair can mitigate the negative effects of backorders. Under the emergency repair, the faulty item is expedited to the repair shop where it is fixed immediately. When fixed, the same item is returned and installed in the down machine. A single-echelon inventory system offering both regular and emergency repair services is depicted in Figure 10.10.

Assume that the base stock level is s and that it operates under a one-for-one replenishment policy. The aggregate demand rate to the stock is λ and may come from one customer or from multiple customers. In the latter, $\lambda = \sum_i^k \lambda_i$, where λ_i is the demand rate of the customer site i for $i = 1, 2, \ldots, k$. The operation of this prioritized repairable inventory system is described as follows:

1) When a demand to the local stock is generated, the availability of on-hand spare parts is checked. If a spare unit is available, the demand is fulfilled by delivering the spare

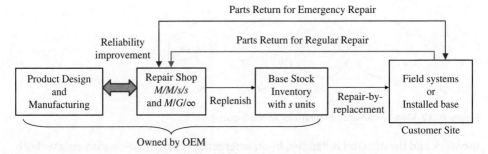

Figure 10.10 A repairable inventory system with emergency repair.

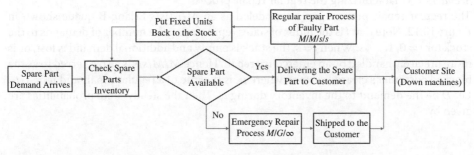

Figure 10.11 Flowchart of the repairable inventory with emergency repair.

item to the customer. In the meantime, a replenishment order is placed to the repair shop and the faulty part is shipped to the repair shop via a regular repair process.

2) If the local stock is empty upon the demand arrival, the faulty item is shipped to the repair shop using expedited transportation. The item is repaired by a dedicated testbed. When fixed, the same item is returned to the customer and put back in the down machine. There is no backorder to the local stock under emergency repair.

Figure 10.11 shows the routing path of faulty items in a repair shop that offers both regular and emergency repair. This prioritized repairable inventory system is equivalent to a tandem queuing network comprised of $M/M/s/s$ and $M/G/\infty$ in series. If the emergency repair shop has limited capacity, $M/G/m/\infty$ is more appropriate than $M/G/\infty$. The former captures the regular repair process and the latter represents the emergency repair process. The regular repair TAT is relatively long, but the resources and cost are lower. The operation cost of an emergency repair is much higher because of dedicated transportation, repair personnel, and tools, but the repair TAT is significantly compressed.

10.6.2 System Downtime and Availability

We are interested in estimating the expected machine downtime based on the inventory system in Figure 10.10. To proceed with that calculation, let us define t_r and t_e as the repair turnaround times corresponding to regular and emergency repairs, respectively. Note that $t_r \gg t_e$. Given that an on-hand spare is available, the hands-on time to replace the faulty item is assumed to be short and negligible compared to t_r and t_e. The stockout probability is calculated by considering two situations, i.e. the demand is fulfilled from

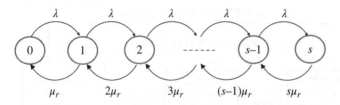

Figure 10.12 Markov transition diagram for *M/M/s/s* queue.

the stock and the demand is handled by an emergency repair. Below we investigate both scenarios, respectively.

Scenario 1: Characterizing the regular repair process

The regular repair process can be modeled as an *M/M/s/s* Erlang-B queue shown in Figure 10.12. Note that *j* is the index of states representing the number of demands to the stock for $j = 0, 1, \ldots, s$. When $j = s$, the stock is empty and additional demand is lost, or is more precisely handled by an emergency repair. Hence *M/M/s/s* is also referred to as the blocked customer's cleared system, where μ_r is the rate of a regular repair and $\mu_r = 1/t_r$. Let *D* be the demand to the inventory during t_r. Then the steady-state probabilities are given by

$$\pi_j = P\{D = j\} = \frac{\dfrac{\rho_r^j}{j!}}{\displaystyle\sum_{k=0}^{s} \dfrac{\rho_r^k}{k!}}, \quad \text{for} \quad 0 \le j \le s \tag{10.6.1}$$

where $\rho_r = \lambda/\mu_r$, which is also called the offered load (Winston 2004). The demand is blocked if and only if $D \ge s$. Hence, the probability that the inventory blocks the demand is

$$\pi_s = P\{D = s\} = \frac{\dfrac{\rho_r^s}{s!}}{\displaystyle\sum_{k=0}^{s} \dfrac{\rho_r^k}{k!}} \tag{10.6.2}$$

Let N_r be the number of parts under regular repair. Then its expected value is

$$E[N_r] = \sum_{j=0}^{s} j\pi_j = \frac{\lambda}{\mu}(1 - \pi_s) = \rho_r(1 - \pi_s) \tag{10.6.3}$$

Scenario 2: Characterizing the emergency repair process

Under an emergency repair, the faulty item is handled in an expedited mode in both the shipping and the repair. There always exist resource constraints in the repair shop. Hence it is more realistic to assume only *m* dedicated servers are available in handling all emergency repair jobs as opposed to the infinite capacity queue in Figures 10.10 and 10.11. This can be modeled as an *M/G/m/∞* queue or Erlang-C model. Figure 10.13 shows the transition diagram where λ_e is the demand rate of emergency repair and μ_e is the repair rate with $\mu_e = 1/t_e$ with t_e being the emergency repair TAT.

The demands for the emergency repair are those that are blocked from a regular repair because of the empty inventory. Therefore the demand rate of an emergency repair λ_e

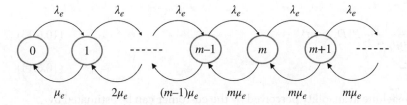

Figure 10.13 Markov transition diagram for $M/G/m/\infty$ queue.

can be estimated by

$$\lambda_e = \lambda \pi_s \tag{10.6.4}$$

where λ is the total demand rate to the local stock and π_s is the block probability given in Eq. (10.6.2). Let D_e be the demand for an emergency repair. Then the steady-state probabilities for the $M/G/m/\infty$ queue are given as

$$\pi_0 = P\{D_e = 0\} = \frac{1}{\displaystyle\sum_{k=0}^{m-1} \frac{(m\rho_e)^k}{k!} + \frac{(m\rho_e)^m}{m!(1 - \rho_e)}}, \qquad \text{for } j = 0 \tag{10.6.5}$$

$$\pi_j = P\{D_e = j\} = \frac{(m\rho_e)^j}{j!} P\{D_e = 0\}, \qquad \text{for } j = 1, \ 2, \ \ldots, \ m \tag{10.6.6}$$

$$\pi_j = P\{D_e = j\} = \frac{(m\rho_e)^j}{\dfrac{(m\rho_e)^m}{m!m^{j-m}}} P\{D_e = 0\}, \qquad \text{for } j \geq m + 1, \ \ldots \tag{10.6.7}$$

where $\rho_e = \lambda_e/(m\mu_e)$. The probability that a faulty part has to wait prior to being served is

$$P\{D_e \geq m\} = 1 - \sum_{j=0}^{m-1} P\{D_e = j\} = \frac{(m\rho_e)^m}{m!(1 - \rho_e)} P\{D_e = 0\} \tag{10.6.8}$$

Let N_e be the number of items in the emergency repair pipeline. Let N_q and N_s be the number of items in waiting and in servicing, respectively. Then $E[N_e]$ is obtained by

$$E[N_e] = E[N_q] + E[N_s] = \frac{\rho_e}{1 - \rho_e} P\{D_e \geq m\} + \frac{\lambda_e}{\mu_e} \tag{10.6.9}$$

where

$$E[N_q] = \frac{\rho_e}{1 - \rho_e} P\{D_e \geq m\} \tag{10.6.10}$$

$$E[N_s] = \frac{\lambda_e}{\mu_e} \tag{10.6.11}$$

Similarly, let W_e be the total time that a part spends in an emergency repair process. Let W_q and W_s be the waiting and actual servicing times, respectively. According to Little's law, we have

$$W_e = W_q + W_s = \frac{1}{m\mu_e - \lambda_e} P\{D_e \geq m\} + \frac{1}{\mu_e} \tag{10.6.12}$$

where

$$W_q = \frac{1}{m\mu_e - \lambda_e} P\{D_e \geq m\} \tag{10.6.13}$$

$$W_s = \frac{1}{\mu_e} \tag{10.6.14}$$

Finally, the machine availability perceived by the customer can be estimated by

$$A_s = \frac{\text{Uptime}}{\text{Uptime} + \text{Downtime}} = \frac{1/\lambda_s}{1/\lambda_s + W_e \pi_s} \tag{10.6.15}$$

where λ_s is the failure rate of the machine and π_s is the probability of being blocked given in Eq. (10.6.2). In other words, the downtime happens and is prolonged if and only if the demand for the spare part is blocked and processed through the emergency repair.

10.7 Repairable Inventory Planning under Fleet Expansion

10.7.1 Non-stationary Parts Demand

The repairable inventory models in Sections 10.3–10.6 are developed based upon the following assumptions: (i) the demand is stationary or follows a Poisson process with a constant rate and (ii) the installed base is either static or infinite. In general these assumptions hold when the fleet size reaches the steady-state condition. However, during a new product introduction, the installed base and the systemt reliability are likely to change because of the market expansion and corrective actions (CA). As a result, the spare parts demand stream tends to be non-stationary with time-varying mean or variance. Therefore, inventory models based on stationary demand may not be effective in managing the parts supply during a new product introduction. In this section we design and optimize a repairable inventory system to supply spare parts to a randomly changing (i.e. increasing) fleet size. When an installed unit fails, it is replaced by a good item from the inventory. We propose a multiresolution inventory control policy that enables the planner to periodically adjust the stock level against the time-varying demand rate. An explicit demand forecasting model is derived when the growth of the fleet size follows a Poisson process.

Figure 10.14 depicts a repairable inventory system facing a time-varying demand rate. New products or machines are continuously shipped and installed at the customer sites driven by the market. It is anticipated that more spare parts are needed for failure replacement because of the growing fleet size. To handle the uncertain growth of the spares demand issue, a multiresolution (Q, r) inventory model is formulated to minimize the total cost associated with materials, ordering, holding, and shortage by periodically adjusting the inventory parameters. The notation and the assumptions of the model are highlighted below.

Notation for the inventory model
  λ = product installation rate
  a = product failure rate
  $N(t)$ = the number of new installations in $(0, t]$
  $H(t)$ = the expected number of renewals in $[0, t]$

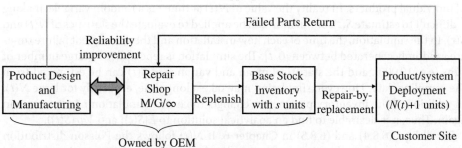

Figure 10.14 Repairable inventory system under fleet expansion.

$S(t)$ = aggregate fleet demand for spare parts in $[0, t]$
m_{max} = the maximum number of setups in $[0, t]$
c_1 = unit holding cost per unit time
c_2 = unit shortage cost
c_3 = ordering cost per replenishment order
c_4 = updating cost per inventory setup
L_i = length of the ith time interval for $i = 1, 2, ..., m$
l = inventory replenishment lead time
D_i = random demand during lead time l in the ith time interval
θ_i = expected demand during lead time l in the ith time interval, $\theta_i = E[D_i]$
β= service level or stockout probability
TC = total inventory cost

The decision variables
m = number of intervals for $m \in [1, m_{max}]$
r_i = reorder point for the ith time interval for $i = 1, 2, ..., m$
Q_i = order quantity for the ith time interval $i = 1, 2, ..., m$

The following assumptions are made:

(1) New installations occur according to the homogeneous Poisson process with rate λ. Let $N(t)$ be the cumulative number of installed products in $(0, t]$. Then the probability mass function (PMF) of $N(t)$ is

$$P\{N(t) = n\} = \frac{(\lambda t)^n e^{-\lambda t}}{n!}, \qquad \text{for } n = 0, \ 1, \ 2, \ ... \qquad (10.7.1)$$

(2) The total installations during $[0, t]$ is $N(t) + 1$, including the first installed system at $t = 0$.
(3) The (Q, r) inventory policy is adopted. Backorders are allowed but penalized.
(4) Repair TAT is a random variable with known distribution.

10.7.2 Multiresolution Inventory Control

To manage an inventory system, the first step is to characterize the demand profile. Let $S(t)$ be the cumulative spare parts demand from $N(t) + 1$ systems over the interval $[0, t]$. If the installed base is fixed, say n, and all are installed at $t = 0$, the expected cumulative demand can be simply calculated as $E[S(t)] = nH(t)$, where $H(t)$ is the renewal function of individual products. In reality the value of $N(t)$ at time t is a random variable, making it difficult to estimate $S(t)$. Simulation can be applied to evaluate the statistics of $N(t)$ and $S(t)$. In the simulation, the time of each new installation and the subsequent failure times are randomly generated between $[0, t]$. The simulation is repeated for a large number of runs (e.g. 10 000), and the statistical mean and variance of $S(t)$ can be estimated from the sampling data. However, simulation may take a long time, especially for large $N(t)$. In addition, if the inventory parameters change, the entire simulation needs to be run again. Thus, it is desirable to find an analytical solution to $E[S(t)]$ and $Var(S(t))$.

Recall Eqs. (6.8.4) and (6.8.5) in Chapter 6. If $N(t)$ follows the Poisson distribution and the lifetime of product is exponential, the closed-form solutions to the mean and variance of $S(t)$ are available. The expressions of $E[S(t)]$ and $Var(S(t))$ are duplicated below, and the detailed derivation is available in Jin and Liao (2009):

$$E[S(t)] = at + \frac{1}{2}a\lambda t^2 \tag{10.7.2}$$

$$Var(S(t)) = at + \frac{1}{2}a\lambda t^2 + \frac{1}{3}a^2\lambda t^3 \tag{10.7.3}$$

where λ is the Poisson installation rate and α is the product failure rate. It shows that the demand is non-stationary because $E[S(t)]$ is a quadratic function and $Var(S(t))$ is a cubic function.

The idea of the multiresolution inventory control is to divide the planning horizon into several intervals, say m, and the demand in each interval can be treated as a stationary process. Let t_i and t_{i-1} be the end and starting points for the ith time interval with $t_0 = 0$. The demand rate within the ith interval is estimated by

$$v_i = \frac{E[S(t_i)] - E[S(t_{i-1})]}{t_i - t_{i-1}} = a + \frac{1}{2}a\lambda(t_i + t_{i-1}), \quad \text{for } i = 1, 2, \ldots, m \tag{10.7.4}$$

where $E[S(t_i)]$ and $E[S(t_{i-1})]$ are given in Eq. (10.7.2). The accuracy of this approximation depends on the length of the time intervals. When a shorter time interval (i.e. higher resolution) is adopted, the approximation to the Poisson demand is more accurate, but it requires more inventory setups. Therefore, a trade-off between the approximation quality and the computational efficiency must be made through the optimization procedure, which will be discussed next.

Figure 10.15 illustrates the operation of the multiresolution inventory system where the entire period, say t, is divided into $m = 3$ intervals with lengths of L_i for $i = 1, 2,$ and 3, namely $t = L_1 + L_2 + L_3$. Since the value of m determines L_i, which further influences v_i and stocking policy, inventory parameters including m and (Q_i, r_i) for $i = 1, 2, \ldots, m$ need to be optimized. Without loss of generality, an equal length of time intervals is adopted in Figure 10.15, yet the proposed multiresolution concept can be extended to variable intervals.

10.7.3 Optimization Model and Algorithm

The cost associated with this inventory system consists of four cost items, i.e. the expenses for holding, backorders, ordering, and update of inventory parameters. The models for each cost item are presented below.

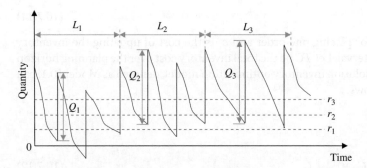

Figure 10.15 A three-interval (Q, r) inventory policy.

The inventory holding cost during the ith interval is given as follows:

$$C_{hi} = c_1 L_i \left(r_i - \theta_i + \frac{Q_i}{2} \right), \quad \text{for } i = 1, 2, \ldots, m \tag{10.7.5}$$

where c_1 be the unit holding cost per unit time and L_i is the interval length. The total inventory holding cost in $[0, t]$ is the sum of the C_{hi} for all i. That is,

$$C_h = \sum_{i=1}^{m} C_{hi} = c_1 \sum_{i=1}^{m} L_i \left(r_i - \theta_i + \frac{Q_i}{2} \right) \tag{10.7.6}$$

The total backorder cost is the sum of the backorder cost incurred in m intervals. Since multiple replenishment cycles may take place within one interval, we need to estimate the expected number of backorders per replenishment cycle. The expected number of backorders during replenishment lead time l can be obtained by

$$E[\text{backorders/cycle}] = \sum_{k=r_i}^{\infty} \frac{e^{-\theta_i} \theta_i^k}{k!} (k - r_i), \quad \text{for } i = 1, 2, \ldots, m \tag{10.7.7}$$

where $\theta_i = \bar{l} v_i$ is the expected lead time demand in the ith interval and k is the number of demands during l. Let c_2 be the unit backorder cost. Now the expected backorder cost per replenishment cycle can be expressed as

$$E[\text{backorder cost/cycle}] = c_2 \left(\theta_i - r_i - \sum_{k=0}^{r_i} \frac{e^{-\theta_i} \theta_i^k}{k!} (k - r_i) \right),$$

$$\text{for } i = 1, 2, \ldots, m \tag{10.7.8}$$

For a given interval L_i, the expected number of replenishments would be $L_i v_i / Q_i$. Since there are m intervals in $[0, t]$, now the total backorder cost can be estimated as

$$C_s = c_2 \sum_{i=1}^{m} \left(\frac{L_i v_i}{Q_i} \right) \left(\theta_i - r_i - \sum_{k=0}^{r_i} \frac{e^{-\theta_i} \theta_i^k}{k!} (k - r_i) \right) \tag{10.7.9}$$

Finally, the cost models for spare parts ordering C_o and inventory updating C_u are given as follows:

$$C_o = c_3 \sum_{i=1}^{m} \frac{L_i v_i}{Q_i} \tag{10.7.10}$$

$$C_u = mc_4 \tag{10.7.11}$$

where c_3 is the cost for placing one order and c_4 is the cost of updating the inventory parameters in each interval. Let TC be the total inventory cost over the planning horizon from $[0, t]$. A multiresolution inventory optimization model, denoted as Model 10.1, can be formulated as follows.

Model 10.1

Min:

$$E[TC] = C_h + C_s + C_o + C_u \tag{10.7.12}$$

Subject to:

$$1 - \sum_{k=0}^{r_i} \frac{e^{-\theta_i} \theta_i^k}{k!} \leq \beta, \qquad \text{for all } i = 1, \ 2, \ \ldots, \ m \tag{10.7.13}$$

$$1 \leq m \leq m_{max} \tag{10.7.14}$$

$$Q_i \geq r_i \geq 0, \qquad \text{for all } i = 1, \ 2, \ldots, \ m \tag{10.7.15}$$

where C_h, C_s, C_o, and C_u are given in Eqs. (10.7.6) and (10.7.9) to (10.7.11). In Model 10.1, the decision variables are the number of time intervals m, the order quantity Q_i, and the reorder point r_i in interval i. The objective function in Eq. (10.7.12) minimizes the expected inventory cost over the planning horizon. Constraint (10.7.13) is the service level criterion, stating that the probability of stockout should not exceed β. Constraint (10.7.14) prescribes the upper and lower bounds of updates. Constraint (10.7.15) indicates that the batch size should be larger than the re-order point.

Model 10.1 is a nonlinear integer programming problem. Many optimization techniques can be used to solve this problem such as the genetic algorithm used by Marseguerra et al. (2005). We propose a bisection search algorithm to seek the optimal values of m, Q_i, and r_i. Let m_L and m_H be the lower and upper limits of the intervals, respectively. The procedure of the algorithm is given as follows:

Step 1: Let $j = 1$, $m_L = 1$, and $m_H = m_{max}$.

Step 2: For given m_L and m_H, determine $TC^*(m_L, (Q_i^*, r_i^*)_{\text{all } i})$ and $TC^*(m_H, (Q_i^*, r_i^*)_{\text{all } i})$ by solving the nonlinear optimization problems and let $m_m^j = \lfloor (m_L + m_H)/2 \rfloor$, where $\lfloor \cdot \rfloor$ is the floor operator to take the integer part of the original value.

Step 3: Let $j = j + 1$. If $TC^*(m_L, (Q_i^*, r_i^*)_{\text{all } i}) \geq TC^*(m_H, (Q_i^*, r_i^*)_{\text{all } i})$, then let $m_L = m_m^{j-1}$ and $m_m^j = \lfloor (m_L + m_H)/2 \rfloor$; otherwise let $m_H = m_m^{j-1}$ and $m_m^j = \lfloor (m_L + m_H)/2 \rfloor$.

Step 4: If $|m_m^j - m_m^{j-1}| \leq 1$, then stop and choose $\min\{TC(m_L, (Q_i^*, r_i^*)_{\text{all } i}), TC(m_H, (Q_i^*, r_i^*)_{\text{all } i})\}$ as the overall optimal solution; otherwise go to Step 2.

Note that, in this algorithm, the optimal $(Q_i^*, r_i^*)_{\text{all } i}$ for each value of m can be obtained using the off-the-shelf optimization tools, such as AMPL, Cplex, and Microsoft Excel Solver. For instance, Excel Solver is widely used in industry and the multiresolution inventory model can be implemented in the Excel spreadsheet to solve spare parts allocation problems arising during the new product introduction.

10.8   Multi-echelon, Multi-item Repairable Inventory

10.8.1   Basic Assumptions of the METRIC Model

The overarching goal of a multi-echelon inventory system is to ensure that the stocks in different echelons hold the right parts at the right time to deliver a committed service with reduced operational costs. Since the inception of the multi-echelon technique for recoverable item control (METRIC) model (Sherbrooke 1968), different variations have been proposed in the operations management community. For instance, Muckstadt (1973) investigated the multi-indenture stocking policy where a LRU unit failed in the field and found that the root case could be caused at the subassembly level within the original LRU. Hence, it is relevant to keep spare parts in stock at different levels, ranging from assemblies, subassemblies, and down to a discrete device level. Other extensions of the METRIC model include transshipment, capacitated repair, variable or stochastic lead time, cost variations between locations, and reliability growth. Figure 10.16 depicts a two-echelon METRIC system comprised of one depot and J bases. The upstream echelon or the depot has a repair center and a central warehouse. The downstream echelon, also known as the base, consists of multiple locations where local stocks are created to meet customer demand. Each base also possesses a limited capability for repairing field returns.

The goal of METRIC is to minimize the sum of the backorders at the bases subject to cost constraint. METRIC adopts the marginal analysis technique to allocate the stock level in the base and the depot. It seeks to maximize system level performance for each dollar spent (a.k.a. "biggest bang for the buck"). Particularly it examines the trade-off between the expensive and inexpensive items in a specific stock, as well as across the supply chain hierarchy for the same item, unlike the single-echelon inventory system, where replenishment decisions are made in the local stock site. METRIC treats the network as an interconnected one where the stock level at the upstream echelon impacts the resupply lead time of downstream bases. This holistic view enables the inventory planners to maximize the overall stocking efficiency along the distribution channels to benefit all the downstream locations. This often leads to significant inventory cost

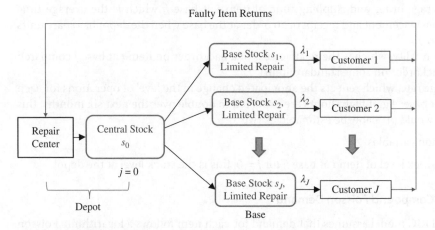

Figure 10.16 The repairable inventory system in the METRIC model.

reductions in addition to improved service levels resulting from the aggregate view of the distribution planning.

A. The following assumptions are made in the METRIC model:

(1) Demand is stationary and follows a logarithm compound Poisson process.
(2) Demands from bases are mutually independent.
(3) Backorders are allowed.
(4) No lateral transshipment.
(5) Base stocking policy with one-for-one replenishment.
(6) Infinite repair capacity in depot.
(7) No part condemnation or scrapping.
(8) No cannibalization, which refers to the practice of removing subsystems necessary for repair from another similar part.
(9) Repairable item may have different essentialities. An essentiality is the relative cost of a backorder on an item at one location compared to a backorder on some standard items.

B. Parameters and variables of the METRIC model:

i = index for item or part type
j = index for base for $j = 1, 2, …, J$, where $j = 0$ is for the depot
T_i = part repair turn-around time, which is the average time for a faulty part to be shipped from the base to the depot, get repaired at the depot, and put back to the central stock
c_i = unit cost
M_i = initial estimate of mean demand for item i.
w_i = uncertainty of the initial estimate M_i
n_i = number of squadron-months or planning horizons of demand data
u_i = demand observed over time period n_i
q = the variance to mean demand ratio and $q = \alpha + \beta\theta$, where α and β are parameters to be estimated by maximum likelihood estimation, and θ is the mean demand
r_{ij} = average faction of units that are base repairable, for $0 < r_{ij} < 1$
τ_{ij} = average repair turn-around time if repaired at base j for item i
φ_{ij} = average order and shipping time of item i at base j, which is the average time between placement and receipt of an order at the base when the depot has spare units in hand
E_{ij} = essentiality, which is the relative cost of a backorder on item i at base j compared to a backorder on some standard item
a_{ij} = usage rate, which reflects the anticipated change in the level of operations for each item at base; e.g. if flight hours are expected to double over the next six months, this factor would probably be estimated as two

C. Decision variables

s_{ij} = the stock level of item i at base j; for $j = 0$, this is the stock level at the depot

10.8.2 Compound Poisson Demand

The METRIC model assumes that demand for each item follows a logarithmic Poisson distribution, which belongs to the compound Poisson process family. This compound

Poisson process is formed by two independent distributions: (i) the order from the customer that arrives at the base is a homogeneous Poisson process and (ii) the number of requested parts per order (or the batch size) has a logarithmic distribution. The PMF of the logarithmic random variable K with parameter p for $0 < p < 1$ is given by

$$P\{K = k\} = \frac{-p^k}{k \ln(1 - p)}, \quad \text{for } k = 1, \ 2, \ 3, \ \ldots \tag{10.8.1}$$

and the cumulative distribution function is

$$P\{K \le k\} = \sum_{j=1}^{k} \frac{-p^j}{j \ln(1 - p)}, \quad \text{for } k = 1, \ 2, \ 3 \ \ldots \tag{10.8.2}$$

The mean and variance of K can be estimated by

$$E[K] = \frac{-p}{(1 - p) \ln(1 - p)} \tag{10.8.3}$$

$$Var(K) = \frac{-p(p + \ln(1 - p))}{(1 - p)^2 [\ln(1 - p)]^2} \tag{10.8.4}$$

To illustrate the distribution of K, Figure 10.17 plots the logarithmic PMF for $p = 0.2$, 0.5, and 0.9, respectively. It shows that for a larger p, there is a higher chance for K to take a larger integer value. This is also reflected by their mean value $E[K] = 1.12$, 1.44, and 3.91, which corresponds to $p = 0.2$, 0.5, and 0.9, respectively.

Let $N(t)$ be the number of customer orders in a given time interval, say $[0, t]$, which follows a Poisson distribution with rate λ. Let K_i, $i = 1, 2, 3, \ldots$ be a sequence of independently and identically distributed random variables (i.e. batch size) each having a log (p) distribution. Let X be the total amount of spare parts requested in $[0, t]$. Then

$$X = \sum_{k=1}^{N(t)} K_i \tag{10.8.5}$$

Since a Poisson process compounded with a log (p)-distributed random variable is equivalent to the negative binomial distribution (Feller 1950), where X is a discrete random variable taking the positive integer values with $x = 1, 2, 3, \ldots$. Thus the PMF of X is given by

$$p\{x \mid \lambda\} = \binom{x + b - 1}{x} (1 - p)^k p^x, \quad \text{for } x = 0, \ 1, \ 2, \ \ldots \tag{10.8.6}$$

Figure 10.17 Probability mass functions of a logarithmic distribution.

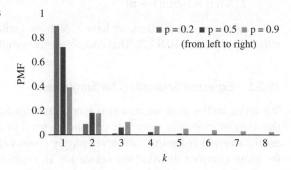

where p is the probability of success in each trial and b is the number of failures in the initial negative binomial random variable. The mean and variance of X are given as follows:

$$E[X] = \frac{pb}{1-p} \tag{10.8.7}$$

$$Var(X) = \frac{pb}{(1-p)^2} \tag{10.8.8}$$

and the variance to mean ratio is also obtained as

$$q = \frac{Var(X)}{E[X]} = \frac{1}{1-p} \tag{10.8.9}$$

Substituting p with q in Eq. (10.8.7), the PMF of X can also be expressed as

$$p\{x \mid \lambda\} = \binom{x+b-1}{x} \frac{(q-1)^x}{q^{k+x}}, \quad \text{for } x = 0, \ 1, \ 2, \ \ldots \tag{10.8.10}$$

where q and b are the distribution parameters with $q > 1$ and $b > 0$. Now the value of $E[X]$ can be expressed as the function of q and b as follows:

$$E[X] = b(q-1) \tag{10.8.11}$$

with λ being the rate of the customer order rate, which is given by Feller (1950) as follows:

$$\lambda = b \ln(q) = -b \ln(1-p) \tag{10.8.12}$$

In the actual deployment of the METRIC system, parameters b and p (or q) are estimated using Eqs. (10.8.11) and (10.8.12) based on observed demand rate λ and the batch size K_i for $i = 1, 2, \ldots, N(t)$. Below we give the proof of Eq. (10.8.12).

Proof:
From Eq. (10.8.5), taking the expectation yields

$$E[X] = E\left[\sum_{k=1}^{N(t)} K_i\right] = E[N(t)]E[K] \tag{10.8.13}$$

This result is due to the fact that K_i for all i are identically and independently distributed. Substituting Eqs. (10.8.3) and (10.8.7) into Eq. (10.8.13) and rearranging, we have

$$E[N(t)] = -b \ln(1-p) \tag{10.8.14}$$

For a Poisson distribution, we have $E[N(t)] = \lambda$ where λ is the distribution parameter, which implies Eq. (10.8.12). This completes the proof.

10.8.3 Expected Backorders for Single Item

We focus on the analysis on a two-level, base-depot hierarchy in Figure 10.16 because this type of inventory system is commonly used in the after-sales service market. The results derived from this two-level supply chain can also serve as the building block for more complex distribution networks. To simplify the notation, index i is dropped because we deal with a single item in this section. The arrival rate of demand order at base j is λ_j for $j = 1, 2, \ldots, J$. Therefore, the pooled demand rate at the depot is

$$\lambda_0 = \sum_{j=1}^{J} (1 - r_j)\lambda_j \tag{10.8.15}$$

where r_j is the fraction of the demand that can be repaired at base j, instead of shipping to the depot for repair. Let $\bar{d}_j$ be the mean batch size per order at base j. Then the aggregate demand is

$$\theta_0 = \sum_{j=1}^{J} \lambda_j \bar{d}_j (1 - r_j) = \sum_{j=1}^{J} \theta_j (1 - r_j) \tag{10.8.16}$$

with

$$\theta_j = \lambda_j \bar{d}_j \tag{10.8.17}$$

Note that θ_j is the mean demand of parts at base j. Let s_0 be the base stock level at the depot. The expected number of backorders at the depot is

$$E[B_0(s_0)] = \sum_{x=s_0+1}^{\infty} (x - s_0)p(x \mid \lambda_0 T_0) \tag{10.8.18}$$

where T_0 is the repair TAT at the depot repair shop and λ_0 is the total parts demand rate to the depot given in Eq. (10.8.15). Note that $E[B_0(s_0)]$ is a convex function of s_0 because

$$E[B_0(s_0 + 1)] - E[B_0(s_0)] = - \sum_{x=s_0+1}^{\infty} p(x \mid \lambda_0 T_0) < 0 \tag{10.8.19}$$

Let $\bar{d}_0$ be the mean spares demand per order at the depot. The average delay per order per unit time perceived by the base can be expressed as

$$\frac{\displaystyle\sum_{x=s_0+1}^{\infty} (x - s_0)p(x \mid \lambda_0 T_0)}{\lambda_0 \bar{d}_0} = \delta(s_0)T_0 \tag{10.8.20}$$

where

$$\delta(s_0) = E[B_0(s_0 \mid \lambda_0 T_0)]/E[B_0(s_0 = 0 \mid \lambda_0 T_0)] \tag{10.8.21}$$

For each $s_0 > 0$, we can also compute the expected number of backorders at base j. This can be accomplished by the following equation:

$$E[B_j(s_j)] = \sum_{x=s_j+1}^{\infty} (x - s_j)p(x \mid \lambda_j T_j), \quad \text{for } j = 1, 2, \ldots, J \tag{10.8.22}$$

where

$$T_j = r_j \tau_j + (1 - r_j)[\varphi_j + \delta(s_0)T_0] \tag{10.8.23}$$

Note that T_j is the weighted repair TAT perceived by base j. It is comprised of the repair TAT of the parts at the base j and the repair TAT for the parts at the depot. τ_j is the average TAT at base j given that a part is repaired at the base and φ_j is the average order and shipping time of an item at base j given that the depot has spare items. For given s_0, it is noticed that $E[B_j(s_j)]$ is also a convex function of s_j for $j = 1, 2, ..., J$. To minimize the expected backorders at the base, a marginal analysis was adopted to determine s_0 and s_j by Sherbrooke (1968). Using the expected item backorder functions in Eqs. (10.8.18) and (10.8.22), the next investment is allocated to that item that produces the maximum decrease in expected backorders divided by unit cost.

10.8.4 Multi-echelon, Multi-item Optimization Model

Now we extend the backorder analysis to a two-echelon, multi-item system that stores and distributes I types of items. Based on Eqs. (10.8.18) and (10.8.21), the aggregate backorders at the depot $j = 0$ and base j are given as

$$E[B_0] = \sum_{i=1}^{I} E[B_{i0}(s_{i0})] = \sum_{i=1}^{I} \sum_{x_i=s_{i0}+1}^{\infty} (x_i - s_{i0}) p_i(x_i \mid \lambda_{i0} T_{0i}), \quad \text{for } j = 0 \quad (10.8.24)$$

$$E[B_j] = \sum_{i=1}^{I} E[B_{ij}(s_{ij})] = \sum_{i=1}^{I} \sum_{x_i=s_{ij}+1}^{\infty} (x_i - s_{ij}) p_i(x_i \mid \lambda_{ij} T_{ij}), \quad \text{for } j = 1, 2, ..., J$$

$$(10.8.25)$$

where T_{0i} represents the repair TAT of item i at the depot and T_{ij} represents the average repair TAT of item i at base j, which is shown in Eq. (10.8.23). Similar to a single item, the objective of the multi-item METRIC model is to minimize the sum of an expected number of backorders at the base. The expected number of backorders varies by base, unless all bases have an identical number of systems in operation. The following optimization model, denoted as Model 10.2, is formulated to minimize the sum of expected backorders at the base.

Model 10.2

Min:

$$f(\mathbf{s}_j, \mathbf{s}_0) = \sum_{i=1}^{I} \sum_{j=1}^{J} E[B_{ij}(s_{ij}, s_{i0})] \quad (10.8.26)$$

Subject to:

$$\sum_{i=1}^{I} c_i \sum_{j=0}^{J} s_{ij} \leq C \quad (10.8.27)$$

$$s_{ij} \in \{0, 1, 2, ...\} \text{ for all } i \text{ and } j \quad (10.8.28)$$

The decision variables at the depot are $\mathbf{s}_0 = [s_{10}, s_{20}, ..., s_{I0}]$ and at base j are $\mathbf{s}_j = [s_{1j}, s_{2j}, ..., s_{Ij}]$ for $j = 1, 2, ..., J$. Note C is the total available cost budget. Marginal analysis again can be employed to determine s_0 and s_j. Using the item backorder functions computed in Eq. (10.8.22), the next investment is allocated to the item that produces the maximum decrease in expected backorders divided by unit cost. This is similar to handling the single-item system, except that now we are dealing with I part types so that the unit cost becomes a variable and not a constant.

On the other hand, the METRIC system can also be designed in such a way as to minimize the overall cost subject to the constraints that the expected backorders at base j do not exceed b_j. Then the problem can be stated as follows.

Model 10.3

Min:

$$f(s_{ij}) = \sum_{i=1}^{I} c_i \sum_{j=0}^{J} s_{ij} \tag{10.8.29}$$

Subject to:

$$\sum_{i=1}^{I} E[B_{ij}(s_{ij}, s_{i0})] \leq b_j, \quad \text{for } j = 1, \ 2, \ \ldots, \ J \tag{10.8.30}$$

According to Everett (1975), Model 10.3 can also be reformulated with the Lagrange multiplier γ_j as follows:

$$\text{Min}: \qquad F(s_{ij}) = \sum_{i=1}^{I} c_i \sum_{j=0}^{J} s_{ij} + \sum_{j=1}^{J} \gamma_j \left(b_j - \sum_{i=1}^{I} E[B_{ij}(s_{ij}, s_{i0})] \right) \tag{10.8.31}$$

Since this is a separable cell integer programming problem, we restrict our attention to a single item and Model 10.3 can be rewritten as follows.

Model 10.4

Min:

$$f(s_j) = c \sum_{j=0}^{J} s_j \tag{10.8.32}$$

Subject to:

$$E[B_j(s_j, s_0)] \leq \hat{b}_j, \quad \text{for } j = 1, \ 2, \ \ldots, \ J \tag{10.8.33}$$

where c is the unit cost of the item and $\hat{b}_j$ is the upper limit of backorders. According to Everett (1975), this problem can also be reformulated with the Lagrange multiplier γ_j as

$$\text{Min}: \quad F(s_j) = c \sum_{j=0}^{J} s_j + \sum_{j=1}^{J} \gamma_j (\hat{b}_j - E[B_j(s_j, s_0)]) \tag{10.8.34}$$

In summary, Model 10.2 is solved using a marginal allocation to minimize the expected backorders in the base. Namely, allocate one spare unit in a base to identify the maximum backorder reduction effect. For multiple items, the criterion is to allocate the item that produces the maximum decrease in expected backorders divided by the cost of that item. Models 10.3 and 10.4 aim to minimize the total system cost for multi-item and single items, respectively, and both problems can be solved using the Lagrange multiplier method.

From a management point of view, minimizing the expected backorders is preferred to the minimization of the backorder rate (or the maximization of the service rate). This is because, given the depot stock level s_0, backorders are a convex function of the base stock level s_j. The service rate (SR) and the backorder rate (BR) are correlated as follows:

$$SR = 1 - BR = 1 - \frac{\sum\limits_{i=1}^{I} E[B_i \mid s_i]}{\sum\limits_{i=1}^{I} E[B_i \mid s_i = 0]} \tag{10.8.35}$$

METRIC chooses backorders instead of the backorder rate as the performance measure because the service rates of interest are often so close with each other given that $s_i > 0$.

10.9 Case Study: Teradyne's Spare Parts Supply Chain

10.9.1 Distributed Product-Service Operation

Teradyne is a world leader in designing and marketing automated test equipment (ATE). ATE are widely used for testing the function and performance of integrated chips in the back-end process. An ATE machine usually costs $1–3 million depending on the performance, configuration, and capacity. To facilitate the repair and maintenance, a modularity design is adopted, and each machine is built with 20–40 swappable instruments made of printed circuit boards (PCBs). Nowadays the design, manufacturing, and support of capital goods are carried out in a distributed supply chain network. The driver behind this distributed product-service operation lies in the pursuant of cost reduction, access to global resources, and closeness to local customers. In Teradyne the design of a new ATE is undertaken by the engineers in Boston, MA, and San Jose, CA, USA. The software development is outsourced to India. The production of PCB is carried out by subcontractors in Charlotte, NC. The finished PCB modules are then shipped to Teradyne's factory in Shanghai, China, where the machine is finally assembled and shipped to the customers. Figure 10.18 depicts the Teradyne distributed product-service network that offers ATE machines and spare parts to its customers in North America, Asian-Pacific, and European regions.

After-sales support plays a vital role in sustaining the operation of the worldwide ATE fleet. Equipment downtime is costly to customers, resulting in delayed device testing, idle labor, and material waste. Thus spare part stocks are created in different regions to meet global customer demands. As shown in Figure 10.18, the after-sales service network consists of two repair facilities, one in the Philippines and the other in Costa Rica. In addition, a central warehouse in Memphis, TN, and 17 local stocks are created worldwide for a type of AT product. The facility located in Cebu, Philippines, handles the repair of PCB modules. The one located in Costa Rica repairs all non-PCB parts such as cables and computers. When fixed, all recovered items, both PCB and non-PCB items, are sent back to the central warehouse located in Memphis, TN. Memphis is chosen for the location of the central warehouse mainly because it is the headquarters of FedEx. Through an air transport service, spare parts are guaranteed to reach the customer anywhere in the world within 24 hours. If an item has been returned and repaired for three times, it is usually scrapped because excessive thermal or mechanical stresses have accumulated in previous rework jobs.

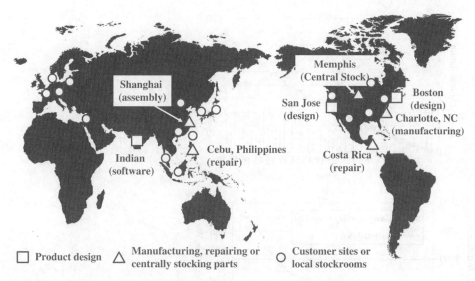

Figure 10.18 Distributed design, manufacturing, and support of ATE machines.

10.9.2 Global Customer Service

The repair and supply of spare parts are handled by the Global Customer Service (GCS) division. The role of the GCS includes: (i) tracking the installed base of ATE machines and parts; (ii) monitoring and predicting the reliability of ATE and parts; (iii) determining and implementing corrective actions against critical failure modes; and (iv) forecasting demand and adjusting the stock level to meet the time-varying demand rate. It is worth mentioning that the data provided in this section are normalized for confidentiality reasons.

10.9.2.1 Tracking the Installed Base of Systems and Parts

An installed base represents the total number of machines that are shipped and in operation at customer sites. Figures 10.19a shows the weekly and cumulative installations of a new product in one year period. Following the first customer shipment, the installed base increases over time due to market expansion. The number of weekly or monthly installations may vary due to market uncertainty. In the meantime, the installed base of parts is also tracked because it allows the inventory planner to predict the amount of spare parts needed for failure replacement. Figure 10.19b shows the weekly or monthly installation and cumulative installation of a part over time. The installed parts information can be estimated by counting the number of parts in each shipped system.

10.9.2.2 Monitoring and Predicting Reliability of System and Parts

The cumulative operating hours of the fleet and parts can be estimated based on the installed base size. Equipment in the semiconductor manufacturing industry typically operates in 24/7 mode. Since machine downtime is small compared to the uptime, each machine is assumed to operate 178 hours a week. If the customer is able to release the machine utilization rate, the actual operating hours can be estimated more precisely.

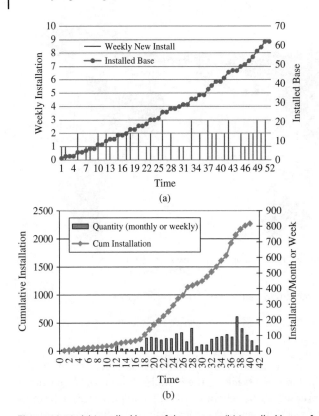

Figure 10.19 (a) Installed base of the system; (b) installed base of a PCB item.

For instance, if a customer owns ten machines and their average utilization rate is 0.9, then the weekly operating hours of the fleet is $10 \times 178 \times 0.9 = 1512$ hours.

Table 10.3 shows the installation data and estimated operation hours of individual parts for one particular week. The machine consists of 14 different part types, of which nine parts are PCB items and five parts are non-PCB items. GCS is responsible for tracking the installed base of 14 part types on a weekly basis. For instance, if a machine has 12 PCB1 units, the weekly operating hours of the PCB1 fleet is $10 \times 12 \times 178 \times 0.9 = 18\,144$ hours. These data are updated weekly as new installations are added to the fleet. Figure 10.20a and b plot the run charts of mean-time-between-failures (MTBF) for the system and PCB1. The rolling MTBF is defined as accumulated hours in 13 weeks divided by the failures within 13 weeks. The cumulative MTBF is defined as the total operating hours divided by the total failures since the first shipment.

10.9.2.3 Implementing Corrective Actions Against Critical Failure Modes

Teradyne implemented a Failure Report and Correctives Action (FRACA) database that tracks the failure mode, root cause analysis, and corrective actions status of each returned part. The Pareto charts in Figure 10.21 are plotted for one part type based on a 13-week data that allows GCS to monitor the failure mode trend, evaluate the effectiveness of corrective actions (CA), and the failures per customer site. Figure 10.21a shows that nearly 50% of the failures will be gone after the new CA program, yet 27% of

Table 10.3 Data for installed base and operating hours in a particular week.

No.	Part type (LRU)	New installation in the week	Cumulative installation	Operation hours in the week	Cumulative operation hours	Failures in the week	Cumulative failures
1	PCB1	37	504	7048	1 270 080	2	24
2	PCB2	7	84	1008	211 780	0	2
3	PCB3	9	127	1512	317 520	1	3
4	PCB4	7	84	1008	211 780	0	1
5	PCB5	12	178	2017	423 370	1	4
6	PCB6	7	84	1008	211 780	1	2
7	PCB7	9	127	1512	317 520	1	3
8	PCB8	12	178	2017	423 370	0	2
9	PCB9	9	127	1512	317 520	0	1
10	Non-PCB1	3	42	504	105 840	0	0
11	Non-PCB2	7	84	1008	211 780	2	3
12	Non-PCB3	7	84	1008	211 780	0	1
13	Non-PCB4	9	127	1512	317 520	0	0
14	Non-PCB5	3	42	504	105 840	1	1
	Machine	3	42	504	105 840	9	47

failures due to no-fault found (NFF) will continue to occur in the future. Figure 10.21b indicates that Customer A accounts for 62% of the returns and specific attention should be directed to this customer.

10.9.2.4 Forecasting Demand and Adjusting Safety Stock Level

The demand for spare parts is forecasted based on the installed base of the part and the MTBF value. This forecasting approach is referred to as installation-based forecasting, as discussed in Chapter 9. It takes into account the interactions between reliability growth and the fleet expansion effect. Figure 10.22a shows the forecasted failures (i.e. spares demand) versus the actual weekly failures of PCB1. The forecasting is made by using the Duane model with the MTBF growth rate of 0.3. Among 52 weeks, there are about 50% weeks in which the forecasting matches exactly with the actual weekly demand and there are nine weeks in which the forecasting underestimates the actual failures. However, if we compare the forecast with the cumulative demand, Figure 10.22b shows that the forecasting is quite accurate with a relative error of only 5.1%, i.e. (82−78)/78. It is interesting to see that for this item, the cumulative demand almost linearly increases with time, indicating the CA effectiveness. Otherwise, the demand would increase quadratically if the reliability is fixed. This example shows that for intermittent demand, it is more precise to forecast the cumulative demand rather than the short-term demand.

10.9.3 Repair Turn-Around Time and Failure Mode

In a multi-echelon inventory, part repair turn-around time, denoted as T_d, is the sum of the forward shipping time, the actual repair time in the shop, and the backward shipping

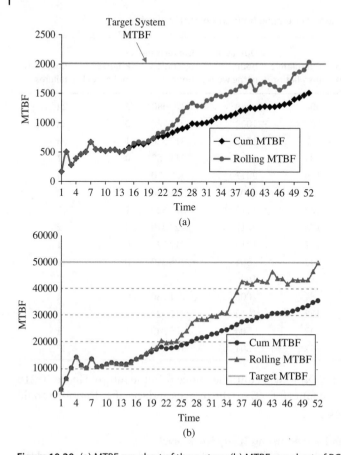

Figure 10.20 (a) MTBF run chart of the system; (b) MTBF run chart of PCB1.

time to the central stock. Let T_s be the total shipping time in forward and backward transits and T_r be the actual repair time in the repair center. Then

$$T_d = T_s + T_r \qquad (10.9.1)$$

In practice, both T_s and T_r are random variables because they are influenced by several uncertain factors, including the shipping mode, customer location and technician's skill, and failure mode, among others. The mean and the variance of T_d are given as

$$E[T_d] = E[T_t] + E[T_r] \qquad (10.9.2)$$

$$Var(T_d) = Var(T_t) + Var(T_r) \qquad (10.9.3)$$

The shipping time varies depending on the geographical distance between the local stock and the repair center and between the repair center and the central stock. Based on the actual measurements, Jin et al. (2006) found that T_s can be modeled by a normal distribution $T_s \sim N(\mu_s, \sigma_s^2)$, where μ_s is the mean and variance of T_s.

Once the part arrives at the repair center, it enters the repair process. For the same type of part, the repair time T_r (i.e. touch labor) may vary from one unit to another. This is because the time to diagnose and repair a part depends on the failure mode.

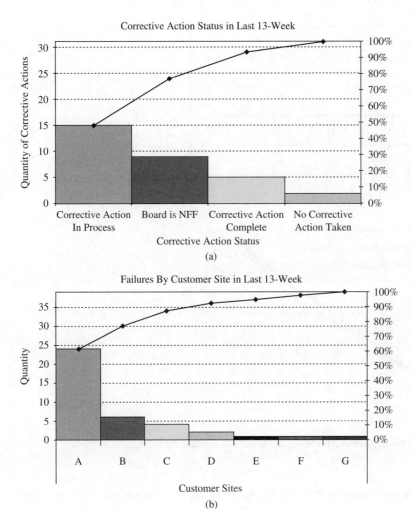

Figure 10.21 (a) Pareto chart of corrective actions; (b) Pareto chart of customer site failures.

Figure 10.23a shows the relation between different failure modes and the actual repair time based on 22 units that are randomly selected from the FRACA database. There are 10 units failed due to cold solder, six due to bad ASIC, five due to bad relays, and four due to corrupted EEPROM. On average it takes 120 minutes to troubleshoot, remove, and replace a bad ASIC, yet it takes only 20 minutes to remove and replace a bad relay. Let n be the number of failure modes; then the mean repair time is

$$E[T_r] = \sum_{i=1}^{n} w_i \tau_i \qquad (10.9.4)$$

where

τ_i = time to fix failure mode i, for $i = 1, 2, \ldots, n$

w_i = percentage of failure mode i and $\sum_{i=1}^{n} w_i = 1$

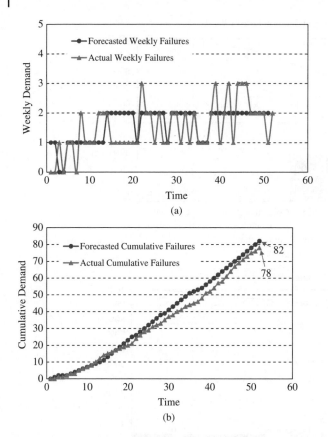

Figure 10.22 (a) Weekly demand for PCB1; (b) cumulative demand for PCB1.

For example, according to the data in Figure 10.23, the average time to repair this part is

$$E[T_r] = \frac{10}{22} \times 30 + \frac{6}{22} \times 120 + \frac{4}{22} \times 20 + \frac{2}{22} \times 70 = 61 \ \text{(minutes)}$$

The value of $E[T_r]$ often changes due to changes in failure modes. For instance, products in early shipment are more likely to fail because of a cold solder joint and ASIC failures. As the corrective actions are implemented, both failure modes decline steadily during six months. Figure 10.23b shows that the MTBF of this part consistently increases at a rate of 0.41. Issues like corrupted EEPROM and relays continue to emerge and become dominant failure modes. Table 10.4 lists the percentage (w_i) of failure modes from months 1 and 6. For instance, the percentage of cold solder failures declines from 45% to 5%, while the relay failure increases from 9% to 64%. Using Eq. (10.9.4), the monthly mean repair time of the part is estimated and shown in the last row of Table 10.4. The value of $E[T_r]$ decreases from 61 minutes in month 1 to 38 minutes in month 6 because of the percentage change in failure modes.

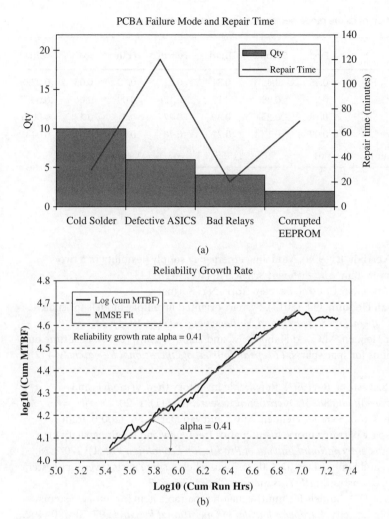

Figure 10.23 (a) Repair time versus failure modes; (b) reliability growth rate. Source: Reproduced with permission of IEEE.

Therefore, the variance of T_r can be used to monitor the variation of the repair TAT. That is,

$$Var(T_r) = Var\left(\sum_{i=1}^{n} w_i \tau_i\right) = \sum_{i=1}^{n} \tau_i^2 Var(w_i) \tag{10.9.5}$$

where $Var(w_i)$ is the variance of w_i. Substituting data of Table 10.4 into Eq. (10.9.3), we obtain $Var(T_r) = 129.4$. In summary, two primary factors cause the change of w_i, and hence the mean and variance of T_r: corrective action and latent failure mode. Corrective action systematically eliminates the critical failure modes over the time. Latent failure is a new failure mode that emerges only after the products have operated for months or even a year.

Table 10.4 The variation of failure mode percentage from first to sixth month.

i	Failure Mode	First	Second	Third	Fourth	Fifth	Sixth	$Var(w_i)$
1	Cold solder	0.45	0.35	0.31	0.21	0.13	0.05	0.0223
2	Defective ASICs	0.27	0.23	0.17	0.12	0.09	0.04	0.0077
4	EEPROM	0.18	0.25	0.30	0.27	0.31	0.27	0.0021
5	Bad relays	0.09	0.17	0.22	0.40	0.47	0.64	0.0431
Mean repair time (min)		61	59	55	48	46	38	N/A

References

Alfredsson, P. and Verrijdt, J. (1999). Modeling emergency supply flexibility in a two echelon inventory system. *Management Science* 45 (10): 1417–1431.

Axsater, S. (2015). *Inventory Control*, 3e. New York, NY: Springer.

Basten, R.J.I. and Van Houtum, G.J. (2014). System-oriented inventory models for spare parts. *Surveys in Operations Research and Management Science.* 19 (1): 34.

Behfard, S., van der Heijden, M.C., Al Hanbali, A., and Zijm, W.H.M. (2015). Last time buy and repair decisions for spare parts. *European Journal of Operational Research.* 244 (2): 498–510.

Guide, D.V. Jr. and Srivastava, R. (1997). Repairable inventory theory: models and applications. *European Journal of Operational Research* 102 (1): 1–20.

De Smidt-Destombes, K.S., van der Heijden, M.C., and van Harten, A. (2009). Joint optimisation of spare part inventory, maintenance frequency and repair capacity for k-out-of-N systems. *International Journal of Production Economics.* 118 (1): 270–278.

Caglar, D., Li, C.-L., and Simchi-Levi, D. (2004). Two-echelon spare parts inventory system subject to a service constraint. *IIE Transactions* 37 (7): 755–777.

Diaz, A. and Fu, M. (1997). Models for multi-echelon repairable item inventory systems with limited repair capacity. *European Journal of Operational Research.* 97 (3): 480–492.

Everett, H. (1975). Generalized Lagrange multiplier method for solving problems of optimal allocation of resources. *Operations Research* 11 (3): 399–417.

Feller, W. (1950). *An Introduction to Probability Theory and Its Applications*, 2e, vol. 1, 271. New York: Wiley.

Houtum, G.-J.V. and Kranenburg, B. (2015). *Spare Parts Inventory Control under System Availability Constraints*, Chapter 2. New York, NY: Springer.

Graves, S.C. (1985). A multi-echelon inventory model for a repairable item with one-for-one replenishment. *Management Science* 31 (10): 1247–1256.

Guide, V.D.R. Jr., Jayaraman, V., Srivastava, R., and Benton, W.C. (2000). Supply-chain management for recoverable manufacturing systems. *Interfaces* 30 (3): 125–142.

Jin, T. and Liao, H. (2009). Spare parts inventory control considering stochastic growth of an installed base. *Computers and Industrial Engineering* 56 (1): 452–460.

Jin, T., H. Liao, Z. Xiong, C. Sung, (2006), "Computerized reparable inventory management with reliability growth and increased product population," in *Proceedings of IEEE Conference on Automation Science and Engineering*, pp. 337–341.

Jin, J. and Tian, Y. (2012). Optimizing reliability and service parts logistics for a time-varying installed base. *European Journal of Operational Research* 218 (1): 152–162.

Jin, T., Tian, Z., and Xie, M. (2015). A game-theoretical approach for optimizing maintenance, spares and service capacity in performance contracting. *International Journal of Production Economics* 31 (3): 31–43.

Jin, T. and Wang, P. (2012). Planning performance based contracts considering reliability and uncertain system usage. *Journal of the Operational Research Society* 63 (2): 1467–1478.

Jin, T., Y. Xiang, and R. Cassady, (2013), "Understanding operational availability in performance-based logistics and maintenance services," in *Proceedings of Annual Reliability and Maintainability Symposium (RAMS)*, pp. 1–6.

Jung, W. (1993). Recoverable inventory systems with time-varying demand. *Production and Inventory Management Journal* 34 (1): 71–81.

Jung, B., Sun, B., Kim, J., and Ahn, S. (2003). Modeling lateral transshipments in multiechelon repairable-item inventory systems with finite repair channels. *Computers and Operations Research* 30 (2003): 1401–1417.

Kennedy, W.J., Patterson, J.W., and Fredenhall, L.D. (2002). An overview of recent literature on spare parts inventories. *International Journal of Production Economics.* 77 (2): 210–215.

Kim, S.H., Cohen, M.A., and Netessine, S. (2007). Performance contracting in aftersales service supply chains. *Management Science* 53 (12): 1843–1858.

Kutanoglu, E. and Mahajan, M. (2009). An inventory sharing and allocation method for a multi-location service parts logistics network with time-based service levels. *European Journal of Operational Research* 194 (3): 728–742.

Lau, H.C., Song, H., See, C.T., and Cheng, S.Y. (2007). Evaluation of time-varying availability in multi-echelon spare parts systems with passivation. *European Journal of Operational Research* 170 (1): 91–105.

Lau, H.C. and Song, H. (2008). Multi-echelon repairable item inventory system with limited repair capacity under nonstationary demands. *International Journal of Inventory Research* 1 (1): 67–92.

Lee, H.L. (1987). A multi-echelon inventory model for repairable items with emergency lateral transhipments. *Management Science* 33 (10): 1302–1316.

Liu, X. and Tang, L.C. (2016). Reliability analysis and spares provisioning for repairable systems with dependent failure processes and time-varying installed base. *IIE Transactions* 48 (1): 48–56.

Louit, D., Pascual, R., Banjevic, D., and Jardine, A.K.S. (2011). Optimization models for critical spare parts inventories-a reliability approach. *Journal of the Operational Research Society* 62 (1): 992–1004.

Marseguerra, M., Zio, E., and Podofillini, L. (2005). Multi-objective spare parts allocation by means of genetic algorithms and Monte Carlo simulations. *Reliability Engineering and System Safety* 87 (3): 325–335.

Muckstadt, J. (1973). A model for a multi-item, multi-echelolon multi-indenture inventory system. *Management Science* 20 (4): 472–481.

Muckstadt, J.A. and Joseph Thomas, L. (1980). Are multi-echelon inventory methods worth implementing in systems with low-demand-rate items? *Management Science* 483–494.

Öner, K.B., Scheller-Wolf, A., and van Houtum, G.J. (2013). Redundancy optimization for critical components in high-availability technical systems. *Operations Research* 71 (2): 244–274.

Öner, K.B., Kiesmüller, G.P., and van Houtum, G.J. (2010). Optimization of component reliability in the design phase of capital goods. *European Journal of Operational Research.* 205 (3): 715–724.

Palm, C. (1938). Analysis of the Erlang traffic formula for busy-signal arrangements. *Ericsson Technics* 5: 39–58.

Patriarca, R., Costantino, F., and Di Gravio, G. (2016). Inventory model for a multi-echelon system with unidirectional lateral transhipment. *Expert Systems with Applications* 65 (C): 372–382.

Saranga, H. and Kumar, U.D. (2007). Optimization of aircraft maintenance/support infrastructure using genetic algorithms – level of repair analysis. *Annals of Operations Research* 143 (1): 91–107.

Selçuk, S. and Ağralí, S. (2013). Joint spare parts inventory and reliability decisions under a service constraint. *Journal of the Operational Research Society* 74: 447–458.

Sherbrooke, C.C. (1992). Multiechelon inventory systems with lateral supply. *Naval Research Logistics* 39 (1): 29–40.

Sherbrooke, C.C. (1968). Metric: a multi-echelon technique for recoverable item control. *Operations Research* 17 (1): 122–141.

Sleptchenko, A. and Van der Heijden, M. (2017). Joint optimization of redundancy level and spare part inventories. *Reliability Engineering and System Safety* 153 (2017): 74–74.

Topan, E., Bayíndír, Z.P., and Tan, T. (2017). Heuristics for multi-item two-echelon spare parts inventory control subject to aggregate and individual service measures. *European Journal of Operational Research* 256 (1): 126–138.

Van Jaarsveld, W., Dollevoet, T., and Dekker, R. (2015). Improving spare parts inventory control at a repair shop. *Omega* 57 (Part B): 217–229.

Van Jaarsveld, W. and Dekker, R. (2011). Spare parts stock control for redundant systems using reliability centered maintenance data. *Reliability Engineering and System Safety* 97 (11): 1577–1587.

Van Horenbeek, A., Scarf, P., Cavalcante, C., and Pintelon, L. (2013). The effect of maintenance quality on spare parts inventory for a fleet of assets. *IEEE Transactions on Reliability* 62 (3): 596–607.

Winston, W.L. (2004). *Operations Research: Application and Algorithm*, Chapter 20, 4e. Belmont, CA: Brooks/Cole.

Xie, W., Liao, H., and Jin, T. (2014). Maximizing system availability through joint decision on redundancy allocation and spares inventory. *European Journal of Operational Research* 237 (1): 174–177.

Problems

Problem 10.1   For each category of repairable inventory model in Table 10.1, give an example that is used in private or public sectors, including manufacturing, energy, transportation, and defense industries (*Hint*: you may use on-line systems like Google to search for the related information).

Problem 10.2   The demand for a particular part is a Poisson process with $\lambda = 3$ items/week. If the inventory lead time is three weeks with one-for-one replenishment, calculate the probability of stockout given the base stock level $s = 5$, 10, and 20, respectively. What are the expected backorders in each case?

Problem 10.3   Based on the situation in Problem 10.2, estimate the fill rate for the base stock level $s = 5$, 10, and 20, respectively. What conclusion you can draw between the stockout probability and the fill rate?

Problem 10.4   Can you give one or two examples to show that the part availability is not closely related to the system availability? In other words, the system availability is still high even if the part availability is low. If feasible, justify your statement using appropriate mathematical equations.

Problem 10.5   Suppose that the fleet consists of 10 machines on the production floor. To meet the customer demand, it is required that at least 7 machines should always be good and available for production with 95% confidence. What is the minimum availability of each machine?

Problem 10.6   A single echelon repairable inventory is established to support the operation of a system fleet (see Figure 10.3). Repair capacity is unlimited and the demand to the inventory is Poisson with $\lambda = 0.4$ item/day. The repair TAT is 15 days including the transportation time. Do the following: (1) What is the expected number of failures during a TAT cycle? (2) What is the average number of parts under repair? (3) To ensure that the stockout probability is less than 10%, what is the minimum base stock level? (4) If 20% of returned parts have to be scrapped and new parts are added by the OEM with average lead time 30 days, how do we set the base stock level such that the stockout probability is ensured with less than 10%?

Problem 10.7   Based on the data in Problem 10.6, do the following: estimate (1) the expected number of backorders, (2) the variance of the expected backorders, and (3) the average machine downtime. Your computations should be performed separately based on whether returned parts are scrapped or not.

Problem 10.8   A single echelon inventory with limited repair capacity is established to support the fleet operation (see Figure 10.5). The demand to the inventory is Poisson with $\lambda = 0.2$ item/day. There are $m = 3$ parallel servers in the repair shop and the repair rate of each server is $\mu = 0.1$ item/day, assuming that the transportation time is small and can be ignored. Do the following: (1) In the steady state, how long does it take for the part to be fixed? (2) What is the average number of parts in the repair shop? (3) How many are actually in repair and how many are actually in waiting? (4) If you want to reduce the waiting time by 50%, how many additional servers should be added?

Problem 10.9 Based on the data in Problem 10.8, do the following: (1) the probability of backorders; (2) estimate the expected number of backorders, (3) the variance of the expected backorders; (4) the on-hand inventory given the base stock is $s = 15$; (4) the average machine downtime, and (5) the machine availability given its MTBF $= 300$ days.

Problem 10.10 A machine-inventory system is created to support the fleet operation with 12 machines (see Figure 10.8). The machine failure rate $\lambda = 0.02$ failure/day. The repair TAT of a part is 20 days, assuming the touch labor time is short and can be ignored. Given the initial inventory level $s = 3$, do the following: (1) In the steady state, what is the average number of parts in the repair shop? (2) How many machines are in the down state on average? (3) What is the machine availability? (4) What is the minimum s such that at least nine machines are always available for operation at a random point in time? (5) What is the fleet readiness if at least nine machines are required in operation.

Problem 10.11 A machine-inventory system is created to support the fleet operation with six machines (see Figure 10.8). The machine consists of two identical parts in parallel. Failure of one part still allows the machine to operate. The machine fails only when two components are malfunctioning. Components fail independently with a failure rate $\lambda = 0.02$ failure/day. The repair TAT of a part is 20 days, assuming that the touch labor time is short and can be ignored. Given the initial inventory level $s = 3$, do the following: (1) In the steady state, what is the average number of parts in the repair shop? (2) How many machines are in the down state on average? (3) What is the machine availability? (4) What is the minimum s such that it ensures that at least five machines are available in operation at any time?

Problem 10.12 The purpose of emergency repair is to directly fix the faulty item and put it back in the downtime in an expedited mode (see Figure 10.10). Assume that the demand to the inventory is $\lambda = 3$ items/week. The regular repair TAT of a part is three weeks, the emergency repair TAT is 0.3 week, and there are $m = 3$ servers dedicated for the emergency service. Given that the initial inventory level $s = 15$, do the following: (1) Estimate the probability of an inventory shortage (i.e. the block rate). (2) In the steady state, what is the average number of parts in regular repair? (3) What is the average number of parts in emergency repair? (4) What is the average machine downtime? (5) What is the machine availability given its MTBF is 30 weeks? (6) Can you find the relation between the parts availability and the system availability?

Problem 10.13 Develop a simulation program to verify the mean and variance of cumulative demand in Eqs. (10.7.2) and (10.7.3) in the following cases: (1) $\alpha = 0.1$ failures/month, $\lambda = 10$ systems/month between 0 and 12 months, (2) $\alpha = 0.1$ failures/month, $\lambda = 30$ systems/month between 0 and 12 months; and (3) $\alpha = 0.1$ failures/month, $\lambda = 10$ systems/month between 0 and 24 months.

Problem 10.14 Provide the detailed derivation to prove the mean and variance of cumulative demand in Eqs. (10.7.2) and (10.7.3).

Problem 10.15 A product has a constant failure rate $\alpha = 0.01$ failures/week and the installation rate $\lambda = 3$ systems/week. The spares inventory planner decides to implement a multiresolution control policy in weeks 1 to 52. The following parameters are available: $m_{max} = 12$, $c_1 = \$50$/week/item, $c_2 = \$4000$/item, $c_3 = \$2000$/order, $c_4 = \$1500$/setup, $l = 2$ weeks, and $\beta = 0.1$. If fixed (Q, r) policy is adopted across the planning horizon, find the best (Q, r) that minimizes the total cost.

Problem 10.16 Given the same setting in Problem 10.15 and assume that $m = 5$ is adopted. Instead of using a fixed (Q, r) policy, a variable (Q_i, r_i) is adopted where Q_i and r_i may differ in a different phase for $i = 1, 2, ..., m$. Find the best (Q_i, r_i) that minimizes the total cost.

Problem 10.17 For the multiresolution inventory model, instead of using the (Q, r) inventory policy, the base stock policy $(s, s-1)$ is adopted. Derive the corresponding cost models, including holding, backorders, ordering, and setup costs.

Problem 10.18 Plot the PMF for the logarithm random variable K with the following parameters: $p = 0.1$, $p = 0.4$, and $p = 0.8$, respectively. Also calculate the mean and the variance of K for each case and compute the variance to mean ratio.

Problem 10.19 Show analytically that a compound Poisson with a logarithm random variable is equivalent to the negative binominal distribution.

Problem 10.20 Develop a simulation program to verify that a compound Poisson with a logarithm random variable is equivalent to the negative binominal distribution.

Problem 10.21 Assume that the parameters of a multi-echelon inventory system are: $J = 2$, $T_1 = 2$ weeks, $c = \$2000$/item, $\lambda_1 = 2$ orders/week, $\lambda_2 = 3$ orders/week, $p_1 = 0.5$, $p_2 = 0.3$, $\tau_1 = 1$ week, $\tau_2 = 1.5$ weeks, $r_1 = 0.2$, $r_2 = 0.3$, $a_1 = a_2 = 1$. Do the following: (1) What is the expected backorders at base and depot if $s_0 = s_1 = s_2 = 0$. (2) What is the expected backorders at base and depot if $s_0 = 0$, $s_1 = s_2 = 5$. (3) What is the expected backorders at base and depot if $s_0 = 10$, $s_1 = s_2 = 0$. (4) What is the expected backorders at base and depot if $s_0 = 2$, $s_1 = s_2 = 4$.

11

Reliability and Service Integration

11.1 Introduction

The product-service system (PSS) is emerging as a new business model in designing and marketing capital equipment as the economy is moving from fragmented products and services to integrated operations. The fusion of product and service markets will have profound impacts on how firms design, manufacture, and support capital goods in this fast-paced, technology-driven world. Under the umbrella of Industry 4.0, the technologies underpinning product-service integration are digital industrialization, the Internet of Things, Big Data analytics, condition-based monitoring, and performance contracting. In this chapter, we design integrated PSSs from two aspects: minimizing the product lifecycle cost and maximizing the service profit margin. The decision variables include reliability, spare parts inventory, and the maintenance strategy. Particularly, Section 11.2 discusses the technological factors and theoretical advances in PSS. Section 11.3 presents a reliability-inventory allocation model to minimize the product life cycle cost across design, manufacturing, and after-sales support. Section 11.4 extends the reliability-inventory optimization model to a variable size of fleet. In Section 11.5 we further incorporate preventive maintenance (PM) into the reliability-inventory allocation model. Section 11.6 presents a case study on designing an integrated PSS for the wind power industry.

11.2 The Rise of Product-Service System

11.2.1 Blurring Between Product and Service

Digitization, globalization, and sociocultural transformations are combining to create a more versatile, dynamic, and interactive economy where value increasingly lies in the synergy between products and services. For industries where capital equipment is used for daily operation, high system availability and lower downtime are crucial for ensuring productivity, competitiveness, and national security. Equipment downtime cost consists of tangible and intangible costs. Tangible cost includes losses in production, capacity, labor, and inventory. Intangible cost includes responsiveness, stress, and innovation. A recent study by the ARC Advisory Group reveals that the global process industry loses up to $20 billion, almost 5% of annual production, due to unscheduled machine downtime (Chantepy 2017). During one Thanksgiving holiday weekend, Amazon suffered a

Reliability Engineering and Services, First Edition. Tongdan Jin.
© 2019 John Wiley & Sons Ltd. Published 2019 by John Wiley & Sons Ltd.
Companion website: www.wiley.com/go/jin/serviceengineering

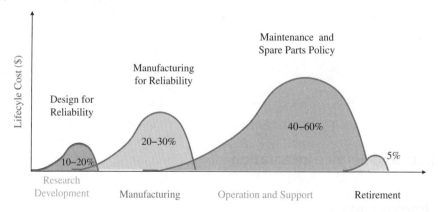

Figure 11.1 Cost distribution across a product lifecycle.

series of power outages due to electricity distribution equipment failure, losing roughly 20 000 product orders (equivalent to $500 000 of revenue loss) during the 20-minute outage (CNET News 2001).

Figure 11.1 shows the typical lifecycle cost distribution for capital equipment in many private and public sectors. It consists of four cost elements: design/development, manufacturing, operations and support, and retirement. The largest portion of the cost lies in support and maintenance of the product during its useful lifetime, which accounts for 40–60% of the lifecycle cost. It can reach as high as 70% as far as many military and defense equipment are concerned (Berkowitz et al. 2004). The second largest expenditure is the manufacturing cost that represents 20–30% of total cost. Design and development cost is relatively small with 10–20%. Finally, the cost of equipment retirement or decommission is about 5%.

In fact, maintenance, repair, and overhaul (MRO) business represents 8–10% of the gross domestic product in the US. These after-sales markets span a wide range of industries including automobiles, energy, airlines, transportation, healthcare, and defense sectors, among others. Cohen et al. (2006) showed that the after-sales revenue counts for only 20–25% of the total income for a firm, but it actually represents 40–50% of the firm's profit. For instance, the annual operation and sustainment cost for the US military equipment fleet alone is $63 billion, and the MRO service is undertaken by 67 800 internal personnel along with hundreds of private contractors (Smith 2007). In the wind industry, according to Lucintel (2013), the MRO service represents more than 75% of the global wind turbine operation and maintenance market in 2012. Turbine blades, gearbox, and generator account for more than 50% of repair and replacement servicing needs. It is projected that the global MRO market of the wind industry will exceed $14 billion by 2018. This motivates the original equipment manufacturer (OEM) and the third-party logistics (3PL) to incorporate service components into the after-sales market through extended warranty, maintenance contract, and spare parts supply.

11.2.2 Technological Factors

There are four technological factors facilitating the integration of product and services: (i) digital industrialization; (ii) the Internet of Things (IoT); (iii) Big Data analytics; and

(iv) Performance-Based Contracting (PBC). Together they creates the basis for planning and operating integrated manufacturing and service systems. These technological factors are the integral and indispensable parts of Industry 4.0 that creates what has been called a "smart factory" (Ivanov et al. 2016), which will affect many areas, most notably to the integrated PSS. They are

- Reliability and continuous quality improvement,
- Maintenance and prognostics health planning,
- Product lifecycle management,
- Smart supply chain and spare parts logistics.

Digital industrialization creates a cloud operating environment where organizations rely on the information technology infrastructure to create standardizations on technologies and economics across facilities, hardware, software, production, and after-sales support. Within the modular structured smart factories, cyber-physical systems create a virtual copy of the physical world and make decentralized decisions, such as creating visual prototype products, performing early reliability assessment, monitoring manufacturing processes, and improving productivity and yield. This means that the total-cost-of-ownership and time-to-market decrease because of its ability to reuse modules and identify early design and manufacturing issues.

Broadly speaking, IoT is the network of physical devices including machines, vehicles, and home appliances embedded with sensors, software, actuators, and connectivity, which enables these objects to connect and exchange data (Wikipedia 2018). Experts estimate that the IoT will consist of about 30 billion objects by 2020 (Nordrum 2016). Each device is uniquely identifiable through its embedded computing system, but is able to interoperate within the information technology infrastructure. Under an integrated product-service framework, product health, and diagnostic data, operating parameters (e.g. temperature and load) and spares stock size can be collected over the IoT, and this information can be fused to determine an optimal maintenance and repair schedule. Readers are referred to the paper by Xu et al. (2014) for a review on this emerging area.

Big data analytics collects and examines large amounts of data to discover hidden patterns, correlations, and other insights. As part of the big data analytics, predictive analytics uses different techniques such as data mining, statistics, modeling, machine learning, neuron network, and artificial intelligence to analyze current data and to make predictions about the future condition of equipment or remaining lifetime (Lee et al. 2013; Wang et al. 2016). For instance, if bearing vibration exceeds a certain level, the incipient failure of the bearing is a simple form of predictive analytics. This enables the deployment of predictive maintenance to maximize equipment uptime, and components are replaced only if they approach the end of lifetime.

The goal of PBC is to reduce the ownership cost while ensuring the reliability and availability of field equipment. This new service business model is often referred to as "long term service agreements" (descent of "power-by-the-hour"), or is called "performance-based logistics" in the defense industry (Smith, 2004; Richardson and Jacopino, 2006; Rees and Van den Heuvel 2012). PBC plays a vital role in designing and implementing an effective product-service contract. PBC differs from time- and material-based services in that the OEM or 3PL is incentivized to maximize service profits by either lowering the lifecycle cost or achieving pre-defined performance goals (Nowicki et al. 2008; Lin et al. 2013; Xiang et al. 2017). These goals are stipulated as

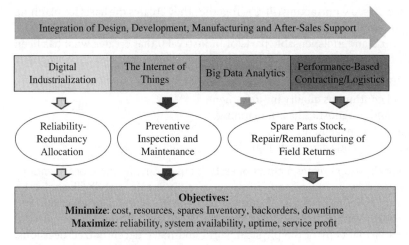

Figure 11.2 Technological factors for product-service integration.

quantifiable measures such as reliability, system availability, inventory fill rates, and logistics response time, among others. Figure 11.2 shows how these technological factors jointly influence the design and operations of integrated product and service systems.

11.2.3 The State of the Art

Introduced by Vandermerwe and Rada (1988), the term "servitization" became nearly synonymous with companies moving from selling products and commodities to selling integrated PSS. Leading manufacturers, including Caterpillar, IBM, and Rolls-Royce, are increasingly garnering benefits from the after-sales services (Huikkola et al. 2016). These PSSs typically include lifecycle reliability and maintenance services that involve PBC programs. This strategic shift has inspired a growing number of publications providing managerial insight and operational guidance to facilitate the paradigm shift (Oliva and Kallenberg 2003; Brax 2005; Gebauer et al. 2012). For instance, Martinez et al. (2010) discuss the cultural and financial challenges in transforming manufacturing-oriented business into a product-service oferings. A comprehensive literature review on PPS up to year 2013 is made by Beuren et al. (2013). In the reliability and operations management community, methods related to product and service integration can be classified into three categories: (i) reliability-inventory allocation model; (ii) redundancy-inventory allocation model; (iii) maintenance-inventory allocation model; and (iv) redundancy-maintenance allocation model. Below we briefly review related literature.

The reliability-inventory allocation model aims to find the trade-off between design for reliability and spare parts stocking policy to achieve the lifecycle cost saving. In Öner et al. (2010), component reliability and spares stock level were jointly optimized to reduce the fleet lifecycle cost based on the $M/M/s/s$ queue. Selçuk and Ağralí (2013) concurrently optimized system reliability and spares inventory under different service measures: aggregate fill rate, average system downtime, and annual number of long downtimes. Jin and Tian (2012) jointly optimized the product failure rate and spares inventory under an expanding system fleet. An adaptive (q, r) stocking policy is proposed to meet the non-stationary service demand. Liu and Tang (2016) took into account component dependent failures and the fleet growth effects. Dynamic programming is used to determine the stocking policy to minimize the total capital cost of a multi-item inventory.

In redundancy-inventory allocation models, instead of improving the item reliability, components of the same type form a parallel configuration to improve the overall system availability. Van Jaarsveld and Dekker (2011) used system maintenance data to optimize spare part inventories for redundant systems assuming a deterministic inventory replacement time. Öner et al. (2013) investigated a one-out-of-two cold standby redundancy scheme to mitigate the system downtime cost in a multi-echelon service network. Xie et al. (2014) and Sleptchenko and Van der Heijden (2016) used a continuous-time Markov chain to allocate active standby components and spares inventory to maximize the availability of a series system connected by redundant subsystems. A common assumption of these models is that the demand for spare parts is stationary.

In maintenance-inventory allocation models, item replacement time and inventory policy are jointly coordinated to achieve the desired performance goal. In early studies (Kabir and Al-Olayan 1996; Vaughan 2005), parts replacement times and inventory ordering policy were jointly optimized in order to minimize the system cost under random failures. Smidt-Destombes et al. (2009) proposed a joint maintenance-inventory allocation model in which the maintenance initiation, the spare parts, and the repair capacity were optimized to minimize the asset ownership cost under availability condition. Van Horenbeek et al. (2013) used stochastic simulation to investigate the effect of fleet size on a joint policy of maintenance and spare parts inventory when spare parts are of variable quality. Bjarnasona and Taghipour (2016) coordinated the inspection time and the (S, s) inventory ordering policy to lower the maintenance cost of k-out-of-m redundant systems. Wu and Ryan (2014) jointly optimized the maintenance interval and inventory level in an integrated PSS. Under a condition-based maintenance (CBM) framework, researchers (Wang et al. 2009; Chen et al. 2013) aimed to coordinate the parts replacement time and the repairable inventory stock so as to minimize the cost. Other works sought to optimize maintenance and spares inventory under a game-theoretic contracting framework between the supplier and the customer (Godoy et al. 2014; Jin et al. 2015), where the service supply chain is coordinated through a time-based PM program, replenishment policy, and utility functions of the supplier and the customer.

Finally, a redundancy-maintenance allocation model aims to coordinate the design with the post-sales maintenance program in order to attain the reliability goal with minimum resources. Nourelfth et al. (2012) formulated a joint redundancy and PM optimization model for series–parallel, multistate systems subject to degradation. The objective was to determine the series–parallel structure and the maintenance actions to maximize the system availability. Bei et al. (2017) proposed a two-stage stochastic programming with recourse to minimize the lifecycle cost of a multi-item system by jointly allocating a component redundancy and maintenance plan with uncertain operating stresses.

11.3 Allocation of Reliability and Inventory for a Static Fleet

11.3.1 Lifecycle Cost Analysis

This section presents a joint reliability-inventory allocation model in the presence of uncertain system usage. The decision variables are the item reliability and the spare stock level. We focus on integrated firms who design and manufacture capital goods and also provide after-sales support through spare parts supply. Two service contracting schemes are investigated under a system availability constraint: (i) minimizing the lifecycle cost of the fleet and (ii) maximizing the service profit margin. A system lifecycle cost consists of expenses associated with design, manufacturing, spare parts inventory, and repair logistics. Below we introduce several methods to characterize these cost terms. All these models are developed for a single item, yet they can be extended to multi-item systems by aggregating the costs of individual items. The notation for the reliability-inventory allocation model is summarized in Table 11.1.

Table 11.1 Notation of reliability-inventory allocation model.

Decision variables

α = item or system failure rate, decision variable

s = base stock level, decision variable

Model parameters

A = system operational availability

T_c = service planning horizon or contractual period in years

N = number of installed systems or items at a customer site

α_{max} = maximum failure rate

α_{min} = minimum failure rate

ρ = system usage rate, and $0 \leq \rho \leq 1$ $\bar{\rho}$ = the mean value of ρ

t_s = time for performing repair-by-replacement given that the spare part is available

t_r = repair turn-around time of a failed part

φ = coefficient to characterize the design difficulty in reliability growth

v = coefficient to characterize the production difficulty in reliability growth

δ = interest rate compounded annually

K = number of item types in a system

B_1 = baseline design cost at $\alpha = \alpha_{max}$

B_2 = baseline unit manufacturing cost at $\alpha = \alpha_{max}$

B_3 = coefficient of the reliability manufacturing cost model

$c(\alpha)$ = unit production cost with the failure rate of α

$D(\alpha), D(r)$ = product design cost

$I(\alpha, s; \rho)$ = inventory and repair cost in contract period

$G(A)$ = reward function

$P(\alpha, s; \rho)$ = service profit of a fleet of systems

$\pi(\alpha, s; \rho)$ = lifecycle cost of a single-item system fleet

11.3.1.1 Design Cost Versus Reliability

It is generally agreed that the design cost increases with the product reliability, but a consensus on how to quantify such a relationship has still not been reached in the literature. Based on empirical data, Mettas (2000) and Yeh and Lin (2009) showed that the product design cost grows exponentially with the reliability. Let $D(\alpha)$ be the design cost for a new product and α be the failure rate. The exponential cost model is given as follows:

$$D(\alpha) = B_1 \exp\left(\varphi \frac{\alpha_{max} - \alpha}{\alpha - \alpha_{min}}\right), \qquad \text{for } \alpha_{min} < \alpha \le \alpha_{max} \qquad (11.3.1)$$

where α_{max} is the maximum acceptable failure rate and α_{min} is the minimum achievable failure rate. B_1 and φ are the model parameters and are always positive. In particular, B_1 is the baseline design cost given $\alpha = \alpha_{max}$ and φ captures the degree of difficulties in reducing α under the technological constraint. To illustrate Eq. (11.3.1), two reliability–cost functions with different φ are depicted in Figure 11.3. Two observations can be made. First, a larger φ implies that it is more costly to increase the reliability (or reduce the failure rate) for the same α. Second, $D(\alpha)$ increases rapidly with the growth of the reliability. For instance, given $\varphi = 0.005$, reducing the failure rate from $\alpha = 5 \times 10^{-5}$ to 4×10^{-5}, the incremental cost is only 3% compared with B_1, but further reducing α to 3×10^{-5}, the incremental cost goes up by almost 22%. A similar conclusion can be made for $\varphi = 0.002$.

Sometimes it is more preferable to use the item reliability r instead of the failure rate α, to capture the reliability design cost relation. For instance, when computing the reliability importance measure, the probability value r, instead of α, is often adopted. Then the cost model becomes

$$D(r) = B_1 \exp\left(\psi \frac{r - r_{min}}{r_{max} - r}\right), \qquad \text{for } r_{min} < r \le r_{max} \qquad (11.3.2)$$

where r_{min} is the minimum acceptable reliability and r_{max} is the maximum achievable reliability with $0 < r_{min} < r_{max} \le 1$. The feasibility of reliability growth is captured by ψ for $0 < \psi < 1$. A smaller ψ implies that it is easier to improve the reliability if two products have the same reliability.

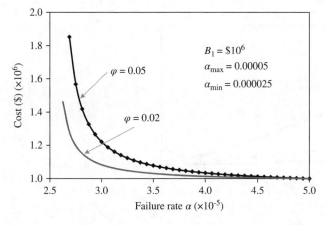

Figure 11.3 Product reliability versus design cost.

11.3.1.2 Manufacturing Cost Versus Reliability

To make a reliable product, it often requires the adoption of advanced manufacturing processes, skilled labor, and total quality management. As a result, the manufacturing cost increases with the reliability. Let $c(\alpha)$ represent the unit manufacturing cost with a failure rate α. According to Öner et al. (2010), the relation between reliability and manufacturing cost can be expressed as

$$c(\alpha) = B_2 + B_3 \left(\frac{1}{\alpha^v} - \frac{1}{\alpha_{max}^v} \right), \quad \text{for } \alpha_{min} \leq \alpha \leq \alpha_{max} \tag{11.3.3}$$

Here B_2 is the baseline unit manufacturing cost at $\alpha = \alpha_{max}$. Both B_3 and v are the model parameters to capture the incremental production cost when α is further reduced, and B_3 and v are always positive. Figure 11.4 depicts the reliability versus the manufacturing cost with $v = 0.4$, 0.5, and 0.6, respectively. It shows that the unit manufacturing cost decreases almost linearly with α. This differs from the design cost in which the cost grows exponentially. In addition, a larger v implies a higher manufacturing cost given the same B_2 and B_3.

Equation (11.3.3) can be augmented to incorporate the learning effects of the manufacturing process. During World War II, US scientists noticed that the manufacturing cost for a given aircraft model declined with increased production in accordance with a fairly predictable formula. In other words, each time the cumulative production quantity doubles, the cost drops by a fixed percentage. For instance, the reduction rate of machine tools for a new module varies between 15% and 25%, and for repetitive electronics production is 5% to 10%. Given that Q units have been produced, the basic learning curve is given as follows:

$$c_Q = c_1 Q^{-\kappa} \tag{11.3.4}$$

where c_Q is the cumulative average cost of producing Q units, c_1 is the cost of making the first unit, and κ is the learning exponent. For example, if $c_1 = \$100$, $\kappa = 0.2$, and $Q = 200$ units, then the average cost of producing the batch is \$34.66 per unit. More information about how to incorporate a learning curve into Eq. (11.3.3) is available in Loerch (1999) and Huang et al. (2007).

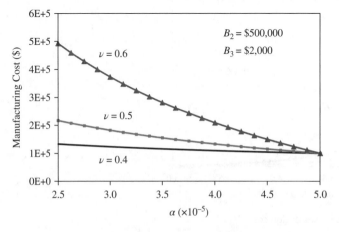

Figure 11.4 Product manufacturing cost vs. reliability.

11.3.1.3 Spare Parts Inventory and Repair Cost

The cost for after-sales support is primarily comprised of the spare parts inventory, maintenance logistics, and repair services. The major portion of the inventory is the capital investment. The repair costs include the part transportation fees, labor, and materials and facilities. Let $I(\alpha, s; \rho)$ be the total post-production support cost incurred during the contractual period of T_c years; then

$$I(\alpha, s; \rho) = sc(\alpha) + c_r \alpha n \rho \phi(\delta, T_c) \tag{11.3.5}$$

where

$$\phi(\delta, T_c) = \frac{(1 + \delta)^{T_c} - 1}{\delta(1 + \delta)^{T_c}} \tag{11.3.6}$$

The first term in Eq. (11.3.5) represents the one-time inventory investment and s is base stock level. The second term captures the present worth of repair costs incurred in $[0, T_c]$ and $\phi(\delta, T_c)$ is the present worth coefficient. For instance, if the interest rate $\delta = 5\%$ and $T_c = 10$ years, then $\phi(\delta, T_c) = 7.72$. In Eq. (11.3.5), c_r is the unit repair cost, n is the fleet size, and ρ is the usage rate for $0 \le \rho \le 1$. It is important to define the unit of α as failures/year in alignment with the years in $\phi(\delta, T_c)$.

11.3.1.4 Fleet Lifecycle Cost

Firms who are responsible for product design, manufacturing, and after-sales support are also referred to as an integrated product-service provider (IPSP). The fleet lifecycle cost during the contract period $[0, T_c]$ perceived from the IPSP is given as follows:

$$\pi(\alpha, s; \rho) = D(\alpha) + nc(\alpha) + I(\alpha, s; \rho) \tag{11.3.7}$$

In reality a system often consists of multiple item types, and each item might be repeatedly used for reliability or redundancy allocation purposes. Thus the lifecycle cost of a multi-item system can be treated as the aggregation of the cost of all items within the system. Let K be the number of item types in a system and m_i be the number of redundant units of type i for $i = 1, 2, \dots, K$. The lifecycle cost for n systems can be estimated as follows:

$$\Pi(\boldsymbol{\alpha}, \mathbf{s}) = \sum_{i=1}^{K} D_i(\alpha_i) + n \sum_{i=1}^{K} m_i c_i(\alpha_i) + \sum_{i=1}^{K} m_i I_i(\alpha_i, s_i; \rho) \tag{11.3.8}$$

where $\boldsymbol{\alpha} = [\alpha_1, \alpha_2, \dots, \alpha_K]$ are the failure rates for the ith item and $\mathbf{s} = [s_1, s_2, \dots, s_K]$ are the base stock level of the ith item. Note that Eq. (11.3.8) is applicable to systems in which redundant units are in the hot-standby mode.

11.3.2 Minimizing Lifecycle Cost of Single-Item Fleet

The IPSP opts for a cost minimization strategy if the product is relatively new or the reliability is uncertain. Meanwhile, customers who are risk-averse usually prefer paying a stable after-sales support fee over T_c years in exchange for guaranteed system reliability and operational availability. Therefore, the IPSP needs to trade off the reliability and the inventory cost in order to minimize the lifecycle cost. Reducing α requires more upfront investment in design and manufacturing, but saves repair and inventory costs. On the other hand, a less upfront investment is needed for making a low reliability product, but more repair and inventory costs will incur due to high failure returns. The following reliability-inventory model originally from Jin and Wang (2012) is presented to minimize the fleet lifecycle cost.

Model 11.1

$$\text{Min:} \quad \pi(\alpha, s; \rho) = D(\alpha) + nc(\alpha) + I(\alpha, s; \rho) \tag{11.3.9}$$

Subject to:

$$\alpha_{min} \leq \alpha \leq \alpha_{max} \tag{11.3.10}$$

$$A(\alpha, s; \rho) \geq A_{min} \tag{11.3.11}$$

$$s \in \{0, 1, 2, \ldots\} \tag{11.3.12}$$

In this formulation, the decision variables are the item failure rate α and spares stock level s. The goal is to determine the optimal α and s such that the fleet lifecycle cost is minimized. Constraint (11.3.10) stipulates the maximum and minimum failure rates. Constraint (11.3.11) ensures that the operational availability meets the customer requirement A_{min}. The analytical formula of $A(\alpha, s; \rho)$ is given in Eq. (7.7.11) in Chapter 7, where λ is used to denote the failure rate. Constraints (11.3.12) simply state that s is a non-negative integer. Figure 11.5 illustrates how the failure rate influences the trade-off between design, manufacturing, and support costs. For a given s, the optimal α is the one that results in the lowest total cost, as shown in the figure.

The value of ρ may vary over the support period. For instance, heavier usages are often observed for military aircraft during mission engagement versus training seasons. Hence Model 11.1 is actually a stochastic optimization model due to the randomness of ρ. A stochastic programming model can be solved based on a simulation-based optimization algorithm or scenario trees, but the computational time could be long. Therefore, we propose to transform Model 11.1 into a deterministic model by minimizing the expected cost with respect to ρ as follows.

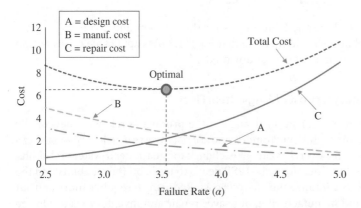

Figure 11.5 Trade-off between design, manufacturing, and support costs.

Model 11.2

Min: $\quad E_\rho[\pi(\alpha, s; \rho)] = D(\alpha) + nc(\alpha) + I(\alpha, s; \overline{\rho})$ $\qquad$ (11.3.13)

Subject to:

$$\alpha_{\min} \leq \alpha \leq \alpha_{\max} \qquad\qquad (11.3.14)$$

$$E_\rho[A(\alpha, s; \rho)] \geq A_{\min} \qquad\qquad (11.3.15)$$

$$s \in \{0, 1, 2, \ldots\} \qquad\qquad (11.3.16)$$

where $\overline{\rho} = E[\rho]$ is the mean value. It is still difficult to compute $E_\rho[A(\alpha, s; \rho)]$ due to the complexity of $A(\alpha, s; \rho)$. To make Model 11.2 tractable, yet without compromising the accuracy, $E_\rho[A(\alpha, s; \rho)]$ can be approximated and substituted by the following estimate:

$$E_\rho[A(\alpha, s; \rho)] \cong \cfrac{1}{1 + \alpha \overline{\rho} t_s + \alpha \overline{\rho} t_r \left(1 - \sum_{j=0}^{s} \cfrac{(\alpha \overline{\rho} n t_r)^j e^{-\alpha \overline{\rho} n t_r}}{j!}\right)} \qquad (11.3.17)$$

One may notice that Eqs. (11.3.17) and (7.7.11) are identical except that ρ is substituted with its mean value. Note that t_r is the average repair turn-around time from when a part failed to when it is fixed and put back to the stockroom. Now the remaining task is to search for α and s such that the objective function in Eq. (11.3.13) is minimized. This will be discussed in Section 11.3.4.

11.3.3 Maximizing Service Profit of Multi-item System Fleet

Under performance-based contracting, the IPSP is incentivized to meet the reliability and operational availability goals in pursuit of monetary reward. It is quite natural for the provider to shift the concentration from purely cost reduction to the pursuit of profit maximization. Both the linear reward function (7.8.1) and the exponential reward functions (7.8.2) in Chapter 7 can be adopted to formulate the profit maximization model. They are depicted in Figure 11.6a and b, respectively. Note that $\{a, b_1, b_2\}$ and $\{\gamma, \rho_1, \rho_2\}$ are corresponding model parameters. For the linear function, a is the fixed payment when the operational availability reaches $A_{\min}$, b_1 is the reward rate if A exceeds $A_{\min}$, and b_2 is the penalty rate if A drops below $A_{\min}$. A larger value of b_1 (or b_2) implies that

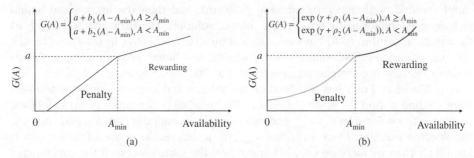

Figure 11.6 (a) Linear reward function. (b) Exponential reward function.

the IPSP receives more compensation (or penalty) given the same amount of availability increment or reduction. A similar interpretation can be applied to the exponential reward function in Figure 11.6b.

Assume that the IPSP is responsible for designing, manufacturing, and supporting a multi-item system fleet. Jin and Wang (2012) proposed a profit-centric maximization model, denoted as Model 11.3, as follows.

Model 11.3

$$\text{Max: } E_{\rho}[P(\boldsymbol{\alpha}, \mathbf{s}; \rho)] = G(A(\boldsymbol{\alpha}, \mathbf{s}; \overline{\rho})) - \sum_{i=1}^{K} D_i(\alpha_i) - n \sum_{i=1}^{K} m_i c_i(\alpha_i) - \sum_{i=1}^{K} m_i I_i(\alpha_i, s_i; \overline{\rho})$$

$$(11.3.18)$$

Subject to:

$$\alpha_{\min,i} \le \alpha_i \le \alpha_{\max,I}, \qquad \text{for } i = 1, 2, \ldots, K \tag{11.3.19}$$

$$A(\boldsymbol{\alpha}, \mathbf{s}; \overline{\rho}) = \prod_{i=1}^{K} (A_i(\alpha_i, s_i; \overline{\rho})^{m_i} \ge A_{\min} \tag{11.3.20}$$

$$s_i \in \{0, 1, 2, \ldots\}, \qquad \text{for } i = 1, 2, \ldots, K \tag{11.3.21}$$

The decision variables are α_i and s_i for item type i for $i = 1, 2, \ldots, K$. The objective function (11.3.18) aims to maximize the service profit over the contractual period $[0, T_c]$, and it consists of two parts: the total revenue $G(A)$ over T_c years and the fleet life-cycle cost associated with design, manufacturing, inventory, and repairs. Constraints (11.3.19), (11.3.20), and (11.3.21) are similar to Model 11.2, specifying the failure rate range for each item, the system operational availability, as well as the integer nature of s_i. Notice that $A(\boldsymbol{\alpha}, \mathbf{s}; \overline{\rho})$ is defined as the multiplication of availability of all items, yet this can be adjusted or modified depending on whether redundant units are in hot or cold standby mode.

11.3.4 Optimization Algorithm

Models 11.2 and 11.3 belong to non-linear mixed integer programming models. These types of problems are in general difficult to solve because they involve the complexity of nonlinearity and the combinatorial natures of integer programming (Gupta and Ravindran 1985). Existing algorithms generally rely on the successive solutions of closely related nonlinear programming problems, and then the branch-and-bound technique is employed to explore the integer solution. Heuristic methods such as the genetic algorithm (GA) have also been shown to be effective in searching for the optimal or near-optimal solution for complex non-linear programming problems within a reasonable computation time (Coit et al. 2004; Marseguerra et al. 2005).

Since Model 11.2 only has two decision variables, α and s, we can apply the enumeration method to find the optimal solution. The detailed procedure is given as follows: starting with $s = 0$, we let $\alpha = \alpha_{\min}$ and increase α by a small step, say $\Delta\alpha$, and compute the objective function. Once α reaches $\alpha_{\max}$, the α that results in the minimum cost is identified. Then we increase $s = s + 1$ and repeat the same process. If the current cost is lower than the previous iteration, choose the current s and α as the optimal solution;

otherwise a new iteration begins by increasing $s = s + 1$. The enumeration terminates until s reaches a sufficient level with no further improvement made.

The enumeration method cannot be de directly applied to Model 11.3 because the number of the decision variables becomes large if $K >> 2$. Therefore, we propose a hybrid method that combines a genetic algorithm with enumeration to search for the optimal or near-optimal α_i and s_i for $i = 1, 2, ..., K$. The algorithm is summarized in the following steps:

Step 1: For each item for $i = 1, 2, ..., K$, usie the enumeration method to find the optimal α_i and s_i locally under the availability constraint $(A_i(\alpha_i, s_i; \overline{p})^{m_i} \geq A_{\min}^{1/K}$. This is equivalent to solving Model 11.2 separately for $i = 1, 2, ..., K$.

Step 2: Create an initial solution $\boldsymbol{\alpha} = [\alpha_1, \alpha_1, ..., \alpha_K]$ and $\mathbf{s} = [s_1, s_2, ..., s_K]$ based on the result from the previous step.

Step 3: Randomly generate a set of feasible solutions based on the initial solution. Then GA is applied to solve Model 11.3 based on the first generation of the feasible solution set.

Step 4: The algorithm stops either because GA reaches the predefined iteration or the result cannot be further improved.

11.4 Allocation of Reliability and Inventory under Fleet Expansion

11.4.1 Multiresolution Inventory Policy

In this section, the reliability–inventory allocation model is extended to a system fleet with variable size. The demand rate of spare parts is likely to change and in fact increase because of the growing installed base. In Chapter 10, we introduced an adaptive, multiresolution (q, r) inventory policy to cope with the time-varying demand rate, where q is the order quantity and r is the reorder point. The idea is to adjust q and r periodically based on the variable mean and variance of the lead-time demand. The decision variables of this joint reliability–inventory allocation model include item reliability, the number of inventory phases, and the (q, r) policy in each phase. The following assumptions are made:

1) Each product is a single-item system and the item failure rate is α.
2) New system installations occur according to the homogeneous Poisson process (HPP) with rate λ. Let $N(t)$ be the cumulative installed systems in $(0, t]$. Then the total installations during $[0, t]$ is $N(t) + 1$, including the first installation at time 0.
3) Backorders are allowed; if they occur, an emergency repair is executed. Hence, it is a lost demand to the spare parts inventory.
4) A continuous review (q, r) control policy is adopted and the inventory replenishment lead time is known and fixed.

It is worth mentioning that both the (q, r) policy and $(s - 1, s)$ policy are often used in repairable inventory management. Our study assumes a system failure corresponds to an item failure that requires one spare part for replacement. Therefore, the spare parts inventory depletes one unit at a time, instead of a batch of units. If $r = q - 1$, the (q, r)

Table 11.2 Notation for reliability-inventory allocation under fleet expansion.

Decision variables

α = item failure rate,

m = number of inventory phase,

q_i = quantity per order for phase i and $\mathbf{q} = \{q_1, q_2, ..., q_m\}$,

r_i = reorder point for phase i and $\mathbf{r} = \{r_1, r_2, ..., r_m\}$,

Model parameters

i = index for inventory phases for $i = 1, 2, ..., m$

j = index for replenishment cycles for $j = 1, 2, ..., n_i$

n_i = expected number of replenish cycles in phase i

t = planning horizon in years

$N(t)$ = the number of installed systems between $(0, t]$

λ = product or system installation rate

$S(t)$ = aggregate fleet failures by time t

$\overline{S}(t), E[S(t)]$ = expected aggregate fleet failures by time t

$m_{\max}$ = maximum inventory setups

c_1 = cost per setup of inventory parameters

c_2 = ordering cost per replenishment

c_3 = unit inventory holding cost per unit time

L = lead time for inventory replenishment

CO_i = expected ordering cost for phase i, for $i = 1, 2, ..., m$

CH_i = expected holding cost for phase i, for $i = 1, 2, ..., m$

L = length of the ith phase, a constant for all i

Y_i = cumulative inventory time prior to the ith phase

t_{ij} = expected end time of the jth cycle in phase i

θ_{ij} = expected lead time demand for the jth cycle in phase i

b_1 = cost for performing a regular repair

b_2 = cost for performing an emergency repair

d_1 = expected downtime cost for a regular repair

d_2 = expected downtime cost for an emergency repair

p_{ij} = probability of stockout for the jth cycle in phase i

p_{out} = average stockout probability during $[0, t]$

$Q(\alpha, m, \mathbf{q})$ = NPV of spare parts capital cost

$I(\alpha, m, \mathbf{q}, \mathbf{r})$ = NPV of inventory cost

$R(\alpha, m, \mathbf{q}, \mathbf{r})$ = NPV of repair cost

$D(\alpha, m, \mathbf{q}, \mathbf{r})$ = NPV of downtime cost

$K(\alpha)$ = NPV of incremental product design cost

$P(\alpha)$ = NPV of incremental fleet manufacturing cost

$\pi(\alpha, m, \mathbf{q}, \mathbf{r})$ = NPV of the fleet lifecycle cost

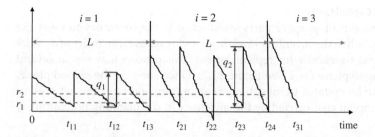

Figure 11.7 A multi-resolution (q, r) control policy. Source: Reproduced with permission of Elsevier.

policy is essentially the same as the $(s - 1, s)$ policy under the one-for-one replenishment policy. Table 11.2 lists the notation of the model. Unless specified, the same notation as in the previous section is also used here.

Figure 11.7 shows a multiresolution (q, r) inventory policy that handles a non-stationary demand process. The planning horizon is divided into three periods of equal length L. In each period, a fixed (q, r) policy is adopted even if the actual demand rate is time-dependent. As the time evolves to the next period, (q, r) is adjusted to meet the new demand rate. Let t_{ij} be the expected ending time for the jth replenishment cycle in period i. According to Jin and Tian (2012), we have

$$t_{ij} = \frac{-\alpha + \sqrt{\alpha^2 + 2\alpha\lambda(jq_i + S((i-1)L))}}{\alpha\lambda}, \quad \text{for } j = 1, 2, \ldots, n_i \tag{11.4.1}$$

where

$$S((i-1)L) = a(i-1)L + \frac{1}{2}a\lambda[(i-1)L]^2 \tag{11.4.2}$$

Note that q_i is the order quantity for phase i and n_i is the expected number of replenishment cycles in phase i. Let θ_{ij} be the expected lead-time demand for the jth cycle in phase i; then we have

$$\theta_{ij} = S(t_{ij}) - S(t_{ij} - l) = \alpha l + \alpha\lambda l t_{ij} - \frac{1}{2}\alpha\lambda l^2 \tag{11.4.3}$$

where l is the constant lead time of inventory replenishment. Upon the partitioning of the planning horizon, the lead-time demand in the jth cycle of the ith phase can be approximated as an HPP with the constant demand rate

$$d_{ij} = \theta_{ij}/l = \alpha + \alpha\lambda t_{ij} - 0.5\,\alpha\lambda l \tag{11.4.4}$$

This assumption is reasonable in practice as long as $t_{ij} \gg l$. As the inventory evolves over time, t_{ij} becomes larger and the condition $t_{ij} \gg l$ is always satisfied. Nevertheless, differentiation must be made between d_{ij} and the average demand rate in Eq. (10.6.4) in Chapter 10, where the lead-time demands in the same phase are treated the same, whereas they are not in Eq. (11.4.4).

11.4.2 Estimating the Inventory Cost

The costs of the multiresolution inventory system include the spare parts capital, holding cost, ordering cost, and cost for updating (q, r) parameters. Though the shortage cost is not considered, it will be treated as part of the emergency repair cost later on.

11.4.2.1 Spare Parts Capital Cost

At time $t = 0$, an amount of q_1 spare parts are stocked in the inventory to meet the failure replacements. When the inventory level drops and reaches r_1, an order of q_1 units is issued to the central warehouse for replenishment. This process may repeat several times before evolving to phase $i = 2$ (see Figure 11.7). Once entering the second phase, both q_i and r_i need to be updated to meet the increasing demand rate. The net present value (NPV) of the capital cost carried by the inventory in the planning horizon $[0, t]$ can be estimated by

$$Q(\alpha, m, \mathbf{q}, \mathbf{r}) = c(\alpha)\left(q_1 + r_1 + \sum_{i=2}^{m}(q_i + r_i - q_{i-1} - r_{i-1})e^{-Y_i\delta} \right) \tag{11.4.5}$$

where

$$Y_i = \sum_{j=1}^{i} L_j - L_i, \qquad \text{for } i = 1, 2, 3, \ldots, m \tag{11.4.6}$$

Note that $c(\alpha)$ is the unit manufacturing cost of items in Eq. (11.3.3). Also δ and Y_i are the annual interest rate and the cumulative inventory time prior to the ith phase, respectively.

11.4.2.2 Ordering Cost

Assuming c_2 is the ordering cost per replenishment cycle, the NPV of the ordering cost in the ith period can be estimated as

$$CO_i = c_2 \frac{\overline{S}(iL) - \overline{S}((i-1)L)}{q_i} e^{-Y_i\delta}, \qquad \text{for } i = 1, 2, \ldots, m \tag{11.4.7}$$

11.4.2.3 Holding Cost

Let c_3 be the unit holding cost per unit time. The NPV of the expected holding cost in the ith period can be estimated as

$$CH_i = c_3 \sum_{j=1}^{n_i}(t_{ij} - t_{ij-1})(0.5q_i + r_i - \theta_{ij})e^{-Y_i\delta}, \qquad \text{for } i = 1, 2, \ldots, m \tag{11.4.8}$$

where θ_{ij} is the expected lead-time demand given in Eq. (11.4.3). If $j = 1$, then $t_{ij-1} = (i - 1)L$ for $i = 1, 2, \ldots, m$.

11.4.2.4 Total Inventory Cost

By summing Eqs. (11.4.5), (11.4.7), and (11.4.8) along with the setup cost, the NPV of the inventory cost in $[0, t]$, denoted as $I(\alpha, m, \mathbf{q}, \mathbf{r})$, is obtained as

$$I(\alpha, m, \mathbf{q}, \mathbf{r}) = c_1 \sum_{i=1}^{m} e^{-Y_i\delta} + Q(\alpha, m, \mathbf{q}, \mathbf{r}) + \sum_{i=1}^{m}(CO_i(q_i, r_i) + CH_i(q_i, r_i)) \tag{11.4.9}$$

where c_1 is the setup cost incurred each time the inventory policy is adjusted. Obviously the inventory cost depends on α, m, $\mathbf{q} = [q_1, q_2, \ldots, q_m]$ and $\mathbf{r} = [r_1, r_2, \ldots, r_m]$.

11.4.3   Lifecycle Cost of a Variable-Size Fleet

Besides the inventory cost, the costs associated with design, manufacturing, repair, and equipment downtime need to be estimated as well. Without loss of generality, the following analysis is developed for a single-item system, but it can be appropriately extended to multi-item systems through the cost aggregation.

11.4.3.1   Incremental Reliability Design Cost

Let $K(\alpha)$ be the incremental design cost with respect to the baseline design cost B_1. Then

$$K(\alpha) = B_1 \exp\left(\varphi \frac{\alpha_{\max} - \alpha}{\alpha - \alpha_{\min}}\right) - B_1, \qquad \text{for } \alpha_{\min} \le \alpha \le \alpha_{\max} \qquad (11.4.10)$$

where, $\alpha_{\max}$ and $\alpha_{\min}$ represent the maximum and the minimum failure rates, respectively. Note that $D(\alpha) = B_1 + K(\alpha)$ by referring to Eq. (11.3.1).

11.4.3.2   Incremental Reliability Manufacturing Cost

Under the assumption of Poisson fleet expansion with a rate of λ, the installed base reaches $N(t) + 1$ at time t. The NPV of the incremental manufacturing cost of the fleet, denoted as $P(\alpha)$, can be estimated as

$$P(\alpha) = (c(\alpha) - B_2)\left(1 + \sum_{i=1}^{N(t)} e^{-\delta W_i}\right) \qquad (11.4.11)$$

where $c(\alpha)$ is the unit production cost given in Eq. (11.3.3) and B_2 is the baseline manufacturing cost. W_i is the installation time for the ith system for $i = 1, 2, \ldots, N(t)$. Manufacturing costs for later installations are converted into the NPV by multiplying $\exp(-\delta W_i)$. Notice that $P(\alpha)$ is a random variable because of the stochastic nature of $N(t)$ and W_i. The expected incremental manufacturing cost for the entire fleet installed in $[0, t]$ is given in the following equation and the detailed derivation is available in Jin and Tian (2012):

$$E[P(\alpha)] = (c(\alpha) - B_2)\left[1 + \frac{\lambda}{\delta}(1 - e^{-\delta t})\right] \qquad (11.4.12)$$

11.4.3.3   Parts Repair Cost

Under the multiresolution $(\mathbf{q}, \mathbf{r})$ inventory control policy, the stockout probability of the jth cycle in the ith phase is given by

$$p_{ij} = 1 - \sum_{x=0}^{r_i} \frac{\theta_{ij}^x e^{-\theta_{ij}}}{x!}, \qquad \text{for } j = 1, 2, \ldots, n_i \qquad (11.4.13)$$

where θ_{ij} is given by Eq. (11.4.3). Since θ_{ij} is time-dependent, the value of p_{ij} slightly differs in different cycles within the same phase. Then the average stockout probability in the ith phase, denoted as p_i, can be estimated by

$$p_i = \frac{1}{n_i} \sum_{j=1}^{n_i} p_{ij}, \qquad \text{for } i = 1, 2, \ldots, m \qquad (11.4.14)$$

To estimate the total repair costs in $[0, t]$, the average stockout probability for the entire planning horizon, denoted as p_{out}, needs to be estimated. That is,

$$p_{out} = \frac{1}{m} \sum_{i=1}^{m} p_i \qquad (11.4.15)$$

Let b_1 and b_2 be the expected cost for performing a regular and an emergency repair, respectively. Then the NPV of the fleet repair cost in $[0, t]$ is the sum of the regular repair cost and the emergency repair cost. That is,

$$E[R(\alpha, m, \mathbf{q}, \mathbf{r})] = (1 - p_{out})E[R_1] + p_{out}E[R_2] \tag{11.4.16}$$

where

$$E[R_1] = \frac{\alpha b_1}{\delta}\left[(1 - e^{-\delta t})\left(1 + \frac{\lambda}{\delta}\right) - \lambda te^{-\delta t}\right] \tag{11.4.17}$$

$$E[R_2] = \frac{\alpha b_2}{\delta}\left[(1 - e^{-\delta t})\left(1 + \frac{\lambda}{\delta}\right) - \lambda te^{-\delta t}\right] \tag{11.4.18}$$

Both $E[R_1]$ and $E[R_2]$ are the functions of α. Since p_{out} is the function of m, $\mathbf{q}$, and $\mathbf{r}$, the repair cost is actually dependent upon the values of α, m, $\mathbf{q}$, and $\mathbf{r}$.

11.4.3.4 System Downtime Cost

Let d_1 and d_2 be the expected customer downtime cost under a regular and an emergency repair, respectively. The total downtime cost between $[0, t]$ is given as follows:

$$E[D(\alpha, m, \mathbf{q}, \mathbf{r})] = (1 - p_{out})E[D_1] + p_{out}E[D_2] \tag{11.4.19}$$

where

$$E[D_1] = \frac{\alpha d_1}{\delta}\left[(1 - e^{-\delta t})\left(1 + \frac{\lambda}{\delta}\right) - \lambda te^{-\delta t}\right] \tag{11.4.20}$$

$$E[D_2] = \frac{\alpha d_2}{\delta}\left[(1 - e^{-\delta t})\left(1 + \frac{\lambda}{\delta}\right) - \lambda te^{-\delta t}\right] \tag{11.4.21}$$

Similar to the repair cost, the downtime cost is also jointly determined by α, m, $\mathbf{q}$ and $\mathbf{r}$. Detailed derivations of Eqs. (11.4.16) and (11.4.19) are available in the work of Jin and Tian (2012).

11.4.3.5 Fleet Lifecycle Cost

By summing Eqs. (11.4.9), (11.4.10), (11.4.12), (11.4.16), and (11.4.19), the NPV of the fleet lifecycle in $[0, t]$ is obtained as follows:

$$\begin{aligned}
&\pi(\alpha, m, \mathbf{q}, \mathbf{r}) \\
&= K(\alpha) + E[P(\alpha)] + E[I(\alpha, m, \mathbf{q}, \mathbf{r})] + E[R(\alpha, m, \mathbf{q}, \mathbf{r})] + E[D(\alpha, m, \mathbf{q}, \mathbf{r})]
\end{aligned} \tag{11.4.22}$$

Next we will show how to jointly coordinate the item reliability (i.e. manifested as the failure rate α) and the inventory policy $\{m, \mathbf{q}, \mathbf{r}\}$ to achieve the maximum cost savings.

11.4.4 Optimization Model and Solution Algorithms

11.4.4.1 The Optimization Model

Under system reliability and availability constraints, the objective of the IPSP is to allocate resources to design, manufacturing, and repairable inventory such that the fleet lifecycle cost over the contractual period is minimized. The can be achieved by solving the following optimization model.

Model 11.4

$$\text{Min: } E[\pi(\alpha, m, \mathbf{q}, \mathbf{r})] = K(\alpha) + E[P(\alpha)] + I(\alpha, m, \mathbf{q}, \mathbf{r}) + E[R(\alpha, m, \mathbf{q}, \mathbf{r})]$$

$$+ E[D(\alpha, m, \mathbf{q}, \mathbf{r})] \tag{11.4.23}$$

Subject to:

$$\alpha_{\min} \le \alpha \le \alpha_{\max} \tag{11.4.24}$$

$$A_i(\alpha, m, q_i, r_i) \ge A_{\min}, \quad \text{for } i = 1, 2, \dots, m \tag{11.4.25}$$

$$1 \le m \le m_{\max} \tag{11.4.26}$$

$$0 < r_i < q_i, \quad \text{for all } i = 1, 2, \dots, m \tag{11.4.27}$$

$$m, r_i, \text{ and } q_i \in \{1, 2, \dots\}, \quad \text{for all } i = 1, 2, \dots, m \tag{11.4.28}$$

where

$$A(\alpha, m, q_i, r_i) = \frac{1}{1 + \alpha t_s + \alpha t_r p_i}, \qquad \text{for } i = 1, 2, \dots, m \tag{11.4.29}$$

Model 11.4 is formulated to minimize the NPV of the fleet cost in $[0, t]$. The decision variables are α, m, $\mathbf{q}$, and $\mathbf{r}$. Except for α, which is a real number, all others are integers. Constraint (11.4.24) stipulates the range of acceptable failure rates. Constraint (11.4.25) ensures that the system availability $A_i(\alpha, m, q_i, r_i)$ in each phase meets the target and p_i in Eq. (11.4.29) is the stockout probability in phase i. Constraint (11.4.27) controls the maximum number of inventory setups. The rest of the constraints are self-explanatory.

11.4.4.2 Optimization Algorithm

Model 11.4 is a nonlinear mixed integer programming model that is difficult to solve. Below we present a heuristic algorithm to seek the optimal or a near-optimal solution. The algorithm is comprised of a bisection search for optimal α and a GA search for $(\mathbf{q}, \mathbf{r})$. Before applying the bisection search, α is divided into multiple levels (e.g. 50) within $[\alpha_{\min}, \alpha_{\max}]$. For each level of α, a bisection search is performed to find the optimal inventory policy $(\mathbf{q}, \mathbf{r})$. Let m_L and m_H be, respectively, the lower and upper limits of inventory phases. The steps of the algorithm are given as follows:

Step 1: Let $k = 1$, $m_L = 1$, $m_H = m_{\max}$.

Step 2: For given m_L and m_H, determine $E[\pi(\alpha, m_L, \mathbf{q}, \mathbf{r})]$ and $E[\pi(\alpha, m_H, \mathbf{q}, \mathbf{r})]$ by solving Model 11.4, respectively. Also let $m_m^k = \lfloor (m_L + m_H)/2 \rfloor$, where $\lfloor \ \rfloor$ is the floor operator to take the integer part of the original value.

Step 3: Let $k = k + 1$. If $E[\pi(\alpha, m_L, \mathbf{q}, \mathbf{r})] \ge E[\pi(\alpha, m_H, \mathbf{q}, \mathbf{r})]$, then let $m_L = m_m^{k-1}$ and $m_m^k = \lfloor (m_L + m_H)/2 \rfloor$; otherwise let $m_H = m_m^{k-1}$ and $m_m^k = \lfloor (m_L + m_H)/2 \rfloor$.

Step 4: If $|m_m^k - m_m^{k-1}| \le 1$, then stop and choose $\min\{E[\pi(\alpha, m_L, \mathbf{q}, \mathbf{r})], E[\pi(\alpha, m_H, \mathbf{q}, \mathbf{r})]\}$ as the best solution at the current value of α; otherwise go to Step 2.

For a given value of α, the GA is applied to search the optimal $(\mathbf{q}, \mathbf{r})$ corresponding to m_L and m_H, respectively. The GA search is repeated for a considerable number of generations (e.g. 100 generations) until the cost cannot be further reduced in the current search. Then the current best solution, denoted as $\{\alpha, m^*, \mathbf{q}^*, \mathbf{r}^*\}$, that minimizes $\{E[\pi(\alpha,$

m_L, **q**, **r**)], $E[\pi(\alpha, m_H, \mathbf{q}, \mathbf{r})]$} is recorded. Now the algorithm moves to the next level of α, and the bisection search process is repeated to find new {α, m^*, $\mathbf{q}^*$, $\mathbf{r}^*$} solutions that minimize the objective function. After all levels of $\alpha \in [\alpha_{min}, \alpha_{max}]$ have been exhausted, the one resulting in the lowest cost is chosen as the optimal solution.

11.5   Joint Allocation of Maintenance, Inventory, and Repair

11.5.1   Inventory Demand under Preventive Maintenance

Under age-based PM, systems are inspected regularly and components are replaced if they have reached the predetermined age τ or fail randomly, whichever comes first. The former is referred to as planned replacement and the latter is called failure replacement. All replaced items are returned to the repair center for reconditioning or repair depending on the failure severity. An item from planned replacement, though degraded, technically is still functional. Hence less time and effort is required to renew a degraded item than repairing a failed part.

Figure 11.8 shows a typical single-echelon repairable inventory system consisting of two service queues: $M_1/G_1/\infty$ and $M_2/G_2/\infty$. The former corresponds to the part reconditioning and the latter is for repairing failed items. Our treatment to the repair center differs from existing repairable inventory literature (Sherbrooke 1992; Kennedy et al. 2002) in that the repair turn-around time in Figure 11.8 is dependent on the health condition of returned items. Table 11.3 lists the notation used in the model.

To jointly allocate the maintenance interval, spares inventory, and repair capacity, it is imperative to characterize the demand profiles of the spare parts. Originally from Jin et al. (2015), the following theorem governing the relationship between the failure replacements and the planned replacements is presented.

Theorem 11.1
Let τ be the predefined replacement interval for an age-based PM. In addition, let Y be the time between two consecutive failure replacements and Z be the time between two consecutive planned replacements. Then both Y and Z are random variables, and are mutually independent, albeit τ is fixed. Let us prove this theorem.

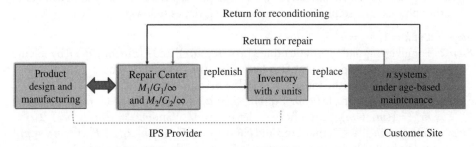

Figure 11.8 A repairable inventory supporting preventive maintenance. Source: Reproduced with permission of Elsevier.

Table 11.3 Notation for allocation of maintenance, spares, and repair capacity.

Decision variables

τ = maintenance interval,

s = base-stock level of spare parts,

t_p, t_r = reconditioning or repair turn-around times,

Model parameters

T_c = number of contract years

A = operational availability

n = fleet size

λ_p, λ_f = planned and failure replacement rates, respectively

λ = aggregate fleet replacement rate under PM

t_s = hands-on time for doing a repair-by-replacement job

$\overline{T}_u$ = mean system uptime

$R(t), F(t)$ = reliability function and cumulative distribution function

O = steady-state inventory on-order

δ = interest rate compounded annually

h = equipment loan payment period in years

ϕ_1 = capital recovery factor for system

ϕ_2 = present worth annuity coefficient for spares capital

C_s = annualized system cost

C_e = system capital cost

C_o = annual operating cost

C_d = annual production losses

C_a = levelized system cost

c_f = production loss of a failure replacement

c_p = production loss of a planned replacement

C_m = annual maintenance and logistics cost

c_1 = spare part unit cost

c_2 = spare part holding cost per unit time

$c_3(t_p), c_4(t_r)$ = costs for reconditioning and repairing a part, respectively

c_{b3}, c_b = baseline reconditioning and repair costs, respectively

$t_p^{\min}, t_r^{\min}$ = shortest reconditioning and repair times, respectively

$t_p^{\max}, t_r^{\max}$ = longest reconditioning and repair times, respectively

γ_1, γ_2 = parameters in the reconditioning/repair cost models

$\tau_{\min}$ = minimum replacement interval

$G(A)$ = service revenue function

π_1, π_2 = profit function and levelized system cost for single-item system, respectively.

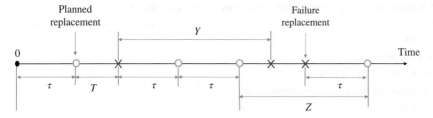

Figure 11.9 Planned and failure replacements in age-based maintenance.

Proof:

Figure 11.9 graphically shows two types of replacements under age-based PM policy: failure replacements and planned replacements. Let Y be the time between two consecutive failure replacements. The renewal period Y is a random variable, which can expressed as

$$Y = J\tau + T \tag{11.5.1}$$

where T is the time-to-failure after J consecutive planned replacements since the last failure replacement. The random number J has the following geometric distribution:

$$P\{J = j\} = R(\tau)^j F(\tau), \qquad \text{for } j = 0, 1, 2 \ldots \tag{11.5.2}$$

Note that $R(\tau)$ and $F(\tau)$ are the reliability and cumulative distribution functions of the item, respectively. The expected number of planed replacements during Y is thus given as

$$E[J] = jP\{J = j\} = \frac{R(\tau)}{F(\tau)} \tag{11.5.3}$$

The distribution of T is a conditional probability that the failure occurred prior to τ, that is

$$P\{T < t \mid T < \tau\} = \frac{P\{T < t, T < \tau\}}{P\{T < \tau\}} = \frac{F(t)}{F(\tau)} \tag{11.5.4}$$

The mean value of T can be estimated by conditioning $T < \tau$. That is,

$$E[T] = \int_0^\tau \left(1 - \frac{F(t)}{F(\tau)}\right) dt = \frac{1}{F(\tau)} \int_0^\tau (F(\tau) - F(t)) dt \tag{11.5.5}$$

By substituting Eqs. (11.5.3) and (11.5.5) into Eq. (11.5.1), the expected value of Y is

$$E[Y] = E[J]\tau + E[T]$$

$$= \frac{\tau}{F(\tau)} \int_0^\tau R(t) dt + \frac{1}{F(\tau)} \int_0^\tau (F(\tau) - F(t)) dt$$

$$= \frac{1}{F(\tau)} \int_0^\tau R(t) dt \tag{11.5.6}$$

In defense and military logistics literature, $E[Y]$ is also referred to as the mean time between unscheduled removals. Next let Z be the time interval between two consecutive planned replacements (see Figure 11.9). It can be estimated as

$$Z = KT + \tau \tag{11.5.7}$$

where K is the number of failures during Z, and the geometric distribution is given as follows:

$$P\{K = k\} = F(\tau)^k R(\tau), \qquad \text{for } k = 0, 1, 2, \dots \qquad (11.5.8)$$

The expected number of failures during Z, denoted as $E[K]$, is given as

$$E[K] = \sum_{k=0}^{\infty} kP\{K = k\} = \frac{F(\tau)}{R(\tau)} \qquad (11.5.9)$$

Now substituting Eqs. (11.5.5) and (11.5.9) into Eq. (11.5.7), the mean time between two consecutive planned replacements is

$$E[Z] = E[K]E[T] + \tau = \frac{1}{R(\tau)} \int_0^\tau R(t)dt \qquad (11.5.10)$$

This completes the proof of Theorem 11.1.

Given $E[Y]$ and $E[Z]$, we can estimate the parts demand rate of failure and planned replacements. Let $\lambda_f(\tau)$ be the expected failure replacement rate and $\lambda_p(\tau)$ be the planned replacement rate. Assume that the time for replacing an item is relatively short compared to its uptime. From Eqs. (11.5.6) and (11.5.10), we have

$$\lambda_f(\tau) = \frac{n}{E[Y]} = \frac{nF(\tau)}{\int_0^\tau R(t)dt} \qquad (11.5.11)$$

$$\lambda_p(\tau) = \frac{n}{E[Z]} = \frac{nR(\tau)}{\int_0^\tau R(t)dt} \qquad (11.5.12)$$

where n is the number of items or fleet size. The aggregate fleet demand rate, denoted as $\lambda(\tau)$, is the sum of both planned and failure replacements. That is,

$$\lambda(\tau) = \lambda_f(\tau) + \lambda_p(\tau) = \frac{n}{\int_0^\tau R(t)dt} \qquad (11.5.13)$$

If $n = 1$, $\lambda(\tau)$ becomes the spare parts demand rate of a single system. The process formed by the union of all the failures/replacements is called a superimposed renewal process (SRP). Cox and Smith (1954) have proved that, as n approaches infinity, the SRP becomes a homogeneous Poisson process. Wang (2012) shows that as long as $n \geq 10$,

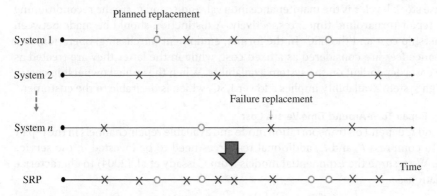

Figure 11.10 Superimposed renewal processes.

the occurrence times between two successive replacements can be approximated by an exponential distribution. This condition usually holds as far as the fleet size in manufacturing, airline, energy, and defense industries is considered. A typical SRP formed by both failure and planned replacements of n systems is shown in Figure 11.10.

11.5.2 Estimating the Cost of Product-Service System

11.5.2.1 Levelized Cost of Single-Item System

The cost models below are developed for single-item systems, but can be extended to a multi-item system through the cost aggregation of individual parts. Let C_a be the annualized system cost including the initial capital, operation expenses, and downtime losses. That is,

$$C_a = \varphi_1 C_e + C_o + C_d \tag{11.5.14}$$

where C_e is the initial capital cost, C_o is the annual operation cost, and C_d is the annual production losses due to maintenance downtime. Note that $\varphi_1 = \delta(1 + \delta)^h / [(1 + \delta)^h - 1]$ is the capital recovery factor where h is the number of years to pay off the loan and δ is the annual interest rate. The annual operation cost C_o includes all the necessary resources such as labor, energy, materials, and facilities that are consumed while operating the system. The annual production loss C_d depends on the maintenance frequency and downtime duration and can be further expressed as

$$C_d = \lambda(\tau)(c_p R(\tau) + c_f F(\tau)) \tag{11.5.15}$$

where $\lambda(\tau)$ is the part replacement rate per year given in Eq. (11.5.13). Parameters c_f and c_p represent the production losses in a failure or a planned replacement, respectively. In general $c_f \gg c_p$, because an unexpected failure usually causes a much higher production loss than a scheduled replacement. Since costs for maintenance and spare parts are borne by the IPSP under the PBC agreement, they are excluded from C_a.

We propose a levelized system cost (LSC) model to measure the ownership cost by taking into account the effect of system availability. LSC is defined as the annualized system cost C_a divided by its operational availability. Let $C_s(s, \tau, t_p, t_r)$ be the LSC; then

$$C_s(s, \tau, t_p, t_r) = \frac{1}{A(s, \tau, t_p, t_r)} [\varphi_1 C_e + C_o + \lambda(\tau)(c_p R(\tau) + c_f F(\tau))] \tag{11.5.16}$$

where $A(s, \tau, t_p, t_r)$ is the system availability given by Eq. (7.7.16) in Chapter 7. Note that s is the base stock level, τ is the maintenance interval, and t_p and t_r are the reconditioning and the repair turnaround times, respectively. A distinction should be made between the ownership cost and the LSC. In the former, equipment purchasing, operation, and maintenance fees are considered as a fixed cost, while in the latter they are treated as variable costs dependent on the system availability. When the annual ownership cost is fixed, high system availability implies a lower LSC, which is desirable to the customer.

11.5.2.2 Repair Turn-Around Time Versus Cost

Both t_p and t_r depend on transportation mode and available repair channels in the repair center. To compress t_p and t_r, additional resources need to be invested in the service pipeline. We resume the exponential models from Cassady et al. (2004) to characterize their relations:

$$c_3(t_p) = c_{b3} \exp\left[\gamma_1 \left(\frac{t_p^{\max} - t_p}{t_p - t_p^{\min}}\right)\right], \quad \text{for } t_p^{\min} < t_p \le t_p^{\max} \tag{11.5.17}$$

$$c_4(t_r) = c_{b4} \exp\left[\gamma_2\left(\frac{t_r^{max} - t_r}{t_r - t_r^{min}}\right)\right], \quad \text{for } t_r^{min} < t_r \leq t_r^{max} \tag{11.5.18}$$

Here c_{b3} and c_{b4} are the baseline costs for reconditioning or repairing a part, t_p^{max} and t_p^{min} represent the longest acceptable and the shortest achievable reconditioning turn-around times, and so do t_r^{max} and t_r^{min} for the repair turn-around times. Finally, γ_1 and $\gamma_2 \in [0, 1]$ characterize the levels of difficulty in compressing t_p and t_r due to transportation, labor skills, and facility limitations. In general, $c_3(t_p) < c_4(t_r)$ because more resources are involved in repairing a failed item.

11.5.2.3 Maintenance, Inventory, and Repair Costs

The budget for maintenance service during the contract period comprises the spares inventory, the holding cost, and all the fees associated with repair and reconditioning tasks. For a fleet of n systems, the annualized maintenance cost can be estimated as

$$C_m(s, \tau, t_p, t_r) = \phi_2 c_1 s + c_2 s + n\lambda(\tau)(c_3(t_p)R(\tau) + c_4(t_r)F(\tau)) \tag{11.5.19}$$

where c_1 is the unit spare part cost, c_2 is the annual holding cost per item, and $c_3(t_p)$ and $c_4(t_r)$ are the costs for reconditioning and repairing an item, respectively; $\phi_2 = \delta(1+\delta)^{T_c}/[(1+\delta)^{T_c} - 1]$ is the inventory capital recovery factor, where T_c is the number of contractual years and δ is the annual interest or discount rate.

11.5.3 Principal–Agent Contracting Model

A principal–agent model belongs to the Stackelberg leadership game theory in which the leader or principal moves first and then the follower or the agent moves sequentially. We propose a principal–agent game model to investigate the supplier–customer interactions in the PBC program. Two players involved in the game are the IPSP and the customer. The customer serves as the principal who defines the reward scheme, system reliability, and availability goals. The IPSP behaves as an agent whose goal is to maximize the service profit by coordinating maintenance, spares inventory, and repair capacities. Without loss of generality, the linear reward function in Figure 11.6a is adopted to formulate the principal–agent model.

Model 11.5 The agent model for IPSP:

Max: $\pi_1(s, \tau, t_p, t_r \mid a, b_1, b_2, \theta)$

$$= G(A(s, \tau, t_p, t_r)) - [\phi_2 c_1 s + c_2 s + n\lambda(\tau)(c_3(t_p)R(\tau) + c_4(t_r)F(\tau))] \tag{11.5.20}$$

Subject to:

$$A(s, \tau, t_p, t_r) \geq A_{min}, \tag{11.5.21}$$

$$\tau \geq \tau_{min} \tag{11.5.22}$$

$$t_p^{min} < t_p \leq t_p^{max} \tag{11.5.23}$$

$$t_r^{min} < t_r \leq t_r^{max} \tag{11.5.24}$$

$$s \in \{0, 1, 2, \dots\} \tag{11.5.25}$$

The principal model for customer:

$$\text{Min: } \pi_2(a, b_1, b_2, \tau_{\min} \mid s, \tau, t_p, t_r)$$

$$= \frac{1}{n} G(A(s, \tau, t_p, t_r)) + \frac{1}{A_s(s, \tau, t_p, t_r)} [\phi_1 C_e + C_o + \lambda(\tau)(c_p R(\tau) + c_f F(\tau))]$$

$$(11.5.26)$$

Subject to:

$$\pi_1(s, \tau, t_p, t_r \mid a, b_1, b_2, \tau_{\min}) > 0 \qquad\qquad (11.5.27)$$

In Model 11.5, the supplier makes a decision on $\{s, \tau, t_p, t_r\}$ such that the service profit in Eq. (11.5.20) is maximized. Constraint (11.5.21) states that the supplier must meet the minimum system availability $A_{\min}$ required by the customer or a penalty is imposed if $A < A_{\min}$. Constraint (11.5.22) states that τ should be larger than $\tau_{\min}$. This criterion is essential to the customer in terms of mitigating the production losses by avoiding excessive maintenance. Constraints (11.5.23) and (11.5.24) define the range of reconditioning and repair turn-around times. Constraint (11.5.25) ensures that s is a non-negative integer and τ is positive.

The objective of the customer is defined in Eq. (11.5.26). The customer needs to specify $\{a, b_1, b_2, \tau_{\min}\}$ so that the sum of the service fee and the levelized cost of system (LCS) is minimized. Note that $G(A(s, \tau, t_p, t_r))/n$ is the amount of service fee paid to the IPSP for each system. Constraint (11.5.27) makes sure that the service profit is positive as it ensures the supplier's participation in the program.

11.5.4 Moral Hazard and Asymmetric Information

A main challenge in implementing a principal–agent model has to do with the so-called moral hazard issue. A moral hazard is a situation where one player is more likely to take risks because the costs that could result will not be borne by the player himself. Asymmetric information is the main reason contributing to the occurrence of a moral hazard issue. Usually the two players are unwilling to exchange the information as each side is concerned that the other will gain the advantage. This is quite common in the protection of critical facilities or infrastructures (Zhang et al. 2018). Nevertheless, the following theorem guarantees the full efficiency of the maintenance contracting service as the supplier's effort levels in terms of τ, s, t_p, and t_r are observable to the customer.

Theorem 11.2

The supplier's action on τ and s are completely observable to the customer. In particular, the supplier must inform the customer of τ in order to perform a planned replacement job. The customer is able to deduce s from the nature and frequency of maintenance events. In addition, the customer can derive the full knowledge of an average part turn-around time, which is $t_p R(\tau) + t_r F(\tau)$, though t_p and t_r are not known.

Proof:

Under an age-based PM, the customer can directly observe τ because this is the replacement interval predetermined by the supplier. In the following, the procedure to estimate $t_p R(\tau) + t_r R(\tau)$ and s are presented from the customer's perspective.

Let $\overline{T}_u$ be the observed mean-time-between-replacements including both planned and failure replacements. Now the operational availability in Eq. (7.7.16) can be rewritten as

$$A = \frac{\overline{T}_u}{\overline{T}_u + t_s + p_{out}(t_p R(\tau) + t_r F(\tau))} \tag{11.5.28}$$

where p_{out} is the probability of an out-of-stock item, which can be deduced as follows: for any replacement, this is indicated if the job is completed within a predefined duration t_s, meaning the stock is not empty. On the other hand, if a replacement job delays and exceeds t_s, this indicates a stock-out signal. Hence, the customer can estimate p_{out} based on the actual duration of a replacement job. By rearranging Eq. (11.5.28), we have

$$t_p R(\tau) + t_r F(\tau) = \frac{1}{p_{out}} \left(\frac{\overline{T}_u}{A_s} - \overline{T}_u - t_s \right) \tag{11.5.29}$$

Equation (11.5.29) indicates that the customer can estimate $t_p R(\tau) + t_r F(\tau)$ based on observations of p_{out}, t_s, A, and $\overline{T}_u$. Next, let us estimate the base stock level s. Under the Poisson demand process, the customer can derive the supplier's base stock level s using the following method:

$$p_{out} = 1 - P\{O \leq s\} = 1 - \sum_{j=0}^{s} \frac{\mu^j e^{-\mu}}{j!} \tag{11.5.30}$$

with

$$\mu \cong \frac{n[t_p R(\tau) + t_r F(\tau)]}{\overline{T}_u + t_s} \tag{11.5.31}$$

where O is the random variable representing the steady-state inventory on-order and μ is the lead-time spare parts demand during a repair or reconditioning turn-around cycle. The value of μ can be derived based on Eq. (11.5.31), where n, $t_p R(\tau) + t_r F(\tau)$, $\overline{T}_u$, and t_s are known to the customer. Based on the observed value of p_{out}, the customer can derive s based on Eq. (11.5.30). This completes the proof of Theorem 11.2.

11.5.5 First-Best Solution

According to Theorem 11.2, the customer has the full observability of s, τ, and $t_p R(\tau) + t_r F(\tau)$. Meanwhile the supplier has all the information of the reward scheme such as $\{a, b_1, b_2, \tau_{min}\}$. Hence the principal–agent game problem in Model 11.5 is immune from the moral hazard issue. Now the principal–agent model can be transformed into a single objective programming model as follows.

Model 11.6

$$\text{Max: } \pi_1(s, \tau, t_p, t_r \mid a, b_1, b_2, \tau_{min})$$

$$= G(A(s, \tau, t_p, t_r)) - [\phi_2 c_1 s + c_2 s + n\lambda(\tau)(c_3(t_p)R(\tau) + c_4(t_r)F(\tau))] \tag{11.5.32}$$

Subject to:

$$\pi_2(s, \tau, t_p, t_r \mid a, b_1, b_2, \theta) \leq \pi_2^{max} \tag{11.5.33}$$

$$A(s, \tau, t_p, t_r) \geq A_{min} \tag{11.5.34}$$

$$\tau \geq \tau_{min} \tag{11.5.35}$$

$$t_p^{min} < t_p \leq t_p^{max} \tag{11.5.36}$$

$$t_r^{min} < t_r \leq t_r^{max} \tag{11.5.37}$$

$$s \in \{0, 1, 2, \ldots\} \tag{11.5.38}$$

The new formulation differs from Model 11.5 in that the customer's objective function (11.5.26) becomes constraint (11.5.33), where π_2^{max} is the maximum LSC acceptable to the customer. It is worth mentioning that Model 11.6 is equivalent to Model 11.5 because of the customer's full observability on the supplier's actions. The decision process is summarized as follows. First, the customer defines the payment scheme (i.e. a, b_1, b_2) and the performance criteria (i.e. A_{min} and τ_{min}). Then the IPSP seeks the optimal s, τ, t_p, and t_r to maximize π_1 subject to the customer's requirements. Finally, the customer can recalculate π_2 based on knowledge of $\{s, \tau, t_p, t_r\}$ to ensure that the LSC does not exceed π_2^{max}. Because of the information symmetry between the supplier and the customer, the first-best solution is guaranteed.

Given four decision variables, Model 11.6 can be tackled by a gradient-based heuristics search or enumeration methods. The procedure of the gradient-based heuristics method is summarized in the following steps, and the detailed derivation of the gradient with respect to τ, t_p, and t_r is available in Jin et al. (2015).

Step 1: Initialize and set $s = 0$, $\tau = \tau_{min}$, $t_p = t_p^{min}$, and $t_p = t_r^{min}$.

Step 2: Set $\tau = \tau_{min} + \Delta\tau$, $t_p = t_p + \Delta t_p$, and $t_p = t_r + \Delta t_r$, where $\Delta\tau$, Δt_p, and Δt_r are incremental steps.

Step 3: Compute the system availability using Eq. (7.716) based on s, τ, t_p, and t_r. If $A < A_{min}$, let $s = s + 1$, and go back to Step 2.

Step 4: Compute the gradients according to Eqs. (C7), (C11), and (C14) in Appendix C in Jin et al. (2015). Update $x = x + \Delta x$ or $x = x - \Delta x$ according to the gradient signs where $x \in \{\tau, t_p, t_r\}$. Here Δx is the incremental step.

Step 5: Check the validity of x based on the constraints from (11.5.33) to (11.5.37). If all constraints are satisfied, compute the objective function $\pi_1(s, \tau, t_p, t_r)$. If any constraint is violated, let $s = s + 1$, and go back to Step 2.

Step 6: The algorithm terminates if $\pi_1(s, \tau, t_p, t_r)$ stops increasing, or it reaches a predefined threshold, say $|\pi_1^c - \pi_1^p| < \varepsilon$, where π_1^c and π_1^p are the current and the previous value, and ε is a small threshold. Finally, π_2 is computed based on the current solution.

For a single-item system with only four decision variables, we can also use the enumeration method to search for the optimum solution for π_1. Starting with $s = 0$, we sequentially increase τ, t_p, and t_r with a small step, and compute the objective function (11.5.32). The iteration resulting in the largest π_1 is recorded at present s. Then we increase $s = s + 1$ and the process is repeated until all the variables reach the upper limits. Sherbrooke (1992) showed that the upper limit of s tends to be very small for critical components with sporadic failures. The iteration that yields the largest π_1 is designated as the best solution for the current setting. The optimality is generally guaranteed as long as the step size for τ, t_p, and t_r is sufficiently small.

11.6   Case Study: Supporting Wind Generation Using PBC

11.6.1   Line Replaceable Units of Wind Turbine

A wind turbine (WT) is a complex electromechanical system comprising multiple subsystems, including rotor, drive chain, nacelle, balance of station, control mechanism, and tower. Each system may consist of multiple items called line replaceable units, such as blades, main shaft/bearing (MS/B), gearbox, generator, AC–DC and DC–AC converters, and hydraulic and cooling systems. A breakdown of a wind turbine system at subsystem and item levels is listed Table 11.4 (NREL 2012).

Table 11.4 Decomposition of a wind turbine system.

	Primary line replaceable units
Subsystem	**1. Rotor**
Item	1.1. Blades
Item	1.2. Hub
Item	1.3. Pitch mechanisms and bearings
Item	1.4. Spinner, nose cone
Subsystem	**2 Drive train, nacelle**
Item	2.1. Low-speed shaft
Item	2.2. Bearings
Item	2.3. Gearbox
Item	2.4. Mechanical brake, high-speed coupling, and associated components
Item	2.5. Generator
Item	2.6. Variable-speed electronics
Item	2.7. Yaw drive and bearing
Item	2.8. Main frame
Item	2.9. Electrical connections
Item	2.10. Hydraulic and cooling systems
Item	2.11 Nacelle cover
Subsystem	**3. Balance of station**
Item	3.1. Foundation/support structure
Item	3.2. Transportation
Item	3.3. Roads, civil work
Item	3.4. Assembly and installation
Item	3.5. Electrical interface/connections
Item	3.6. Engineering permits
Subsystem	**4. Control, safety system, and condition monitoring**
Subsystem	**5. Tower**

High system availability is desirable as wind farm owners are able to harvest more energy yield given the intermittent wind speed. Due to the complexity of technology and difficulties in accessibility, maintenance and repair of wind turbines are usually undertaken by OEM or third party logistics. When a turbine fails, technicians have to access the nacelle and remove the defective item using a crane (if the parts are bulky and heavy). For off-shore wind turbines, ferry boats or even helicopters must be used in order to access the nacelle. Many items like gearboxes and generators are repairable or remanufactured in the repair center for future use.

11.6.2 Maximizing the Service Profitability

We take the case studies from the work of Espiritu et al. (2012) to demonstrate the service profit maximization model based on Model 11.3. The service contract focuses on three items: blades, main shaft/bearing, and the gearbox, because they represent the majority field failures. Parameters related to the design, manufacturing, spare parts, and repairs of these items are listed in Table 11.5. The reliability data and relevant cost information are inferred from the studies in Tavner et al. (2007) and Fingersh et al. (2006).

The IPSP signed a service contract with the wind farmer and agreed to design, manufacture and sustain the operation of blades, MS/B, and gearbox of n WT systems over T_c years. Let α_i and s_i for $i = 1$, 2, and 3 be the failure rate and the spare parts quantity of these items, respectively. If the cost plus incentive fee (CPIF) payment scheme is offered by the wind farmer, the IPSP aims to maximize the service profit while meeting the reliability and availability criteria (see Model 11.7).

Model 11.7

$$\text{Max: } E_\rho[P(\boldsymbol{\alpha}, \mathbf{s}; \rho)] = G(A(\boldsymbol{\alpha}, \mathbf{s}; \overline{\rho})) - \sum_{i=1}^{3}(D_i(\alpha_i) - B_{1i}) - n\sum_{i=1}^{3} m_i(c_i(\alpha_i) - B_{2i})$$

$$- \sum_{i=1}^{3} m_i I_i(\alpha_i, s_i; \overline{\rho}) \tag{11.6.1}$$

Table 11.5 Reliability and cost data.

Index	$i = 1$	$i = 2$	$i = 3$
Item name	Blade	MS/B	Gearbox
m_i	3	1	1
α_{max} (faults/year)	0.2898	0.0312	0.1306
α_{min} (faults/year)	0.1560	0.0168	0.0703
$B_1($\$$)$	3 330 000	675 000	1 936 500
$B_2($\$$)$	333 000	67 500	193 650
$B_3($\$$)$	20 000	7000	12 000
φ	0.02	0.02	0.02
v	0.6	0.6	0.6
$\overline{\rho}$	1	1	1
c_r ($\$$/defective part)	40 000	50 000	60 000
t_r (days)	45	90	120
t_s (days)	3	4	6

Source: Reproduced with permission of World Scientific Publishing.

Subject to:

$$\alpha_{\min,i} \le \alpha_i \le \alpha_{\max,i}, \qquad \text{for } i = 1, 2, 3 \tag{11.6.2}$$

$$A(\alpha, s; \bar{\rho}) = \prod_{i=1}^{3} (A_i(\alpha_i, s_i; \bar{\rho})^{m_i} \ge A_{\min} \tag{11.6.3}$$

$$s_i \in \{0, 1, 2, \ldots\} \quad \text{for } i = 1, 2, 3 \tag{11.6.4}$$

Since wind farms are usually located in windy sites where the turbines are operating 24/7 mode, $\bar{\rho} = 1$ is assumed. Note that $A_i(\alpha_i, s_i; \bar{\rho})$ is the item operational availability, which is given in Eq. (7.7.11). The enumeration technique is adopted to find the optimal values of $\{s_1, s_2, s_3, \alpha_1, \alpha_2, \alpha_3\}$. The results are given in Table 11.6. Under the linear reward function with $\{a = \$3 \times 10^7, b_1 = \$3 \times 10^8, b_2 = 0\}$ the IPSP can reap nearly \$25.06 million profit for a five-year contract by keeping the system availability at 0.9889. The availability outcome is higher than $A_{\min} = 0.97$ because IPSP is motivated to maximize the service profit.

Model 11.7 is also solved under the exponential reward function. To make a fair comparison, we set $\{\gamma = 17.217, \rho_1 = 8.745, \rho_2 = 0\}$ such that the base revenue and maximum revenue are the same as in the linear model. The profit is reduced to \$24.78 million, less than the linear award function. However, the system availability still reaches 0.989, exceeding $A_{\min} = 0.97$. This finding shows that the OEM must spend more effort to gain the same amount of profit under the exponential reward function. The inventory decision on both cases happened to be the same. Seven spare blades are required because of the relatively high failure rate. Five gearboxes are needed in the stock considering the longer repair turn-around time. Spare main shafts/bearings are not needed because the reliability is high and repair time is short. This result is quite interesting as it shows that we are able to achieve a high operational availability if the parts reliability and repair logistics are appropriately coordinated.

Table 11.7 presents the optimal decision on λ_i and s_i under a 10-year service contract. Under the 10-year contract, the failure rates of three items are lower than those under the 5-year contract, both in the linear and the exponential cases. This indicates that the IPSP is willing to invest more resources on product reliability under a long-term contract. The second observation is that the overall system availability under the 10-year

Table 11.6 Performance outcome under a 5-year contract.

| CPIF | $T_c = 5$ years, $A_{\min} = 0.97$, and $n = 50$ systems | | | | | |
	Linear			Exponential		
i	1	2	3	1	2	3
m_i	3	1	1	3	1	1
Item	Blade	MS/B	Gearbox	Blade	MS/B	Gearbox
α	0.180	0.031	0.120	0.179	0.031	0.119
s	7	0	5	7	0	5
A_{item}	0.9981	0.9972	0.9974	0.9981	0.9972	0.9975
A_{sys}		0.9889			0.9890	
Profit		\$25.06 M			\$24.78 M	

Source: Reproduced with permission of World Scientific Publishing.

Table 11.7 Performance outcome under a 10-year contract.

CPIF	Linear			Exponential		
	$T_c = 10$ years, $A_{min} = 0.97$, and $n = 50$ systems					
i	1	2	3	1	2	3
m_i	3	1	1	3	1	1
Item	Blade	MS/B	Gearbox	Blade	MS/B	Gearbox
α	0.172	0.025	0.103	0.172	0.024	0.0990
s	8	0	5	8	0	5
A_{item}	0.9985	0.9981	0.9980	0.9985	0.9982	0.9982
A_{sys}		0.9916			0.9918	
Profit		$57.59 M			$57.11 M	

Source: Reproduced with permission of World Scientific Publishing.

contract is also higher than that of the 5-year contract. Therefore a long-term service contract is more preferable to the IPSP who intends to improve item reliability in design and manufacturing to reap more savings in after-sales support.

Appendix 11.A. Matlab Codes for Casey Study with Linear Reward

```
% Program to find the optimal reliability and inventory of blade, main
% bearing and the gearbox
clear
counter=0
fleet=50              % define the fleet size
beta=1                % define the usage rate (fully used)
fi=[0.02, 0.02, 0.02]   % define the coefficient for reliability
  improvement in design
lambda_max=[0.2898, 0.0312, 0.1306]   % max acceptable inherent
  failure rate in faults/year
lambda_min=[0.1560, 0.0168, 0.0703]   % best achievable failure rate
  in faults/year
m=[3, 1, 1]   % define the quantity per RLU type used in the system
n=fleet*m   % dedine the number of items per type
B1=[3330000, 675000, 1936500]   % define the design cost
B2=[333000, 67500, 193650]    % define the baseline manufacturing cost
B3=[20000, 7000, 12000]   % cost coefficient for improving manufacturing
mu=[0.6, 0.6, 0.6]   % coefficient for manufacturing-reliability
cr=[40000, 50,000, 60000]   % defective part repair cost (labor,
  facility, and transportation)
repair_days=[45, 90, 120]   % define the repair turn-around days
tr=repair_days/365   % compute the average defective part repair time in
  unit of years
ts=[3/365, 4/365, 6/365]   % converting hands-on time into unit of year
tao=5  % contractual planning horizon in years
theta=0.05  % interest rate
As_min=0.97  % minimum system availability criterion by the customer
```

```
Amin=[As_min^(1/3), As_min^(1/3), As_min^(1/3)]  % minimum subsystem
 cluster availability
lambda_op=[0, 0, 0]  % define the inital value of lambda
s_op=[0, 0, 0]  % define the initial value of s
A_LRU_op=[0, 0, 0]
fe=((1+theta)^tao-1)/(theta*(1+theta)^tao);  % compute the PWV
 of unity factor
cost_op=[10^12, 10^12, 10^12]  % initializing the solution

lpsize=300
loop1=lpsize;    % for lambda3
loop2=20;        % for s1
loop3=lpsize;    % for lambda1
loop4=20;        % for s2
loop5=lpsize;    % for lambda2
loop6=20;        % for s3
del_lambda=(lambda_max-lambda_min)/lpsize    % compute the step size
 of the lambda
lambda(1)=lambda_min(1);

% loop1 starts
for k1=1:loop1;
    lambda(1)=lambda(1)+del_lambda(1);
    % compute the incremental design cost;
    D_lambda(1)=B1(1)*(exp(fi(1)*(lambda_max(1)-lambda(1))/
        (lambda(1)-lambda_min(1)))-1);
    % compute the incremental manufacturing cost;
    c_lambda(1)=B2(1)+B3(1)*(1/lambda(1)^mu(1)-1/lambda_max(1)^mu(1));
    M_lambda(1)=n(1)*(c_lambda(1)-B2(1));

    % start the loop 2
    s(1)=loop2;
    for k2=1:loop2;
    s(1)=s(1)-1;

    % compute the cluster availability of Blade, bearing and generator
    temp2=0;
    x=-1;
    for i=1:s(1)+1;
        x=x+1;
        temp2=temp2+(n(1)*beta*tr(1)*lambda(1))^x*exp(-n(1)*beta*tr(1)
*lambda(1))/factorial(x);
    end
    temp3=1+beta*lambda(1)*ts(1)+beta*lambda(1)*tr(1)*(1-temp2);
    A(1)=1/temp3;
    A_LRU(1)=A(1)^m(1)        % compute the cluster availability

    % compute the inventory cost;
    I_spare(1)=s(1)*c_lambda(1);
    I_repair(1)=cr(1)*n(1)*beta*lambda(1)*fe;
    I_lambda(1)=I_spare(1)+I_repair(1);

    % total cost
    LCC(1)=D_lambda(1)+M_lambda(1)+I_lambda(1);

    % identify the best solution with lowest cost
```

```
    if (A_LRU(1)>=Amin(1))&&(LCC(1)<cost_op(1))
        cost_op(1)=LCC(1)
        A_LRU_op(1)=A_LRU(1)
        s_op(1)=s(1)
        lambda_op(1)=lambda(1)
    end
    end % end of loop of loop2
end % end of loop of loop1

 % ******************** Start the loop 3 *****************
 lambda(2)=lambda_min(2);
 for k3=1:loop3;
    lambda(2)=lambda(2)+del_lambda(2);

    % compute the incremental design cost;
    D_lambda(2)=B1(2)*(exp(fi(2)*(lambda_max(2)-lambda(2))/
(lambda(2)-lambda_min(2)))-1);

    % compute the incremental manufacturing cost;
    c_lambda(2)=B2(2)+B3(2)*(1/lambda(2)^mu(2)-1/lambda_max(2)^mu(2));
    M_lambda(2)=n(2)*(c_lambda(2)-B2(2));

    % start the loop 4
    s(2)=loop4;
    for k4=1:loop4;
    s(2)=s(2)-1;

    % compute the cluster availability
    temp2=0;
    x=-1;
    for i=1:s(2)+1;
        x=x+1;
        temp2=temp2+(n(2)*beta*tr(2)*lambda(2))^x*exp(-n(2)*beta*tr(2)
*lambda(2))/factorial(x);
    end
    temp3=1+beta*lambda(2)*ts(2)+beta*lambda(2)*tr(2)*(1-temp2);
    A(2)=1/temp3;

    % compute the cluster availability
    A_LRU(2)=A(2)^m(2);

    % compute the inventory cost;
    I_spare(2)=s(2)*c_lambda(2);
    I_repair(2)=cr(2)*n(2)*beta*lambda(2)*fe;
    I_lambda(2)=I_spare(2)+I_repair(2);

    % total cost
    LCC(2)=D_lambda(2)+M_lambda(2)+I_lambda(2);

    % identify the best solution with lowest cost
    if (A_LRU(2)>=Amin(2))&&(LCC(2)<cost_op(2))
        cost_op(2)=LCC(2)
        A_LRU_op(2)=A_LRU(2)
        s_op(2)=s(2)
        lambda_op(2)=lambda(2)
    end
```

```
    end    % end for loop4
  end  % end for loop3

  % *********************** Start the loop 5 *****************
    lambda(3)=lambda_min(3);
    for k5=1:loop5;
    lambda(3)=lambda(3)+del_lambda(3);

    % compute the incremental design cost;
    D_lambda(3)=B1(3)*(exp(fi(3)*(lambda_max(3)-lambda(3))/
(lambda(3)-lambda_min(3)))-1);

% compute the incremental manufacturing cost;
    c_lambda(3)=B2(3)+B3(3)*(1/lambda(3)^mu(3)-1/lambda_max(3)^mu(3));
    M_lambda(3)=n(3)*(c_lambda(3)-B2(3));

% start the loop 6
    s(3)=loop6;
    for k6=1:loop6;
    s(3)=s(3)-1;

% compute the cluster availability
    temp2=0;
    x=-1;
    for i=1:s(3)+1;
        x=x+1;
        temp2=temp2+(n(3)*beta*tr(3)*lambda(3))^x*exp(-n(3)*beta*tr(3)
*lambda(3))/factorial(x);
    end
    temp3=1+beta*lambda(3)*ts(3)+beta*lambda(3)*tr(3)*(1-temp2);
    A(3)=1/temp3;

    % compute the cluster availability
    A_LRU(3)=A(3)^m(3);

% compute the inventory cost;
    I_spare(3)=s(3)*c_lambda(3);
    I_repair(3)=cr(3)*n(3)*beta*lambda(3)*fe;
    I_lambda(3)=I_spare(3)+I_repair(3);

% total cost
    LCC(3)=D_lambda(3)+M_lambda(3)+I_lambda(3);

% identify the best solution with lowest cost
    if (A_LRU(3)>=Amin(3))&&(LCC(3)<cost_op(3))
        cost_op(3)=LCC(3)
        A_LRU_op(3)=A_LRU(3)
        s_op(3)=s(3)
        lambda_op(3)=lambda(3)
    end
    end    % end of loop 6
    end    % end of loop 5

% compute the profit margin using linear function
a=3*10^7;
```

```
b=10*a;
Asys=prod(A_LRU_op);
profit_op=a+b*(Asys-As_min)-sum(cost_op)

% using GA to find the optimal solution
% create some variations with 5% of the mean
sigma_lam=0.05*lambda_op

% create some variations with 20% of the mean
sigma_s=0.2*s_op
loops=1000;    % define the generation number of GA

for i=1:loops;
% randomly generate a new solution
    for j=1:3;
    lambda(j)=lambda_op(j)+(2*rand()-1)*sigma_lam(j);
    s(j)=round(s_op(j)+(2*rand()-1)*sigma_s(j));
    % make judgment on the validity of the new potential solution
        if (lambda(j)>=lambda_max(j))||(lambda(j)<=lambda_min(j));
           lambda(j)=lambda_op(j);
        end
        if s(j)<0;
            s(j)=s_op(j);
        end;
    end

    % recomputed the cost and availability of LRU
    for j=1:3;

    % compute the incremental design cost;
    D_lambda(j)=B1(j)*(exp(fi(j)*(lambda_max(j)-lambda(j)))/
(lambda(j)-lambda_min(j)))-1);

    % compute the incremental manufacturing cost;
    c_lambda(j)=B2(j)+B3(j)*(1/lambda(j)^mu(j)-1/lambda_max(j)^mu(j));
    M_lambda(j)=n(j)*(c_lambda(j)-B2(j));

    % compute the cluster availability
    temp2=0;
    x=-1;
    for i=1:s(j)+1;
        x=x+1;
        temp2=temp2+(n(j)*beta*tr(j)*lambda(j))^x*exp(-n(j)*beta*tr(j)
*lambda(j))/factorial(x);
    end
    temp3=1+beta*lambda(j)*ts(j)+beta*lambda(j)*tr(j)*(1-temp2);
    A(j)=1/temp3;

    % compute the cluster availability
    A_LRU(j)=A(j)^m(j);

    % compute the inventory cost;
    I_spare(j)=s(j)*c_lambda(j);
    I_repair(j)=cr(j)*n(j)*beta*lambda(j)*fe;
    I_lambda(j)=I_spare(j)+I_repair(j);
```

```
    % total cost
    LCC(j)=D_lambda(j)+M_lambda(j)+I_lambda(j);
    end;

    % compute the system availability, compare the cost now
    Asys=prod(A_LRU);
    temp=a+b*(Asys-As_min)-sum(LCC);
    if (Asys>=As_min)&&(temp>profit_op);
    % compute the profit margin using linear function
     profit_op=temp;
     lambda_op=lambda;
     s_op=s;
     A_LRU_op=A_LRU;
     Asys_op=Asys
     counter=counter+1
     tempcost(counter)=temp
    end; % end of comparing the original solution
end

% output the results
    profit_op
    lambda_op
    s_op
    A_LRU_op
    Asys_op
    tempcost
```

References

Berkowitz, D., Gupta, J., Simpson, J.T., and McWilliams, J.B. (2004). Defining and implementing performance-based logistics in government. *Defense AR Journal* 11 (3): 254–267.

Bei, X., Chatwattanasiri, N., Coit, D., and Zhu, X. (2017). Combined redundancy allocation and maintenance planning using a two-stage stochastic programming model for multiple component systems. *IEEE Transactions on Reliability* 66 (3): 950–962.

Beuren, F.H., Gomes Ferreira, M.G., and Cauchick Miguel, P.A. (2013). Product-service systems: a literature review on integrated products and services. *Journal of Cleaner Production* 47 (5): 222–231.

Bjarnasona, E.T.S. and Taghipour, S. (2016). Periodic inspection frequency and inventory policies for a k-out-of-n system. *IIE Transactions*. 48 (7): 638–650.

Brax, S.A. (2005). A manufacturer becoming service provider – challenges and a paradox. *Managing Service Quality: An International Journal* 15 (2): 142–155.

Cassady, C.R., Pohl, E.A., and Jin, S. (2004). Managing availability improvement efforts with importance measures and optimization. *IMA Journal of Management Mathematics* 15 (2): 161–174.

Chantepy, L., (2017), "Minimizing downtime: moving from reactive to proactive operations," Forbes Business, available at: https://www.forbesmiddleeast.com/en/ (accessed on February 10, 2018).

Chen, L., Ye, Z.-S., and Xie, M. (2013). Joint maintenance and spare component provisioning policy for k-out-of-n systems. *Asia-Pacific Journal of Operational Research* 30 (6): 1–21.

CNET News, (2001), "California power outages suspended – for now," available at http://news.cnet.com/2100-1017-251167.html (accessed December 21, 2018).

Cohen, M., Agrawal, N., and Agrawal, V. (2006). Winning in the aftermarket. *Harvard Business Review* (May): 129–138.

Coit, D.W., Jin, T., and Wattanapongsakorn, N. (2004). System optimization considering component reliability estimation uncertainty: multi-criteria approach. *IEEE Transactions on Reliability* 53 (3): 369–380.

Cox, D.R. and Smith, W.L. (1954). On the superposition of renewal processes. *Biometrika* 41: 91–99.

Fingersh, L., M. Hand, A. Laxson, (2006), "Wind turbine design cost and scaling model." National Renewable Energy Lab, Technical Report, available at: http://www.phillipselectric.com/pdf/WTG_NREL-Wind-Turbine-Design-Costs-Model-2005.pdf (last accessed on January 23, 2016).

Espiritu, J., C.-H. Sung, T. Jin, H.-Z. Huang, (2012), "Contracting for performance-based maintenance service under profit maximization," in *Proceedings of the ISSAT Conference*, 2012, pp. 335–339.

Gebauer, H., Ren, G., Valtakoski, A., and Reynoso, J. (2012). Service-driven manufacturing. *Journal of Service Management* 23 (1): 120–136.

Godoy, D.R., Pascual, R., and Knights, P. (2014). A decision-making framework to integrate maintenance contract conditions with critical spares management. *Reliability Engineering and System Safety* 131 (11): 102–108.

Gupta, O.K. and Ravindran, A. (1985). Branch and bound experiments in convex nonlinear integer programming. *Management Science* 31 (12): 1533–1546.

Huang, H.-Z., Liu, H.J., and Murthy, D.N.P. (2007). Optimal reliability, warranty and price for new products. *IIE Transactions* 39 (8): 819–827.

Huikkola, T., Kohtamäki, M., and Rabetino, R. (2016). Resource realignment in servitization. *Research-Technology Management* 59 (4): 30–39.

Horenbeek, A., Scarf, P., Cavalcante, C., and Pintelon, L. (2013). The effect of maintenance quality on spare parts inventory for a fleet of assets. *IEEE Transactions on Reliability* 62 (3): 596–607.

Ivanov, D., Dolgui, A., Sokolov, B. et al. (2016). A dynamic model and an algorithm for short-term supply chain scheduling in the smart factory industry 4.0. *International Journal of Production Research* 54 (2): 386–402.

Jin, J. and Tian, Y. (2012). Optimizing reliability and service parts logistics for a time-varying installed base. *European Journal of Operational Research* 218 (1): 152–162.

Jin, T., Tian, Z., and Xie, M. (2015). A game-theoretical approach for optimizing maintenance, spares and service capacity in performance contracting. *International Journal of Production Economics* 31 (3): 31–43.

Jin, T. and Wang, P. (2012). Planning performance based contracts considering reliability and uncertain system usage. *Journal of the Operational Research Society* 63 (2): 1467–1478.

Kabir, A. and Al-Olayan, A. (1996). A stocking policy for spare part provisioning under age based preventive replacement. *European Journal of Operational Research* 90 (1): 171–181.

Kennedy, W.J., Patterson, J.W., and Fredenhall, L.D. (2002). An overview of recent literature on spare parts inventories. *International Journal of Production Economics* 76 (2): 210–215.

Lee, J., Bagheri, B., and Kao, H.-A. (2013). Recent advances and trends in predictive manufacturing systems in big data environment. *Manufacturing Letters* 1 (1): 38–41.

Lin, Y.-K., Lin, J.-J., and Yeh, R.-H. (2013). A dominant maintenance strategy assessment model for localized third-party logistics service under performance-based consideration. *Quality Technology and Quantitate Management* 10 (2): 221–240.

Liu, X. and Tang, L.C. (2016). Reliability analysis and spares provisioning for repairable systems with dependent failure processes and time-varying installed base. *IIE Transactions* 48 (1): 48–56.

Loerch, A.G. (1999). Incorporating learning curve costs in acquisition strategy optimization. *Naval Research Logistics* 46: 255–271.

Lucintel Report, (2013), "Growth opportunities in wind operation and maintenance services market," available at: http://www.lucintel.com/wind-maintenance-services-2018.aspx (accessed on February 10, 2018).

Marseguerra, M., Zio, E., and Podofillini, L. (2005). Multi-objective spare parts allocation by means of genetic algorithms and Monte Carlo simulations. *Reliability Engineering and System Safety* 87 (3): 325–335.

Martinez, V., Bastl, M., Kingston, J., and Evans, S. (2010). Challenges in transforming manufacturing organisations into product-service providers. *Journal of Manufacturing Technology Management* 21 (4): 449–469.

Mettas, A. (2000). Reliability allocation and optimization for complex systems. *Proceedings of Reliability and Maintainability Symposium* 216–221.

Nordrum, A., (2016), "Popular Internet of Things forecast of 50 billion devices by 2020 is outdated," IEEE Spectrum, 2016, available at: https://spectrum.ieee.org/tech-talk/telecom/internet (accessed on February 2018).

Nourelfth, M., Châtelet, E., and Nahas, N. (2012). Joint redundancy and imperfect preventive maintenance optimization for series–parallel multi-state degraded systems. *Reliability Engineering and System Safety, vol.* 103 (July): 51–60.

Nowicki, D., Kumar, U.D., Steudel, H.J., and Verma, D. (2008). Spares provisioning under performance-based logistics contract: profit-centric approach. *Journal of the Operational Research Society* 59 (3): 342–352.

NREL, (2012), "Distributed generation energy technology operations and maintenance costs," Report of National Renewable Energy Laboratory, available at: http://www.nrel.gov/analysis/tech_cost_om_dg.html (last accessed on February 24, 2012).

Oliva, R. and Kallenberg, R. (2003). Managing the transition from products to services. *International Journal of Service Industry Management* 14 (2): 160–172.

Öner, K.B., Kiesmüller, G.P., and van Houtum, G.J. (2010). Optimization of component reliability in the design phase of capital goods. *European Journal of Operational Research* 205 (3): 615–624.

Öner, K.B., Scheller-Wolf, A., and van Houtum, G.J. (2013). Redundancy optimization for critical components in high-availability technical systems. *Operations Research* 61 (2): 244–264.

Rees, J.D. and Van den Heuvel, J. (2012). Know, predict, control: a case study in services management. *Proceedings of Reliability and Maintainability Symposium* 1–6.

Richardson, D. and Jacopino, A. (2006). Use of r&m measures in Australian defense aerospace performance based contracts. *Proceedings of Reliability and Maintainability Symposium* 331–336.

Selçuk, B. and Ağralí, S. (2013). Joint spare parts inventory and reliability decisions under a service constraint. *Journal of the Operational Research Society* 64: 446–458.

Sherbrooke, C.C. (1992). Multiechelon inventory systems with lateral supply. *Naval Research Logistics* 39: 29–40.

Sleptchenko, A. and Van der Heijden, M. (2016). Joint optimization of redundancy level and spare part inventories. *Reliability Engineering and System Safety.* 153: 64–74.

Smidt-Destombes, K.S., Heijden van der, M.C., and Harten van, A. (2009). Joint optimisation of spare part inventory, maintenance frequency and repair capacity for *k*-out-of-*N* systems. *International Journal of Production Economics* 118 (1): 260–268.

Smith, S. J., (2007), "Equipping the American forces with a reliable and robust condition based maintenance plus capability," USAF Report, available at: www.acq.osd.mil/log/mpp/cbm+/_LG301T6_FINAL.PDF (accessed on February 10, 2018).

Smith, T. C., (2004), "Reliability growth planning under performance based logistics," in *Proceedings of Reliability and Maintainability Symposium*, 2004, pp. 418–423.

Tavner, P.J., Xiang, J., and Spinato, F. (2007). Reliability analysis for wind turbines. *Wind Energy* 10 (1): 1–18.

Van Jaarsveld, W. and Dekker, R. (2011). Spare parts stock control for redundant systems using reliability centered maintenance data. *Reliability Engineering and System Safety* 96 (11): 1576–1586.

Vandermerwe, S. and Rada, J. (1988). Servitization of business: adding value by adding services. *European Management Journal* 6 (4): 314–324.

Vaughan, T. (2005). Failure replacement and preventive maintenance spare parts ordering policy. *European Journal of Operational Research* 161 (1): 183–190.

Wang, W. (2012). A stochastic model for joint spare parts inventory and planned maintenance optimization. *European Journal of Operational Research* 216 (1): 127–139.

Wang, L., Chu, J., and Mao, W. (2009). A condition-based replacement and spare provisioning policy for deteriorating systems with uncertain deterioration to failure. *European Journal of Operational Research* 194 (1): 184–205.

Wang, S., Wan, J., Zhang, D. et al. (2016). Towards smart factory for industry 4.0: a self-organized multi-agent system with big data based feedback and coordination. *Computer Networks* 101 (6): 158–168.

Wikipedia, (2018), Internet of Things, available at: https://en.wikipedia.org/wiki/Internet_of_things (accessed February, 2018).

Wu, X. and Ryan, S.M. (2014). Joint optimization of asset and inventory management in a product-service system. *The Engineering Economist* 59 (2): 91–115.

Xiang, Y., Zhu, Z., Coit, D., and Feng, Q. (2017). Condition-based maintenance under performance-based contracting. *Computers and Industrial Engineering* 111 (September): 391–402.

Xie, W., Liao, H., and Jin, T. (2014). Maximizing system availability through joint decision on redundancy allocation and spares inventory. *European Journal of Operational Research.* 237 (1): 164–176.

Xu, L.D., He, W., and Li, S. (2014). Internet of Things in industries: a survey. *IEEE Transactions on Industrial Informatics* 10 (4): 2233–2243.

Yeh, W.-C. and Lin, C.-H. (2009). A squeeze response surface methodology for finding symbolic network reliability functions. *IEEE Transactions on Reliability* 58 (2): 374–382.

Zhang, C., Ramirez, J.E., and Li, Q. (2018). Locating and protecting facilities from intentional attacks using secrecy. *Reliability Engineering and System Safety* 169: 51–62.

Problems

Problem 11.1 Find 2 or 3 industry cases to illustrate the fact that capital equipment manufacturers are now leaning toward the product-service offering. What are the main drivers behind this paradigm change?

Problem 11.2 There are four technological factors fostering the integration of product and services. Take an example from the automobile, wind power, or transportation industries to explain how these technologies facilitate the product-service offering, including design, development, prototyping, manufacturing, and in-service support.

Problem 11.3 Section 11.2.3 surveys the literature related to the modeling and design of a product-service system focused on reliability, spare parts, and maintenance. Particularly, reliability-inventory, redundancy-inventory, and maintenance-inventory models are discussed. Could you expand the literature review scope and find 1 or 2 new modeling frameworks that are also under the PSS framework.

Problem 11.4 Produce a reliability design cost curve given $r_{min} = 0.5$, $r_{max} = 0.95$, $B_1 = 100$ with different values of ψ for $\psi=0.3$, $\psi=0.5$, and $\psi=0.8$. What conclusion can you draw?

Problem 11.5 Show that $D(\alpha)$ in Eq. (11.3.1) is a convex function of α. Also show it is valid for $D(r)$ in Eq. (11.3.2).

Problem 11.6 Find the reliability manufacturing cost function given $\alpha_{min} = 0.01$, $\alpha_{max} = 0.05$, $B_2 = 100$, and $B_3 = 20$ in the following cases: (1) $v = 0.2$; (2) $v = 0.5$; (3) $v = 0.7$. What conclusion can you draw?

Problem 11.7 Show that $c(\alpha)$ in Eq. (11.3.3) is a convex function of α.

Problem 11.8 The initial unit production cost $c_1 = \$300$, and after making 400 pieces, the cumulative average unit cost $c_Q = \$170$. Estimate the learning exponent.

Problem 11.9 A fleet of 50 items are operating in the field and the item failure rate is $\alpha = 0.02$ failures/month. The repair cost of a failed item is $c_r = \$3000/item$. Assume that all items operate in the 24/7 mode. Do the following: (1) Estimate the annual repair cost of supporting this fleet product. (2) If the service contract is 5 years and the base stock level $s = 2$ parts, what is the present worth during five years?

Problem 11.10 Assume that an OEM is responsible for design, manufacturing, and support for a fleet of single-item systems. The following data are given: $\alpha_{max} = 0.3$ failures/year, $\alpha_{min} = 0.15$ failures/year, $B_1 = \$2\,M$, $B_2 = \$0.3\,M$, $B_3 = \$0.02\,M$, $\varphi = 0.03$, $v = 0.55$, and $c_r = \$50\,000/item$

Based on Model 11.1, minimize the cost in the following cases: (1) contract period $T_c = 5$ years and $A_{min} = 0.97$, (2) contract period $T_c = 5$ years and $A_{min} = 0.99$, (3) contract period $T_c = 10$ years and $A_{min} = 0.97$, and (4) contract period $T_c = 10$ years and $A_{min} = 0.99$.

Problem 11.11   For the linear reward function in Figure 11.6a, plot $G(A)$ in the following cases: (1) $a = 1000$, $b_1 = 500$, $b_2 = 400$, $A_{min} = 0.9$; (2) $a = 0$, $b_1 = 500$, $b_2 = 400$, $A_{min} = 0.9$; (2) $a = 2000$, $b_1 = 500$, $b_2 = 400$, $A_{min} = 0.9$; (4) $a = 1000$, $b_1 = 500$, $b_2 = 400$, $A_{min} = 0.85$.

Problem 11.12   For the exponential reward function in Figure 11.6b, plot $G(A)$ in the following cases: (1) $\gamma = 1$, $\rho_1 = 2$, $\rho_2 = 3$, $A_{min} = 0.9$; (2) $\gamma = 0$, $\rho_1 = 2$, $\rho_2 = 3$; $A_{min} = 0.9$; (2) $\gamma = 2$, $\rho_1 = 2$, $\rho_2 = 3$, $A_{min} = 0.9$; (4) $\gamma = 1$, $\rho_1 = 2$, $\rho_2 = 3$, $A_{min} = 0.85$.

Problem 11.13   For a 5-year service contract, use the data in Problem 10.10 to solve the optimization problem based on Model 11.3 assuming the reward function has the following value: (1) $G(A) = 0$, (2) $G(A) = \$30$ M, (3) $G(A) = \$30 \times 10^6 \times A$, where A is the achieved system availability

Problem 11.14   Use a simulation to show the superimposed renewal process in the following cases, where α is the scale parameter and β is the shape parameter of a Weibul lifetime.

Case	Fleet size	α	β	Case	Fleet size	α	β
1	5	1	1	5	10	1	0.5
2	5	1	3	6	20	1	0.5
3	10	1	1	7	50	1	1
4	10	1	3	8	50	1	0.5

Problem 11.15   Assume that preventive maintenance is adopted. Based on α and β in Problem 11.14, use simulation to show the superimposed renewal process. For each case, we assume the maintenance interval $\tau = 0.5$, $\tau = 1$, and $\tau = 1.5$, respectively.

12

Resilience Engineering and Management

12.1 Introduction

Unlike reliability issues, events considered in resilience management possess two unique features: (i) high impact with low probability of occurrence and (ii) catastrophic damages resulting from cascading failure. For instance, the 2012 Hurricane Sandy inflicted nearly $69 billion in damage to the east coast of the US, resulting in power outage for weeks or even a month in certain affected areas. Hence, enhancing the resilience of engineering systems like the electric grid and transportation networks against extreme events (such as hurricanes) becomes a fundamental task for engineers and policy makers. This chapter introduces the resilience concept, modeling method, and design strategies with the focus on the power grid system. Section 12.2 introduces the concept of a resilience curve from which several performance measures are derived. Section 12.3 compares the difference between reliability and resilience perceived from power grid operation, and probabilistic models simulating the hurricane landing characteristics are provided. Section 12.4 elucidates three design aspects of a resilience system, namely prevention, survivability, and recovery. In Section 12.5, we use moment methods and stochastic models to characterize the intermittency of wind and solar generation, and the demand response program. Section 12.6 solves a distributed generation planning problem with two objectives: (i) improving grid protection and survivability via a microgrid operation and (ii) accelerating the recovery of damaged assets through proactive maintenance and multiteam repair.

12.2 Resilience Concept and Measures

12.2.1 Resilience in Different Domains

Resilience has emerged in the last decade as a concept for better understanding the performance of systems, especially their behavior during and after the occurrence of disruptive events, such as natural hazards, technical failures, economic crises, and adverse threats. The word resilience originates from the Latin word "resiliere" meaning "to bounce back." For instance, the disaster resilience of a critical infrastructure (TISP 2006) is characterized as the capability to prevent or protect against significant multihazard threats and incidents, and to restore critical services with minimum devastation to public safety and health. Vugrin et al. (2010) defined system resilience as

Reliability Engineering and Services, First Edition. Tongdan Jin.
© 2019 John Wiley & Sons Ltd. Published 2019 by John Wiley & Sons Ltd.
Companion website: www.wiley.com/go/jin/serviceengineering

follows: given the occurrence of a particular disruptive event, the resilience of a system against that event is its ability to efficiently reduce both the magnitude and duration of deviation from expected performance levels. There are two key elements in this definition: the system impact measured by the difference between the target level and the degraded performance and the recovery efforts measuring the amount of resources involved in the restoration process. Several alternative definitions of resilience are also available, and though similar, they all incorporate a number of existing concepts, such as robustness, survivability, agility, and recoverability, among others. Today resilience generally implies the ability of a system or entity to withstand or return to the normal state after the onslaught of a disruptive event including man-made errors.

Due to its significance, the concept of resilience has been explored in a wide array of disciplines, including psychology, ecology, enterprise, supply chain, and engineering design, among others. The application domains of resilience can be classified into four categories: organizational, social, economic, and engineering (Yodo and Wang 2016). Note that resilience in some applications may overlap due to their mutual dependency or interconnections. For instance, resilience of societal safety depends on the reliability and resilience of engineering infrastructure systems such as transportation and communication networks. Related definitions of resilience in each category is elaborated below.

12.2.1.1 Organizational Resilience

Organizational resilience is referred to as the ability of an organization, particularly business entity, to prepare for, respond, and adapt to sudden disruptions in order to survive and prosper. It goes beyond the traditional risk management and seeks a more holistic approach to business health and success. Excellent product, reliable process, and quality people are generally treated as three essential elements of organizational resilience. In addition, information resilience, operational resilience, and supply chain resilience represent the three functional domains that unleash the potential of organizational resilience within a firm or enterprise system.

12.2.1.2 Social Resilience

Social resilience aims to characterize the capacity of individuals and groups (i.e. family, community, and country) to be more generative during times of stability and to adapt, reorganize, and grow in response to disruptive events. Keck and Sakdapolrak (2013) viewed social resilience as an entity comprised of three dimensions: coping capacity, adaptive capacity, and transformative capacity. The concept of social resilience has been extensively studied in subdomains such as sociology, psychology, and ecology. The core values of social resilience include: (i) promoting survival, respect, engagement, dignity, and livelihood; (ii) creating capacity for creativity and generativity at all levels; (iii) enhancing human and social capital; and (iv) engaging stakeholders at all levels.

12.2.1.3 Economic Resilience

Martin (2012) described economic resilience as "the capacity to reconfigure and adapt its structure, including firms, industries, technologies, and institutions, so as to maintain an acceptable growth path in output, employment, and wealth over time." Economic resilience can be strengthened by implementing policies aimed at mitigating the risks and consequences of severe crises. This requires that the system should be able to monitor economic vulnerabilities, adjust, and implement policy mechanisms so as to absorb the impact of an adverse downturn. "Singapore Paradox," termed by Briguglio (2002), refers to the reality that the island state is highly exposed to external uncertainties and shocks, and yet the country has managed to maintain a high economic growth rate and gross domestic product per capita. This success can be ascribed to the ability of Singapore to build its economic resilience.

12.2.1.4 Engineering Resilience

Compared with the other three domains, the concept of resilience in the engineering domain is relatively new. The engineering domain includes technical systems or critical infrastructures that are designed to facilitate transportation, communication, logistics and material delivery, production, energy generation and use, and national security, among others. The American Society of Mechanical Engineers (ASME) (2009) defined engineering resilience as the ability of a system to sustain internal and external disruptions without discontinuity of performing its intended function, or to rapidly recover the function if it is disconnected. In a military operation, a resilient engineered system is defined as being capable of completing its planned missions in the face of environmental and adversarial threats, and has the capabilities to flexibly adapt to future missions with evolving threats (Small et al. 2017). With respect to the resilience of industrial processes, Dinh et al. (2012) identify six factors for resilience enhancement, including early detection, controllability, flexibility, administrative controls and procedure, limitation of effects, and minimization of failure. In the infrastructure domain, engineering resilience has grown as a proactive approach to enhance its ability to prevent damage prior to disturbance events, mitigate losses during the events, and improve the recovery capability after the event (Woods 2015). These new approaches have gone beyond the concept of traditional reliability engineering.

Our daily life and businesses are heavily dependent on the reliable functioning of critical infrastructures, such as the power grid, water distribution, nuclear plants, transportation and logistics, and communication networks. These critical infrastructures can be considered as a subdomain of the engineering systems, as their construction, operation, and restoration require sophisticated technologies and resources. The National Infrastructure Advisory Council (NIAC) (NIAC 2009) defines the resilience of infrastructure systems as their ability to predict, absorb, adapt, and/or quickly recover from a disruptive event such as a natural disaster. Modern infrastructure systems can greatly improve the economic efficiency and the social welfare of human beings by offering safe and fast transportation, delivering real-time information, and producing clean and affordable energy. As a result, critical infrastructures are also considered as a subdomain of social and economic resilience as their failure causes adverse impacts on a nation's security, economic growth, and risk to human life (MacKenzie and Barker 2013; DiPietro et al. 2014).

A power system is well suited to the resilience engineering domain of critical infrastructure because of its ubiquitousness and interdependency nature. Enhancing the resilience of the electric grid against extreme weather is becoming a fundamental task for all the stakeholders in energy generation, transmission, distribution, and consumption sectors. Panteli and Mancarella (2017) indicated that events considered in power resilience management possess two unique features: (i) high impact with low probability of occurrence and (ii) catastrophic consequence resulting from cascading failures (Dong and Cui 2016). For instance, Hurricane Sandy is an N-90 event with estimated economic losses of up to $50 billion (US DoC 2013). Obviously, a traditional N-1 reliability design is no longer sufficient to withstand such an extreme event. The 2008 Great Ice Storm in Southern China resulted in power loss of 14.66 million households across five provinces. The investigation by Zhou et al. (2011) concluded that the interdependency of coal mining, fuel lifeline, and harsh weather triggers the cascading failure of electricity generation and supply. Both examples indicate that resilience planning and management is becoming a critical issue as the grid is increasingly penetrated with variable energy resources (i.e. wind and solar), distributed energy storage systems, and a consumer trading economy.

12.2.2 Resilience Curves

When an engineering system is exposed to unexpected destructive events such as extreme weather, the system performance will be affected adversely, and indeed is likely to deteriorate during the course of the event. Since resilience is generally associated with the loss of system performance after a destructive event, a resilience curve can be represented as a system performance plotted against time. Let $P(t)$ be the performance index of the system at time t. Figure 12.1 shows how $P(t)$ changes under an adverse attack for systems with different resilience capacities: resilient system, less-resilient system, and nonresilient system.

Prior to the onset of the destructive event, three systems can operate reliably between $[0, t_1]$. Upon the destructive event attack at time t_1, the performance of these systems starts to decline from the initial $P(t) = P_0$ and reaches level P_d at time t_2. The maximum impact level, denoted P_{impact}, is defined as the difference between the initial P_0 and the lower level P_d, namely $P_{impact} = P_0 - P_d$. Three performance scenarios are shown in Figure 12.1. A resilient system is able to be resorted and returned to the normal performance level at time t_3. For a less-resilient system, though $P(t)$ stops declining, its performance cannot be fully restored to its initial level. If a system is non-resilient, its performance continues declining after t_2. In fact, the system completely fails or collapses at time t_4. Thus, the resilient system differs from other two systems in that it is capable

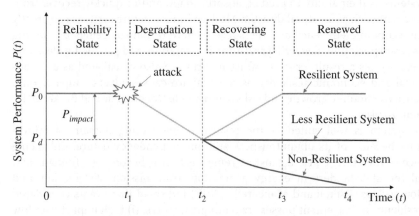

Figure 12.1 Behavior of systems with different resilience capacities.

of surviving or adapting to adverse events, and most importantly recovering itself to its normal state.

The resilience curve in Figure 12.1 is manifested by the system performance $P(t)$, which can be broken down into four states: reliability state, degradation state, recovering state, and renewed state. The interpretations of each state are described in the following (Yodo and Wang 2016):

(1) *Reliability state*: It is the baseline or original state where the system operates normally prior to the attack of disruptive events. This corresponds to the period of $[0, t_1]$.
(2) *Degradation state*: It is a vulnerable state where the system performance is declining from P_0 to P_d following a destructive event. This corresponds to the period $[t_1, t_2]$.
(3) *Recovering state*: In this state, the system performance is improving as a result of restoration efforts. This corresponds to the period of $[t_2, t_3]$.
(4) *Renewed state*: System performance reaches a newly established level after implementing the recovery procedures and $P(t)$ for $t \geq t_3$ is at least as good as prior to the disruptive event.

Apart from the resilience curve illustrated in Figure 12.2, several variations exist in modeling the resilience behavior of a system. These different versions originate from different behavior of system performance during the adaptation and restoration phases. In other words, the variations are mostly due to the differences in degradation, adaptation, and recovery processes for different engineering systems. Since a disruptive event typically varies in terms of severity and duration, both the adaptation path and the recovery path may also change in different scenarios. Figure 12.2 shows some conceptual profiles that lead to various forms of the resilience curve. For instance, the adaptation path could be a straight line AC, a curve AC (dashed line), or an "L" shaped line ABC. Meanwhile, the system performance after a disruptive event could be recovered in a linear fashion as line CD, or the recovery process follows a nonlinear path such as curve CE.

12.2.3 Resilience Measures

Though many resilience performance measures have been proposed in the literature, there still lacks the consensus on the standardization of performance characterization.

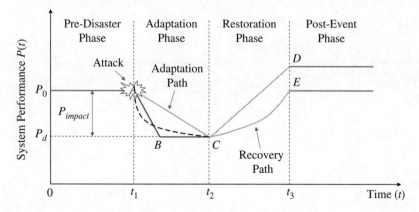

Figure 12.2 Variations of resilience curves.

This section introduces several quantitative measures derived from resilience curves. Other resilience measures are also proposed based on pre- and post-disaster performances, or on reliability and restoration processes. New resilience measures are continuously proposed in the literature, and readers are referred to the recent survey on this subject by Yodo and Wang (2016). In this section we discuss several resilience measures derived based on the performance ratio, impact area, and loss recovery concept. At the end, a five-dimensional view of resilience assessment is also presented.

System resilience can be defined as the ratio of the recovered performance over the loss of the performance. Let $\Psi(t)$ be the resilience measure of an engineering system at time t. Henry and Ramirez-Marquez (2012) and Barker et al. (2013) quantified the resilience as the ratio of recovered performance over the loss at a random point in time:

$$\Psi(t) = \frac{P(t) - P_d}{P_0(t) - P_d}, \quad \text{for } t_1 \le t \le t_3 \tag{12.2.1}$$

where $P_0(t)$ is the initial performance level of the system prior to the destructive event attack at t_1 and $P(t)$ is the actual degraded performance between t_1 and t_3. The value of $\Psi(t)$ varies between 0 and 1. For instance, $\Psi(t) = 1$ if $t = t_3$; this is because $P(t_3) = P_0$, as shown in Figure 12.3. On the other hand, $\Psi(t) = 0$ at $t = t_2$ due to the fact that $P(t_2) - P_d = 0$.

The shaded area in the resilience curve in Figure 12.3 is often referred to as the "impacted area," which defines the cumulative performance loss after the destructive event. Bruneau et al. (2003) used the loss of resilience to capture the performance loss. Denoted as A_{loss}, the resilience loss intends to measure the magnitude of the expected degradation in performance over the recovering time; mathematically P_{loss} is expressed as follows:

$$A_{loss} = \int_{t_1}^{t_3} (P_0(t) - P(t)) dt \tag{12.2.2}$$

If $P_0(t)$ is assumed to be constant with $P_0(t) = P_0$ for $t \in [t_1, t_3]$, then Eq. (12.2.2) can also be expressed as

$$A_{loss} = (t_3 - t_1)P_0 - \int_{t_1}^{t_3} P(t) dt \tag{12.2.3}$$

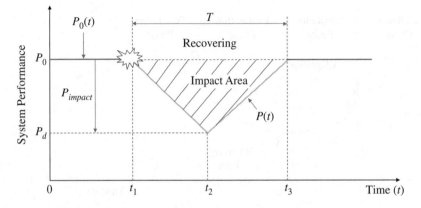

Figure 12.3 Performance loss under a disruptive event.

During the performance loss period from t_1 to t_3, some researchers (Ouyang et al. 2012; Shafieezadeh and Burden 2014) suggested that resilience be quantified by taking the ratio of the areas under the actual resilience curve $P(t)$ over the baseline resilience curve $P_0(t)$ across the time interval $[0, t]$. This metric is also known as the integral resilience (Dessavre et al. 2016) and is given as follows:

$$\Psi_{loss}(t) = \frac{\int_0^t P(x)dx}{\int_0^t P_0(x)dx}, \quad \text{for } t \geq t_1 \tag{12.2.4}$$

where $P_0(t)$ represents the system performance assuming no disruption between $[0, t]$ and $P(t)$ is the actual system performance in the presence of a disruption during the same time period. The value of Ψ_{loss} varies between $[0, 1]$ because $P(t) \leq P_0(t)$. Obviously $\Psi_{loss} = 1$ if $0 \leq t \leq t_1$ because $P(t) = P_0(t)$. If Ψ_{loss} approaches unity, this means the system becomes more resilient.

Instead of measuring the absolute performance, resilience can also be characterized and measured by the performance loss being recovered with respect to its impact level. Let $\Psi_r(t)$ be the resilience of a system at time t for $t \geq t_2$ in the recovering phase. It is used to describe the cumulative system performance that has been restored at time t, normalized by the expected cumulative system performance without being affected by disruption (Fang et al. 2016). That is,

$$\Psi_r(t) = \frac{\int_{t_2}^t [P(x) - P_d]dx}{\int_{t_2}^t [P_0(x) - P_d]dx}, \quad \text{for } t \geq t_2 \tag{12.2.5}$$

Graphically, $\Psi_r(t)$ aims to quantify the ratio of the area with dashed lines (i.e. the triangle BCD) versus the area of the shaded area (i.e. the rectangular $ABCE$), as shown in Figure 12.4.

Note that the formulation in Eq. (12.2.4) focuses mainly on the recoverability dimension and the value of $\Psi_r(t)$ falls in the range of $[0, 1]$. When $P(t) = P_d$, this means that a system resides in its disrupted state (i.e. there has been no "resilience" action); hence $\Psi_r(t) = 0$. As the system is recovering and $P(t)$ is increased to the ideal performance level $P_0(t)$, we have $\Psi_r(t) = 1$. It is worth mentioning that this definition of system resilience

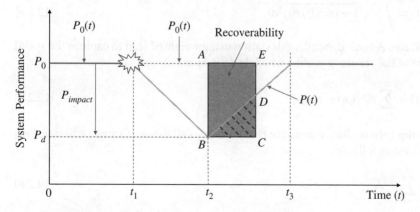

Figure 12.4 Recoverability dimension of resilience performance.

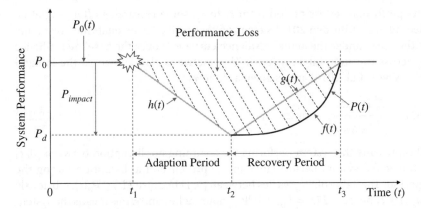

Figure 12.5 Six dimensions of the system resilience.

is not memoryless because it cumulatively calculates the restored system functionality; hence it differs from the Eq. (12.2.1).

In addition to the performance loss metric, other resilience metrics could also be derived from resilience curves. For instance, Munoz and Dunbar (2015) proposed five dimensions of resilience: recovery, impact, performance loss, recovery profiling, and weighted-sum index, as shown in Figure 12.5. The description of each dimension and the corresponding equations are summarized below.

- *Adaptation time*, $t_2 - t_1$: It defines the time duration for the system to reach the lowest performance before restoration action is taken.
- *Recovering time*, $t_3 - t_2$: It defines the time duration for the system to return to the acceptable performance level.
- *Impact size*, $P_0 - P(t_2)$ or $P_0 - P_d$: It measures the maximum loss of the system performance, where P_0 is the initial system performance at $t = 0$.
- *Performance loss*: This is defined in Eq. (12.2.2) or (12.2.3).
- *Profile length*: The length of the recovery profile as it reaches the acceptable performance level, which is given as

$$f(t) = \int_{t_1}^{t_3} \sqrt{1 + (dP(t)/dt)^2} dt \tag{12.2.6}$$

- *Weighted sum*: A time-dependent deviation using weighted sum to capture the speed and shape of the recovery profile, which is defined as

$$g(t) = \sum_{i=1}^{n} a_i[g(a_i) - p(a_i)] \tag{12.2.7}$$

Finally, a composite resilience measure that combines all five performance dimensions can be obtained as follows:

$$\Psi = \sum_{i=1}^{5} w_i \Psi_i \tag{12.2.8}$$

where Ψ_1 stands for recovery, Ψ_2 is for impact, Ψ_3 is for performance loss, Ψ_4 is for profile length, and Ψ_5 is for the weighted sum. Note that $w_1, w_2, \ldots, w_5$ are the weights corresponding to the different dimension of resilience.

12.3   Disaster Resilience Models of Power Grid

12.3.1   Reliability Versus Resilience

Reliability and resilience are both frequently, and often interchangeably, used in electricity generation and services, referred to as "keeping the lights on." The question is what is the major difference between reliability and resilience? For the electric sector, reliability can be defined as the ability of the power grid to deliver electricity to end users with guaranteed quality, quantity, and security in normal operating conditions. Hence reliability is a binary view of grid performance as the system is either functional or failed. In fact N-1 criterion is widely adopted in the power industry for reliability design of the power grid, which implies that the grid is able to meet the load even if the largest generating unit tripped or failed in contingency. There are many reasons that can lead to power failure, compromising the reliability of the electric grid under normal operating conditions, such as hot weather, aging equipment, vegetation, cascading failures, and cybersecurity faults. For example, the Northeast blackout of 2003 was caused by a software bug in the alarm system at a control room, causing an estimated 55 million people across Canada and the US to be without power (Barron 2003).

Grid resilience is concerned with a power system's ability to continue operating and delivering electricity in the event that low probability, high-consequence disruptions such as hurricanes, earthquakes, and cyberattacks occur (Zhang et al. 2015; Vugrin et al. 2017). The focus of grid resilience is on preventing, managing, and minimizing potential consequences resulting from these disruptions. Compared with reliability, resilience emphasizes the survivability, adaptability, and restoration capability when a system (like the power grid) fails in an extreme condition, The National Infrastructure Advisory Council (NIAC 2009) defines critical infrastructure resilience as "…the ability to reduce the magnitude and/or duration of very large scale events (VLSEs). The effectiveness of a resilient infrastructure or enterprise depends upon its ability to anticipate, absorb, adapt to, and/or rapidly recover from a potentially VLSE." The implication of resilience to the power industry is that the electric grid should possess the ability to withstand extraordinary and high-impact, low-probability events that may have never been experienced before (Panteli et al. 2016). Therefore, grid resilience usually prioritizes the concept that systems should be designed to be stronger against disruptive events and bounce back quickly after the disaster strikes.

In Table 12.1 the differences between grid reliability and resilience are compared in seven aspects, including the nature of disturbance, performance measures, assessment metrics, probability of event occurrence, the impact of the event, the controllability of event, and the design criteria. For example, though disturbance pertaining to grid reliability is uncertain, it is often stationary and tends to be predicable in terms of scale and time. A main uncertainty is the electricity demand, of which its peak value reaches the highest in the summer because of the hot weather. Now this spike can be predicted

Table 12.1 Comparisons between grid reliability and resilience.

No.	Features	Reliability	Resilience
1	Disturbance	Uncertain but stationary	Uncertain and non-stationary
2	Performance measure	Static measures	Dynamic measures
3	Assessment metrics	LOLP, LOLE, SAIDI, CAIDI, SAIFI	Impact size, performance loss, adaptation time, and recovery time
4	Probability of event	Medium or high	Very or extremely low
5	Impact of event	Small or medium	Very or extremely large
6	Controllability	Usually controllable	Typically uncontrollable
7	Design criteria	$N-1$	$N-k$ for $k \gg 1$

fairly well given advanced weather forecasting technology and big data analytics. The disturbance associated with resilience is also uncertain, but less predictable in terms of its size (i.e. scale) and occurrence time. For instance, it is generally known that hurricanes are likely to occur in the summer, and snow storms often take place in the winter, but how to precisely predict the actual occurrence time and the scale of the event still remain as a challenge to human beings.

In summary, reliability and resilience are both relevant concepts in relation to critical infrastructures like the power grid. Reliability, with its goal of keeping the lights on in the power grid, can be described as a static performance criterion of attaining the goal. In order to meet this goal in the case of extreme weather or an earthquake, the power grid needs to be resilient, as disastrous events might occur in a manner that gravely compromises the reliability of the electricity supply. Thus, grid resilience can be treated as a dynamic performance measure of which adapting to adverse situations and faster recovery are critical to the reliable operation of a modern power system.

12.3.2 Reliability Measures of Power Grid

Over 90% of electric power interruptions stem from disruptions on the distribution system both in terms of the duration and frequency of outages, which is largely due to weather-related events (Carlson et al. 2012). While much less frequent, damage to the transmission infrastructure results in more widespread power outages of customers, thus causing significant amounts of economic consequences. IEEE Standard 1366 has defined a set of reliability indices, such as System Average Interruption Frequency Index (SAIFI), System Average Interruption Duration Index (SAIDI), Customer Average Interruption Duration Index (CAIDI), Loss of Load Probability (LOLP), and Loss of Load Expectation (LOLE), to measure and characterize the reliability of the electric grid. These customer-based reliability service indexes are briefly reviewed as follows.

SAIFI is the System Average Interruption Frequency Index. This is a measure of the average number of interruptions in a year and is defined as

$$\text{SAIFI} = \frac{\text{Cumulative number of customer interruptions}}{\text{Total number of customers}} \tag{12.3.1}$$

SAIDI is the System Average Interruption Duration Index. It measures the average outage length in a year for each customer and is defined as

$$\text{SAIDI} = \frac{\text{Cumulative number of customer hours of interruptions}}{\text{Total number of customers}} \quad (12.3.2)$$

CAIDI is the Customer Average Interruption Duration Index. It measures the average duration of an interruption experienced by the affected customer and is defined as

$$\text{CAIDI} = \frac{\text{SAIDI}}{\text{SAIFI}} = \frac{\text{Cumulative number of customer hours of interruptions}}{\text{Total number of interrupted customers}} \quad (12.3.3)$$

ASAI is the Average System Availability Index. This is also known as the system reliability index and is defined as

$$\text{ASAI} = \frac{8760 - \text{SAIDI}}{8760} \quad (12.3.4)$$

where 8760 is the total number of hours in a year as the power must be ensured in 24/7 mode.

Example 12.1 In North America, for example, reliability is defined in terms of LOLE or LOLP, and is set at 0.1 day/year or equivalently LOLP = 0.000 274. From Eq. (12.3.2), this can be connected with SAIDI as follows:

$$\text{LOLP} = 1 - \text{ASAI} = \frac{\text{SAIDI}}{8760} \quad (12.3.5)$$

which leads to SAIDI = 8760 × LOLP = 8760 × 0.000 274 = 2.4 hours.

Figure 12.6 shows the SAIDI data of North American utilities between 2005 and 2012. The reliability performance continues to improve from 1.39 interruptions/year in 2005 to 1.08 in 2012. The second and fourth quantiles represent the optimistic and pessimistic values of SAIDI, respectively. Similar interpretations can be applied to SAIFI and CAIDI

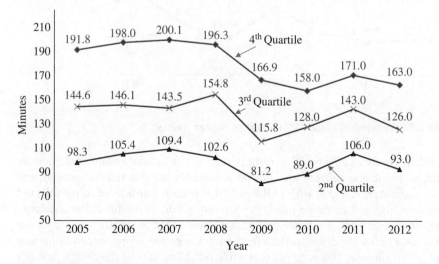

Figure 12.6 SAIDI between 2005 and 2012 for North American utilities.

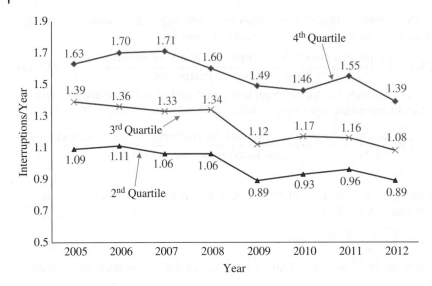

Figure 12.7 SAIFI between 2005 and 2012 for North American utilities.

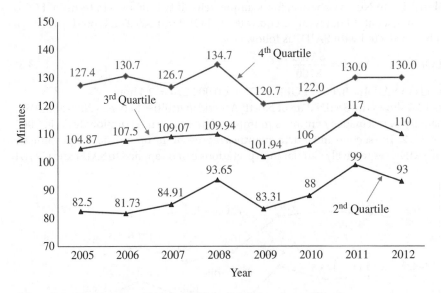

Figure 12.8 CAIDI between 2005 and 2012 for North American utilities.

as depicted in Figures 12.7 and 12.8, respectively. While reliability and resilience are correlated, a grid that is reliable may not be necessarily the one that is resilient, and vice versa. For instance, a grid with LOLP = 0.01 is usually considered unreliable, but its resilience is high if it could be quickly recovered within 30 minutes after a disaster attack. On the other hand, a reliable grid with a lower LOLP = 0.0001 may take a day for a full recovery after the disaster. Therefore, a quick recovery represents an important feature of grid resilience. This suggests that while reliability may be the design goal of a power grid, resilience could be an operational compromise that reflects the rapid change in environment. Therefore, resilience can be an additional yet necessary component of reliability assessment and design. Nevertheless, reliability should remain the end goal of a power system, as it does in other critical infrastructures.

12.3.3 Resilience Measures of Power Grid

At present, there are no formal grid resilience metrics or analytical methods that have been unanimously accepted across the academic and industry community. The metrics and methods described herein combine the results of the Resilience Analysis Process (RAP) developed by Watson et al. (2016) and the Resilience Metrics for the Electric Power Systems proposed by Vugrin et al. (2017). The combination allows for grid planners and operators to leverage system models and historical data to create consequence-based grid resilience metrics and further implement response decision-making strategy.

Broadly speaking, the impacts or consequences of power disruption due to extreme events can be perceived from two stakeholders: utility company and customers. Thus the consequence categories should appropriately reflect the resilience goals of different stakeholders. For the utility, the consequence assessment is directly focused on the impact that is often associated with such metrics as power not delivered, loss of revenue, and cost of repair and restoration. However, direct impacts are only part of the resilience aspect. Power systems provide energy not just for the sake of generating or distributing it, but for creating community and social benefit as well, such as healthcare, transportation, financial services, and education. Therefore, it is realistic and also necessary that resilience analyses should include a broader community perspective by taking into account customer outage consequences, such as business interruption costs, loss of customer goodwill, risk of human life, and vulnerability of national security, among others.

Table 12.2 provides a list of resilience metrics that could serve as the basis to assess the consequence and impact of a disastrous event on the electric grid. These measures are perceived from a utility and customer perspective, respectively. Both stakeholders involve functional loss and monetary cost, but as the utility is also responsible for recovering and bringing the damaged circuit back to a normal condition, the restoration response time and the mean-time-to-recovery are important measures for grid resilience. In addition, the utility may take a proactive approach to hardening an existing grid by adding onsite backup generating units or fortifying the poles or overhead lines.

From a geographic and duration perspective, the resilience measures in Table 12.2 can be classified into spatial and temporal dimensions. For instance, a backup generation belongs to the spatial resilience management because it serves a specific region or zone. The mean-time-to-recovery is a temporal resilience metric as it is characterized over the time horizon. The preventive maintenance cost involves both spatial and temporal dimensions because where and when the assets will be proactively inspected and maintained are associated with geographic and timing decision.

12.3.4 Modeling Hurricane Events

Due to the uncertainty of a natural disaster like hurricane or earthquake, the probability that grid components fail often differs in location and time. This implies that natural

Table 12.2 Grid resilience measures broken down by stakeholders.

Stakeholder	Consequence	Resilience metrics
Utility	Protection	1) Backup generation capacity 2) Percentage of load protected
	Functional loss	1) Cumulative customer-hours of outages 2) Customer energy demand not served 3) Percentage of loads that experience an outage
	Restoration	1) Restoration response time 2) Mean time-to-recovery
	Monetary cost	1) Investment cost for prevention and protection (e.g. backup generator, energy storage, and line redundancy) 2) Preventive maintenance cost 3) Loss of utility revenue 4) Grid damage cost 5) Repair and recovery cost
Customer	Functional loss	1) Percentage of critical load without power (e.g. hospitals, banking) 2) The minimum guaranteed hours for available power in an extreme event 3) The maximum outage duration (hours) of a critical load 4) Risk or vulnerability of national security
	Monetary cost	1) Business or production loss due to interruption 2) Loss of assets and perishable products 3) Indirect cost such as school closing and customer dissatisfaction

disasters must be modeled with spatial–temporal models (Li et al. 2017). First, when disasters occur on power systems, the failure probability of components is different in different geographical zones. Second, natural disasters are non-stationary random processes. As time passes, the destructive consequence of components also varies within the same zone. Since a natural disaster like a hurricane is a spatial–temporal non-stationary random process, the area of the distribution network is usually divided into several zones, and time (the disaster duration) in each zone is further divided into several periods. To perform the disaster assessment, we can assume that, in each time period, the failure probability in a particular zone remains constant, but differs in different zones. Then, for each time step and each zone, assessment tools such as the Monte Carlo method can be used to predict the hurricane strength and further analyze the damage. Results obtained at the current time period can be utilized to predict and evaluate the damage at the next time period.

There are three basic approaches to modeling hurricane characteristics: (i) empirical models, (ii) the sampling approach, and (iii) statistical models using probability distribution functions. Since various uncertainties are involved in a hurricane event, it is necessary to use a mix of the three modeling approaches. Xu and Brown (2008) classify hurricane landing characteristics into eight main features:

A) Annual occurrence
B) Approach angle
C) Translation speed
D) Central pressure difference

E) Radius to maximum wind
F) Wind speed decay rate
G) Central pressure filling rate
H) Maximum wind speed

Below we introduce the hurricane model originally proposed by Xu and Brown (2008). Both statistical and empirical methods describing the hurricane landing characteristics are presented.

12.3.4.1 Annual Occurrence

In literature the annual hurricane occurrence has been successfully modeled parametrically using Poisson distribution and negative binominal distributions (Elsner and Bossak 2001; Elsner and Jagger 2004; Georgiou et al. 1983). Let N be the number of hurricane events occurring in a fixed period (e.g. one year). The Poisson probability of the number of hurricane events can be expressed as

$$P(N = n) = \frac{\lambda^n e^{-\lambda}}{n!}, \quad \text{for } n = 0, 1, 2, \ldots \tag{12.3.6}$$

where λ is the known average occurrence rate. In the Poisson model, the future event happens independently of the time since the last event, and the time between two consecutive events follows the exponential distribution with a mean value of $1/\lambda$. Let x_i be the number of hurricane events in year i for $i = 1, 2, \ldots, K$, where K is the number of sampling years. The maximum likelihood estimate of the occurrences rate is

$$\hat{\lambda} = \frac{1}{K} \sum_{i=1}^{K} x_i \tag{12.3.7}$$

The unit of $\hat{\lambda}$ in Eq. (12.3.7) is the number of events per year. The drawback of the Poisson distribution is that in any two non-overlapping time intervals, the expected numbers of hurricane storms are equal given the same length of interval. Since hurricanes are more likely to happen in the summer than in the winter season, the negative binomial distribution allows for modeling the annual hurricane occurrence with unequal means of two time periods of identical length. The negative binomial distribution is given as

$$P(N = n) = \frac{\Gamma(n + k)}{\Gamma(n + 1)\Gamma(k)} \left(\frac{k}{m + k} \right)^k \left(\frac{m}{m + k} \right)^x \tag{12.3.8}$$

where $\Gamma(\bullet)$ is the gamma function and m and k are parameters of the distribution. The maximum likelihood estimates of m and k are given as follows:

$$\hat{m} = \frac{1}{K} \sum_{i=1}^{K} x_i \quad \text{and} \quad \hat{k} = \frac{\hat{m}^2}{s^2 - \hat{m}} \tag{12.3.9}$$

where s^2 is the sample variance. The difference between the Poisson distribution and the negative binominal distribution in modeling the annual hurricane frequency is negligible. This can be demonstrated by taking the numerical example from Oxenyuk (2014). Figure 12.9 compares the fit in the annual hurricane occurrence between the Poisson and negative binomial distributions.

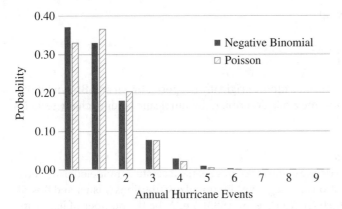

Figure 12.9 Poisson and negative binomial distributions fit for hurricanes.

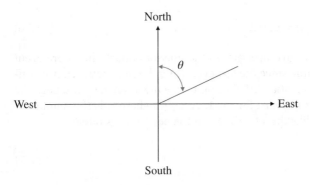

Figure 12.10 Hurricane approach angle.

12.3.4.2 Approach Angle

The approach angle defines the heading direction of a hurricane wind when it comes ashore. As shown in Figure 12.10, it is expressed to the nearest five degrees with North as zero degrees. Let Θ be the random variable for the approach angle and θ its realization. According to Toth and Szentimrey (1990) and Xu and Brown (2008), the approach angle can be modeled as the weighted sum of two normal distributions. Let w for $0 < w < 1$ be the weighting factor; then the probability density function (PDF) is

$$f_\Theta(\theta) = \frac{w}{\sigma_1\sqrt{2\pi}}e^{-\frac{(\theta-\mu_1)^2}{2\sigma_1^2}} + \frac{1-w}{\sigma_2\sqrt{2\pi}}e^{-\frac{(\theta-\mu_2)^2}{2\sigma_2^2}}, \quad \text{for } 0 \le \theta \le 360 \qquad (12.3.10)$$

where μ_1 and μ_2 are the means for two normal distributions, respectively, and σ_1 and σ_2 are their standard deviations. The actual values of mean, variance, and the weight can be estimated from historical data.

12.3.4.3 Translation Speed

Translation speed is also referred to as forward speed. Let V be the random variable representing the translation speed (m/s) of the hurricane upon landfall. According to

Georgiou et al. (1983) and Oxenyuk (2014), the value of V can be modeled as the log-normal distribution as follows:

$$f_V(v) = \frac{1}{\sigma_v v \sqrt{2\pi}} \exp\left[-\frac{1}{2}\left(\frac{\ln v - \mu_v}{\sigma_v}\right)^2\right], \qquad v \geq 0 \qquad (12.3.11)$$

where μ_v and σ_v are the parameters of the distribution and both can be estimated from the historical wind speed data.

12.3.4.4 Central Pressure Difference

The maximum wind speed is largely determined by the difference between atmospheric pressures at the center and at the periphery of a hurricane. Let P_d be the random variable representing the pressure difference between the periphery and the center, and p_d is its realization. Then P_d can be modeled as the Weibull distribution as follows (Georgiou et al. 1983; Oxenyuk 2014):

$$f_{P_d}(p_d) = \frac{k}{c}\left(\frac{p_d}{c}\right)^{k-1} \exp\left[-\left(\frac{p_d}{c}\right)^k\right], \qquad p_d \geq 0 \qquad (12.3.12)$$

where c and k are the scale and shape parameter of the Weibull distribution. If the absolute central pressure p_c instead of the pressure difference is recorded, then p_c can be converted to the central pressure difference as follows:

$$p_d = p_0 - p_c \qquad (12.3.13)$$

where $p_0 = 101.3$ kPa is a constant, representing the normal atmospheric pressure at a location beyond the hurricane's effect.

12.3.4.5 Radius to Maximum Wind

The radius to maximum wind describes the range of the most intensive hurricane wind speed. The radius of the maximum wind speed, denoted as R_{max}, can be empirically estimated as (FEMA 2017)

$$\ln R_{max} = 2.556 - 0.000\,050\,255 p_d^2 + 0.042\,243\,032\phi \qquad (12.3.14)$$

where p_d is the center pressure difference and ϕ is the storm latitude.

12.3.4.6 Wind Speed Decay Rate

A hurricane's intensity decays and dissipates after landfall because of friction and cut-off of the hurricane's circulation caused by land and terrain. The empirical model by Kaplan and Demaria (1995) and Demaria et al. (2006) is among the most widely used approach for simulating the decay of the hurricane's maximum wind speed inland. It is based on the assumption that a hurricane decays exponentially at a rate proportional to its landfall intensity after landfall. That is,

$$v(t) = v_b + (rv_0 - v_b)e^{-at} \qquad (12.3.15)$$

where $r = 0.9$ is a factor used to account for the sea–land wind speed reduction, $v_b = 13.75$ m/s, $a = 0.095$/hour, and v_0 is the maximum sustained one-minute surface wind speed at the time of landfall. Figure 12.11 shows how the wind speed decays over time given different initial wind speed v_0.

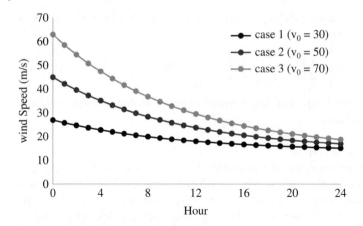

Figure 12.11 Wind speed decays over time.

12.3.4.7 Central Pressure Filling Rate

Vickery and Twisdale (1995a) proposed an empirical model to estimate the filling rate of the central pressure. Let $p_d(t)$ be the minimum central pressure difference at time t after the hurricane landfall. The evolution of $p_d(t)$ can be modeled as follows:

$$p_d(t) = p_{d0}e^{-(a_4+a_5p_0+\varepsilon)t} = p_{d0}e^{-bt} \qquad (12.3.16)$$

with

$$b = a_4 + a_5p_0 + \varepsilon \qquad (12.3.17)$$

Note that p_{d0} is the minimum central pressure difference at $t = 0$ upon the hurricane landfall and a_4 and a_5 are model parameters, and their values are location-dependent. For instance, for the Florida peninsula, $a_4 = 0.006$, $a_5 = 0.000\ 46$, and ε is a normally distributed error term with $\varepsilon \sim N(\mu = 0, \sigma^2 = 0.025^2)$ according to the study by Vickery and Twisdale (1995b).

Figure 12.12 plots the four cases under different values of a_4, a_5, and p_{d0} for comparison purposes. Case 1 serves as the baseline with $a_4 = 0.006$, $a_5 = 0.000\ 46$,

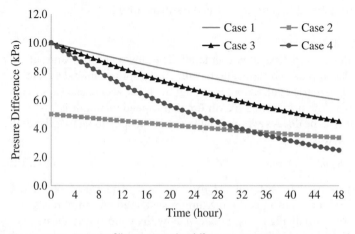

Figure 12.12 Pressure filling rate under different parameters.

and $p_{d0} = 10\,\text{kPa}$. The parameters for other three cases are: Case 2 with $a_4 = 0.006$, $a_5 = 0.000\,46$, and $p_{d0} = 5\,\text{kPa}$; Case 3 with $a_4 = 0.012$, $a_5 = 0.000\,46$, and $p_{d0} = 10\,\text{kPa}$; and Case 4 with $a_4 = 0.006$, $a_5 = 0.0023$, and $p_{d0} = 10\,\text{kPa}$. In addition to the time-based model in Eq. (12.3.16), Georgiou (1985) observed that the filling rate also depends on the translation speed at the time of hurricane landfall. He suggested that the distance inland could be a better indicator of the filling rate than the time since landfall.

12.3.4.8 Maximum Wind Speed

The maximum wind speed is defined as the sustained speed in one minute interval above 10 m on ground. The maximum wind speed can be estimated and approximated based on its minimum central pressure p_c at its landfall. Xu and Brown (2008) proposed a linear interpretation model to estimate the maximum wind speed at landfall, which involves two steps.

Step 1: Using p_c at landfall determines the Saffir–Simpson category of the corresponding hurricane.

Step 2: The maximum wind speed is proportionally extrapolated within the specific Saffir–Simpson speed category.

Example 12.2 Assume the simulated central pressure $p_c = 95.8\,\text{kPa}$. According to the Saffir–Simpson scale in Table 12.3, it is a Category 3 hurricane; then the maximum sustained wind speed for this hurricane, denoted as $v_{\max}$, upon landfall is calculated as

$$v_{\max} = 49.2 + \frac{58.1 - 49.2}{96.4 - 94.5}(96.4 - 95.8) = 52 \ \text{m/s} \tag{12.3.18}$$

In reality, hurricane features and effects may be highly idiosyncratic. For example, the center of a hurricane may not make landfall but it can still affect the state with its outer cloud cluster, the trajectory may not follow a straight line path across land, or some hurricanes come back to the same place after they leave. Nevertheless, the probabilistic hurricane model has been developed to simulate a statistically general hurricane year in a region; in other words, it is designed to simulate expected effects based on the average impact of a large number of simulations rather than track every single possible hurricane scenario.

Table 12.3 Saffir–Simpson scale in Xu and Brown (2008) (N/A = not applicable).

Category	Minimum central pressure (kPa)	Maximum sustained wind speed (m/s)	Storm surge (m)
Tropical Depression	N/A	0–17.4	0
Tropical storm	N/A	17.4–33.1	0–0.9
1	≥ 98	33.1–42.0	1.2–1.5
2	96.5–97.9	42.0–49.2	1.8–2.4
3	94.5–96.4	49.2–58.1	2.7–3.7
4	92–94.4	58.1–69.3	4–5.5
5	<92	≥ 69.3	$> = 5.5$

12.4   Prevention, Survivability, and Recovery

12.4.1   Three Elements of Power Resilience

According to Electric Power Research Institute (EPRI) (2017), grid resiliency is manifested in three elements: prevention, survivability, and recovery.

Grid prevention aims to protect transmission and distribution systems from being damaged by extreme weather, intentional attacks, or other unexpected disruptions. To that end, innovations are required to be developed and included in existing design standards, construction guidelines, and maintenance and inspection procedures. Particularly, hardening, decentralized power generation, and preventive maintenance (PM) are considered as viable means to strengthen the physical components of the grid against extreme events.

Grid survivability refers to the ability to maintain the basic level of power supply to consumers in the event of a complete loss of electrical service after the disaster event. The key elements of survivability include the use of defensive microgrid systems, backup or mobile generators, and energy storage systems to power critical infrastructures, including traffic signals, banks, hospitals, schools, and telecommunication equipment.

Grid recovery means that the resiliency planner ought to provide a quick assessment on a damaged grid, promptly deploy crew to damaged assets, and timely replace and repair failed circuits or equipment. In the 2017 Hurricane Harvey in Houston, TX, accessing affected areas was problematic because it was difficult to dispatch repair crews through streets that were blocked by fallen trees and flooded zones (ERCOT 2017).

In this chapter, we will elaborate difference means to enhance the resilience performance in prevention, survivability, and recovery. In particular, our attention is focused on line hardening, PM, decentralized resource allocation, and multiteam restoration. Other approaches to enhancing the grid resilience prior and post disaster events are also discussed by Kwasinski et al. (2012) and Wang et al. (2016) in their review papers.

12.4.2   Grid Hardening

Line hardening aims to strengthen certain distribution circuits that will be otherwise vulnerable to outages during extreme conditions. Line hardening is a simple resilient solution because it does not require sophisticated technologies and is commonly used in a radial or meshed distribution system (Wang et al. 2018). For overhead circuits, the majority of power outages happen because trees are falling on power lines or gusting winds directly blow down poles during hurricane storms or snow storms. Distribution circuits buried underground can avoid the direct damage due to gusting wind and storms, but the majority of the underground cables are susceptible to flooding, which is part of the hurricane effects. In this section, we introduce two fragility models corresponding to overhead and underground distribution systems, respectively. Both models allow the resilience management to estimate and predict the probability of line damage in a different scale of weather events.

12.4.2.1   Fragility Model of Overhead Lines

Overhead distribution lines consist of poles, conductor wires, and other types of components. The breakdown of a single pole or a wire results in the disconnection of the entire distribution line. Hence, the reliability of a distribution line can be treated as a series system with the fragility analysis of each pole and wire within that line (Ouyang et al. 2012). If we assume that the fragility of different components of an overhead line is independent, the failure probability of an overhead line under an extreme weather event can be expressed as follows (Ouyang et al. 2012; Ma et al. 2017):

$$p_{l,ij}(v(t)) = 1 - \prod_{k=1}^{m}(1 - p_k(v(t)))\prod_{k=1}^{n}(1 - p_{fc_{,k}}(v(t)))$$

(12.4.1)

where $p_{l,ij}(v(t))$ is the failure probability of the overhead line (i, j) with i and j being the line nodes; m is the number of poles supporting line (i, j); n is the number of wires between two adjacent poles at line (i, j); and p_k is defined as the conditional failure probability of the kth pole at line (i, j) as a function of the wind speed $v(t)$, which can be further modeled as a lognormal distribution (Salman et al. 2015) as

$$p_k(v(t)) = \Phi\left(\frac{\ln(v(t)/m_R)}{\xi_R}\right)$$

(12.4.2)

where $v(t)$ is the local three-seconds gusting wind speed, m_R is the median capacity or resistance against the wind, and ξ_R is the lognormal standard deviation of the intensity measurement of the wind speed.

In Eq. (12.4.1), $p_{fc,k}(v(t))$ represents the failure probability of wire k for $k = 1, 2, ..., n$ between two adjacent poles. That is,

$$p_{fc,k}(v(t)) = (1 - f_u)\max\{p_{fw,k}(v(t)), \alpha p_{ftr,k}(v(t))\}$$

(12.4.3)

where

$p_{fw,k}(v(t))$ =the direct wind-induced failure probability of wire k
$p_{ftr,k}(v(t))$ =the fallen tree-induced failure probability of wire k
f_u =the probability that wire k is placed underground (invulnerable to weather events)
α =the average tree-induced damage probability of overhead wires

According to Ouyang and Dueñas-Osorio (2014), the damage probability of wire k directly induced by wind can be estimated as

$$p_{fw,k}(v(t)) = \min\{F_{fw,k}(v(t))/F_{fo,k}(v(t)), 1\}$$

(12.4.4)

where

$F_{fw,k}(v(t))$ = represents the wind force imposed on wire k and
$F_{fo,k}(v(t))$ = represents the maximum perpendicular force that wire k can hold

Finally, the failure probability of wire k induced by fallen trees can be modeled as follows (Canham et al. 2001):

$$p_{ftr,k}(v(t)) = \frac{\exp(a + ck_s s_k d_h^b)}{1 + \exp(a + ck_s s_k d_h^b)}$$

(12.4.5)

where s_k is the wind speed intensity in a range between 0 and 1 on wire k and can be calculated by dividing the local three-second wind speed by the maximum wind speed in the area; d_h is the tree diameter at the height of 1.35 m; k_s is a factor that captures the local terrain effects near wire k; and finally a, b, and c are the model parameters depending on the tree species or types.

12.4.2.2 Fragility Model of Underground Lines Due to Flood

As a hurricane approaches the land, a storm surge creates a wall of water that floods the shore and the adjacent land. As a result, underground wires and other electric components are subject to damage due to water exposure, debris, and salt residuals. Xu and Brown (2008) proposed a linear mathematical model to estimate underground line damages based on categories of hurricanes and storm surge zones. The failure probability of underground line (i, j), denoted as $\lambda_{ij,t}$, can be estimated as

$$\lambda_{ij,t} = [a_u + b_u(H - Z)] \times I(H - Z) \tag{12.4.6}$$

where H is the hurricane category 1 to 5 and Z is the storm surge zone category 1 to 5. Note that the indicator function $I(H - Z)$ shows whether the area is affected by an incoming hurricane or not, and it is 1 if $H - Z \geq 0$ and 0 if $H - Z < 0$. Finally, a_u and b_u are the tuning parameters.

Example 12.3   Assuming $a_u = 0.02$ and $b_u = 0.05$, estimate the value of λ_{ij} under different combinations of hurricane categories and storm surge zones.

Solution:
We substitute $a_u = 0.02$ and $b_u = 0.05$ into Eq. (12.4.6). The value of λ_{ij} associated with different combination of H and Z are shown in Table 12.4.

12.4.3 Proactive Maintenance

Grid resilience can be enhanced through PM with which aging components are proactively replaced prior to its failure. An age-based PM policy has been widely used to support the system operation in power systems (Bertling et al. 2005). Recently, condition-based maintenance (CBM) receives much attention as it has the potential to achieve the just-in-time maintenance (Ye et al. 2014; Byon et al. 2010). Below we present an age-based PM policy to minimize the maintenance cost rate of individual components or grid assets in a steady-state operating condition. However, the result derived from an age-based PM policy can be extended to systems or equipment where CBM is adopted. Components in an electric grid represents independent items, such as conductor wires, transformers, generating units, circuit breakers, poles, and battery systems. Assuming component type i is maintained under a PM policy, the optimal

Table 12.4 Underground line failure probability under $a_u = 0.02$ and $b_u = 0.05$.

Hurricane category (H)	Storm surge zone (Z)				
	1	2	3	4	5
1	2%	0	0	0	0
2	7%	2%	0	0	0
3	12%	7%	2%	0	0
4	17%	12%	7%	2%	0
5	22%	17%	12%	7%	2%

maintenance interval, denoted as $\tau_i{}^*$, can be determined by minimizing the cost rate as follows (Jin et al. 2013):

$$\min_{\tau_i} : u_i(\tau_i) = \frac{c_{fi}F_i(\tau_i) + c_{pi}R_i(\tau_i)}{\int_0^{\tau_i} R_i(t)dt + t_{fi}F_i(\tau_i) + t_{pi}R_i(\tau_i)}, \quad \text{for } i = 1, 2, \ldots, m \qquad (12.4.7)$$

where m is the number of component types in a grid, $R_i(t)$ is the reliability of component type i, and $F_i(t)$ is the cumulative distribution function (CDF) and $F_i(t) = 1 - R_i(t)$. Note that c_{pi} and c_{fi} are the costs for performing a planned replacement and a corrective replacement, respectively. In addition, t_{fi} and t_{pi} are the hands-on time involved in performing a failure and planned replacement, respectively. The numerator represents the average uptime of the component and the denominator is the sum of the uptime and the downtime in a maintenance cycle. The objective is to determine the optimal value $\tau_i{}^*$ to minimize the cost per unit time (i.e. maintenance cost rate). In addition, the steady-state availability of component type i under PM can be estimated as

$$A_i(\tau_i) = \frac{\int_0^{\tau_i} R_i(t)dt}{\int_0^{\tau_i} R_i(t)dt + t_{fi}F_i(\tau_i) + t_{pi}R_i(\tau_i)}, \quad \text{for } i = 1, 2, \ldots, m \qquad (12.4.8)$$

where the numerator represents the mean time between replacements, including scheduled and unscheduled replacements.

Recently distributed generation (DG) emerged as new power production and delivery technology to meet the growing needs of electricity. Unlike the central generation, a DG system is comprised of multiple distributed energy resources (DERs) that are installed in close proximity to the end consumers. Typical DER units include wind turbines, solar PV, diesel generators, battery storage, combined heat and power, and thermal generators. Let $i = 1, 2, \ldots, m$ be the index denoting the DER types including the substation. For instance, if the generation pool consists of 1 MW WT, 1.5 MW WT, and 0.5 MW PV at 10 MW substations, then $i = 1, 2, 3$, and 4. Then the availability of the entire DG pool can be expressed as

$$A_{DG}(\tau) = \prod_{i=1}^{m} \prod_{j=1}^{n} (A_i(\tau_i))^{x_{ij}} \qquad (12.4.9)$$

where $A_i(\tau_i)$ is the availability of DER type i given in Eq. (12.4.8) and x_{ij} is the number of DER type i installed at node j. For instance, if $x_{ij} = 1$, this means that one unit of DER type i is installed at node j, or is $x_{ij} = 0$ otherwise. In Eq. (12.4.9), it is assumed that DER units of the same type adopt the identical maintenance policy regardless of their location and usage variation. This assumption, however, can be relaxed if τ_i is location-dependent in the distribution network.

12.4.4 Decentralized Resource Allocation

The traditional hierarchical power system is undergoing a paradigm change due to the growing integration of DER. Typical DER units include microgrids, microturbines, combined heat and power, bio fuel cells, distributed energy storages, and electric vehicles. Cogeneration combining gas-fired electricity and heat becomes popular because of a lower fuel cost and carbon emissions (Shao et al. 2018). Distributed generation (DG) technologies such as microgrids can enhance the electric power resilience and serve as critical loads in contingency. A good example of resilience performance is the Roppongi Hills microgrid in Tokyo. During the Great East Japan earthquake in 2011, the natural gas based Roppongi Hills microgrid was able to maintain the electricity supply even though the main grid was broken down for several days, highlighting the microgrid survivability in adverse conditions (Lin and Bie 2016). Microgrid systems are distributed electric generation resources incorporating storage, load control, and energy management systems. They are normally integrated with the grid, but can operate independently of the main grid. Below we discuss several types of DER technologies that are capable of enhancing grid survivability and adaptability during the course of extreme weather attacks.

12.4.4.1 Generation Backup

DER units as generation backups are widely used in critical infrastructures or service systems such as banking, e-commerce, network computing, telecommunication, hospitals, and military bases, where power outage or electricity shortage may cause detrimental consequences. Using DG as a backup unit to provide electricity in the case of utility supply interruption or curtailment is perhaps among the early implementations of this technology. Recently Lei et al. (2018) developed an optimization model to pre-allocate a mobile generator fleet for a resilient response to natural disasters. In a local distribution system with DG interconnected, load transfer to other feeders via switches can be performed in an interruption situation in order to maintain the balance between load and supply. In addition, large manufacturing companies usually participate in demand response programs by signing a contract with the utility company and they agree to reduce the load in contingency events like peak hours. To maintain normal production, the manufacturer can leverage a backup DER unit to fill the electricity gap when the supplies from the main grid are curtailed (Jia et al. 2017). If the DER units are correctly coordinated in contingency events, they can enhance the reliability and power quality of the local distribution system (Brown and Freeman 2001).

12.4.4.2 Onsite Generation

In onsite generation, DER units are installed at the consumer site to supply the electric energy along with the main grid. Typical examples include roof-top solar PV or wind turbines (WT). The capacity of onsite generators usually is less than the local demand, and they serve as complementary energy sources for powering large industrial or business facilities. There are two main benefits associated with onsite wind and solar generation: (i) reducing the carbon footprint and (ii) saving the energy losses otherwise incurred during the transmission and distribution. In terms of power resilience, Kwasinski et al. (2012) showed that onsite WT and PV exceed fuel-based generators in terms of resilience operation. This is because WT and PV harness natural resources instead of using fossil fuels for energy generation, as the fuel supply or lifeline is likely to be damaged during natural disasters. At present, the main obstacle impeding large-scale renewable-based onsite generation has to do with the power intermittency. Therefore, better models need to be developed in order to characterize and predict the output of variable generation systems as well as understanding the influence of such units on the grid reliability and power quality.

Figure 12.13 shows a typical onsite generation system installed in a large manufacturing facility to co-supply the power along with the main grid. When the output power of

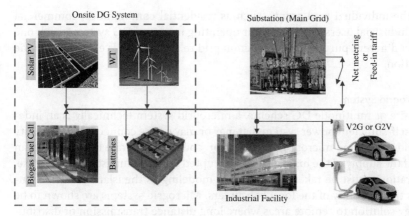

Figure 12.13 Onsite DG for industrial users.

WT, PV, and biogas fuel cell exceeds the demand, the surplus energy can charge employees' electric vehicles (EVs) or be returned to the grid via net metering or a feed-in tariff. The EV fleet can provide frequency and voltage regulation during peak hours via an intelligent vehicle-to-grid (V2G) operations (Villarreal et al. 2013).

12.4.4.3 Active Distribution Network

In an active distribution network, several DER units are injected into different nodes to form a cluster of DG sources. These dispersed units are coordinated via a networked and centralized control platform and jointly supply electricity to a group of points of use through one feeder or via a multifeeder network (see Figure 12.14). To reduce the thermal losses or line congestion, DER power is encouraged to be consumed locally. If the output of a DER unit exceeds the local demand, the surplus energy can be transferred to its adjacent nodes, realizing a two-way power flow. Such a multinode DG architecture is quite appealing to the utility industry as it defers the transmission expansion investment and also relieves the distribution bottleneck in peak hours. If the local consumer is the co-owner of the DER units, the energy that is injected into the utility grid is measured and the consumer has to pay only for the difference between the energy used from the distribution utility and the amount injected into the network.

Figure 12.14b shows an active distribution network integrated with multiple WT and PV units to co-supply the electricity to a local user. Note that D_1 through D_9 are the

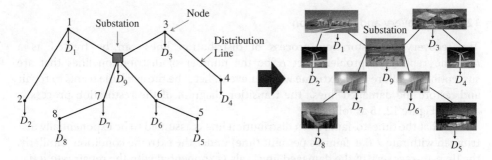

Figure 12.14 (a) Passive distribution network; (b) active distribution network.

demands of the individual load points, such as residential communities, commercial buildings, or industrial users. There are four operating modes of DG systems, each one of which has a different purpose and impact on grid reliability and is elaborated in the following section.

12.4.4.4 Microgrid System

Unlike an onsite or multinode DG scheme, a microgrid system technically is an independent and self-sufficient power system that may or may not be connected to the utility grid. For a grid-connected microgrid, the user can choose to operate the system in an island mode if the supply of the main grid is interrupted or in a failure state. In this case, new considerations must be taken for reliability modeling as the island model essentially benefits the reliability of the local consumers. Microgrid systems are shown to be a viable energy solution to remote areas where long distance transmission or distribution lines are too costly to be constructed. Architecturally, both onsite generation and microgrid systems adopt one or multiple DER units to supply the electricity to meet local needs. The main difference is that the microgrid is capable of maintaining an independent and sustainable supply while onsite generation usually co-supplies the power along the main grid. Hence microgrid systems possess the unique capability of enhancing the grid resilience via island operations in extreme weather conditions (Liu et al. 2017). Technically, a microgrid system is a special DG system that can actively control the regulated power flow between the main grid and local consumer.

12.4.4.5 Dual Feeds

Dual feeds is a common service offered by utilities for customers with critical loads, such as data centers, hospitals, and industrial facilities, among others. It provides a customer with a normal power source and a back-up source. Dual feeds may not eliminate outages if both feeders are disrupted or damaged, but dramatically reduce the likelihood of an outage, even during a VLSE. Damage to both electrical feeders could, of course, occur during a VLSE, but they provide for alternative restoration strategies. For instance, a customer powered by two different circuits from different substations would give system operators greater flexibility in prioritizing crews for repairs. The concept of dual feeds can be applied beyond the electric grid. For example, raw materials can be sourced from two different vendors in production and supply chain operations, hence increasing the supply resilience.

12.4.5 Recovery and Restoration

The recovery or restoration process of distribution lines can be treated as a machine–repairman problem. Let K be the number of distribution lines that are susceptible to failure in an extreme weather event. Let R be the available teams to repair and restore the damaged lines. The transition diagram of the restoration process is given in Figure 12.15 as follows.

Note that the time-to-failure of a distribution line is assumed to be exponentially distributed with rate λ (i.e. failures per unit time) under the extreme condition. Similarly, the time-to-recovery of the damaged line is also exponential with the repair rate μ (i.e. number of lines repaired per unit time). Here failure means a line is disconnected and

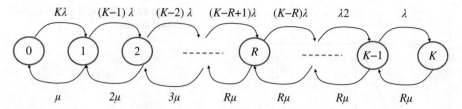

Figure 12.15 Transition diagram of repairing distribution lines.

repair means a disconnected line is recovered. Let L be the expected number of damaged lines and L_q be the number of lines waiting for repair. Then

$$L = \sum_{k=0}^{K} k\pi_k \tag{12.4.10}$$

$$L_q = \sum_{k=R}^{K} (k - R)\pi_k \tag{12.4.11}$$

where π_k is the steady-state probability of k lines that are damaged for $k = 0, 1, \ldots, K$. Then we have

$$\pi_0 = \frac{1}{1 + \sum_{k=1}^{K} c_k}, \quad \text{for } k = 0 \tag{12.4.12}$$

$$\pi_k = c_k \pi_0, \quad \text{for } k = 1, 2, \ldots, K \tag{12.4.13}$$

and

$$c_k = \binom{K}{k} \left(\frac{\lambda}{\mu}\right)^k, \quad \text{for } k = 1, 2, \ldots, R \tag{12.4.14}$$

$$c_k = \binom{K}{k} \left(\frac{\lambda}{\mu}\right)^k \frac{k!}{R! R^{k-R}}, \quad \text{for } k = R + 1, R + 2, \ldots, K \tag{12.4.15}$$

We can also estimate the duration of recovering a damaged line. Let W be the duration from when the line is damaged to when it is restored. According to Little's law, W is given by

$$W = \frac{L}{\bar{\lambda}} = \frac{L}{\sum_{k=0}^{K} (K - k)\lambda\pi_k} \tag{12.4.16}$$

where L is the number of damaged lines given in Eq. (12.4.10) and $\bar{\lambda}$ is the weighted failure rate of lines.

There exist other types of recovery models in the literature. For instance, Figueroa-Candia et al. (2018) considered a multidimensional resiliency measure, such as response-time components, information availability, and quality of service, in evaluation and optimization of restoration policies for distribution grids subject to extreme weather.

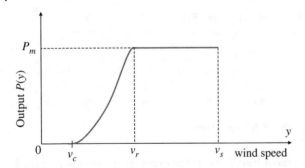

Figure 12.16 A typical wind turbine power curve.

12.5 Variable Generation System Model

12.5.1 Wind Turbine System

The output power from a WT is uniquely determined by its power curve, shown in Figure 12.16. Depending on the wind speed, the power curve consists of four operating phases: standby phase with no output power when $y < v_c$; nonlinear power output for $v_c < y < v_r$; rated or constant power phase $v_r < y < v_s$; and the cut-off phase if $y > v_s$ (for protection if the speed is too large). Note that v_c is the cut-in wind speed, v_r is the rated wind speed, v_s is the cut-off wind speed, and P_m is the rated power or power capacity of the turbine.

Two types of models are available to estimate the power output during the nonlinear production phase. One approach uses the quadratic function to characterize the non-linear relation between y and $P_w(y)$, which is the output power (Giorsetto and Utsurogi 1983; Karki et al. 2006). That is,

$$P_w(y) = \begin{cases} 0, & y < v_c, \ y > v_s \\ P_m(a + by + cy^2), & v_c \le y \le v_r \\ P_m, & v_r \le y \le v_s \end{cases} \tag{12.5.1}$$

where a, b, and c are model parameters of the quadratic function when y falls between v_c and v_r.

Cubic power curve models are developed based on the kinetic theory of the air flow dynamics (Thiringer and Linders 1993; Ehsani et al. 2005). Namely, WT output power is proportional to the cube of the wind speed when $v_r \le y \le v_s$. That is,

$$P_w(y) = \begin{cases} 0, & y < v_c, \ y > v_s \\ 0.5\eta_{max}\rho A y^3, & v_c \le y \le v_r \\ P_m, & v_r \le y \le v_s \end{cases} \tag{12.5.2}$$

where ρ is the air density and A is the area covered by the turbine blades. When the total wind energy is converted into electricity, some energy is lost due to bearings and gearbox friction and the efficiency limitation of induction generators. Thus η_{max} represents the maximum conversion rate, which is 0.5926. The actual conversion rate is often lower and varies between 0.3 and 0.5.

Let Y be the random wind speed and $P_w(Y)$ be the output power at Y. The mean, the second moment, and the variance of $P_w(Y)$ can be estimated as follows (Jin and Tian 2010):

$$E[P_w(Y)] = \gamma \int_{v_c}^{v_t} y^3 f_w(y) dy + P_m(F_w(v_s) - F_w(v_r)) \tag{12.5.3}$$

$$E[P_w^2(Y)] = \gamma^2 \int_{v_c}^{v_r} y^6 f_w(y) dy + P_m^2(F_w(v_s) - F_w(v_r)) \tag{12.5.4}$$

$$\text{var}(P_w(Y)) = E[P_w^2(Y)] - (E[P_w(Y)])^2 \tag{12.5.5}$$

where $\gamma = 0.5 \eta_{max} \rho A$ and $F_w(y)$ and $f_w(y)$ are the cumulative distribution and the PDF of wind speed Y, respectively. Many meteorological data show that Y can be modeled by normal or Weibull distributions (Karki et al. 2006; Seguro and Lambert 2000).

12.5.2 Solar Photovoltaic System

The output power of a PV system depends on multiple factors, including the panel size, the conversion efficiency, the operating temperature, the panel orientation, the tilt angle, the calendar date, the daily hour, the latitude, and the weather condition (Lave and Kleissl 2011). A generic PV generation model incorporating all these factors is given by

$$P_s(s) = \eta_s A_s s(1 - 0.005(T_o - 25)) = Bs \tag{12.5.6}$$

with $\qquad B = \eta_s A_s (1 - 0.005(T_o - 25)) \tag{12.5.7}$

Here, $P_s(s)$ is the instantaneous PV power output, η_s is the PV efficiency with $\eta_s = 15$–20%, A_s is the panel size, T_o is the PV operating temperature (°C), and s is the solar irradiance (W/m^2) incident on the panel surface.

Let S be the random variable representing the solar irradiance during the course of a year and s is its realization. Then S can be modeled as a beta distribution as follows (Sulaiman et al. 1999; Atwa et al. 2009):

$$f_s(s) = \frac{\Gamma(a+b)}{s_m \Gamma(a) \Gamma(b)} \left(\frac{s}{s_m} \right)^{a-1} \left(1 - \frac{s}{s_m} \right)^{b-1}, \quad \text{for } 0 \le s \le s_m \tag{12.5.8}$$

where a and b are the shape parameters of the beta distribution and s_m is the maximum irradiance. Based on Eqs. (12.5.6) and (12.5.8), the mean and the variance of $P_s(s)$ are obtained as

$$E[P_s(S)] = \int_0^{s_m} P_s(s) f_s(s) ds = \frac{a s_m B}{a + b} \tag{12.5.9}$$

$$\text{var}(P_s(S)) = \frac{a b s_m^2 B^2}{(a+b)^2 (a+b+1)} \tag{12.5.10}$$

Figure 12.17 shows various types of beta probability density functions to mimic different solar irradiance values during the course of a year. Note that the maximum solar irradiance values are $s_m = 1000, 1100,$ and 1200 W/m^2, respectively. In general, Case I represents an area where less sunshine is received, while Case III represents the location with strong sunshine across the year.

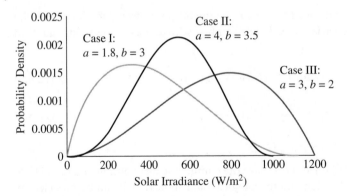

Figure 12.17 Annual solar irradiance distributions.

12.5.3 Battery Charge and Discharge Model

The battery system plays a vital role in streamlining the intermittent wind and solar generation because it has a dual function: storing and discharging energy. When surplus energy is generated from WT and PV, it can be temporally stored in the battery. If the power from WT and PV is less than the demand, the stored energy can be discharged from the battery to meet the generation gap. Let S_{cap} be the battery capacity in units of kW h or MWh and $S(t)$ be the energy storage level at time t. Under the assumption of linear process charging, $S(t)$ can be estimated as

$$S(t + \Delta t) = \max \left\{ S_{cap}, \quad S(t) + \frac{S_{cap}}{T_c} \Delta t \right\} \tag{12.5.11}$$

where Δt is the charging duration in hours and T_c is the full charging cycle in hours. Eq. (12.5.11) shows that the energy level will be saturated once reaching S_{cap}. Similarly, the discharging process can be expressed as

$$S(t + \Delta t) = \max \{0, \quad S(t) - P_d \Delta t\} \tag{12.5.12}$$

where P_d is the battery discharging power. Equation. (12.5.12) shows that the battery energy level will reach zero after a sufficiently long period of discharge. Battery technologies like lithium-ion usually require a minimum of 20% of energy to be retained before recharging. The main reason is to extend the battery lifetime because deep cycling of charge and discharge will significantly shrink the lifetime of a lithium-ion battery.

12.5.4 Demand Response Model

In demand response (DR), the customer changes the power consumption profile to better match demand with the power supply, which often occurs in peak time. DR programs can be divided into two categories: time-based programs and incentive-based programs.

Time-based programs give the customers the choice to curtail or change their consumption habit based on the change of the electricity price during a 24-hour period (Khajavi et al. 2011). Programs under the time-based category include time-of-use, real-time pricing, and critical peak pricing. An incentive-based program is also called a contract-based demand response, and establishes an agreement between the customer and the utility company to curtail the power consumption during peak hours, in exchange for incentives or financial compensation to the customer. Typical incentive-based programs are direct load control, an emergency demand response program, a capacity market program, demand bidding/buy-back, ancillary service markets, and interruptible/curtailable load service. Below, we discuss the interruptible/curtailable program as it is perhaps the most widely used contract-based demand response for large industrial or commercial users.

Four variables are involved in an interruptible/curtailable demand response (I/C DR) program: (i) the number of curtailments in a year; (ii) the start time of each curtailment; (iii) the duration of a curtailment; and (iv) the curtailment level. Probability models are used to characterize the behavior of these random variables.

12.5.4.1 Curtailment Number and Start Time

Under the I/C DR contract, the actual start time of a curtailment event is not known to the customer until being called 30–60 minutes in advance (Baldick et al. 2006). Let N be the number of curtailment calls during a year. The occurrence of I/C DR events can be modeled as a Poisson process as follows:

$$P\{N = n\} = \frac{(\lambda t)^n e^{-\lambda t}}{n!}, \quad \text{for } n = 0, 1, 2, \ldots \tag{12.5.13}$$

where λ is the curtailment call rate. For instance, if the average curtailments in a year are five calls, then $\lambda = 5$ calls/year. Equation (12.5.13) allows us to simulate the DR start time and the number of annual curtailments.

12.5.4.2 Curtailment Duration

The maximum and the minimum curtailment durations, denoted as T_{max} and T_{min}, respectively, are usually written into the contract. In general, we have $1 \leq T_{min} \leq T_{max} \leq 8$ hours for a typical I/C DR program (Aalami et al. 2010). However, the customer does not have the advance information about the actual curtailment duration until being called. Let T be the random duration of a curtailment event and τ be its realization. Then T can be treated as a uniform random variable with the following probability density function:

$$f_T(\tau) = \begin{cases} \frac{1}{T_{max} - T_{min}}, & T_{min} \leq \tau \leq T_{max} \\ 0, & \text{otherwise} \end{cases} \tag{12.5.14}$$

12.5.4.3 Curtailment Levels

The curtailment level may differ in each contingency call, yet the maximum power reduction usually does not exceed 20% of the customer's mean load (Aalami et al. 2010). We use the discrete probability model to characterize the variation of curtailment levels. Let l_j for $j = 1, 2, \ldots, q$ represent the actual realization of possible curtailment levels with $l_1 < l_2 < \cdots < l_q$. The probability mass function can be expressed as

$$P\{L_t = l_j\} = p_j, \quad \text{for } j = 1, 2, \ldots, q \tag{12.5.15}$$

where L_t is the random variable representing the load curtailment level in time t. Obviously, the sum of p_j for all j should be equal to unity.

12.6   Case Study: Design for Resilient Distribution Systems

12.6.1   Background Information

In this section, we use the nine-node distribution network in Figure 12.14a to demonstrate the design for resilience in three elements, namely, prevention, survivability, and recovery. Node 9 is located in the central part of the network and is reserved for the location of a substation that handles bulk power from the transmission network. Nodes 1 through 8 are the sites where WT and PV units can be integrated to form the active distribution network. Five different DER units are available for integration: 1 MW WT, 2 MW WT, 3 MW WT, 0.25 MW PV, and 0.5 MW PV. Without loss of generality, all line resistances between two adjacent nodes are assumed to be 1 Ω. In addition, a DC power flow model is used to determine the location and quantity of DER units in the distribution network because the allocation is made at the strategic level with less influence on reactive power.

Table 12.5 lists the power demand of year 1 and the predicted demand growth in year 3. In this study, it is assumed that both the mean and the variance increase with the time. The LOLP criterion is set with $\alpha_1 = 0.01$ and the power quality confidence is $\alpha_2 = 0.9$. The nominal voltage is $V_{DG} = 33$ KV and the upper and lower limits of nodal voltage is $V_{max} = 1.95 V_{DG}$ and $V_{min} = 1.95 V_{DG}$, respectively.

Table 12.6 provides the costs of equipment, maintenance, and carbon credits which are negative in terms of cost. The capacity factor is the ratio of the mean output power over its nameplate capacity. Though the values of the capacity factor vary with the local wind speed and solar radiation, it has less impact on the justification of the survivability of the distribution power via a defensive microgrid operation.

The goal of DG planning is to determine the sizing and siting of DER units in the distribution network such that the annualized system cost is minimized (El-Khattam et al. 2004; Jin et al. 2015). The optimal DER types and quantity in each node are summarized in Table 12.7 and the detailed solution is available in Jin et al. (2018). For instance, in Year 3, one WT2 and one WT3 will be installed at node 2. The total capacity would be $2 + 3 = 5$ MW. As another example, two WT3 units will be installed at node 8 with a total capacity of 6 MW. As for node 9, it is placed with a 50 MW substation. Since the PV capacity cost is higher than that of WT, a smaller amount of PV capacity is chosen rather than wind generators in this case.

Table 12.5 Mean and variance of load in years 1 and 3 (unit: MW).

Node *j*		1	2	3	4	5	6	7	8	9	System
Notation		D_1	D_2	D_3	D_4	D_5	D_6	D_7	D_9	D_9	D
Year 1	$E[D_j]$	7.64	8.72	4.58	4.00	5.14	6.11	7.64	7.27	0	51.1
	$Var(D_j)$	0.146	0.19	0.052	0.04	0.066	0.093	0.146	0.132	0	0.865
Year 3	$E[D_j]$	8.105	9.251	4.859	4.244	5.453	6.482	8.105	7.713	0	54.212
	$Var(D_j)$	0.164	0.213	0.058	0.045	0.074	0.104	0.164	0.148	0	0.972

Source: Reproduced with permission of IEEE.

Table 12.6 Power capacity and costs for DER units (SS = Substation).

i	DER Type	DER capacity (MW)	Installation ($\times 10^3$ $/MW)	Maintenance ($/MWh)	Carbon credits ($/MWh)	Capacity factor
1	WT1	1	910	10	−5	0.4
2	WT2	2	773.5	9	−5	0.35
3	WT3	3	637	8	−5	0.3
4	PV1	0.25	2000	3	−10	0.3
5	PV2	0.5	1750	2	−10	0.3
6	SS1	45	273	16	10	1.0
7	SS2	50	227.5	16	10	1.0

Source: Reproduced with permission of IEEE.

Table 12.7 Cumulative Allocated DER units by the end of Year 3.

DER\Node	$j = 1$	2	3	4	5	6	7	8	9
WT1									
WT2		1		1		1			
WT3	1	1					1	2	
PV1									
PV2				1	1	1	1		
SS1									
SS2									1

12.6.2 Prevention via Distributed Power Allocation

Protection for the power supply is achieved by progressively integrating WT and PV units into nodes 1 to 8. Table 12.8 summarizes the amount of power being protected in Year 3 by estimating the maximum DER power output of individual nodes. For instance, in Year 3, node 8 installed a total of 6 MW wind capacity, thus ensuring that 6/7.713 = 77.8% of the local demand is protected. At the system level, the protected power is 42.4% in Year 3 as opposed to zero if there are no DER units installed. This clearly shows that distributed resource allocation can effectively protect a power shortage when the substation or the distribution lines are damaged in extreme weather.

12.6.3 Survivability via Microgrid Operation

A key criterion to measure the grid survivability is to assess the robustness of the power supply when it is attacked by extreme events like hurricanes. Let us consider the following extreme case: a substation is totally lost although all the distribution lines are still connected across nine nodes. Since it is in the contingent mode, we also assume that the power demand of each node is reduced by 50% to maintain the operation of critical loads. We compute the grid survivability in Year 3 and the results are summarized

Table 12.8 The amount of protected power of each node in Year 3.

Node j	1	2	3	4	5	6	7	8	9
Mean load (MW)	8.105	9.251	4.859	4.244	5.453	6.482	8.105	7.713	0
DER power (MW)	3	5	0	2.5	0.5	2.5	3.5	6	0
Protection (%)	37.0	54.0	0.0	58.9	9.2	38.6	43.2	77.8	0

Table 12.9 Survivability in Years 1 to 3 with lost substation.

Node j	1	2	3	4	5	6	7	8	9
Critical load (MW)	4.053	4.626	2.429	2.122	2.727	3.241	4.053	3.856	0
DER power (MW)	3	5	0	2.5	0.5	2.5	3.5	6	0
Survivability (%)	74.0	108.1	0.0	117.8	18.3	77.1	86.4	155.6	0

in Table 12.9. It shows that nodes 2, 4, and 8 have surplus power because their maximum generation exceeds the critical load. Obviously the surplus power from these nodes enters their adjacent nodes 1, 3, and 7, respectively. This indeed increases the survivability all the nodes.

12.6.4 Recovery under Multiteam Repair

The repair and recovery process depends on the failure rate of the distribution lines and the availability of repair teams. We compare the recovery time under two different damage severity conditions manifested by $\lambda = 1$ line/hour and $\lambda = 0.5$ line/hour. Obviously a higher failure rate implies a harsher weather condition. $K = 8$ is because there are eight distribution branches in the 9-node network and $R = 2$ means two teams are available for performing the recovery job. We compute the expected number of damaged branches L and the expected recovery time W per branch based on Eqs. (12.4.10) and (12.4.16), respectively. Both L and W are calculated as the repair rate μ increases from 0.1 to 5 line/hour. The results are shown in Figures 12.18 and 12.19.

Two observations can be made from these two figures. First, μ plays a decisive role in reducing the recovery time. Regardless of the damage severity, the repair rate plays a dominant role in the downtime duration. For $\mu = 0.1$ line/hour, it takes on average 39 hours to restore a damaged branch under a severe weather condition. However, if μ increases to 0.5, the line downtime drops dramatically from 39 hours to 6 hours. Second, the weather severity has less impact on the line downtime given the same repair rate. For instance, if the weather severity is reduced by a half (from $\lambda = 1$ to $\lambda = 0.5$), it still needs 38 hours to recover a damaged line, as shown in Figure 12.19.

In summary, this section presents a design case for enhancing distribution grid resilience via wind and solar power integration. We approach the grid resilience

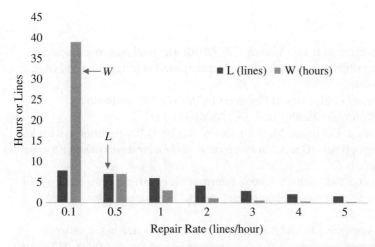

Figure 12.18 Recovery time and disconnected branches with $\lambda = 1$, $R = 2$. Source: Reproduced with permission of IEEE.

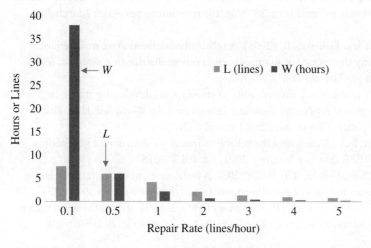

Figure 12.19 Recovery time and disconnected branches with $\lambda = 0.5$, $R = 2$. Source: Reproduced with permission of IEEE.

from three elements, namely, prevention, survivability, and recovery. Prevention aims to increase the supply robustness through integration of distributed microgrid power. Survivability is attained using a defensive microgrid operation or topological reconfiguration in contingency. Recovery is assessed and planned based on the machine–repairman Markov model. In this chapter, the modeling and design of a resilient engineering system is built upon the power grid system, but the idea and method can be appropriately extended to other critical infrastructure systems, such as public transportation, supply chain management, communication network, healthcare delivery, and water distribution systems.

References

Aalami, H.A., Moghaddam, M.P., and Yousefi, G.R. (2010). Demand response modeling considering interruptible/curtailable loads and capacity market programs. *Applied Energy* 87 (1): 243–250.

ASME, American Society of Mechanical Engineers (ASME) (2009). *Innovative Technological Institute (ITI)*. Washington, DC: ASME ITI, LLC.

Atwa, Y. M., El-Saadany, E.F., Salama, M., Seethapathy, R., (2009) "Distribution system loss minimization using optimal DG mix," in *Proceedings of IEEE Power and Energy Society General Meeting*, pp. 1–6.

Baldick, R., Kolos, S., and Tompaidis, S. (2006). Interruptible electricity contracts from an electricity retailer's point of view: valuation and optimal interruption. *Operations Research* 54 (4): 627–642.

Barker, K., Ramirez-Marquez, J.E., and Rocco, C.M. (2013). Resilience-based network component importance measures. *Reliability Engineering and Systems Safety.* 117 (September): 89–97.

Barron, J., (2003), "The blackout of 2003: The overview," *New York Times*, August 15, 2003, available at: http://www.nytimes.com/2003/08/15/nyregion (assessed on March 20, 2018).

Bertling, L., Allan, R., and Eriksson, R. (2005). A reliability-centered asset maintenance method for assessing the impact of maintenance in power distribution systems. *IEEE Transactions on Power Systems* 20 (1): 75–82.

Briguglio, L. (2002). The economic vulnerability of small island developing states. In: *Sustainable Development for Island Societies: Taiwan and the World* (ed. H.H. Hsiao, C.H. Liu and H.M. Tsai). Taiwan: National Central University.

Brown, R.E., Freeman, L.A., "Analyzing the reliability impact of distributed generation," in *Proceedings of the IEEE Summer Meeting*, 2001, pp. 1013–1018.

Bruneau, M., Chang, S.E., Eguchi, R.T. et al. (2003). A framework to quantitatively assess and enhance the seismic resilience of communities. *Earthquake Spectra* 19 (4): 733–752.

Byon, E., Ntaimo, L., and Ding, Y. (2010). Optimal maintenance strategies for wind turbine systems under stochastic weather conditions. *IEEE Transactions on Reliability* 59: 393–404.

Canham, C.D., Papaik, M.J., and Latty, E.F. (2001). Interspecific variation in susceptibility to windthrow as a function of tree size and storm severity for northern temperate tree species. *Canadian Journal of Forest Research* 31 (1): 1–10.

Carlson, L., Bassett, G., Buehring, M. et al. (2012). Resilience: theory and applications. *Argonne National Laboratory Report* 1–64, available at: http://www.ipd.anl.gov/anlpubs/2012/02/72218.pdf. accessed on November 2, 2017.

DeMaria, M., Knaff, J., and Kaplan, J. (2006). On the decay of tropical cyclone winds crossing narrow landmass. *Journal of Applied Meteorology and Climatology* 45: 491–499.

Dessavre, D.G., Ramirez-Marquez, J.E., and Barker, K. (2016). Multidimensional approach to complex system resilience analysis. *Reliability Engineering & Systems Safety* 149: 34–43.

Dinh, L.T.T., Pasman, H., Gao, X., and Mannan, M.S. (2012). Resilience engineering of industrial processes: principles and contributing factors. *Journal of Loss Prevention in the Process Industries* 25 (2): 233–241.

DiPietro, G.S., Matthews, H.S., and Hendrickson, C.T. (2014). Estimating economic and resilience consequences of potential navigation infrastructure failures: a case study of the Monogahela River. *Transportation Research, Part A* 69: 142–164.

Dong, H. and Cui, L. (2016). System reliability under cascading failure models. *IEEE Transactions on Reliability* 65 (2): 929–940.

Ehsani, A., Fotuhi, M., Abbaspour, A., Ranjbar, A. M, (2005), "An analytical method for the reliability evaluation of wind energy systems," in *Proceedings of TENCON 2005 IEEE Region 10*, 21–24 November, pp. 1–7.

El-Khattam, W., Bhattacharya, K., Hegazy, Y., and Salama, M. (2004). Optimal investment planning for distributed generation in a competitive electricity market. *IEEE Transactions on Power Systems* 19 (3): 1674–1684.

Elsner, J.B. and Bossak, B.H. (2001). Bayesian analysis of U.S. hurricane climate. *Journal of Climate* 14: 4341–4350.

Elsner, J.B. and Jagger, T.H. (2004). A hierarchical Bayesian approach to seasonal hurricane modeling. *Journal of Climate* 17: 2813–2827.

EPRI, (2017), Grid Resiliency, available at: https://www.epri.com/#/pages/sa/grid_resiliency (accessed on November 11, 2017).

ERCOT, (2017), "ERCOT responds to Hurricane Harvey," September 6, 2017, available at: http://www.ercot.com/help/harvey (accessed on December 4, 2017).

Fang, Y.-P., Pedroni, N., and Zio, E. (2016). Resilience-based component importance measures for critical infrastructure network systems. *IEEE Transactions on Reliability* 65 (2): 502–512.

FEMA, (2017), HAZUS-MH MR2 Technical Manual, Chapter 2, Report of Federal Emergency Management Agency, available at: https://www.fema.gov/media-library/assets/documents/24609 (accessed on March 29, 2018).

Figueroa-Candia, M., Felder, F., and Coit, D. (2018). Resiliency-based optimization of restoration policies for electric power distribution systems. *Electric Power Systems Research*. 161 (August): 188–198.

Georgiou, P., Davenport, A.G., and Vickery, B.J. (1983). Design wind speeds in regions dominated by tropical cyclones. *Journal of Wind Engineering and Industrial Aerodynamics* 3: 139–152.

Georgiou, P.N., (1985), Design Windspeeds in Tropical Cyclone-Prone Regions. PhD thesis, Faculty of Engineering Science, University of Western Ontario, London, Ontario, Canada.

Giorsetto, P. and Utsurogi, K.F. (1983). Development of a new procedure for reliability modeling of wind turbine generators. *IEEE Transactions on Power Apparatus and Systems* PAS-102 (1): 134–143.

Henry, D. and Ramirez-Marquez, J.E. (2012). Generic metrics and quantitative approaches for system resilience as a function of time. *Reliability Engineering and Systems Safety* 99 (1): 114–122.

Jia, H., Ding, Y., Peng, R., and Song, Y. (2017). Reliability evaluation for demand-based warm standby systems considering degradation process. *IEEE Transactions on Reliability* 66 (3): 795–805.

Jin, T., Tian, Y., Zhang, C., and Coit, D. (2013, 2013). A multi-criteria planning for distributed wind generation under strategic maintenance. *IEEE Transactions on Power Delivery* 28 (1): 357–367.

Jin, T., Tian, Z., (2010), "Uncertainty analysis for wind energy production with dynamic power curves," in *Proceedings of IEEE Conference on Probabilistic Methods Applied to Power Systems (PMAPS)*, pp. 745–750.

Jin, T., Yu, Y., and Elsayed, E. (2015). Reliability and quality control for distributed wind-solar energy integration: a multi-criteria approach. *IIE Transactions* 47 (10): 1122–1138.

Jin, T., Mai, N., Ding, Y., Vo, L., Dawud, R., (2018), "Planning for distribution resilience under variable generation: Prevention, survival and recovery," in *Proceedings of 2018 IEEE Annual Green Technologies Conference*, April 4–6, 2018, Austin, TX (in press).

Kaplan, J. and DeMaria, M. (1995). A simple empirical model for predicting the decay of tropical cyclone winds after landfall. *Journal of Applied Meteorology* 34: 2499–2512.

Karki, R., Hu, P., and Billinton, R. (2006). A simplified wind power generation model for reliability evaluation. *IEEE Transaction on Energy Conversion* 21 (2): 533–540.

Keck, M. and Sakdapolrak, P. (2013). What is social resilience? Lessons learned and ways forward. *Erdkunde* 67 (1): 5–19.

Khajavi, P., Abniki, H., Arani, A. B., (2011), "The role of incentive based demand response programs in smart grid," in *Proceedings of 10th International Conference on Environment and Electrical Engineering (EEEIC)*, page 1–4.

Kwasinski, A., Krishnamurthy, V., Song, J., and Sharma, R. (2012). Availability evaluation of micro-grids for resistant power supply during natural disasters. *IEEE Transactions on Smart Grid* 3 (4): 2007–2018.

Lave, M. and Kleissl, J. (2011). Optimum fixed orientations and benefits of tracking for capturing solar radiation. *Renewable Energy* 36 (3): 1145–1152.

Lei, S., Wang, J., Chen, C., and Hou, Y. (2018). Mobile emergency generator pre-positioning and real-time allocation for resilient response to natural disasters. *IEEE Transactions on Smart Grid*, available at: http://ieeexplore.ieee.org/document/7559799.

Li, B., Roche, R., Miraoui, A., (2017), "A temporal-spatial natural disaster model for power system resilience improvement using DG and lines hardening," in *Proceedings of IEEE PowerTech Conference*, Manchester, NH, 18–22 June 2017, pp. 1–6.

Lin, Y. and Bie, Z. (2016). Study on the resilience of the integrated energy system. *Energy Procedia* 103 (December issue): 171–176.

Liu, X., Shahidehpour, M., Li, Z. et al. (2017). Microgrids for enhancing the power grid resilience in extreme conditions. *IEEE Transactions on Smart Grid* 8 (2): 589–597.

Ma, S., Chen, B., and Wang, Z. (2017). Resilience enhancement strategy for distribution systems under extreme weather events. *IEEE Transactions on Smart Grid* 9 (2): 1442–1451.

MacKenzie, C.A. and Barker, K. (2013). Empirical data and regression analysis for estimation of infrastructure resilience with applications to electric power outages. *Journal of Infrastructure Systems* 19 (1): 25–35.

Martin, R.L. (2012). Regional economic resilience, hysteresis and recessionary shocks. *Journal of Econ Geography* 12 (1): 1–32.

Munoz, A. and Dunbar, M. (2015). On the quantification of operational supply chain resilience. *International Journal of Production Research* 53 (22): 6736–6751.

NIAC, (2009), Critical Infrastructure Resilience: Final Report and Recommendations, National Infrastructure Advisory Council, available at: https://www.dhs.gov/publication/niac-critical-infrastructure-resilience-final-report (accessed on April 1, 2018).

Ouyang, M. and Dueñas-Osorio, L. (2014). Multi-dimensional hurricane resilience assessment of electric power systems. *Structural Safety* 48: 15–24.

Ouyang, M., Duenas-Osorio, L., and Min, X. (2012). A three-stage resilience analysis framework for urban infrastructure systems. *Structural Safety* 36–37: 23–31.

Oxenyuk, V., (2014), Distribution Fits for Various Parameters in the Hurricane Model, Chapter 3, Master Thesis, Florida International University, available at: http://digitalcommons.fiu.edu/cgi/viewcontent.cgi?article=2301&context=etd, (accessed on March 18, 2018).

Panteli, M., Trakas, D.N., Mancarella, P., and Hatziargyriou, N.D. (2016). Boosting the power grid resilience to extreme weather events using defensive islanding. *IEEE Transactions on Smart Grid* 7 (6): 2913–2922.

Panteli, M. and Mancarella, P. (2017). Modeling and evaluating the resilience of critical electrical power infrastructure to extreme weather events. *IEEE Systems Journal* 11 (3): 1733–1742.

Salman, A.M., Li, Y., and Stewart, M.G. (2015). Evaluating system reliability and targeted hardening strategies of power distribution systems subjected to hurricanes. *Reliability Engineering and Systems Safety* 144 (December): 319–333.

Seguro, J.V. and Lambert, T.W. (2000). Modern estimation of the parameters of the Weibull wind speed distribution for wind energy analysis. *Journal of Wind Engineering and Industrial Aerodynamics* 85 (1): 75–84.

Shafieezadeh, A. and Burden, L.I. (2014). Scenario-based resilience assessment framework for critical infrastructure systems: case study for seismic resilience of seaports. *Reliability Engineering and Systems Safety* 132: 207–219.

Shao, C., Ding, Y., Wang, J., and Song, Y. (2018, 2018). Modeling and integration of flexible demand in heat and electricity integrated energy system. *IEEE Transactions on Sustainable Energy* 9 (1): 361–370.

Small, C., Parnell, G., Pohl, E. et al. (2017). Engineering resilience for complex systems. In: *Disciplinary Convergence in Systems Engineering Research* (ed. A. Madni, B. Boehm, R. Ghanem, et al.), 3–15. Cham: Springer.

Sulaiman, M.Y., Oo, W.M.H., Wahab, M.A., and Zakaria, A. (1999). Application of beta distribution model to Malaysian sunshine data. *Renewable Energy* 18 (4): 573–579.

Thiringer, T. and Linders, J. (1993). Control by variable rotor speed of a fixed-pitch wind turbine operating in a wide speed range. *IEEE Transaction on Energy Conversion* 8 (3): 520–526.

TISP (2006). *Regional Disaster Resilience: A Guide for Developing on Action Plan,".* The Infrastructure Security Partnership. Reston, VA: American Society of Civil Engineers.

Toth, Z. and Szentimrey, T. (1990). The binormal distribution: a distribution for representing asymmetrical but normal-like weather elements. *Journal of Climate* 3: 128–136.

US DOC, (2013), Economic Impact of Hurricane Sandy, prepared by Economics and Statistics Administration Office of the Chief Economist, US Department of Commerce, available at: http://www.esa.gov/sites/default/files/sandyfinal101713.pdf (accessed on May 1, 2017).

Vickery, P.J. and Twisdale, L.A. (1995a). Wind-field and filling models for hurricane wind-speed predictions. *Journal of Structural Engineering* 121 (11): 1700–1709.

Vickery, P.J. and Twisdale, L.A. (1995b). Prediction of hurricane wind speeds in the United States. *Journal of Structural Engineering* 121 (11): 1691–1699.

Villarreal, S., Jimenez, J.A., Jin, T., and Cabrera-Rios, M. (2013). Designing a sustainable and distributed generation system for semiconductor wafer fabs. *IEEE Transactions on Automation Science and Engineering* 10 (1): 10–16.

Vugrin, E.D., Warren, D.E., Ehlen, M.A., and Camphouse, R.C. (2010). A framework for assessing the resilience of infrastructure and economic systems. In: *Sustainable Infrastructure Systems: Simulation, Modeling, and Intelligent Engineering* (ed. K. Gopalakrishnan and S. Peeta), 77–117. Berlin Heidelberg: Springer-Verlag.

Vugrin, E. D., Castillo, A., and Silva-Monroy, C. (2017), "Resilience Metrics for the Electric Power System: A Performance-Based Approach," Technical Report SAND2017–1493, Sandia National Laboratories, USA, available at: http://prod.sandia.gov/techlib/access-control.cgi/2017/171493.pdf (accessed on April 30, 2018).

Wang, X., Li, Z., Shahidehpour, M., and Jiang, C. (2018). Robust line hardening strategies for improving the resilience of distribution systems with variable renewable resources. *IEEE Transactions on Sustainable Energy*, available at: http://ieeexplore.ieee.org/document (in early access).

Wang, Y., Chen, C., Wang, J., and Baldick, R. (2016). Research on resilience of power systems under natural disasters – a review. *IEEE Transactions on Power Systems* 31 (2): 1604–1613.

Watson, J. P., R. Guttromson, C. Silva-Monroy, et al., (2016), "Conceptual framework for developing resilience metrics for the electricity, oil, and gas sectors in the United States," Technical Report SAND2014–18019, Sandia National Laboratories, USA, available at: http://energy.gov/sites/prod/files/2015/09/f26/EnergyResilienceReport_Final_SAND2014-18019.pdf (accessed April 28, 2018).

Woods, D.D. (2015). Four concepts for resilience and the implications for the future of resilience engineering. *Reliability Engineering and System Safety* 141 (September): 5–9.

Xu, L., Brown, R., (2008), "Undergrounding Assessment Phase 3 Report: Ex ante cost and benefit modeling," Quanta Technology, Raleigh, Technical Report PSC-06-035-PAA-EI Final Report, available at: http://www.feca.com/PURCPhase3FinalReport.pd (accessed on March 2, 2018).

Ye, Z.S., Xie, M., Tang, L.C., and Chen, N. (2014). Semiparametric estimation of gamma processes for deteriorating products. *Technometrics* 56 (4): 504–513.

Yodo, N. and Wang, P. (2016). Engineering resilience quantification and system design implications: a literature survey. *Journal of Mechanical Design* 138 (11): 111408-1–111408-13.

Zhang, C., Ramirez-Marquez, J.E., and Wang, J. (2015). Critical infrastructure protection using secrecy – a discrete simultaneous game. *European Journal of Operational Research* 242 (1): 212–221.

Zhou, B., Gu, L., Ding, Y. et al. (2011). The great 2008 Chinese ice storm: its socioeconomic–ecological impact and sustainability lessons learned. *Bulletin of the American Meteorological Society* 92 (1): 47–60.

Problems

Problem 12.1 State what are the four main domains in which the concept of resilience is adopted and assessed.

Problem 12.2 Why in the last 10 years has there been growing attention paid to the resilience of the engineering system? In other words, what are the key drivers behind this motivation?

Problem 12.3 Describe the four states associated with a resilience curve. Use the example of a power grid to explain this.

Problem 12.4 Indicate at least three major differences between reliability and resilience concepts. Again, you can take an engineering system to describe their differences.

Problem 12.5 Based on Figure 12.3, estimate the resilience using Eq. (12.2.1) by considering three different cases: (1) $t_1 = 10$ hours, $t_2 = 15$ hours, $t_3 = 30$ hours, $P_0 = 100$, and $P_d = 60$; (2) $t_1 = 10$ hours, $t_2 = 15$ hours, $t_3 = 30$ hours, $P_0 = 100$, and $P_d = 40$; (3) $t_1 = 10$ hours, $t_2 = 20$ hours, $t_3 = 30$ hours, $P_0 = 100$, and $P_d = 60$.

Problem 12.6 Based on Figure 12.3, estimate the resilience using Eq. (12.2.2) using the same datasets as in Problem 12.5.

Problem 12.7 Based on Figure 12.3, estimate the resilience using Eq. (12.2.4) using the same datasets as in Problem 12.5.

Problem 12.8 Based on Figure 12.3, estimate the resilience using Eq. (12.2.5) using the same datasets as in Problem 12.5.

Problem 12.9 Based on Figure 12.5, estimate the six dimensions of the resilience with the following different datasets: (1) $t_1 = 10$ hours, $t_2 = 30$ hours, $t_3 = 50$ hours, $P_0 = 100$, and $P_d = 60$; (2) $t_1 = 10$ hours, $t_2 = 30$ hours, $t_3 = 50$ hours, $P_0 = 100$, and $P_d = 20$; (3) $t_1 = 10$ hours, $t_2 = 20$ hours, $t_3 = 50$ hours, $P_0 = 100$, and $P_d = 60$.

Problem 12.10 Five-year weather data of a hurricane are collected and shown in the table below. Do the following: (1) estimate the annual occurrence rate and distribution using the Poisson process; (2) estimate the occurrence rate and distribution function using the negative binominal distribution; (3) simulate the hurricane arrival time.

Event	Year 1	2	3	4	5
1	June 3	May 10	April 25	June 28	March 25
2	September 20	August 5	July 17	October 2	July 8
3		November 25	October 24		December 10

Problem 12.11 The approach angle of a hurricane can be modeled as the random variable with a binormal distribution, i.e. the sum of two independent

normal distributions with weights of w and $1 - w$, as shown in Eq. (12.3.10). Assuming $\mu_1 = 1$ and $\mu_2 = 2$, and further assuming $\sigma_1 = 0.2$ and $\sigma_2 = 0.3$, plot the binormal distribution as w varies from 0 to 1 as follows: $w = 0$, $w = 0.3$, $w = 0.5$, $w = 0.7$, and $w = 1$.

Problem 12.12 Based on the data in Problem 12.11, develop a computer program to simulate the hurricane approach angles under different values of weights: $w = 0$, $w = 0.3$, $w = 0.5$, $w = 0.7$, and $w = 1$.

Problem 12.13 The radius to maximum winds describes the range of the most intensive hurricane wind speed. Denote as R_{max}; it can be empirically estimated as follows (also see Eq. (12.3.14)):

$$\ln R_{max} = 2.556 - 0.000\,050\,255p_d^2 + 0.042\,243\,032\phi$$

where p_d is the center pressure difference and ϕ is the storm latitude. Calculate R_{max} considering two cases: (1) $P_d = 10\,\text{kPa}$ and (2) $P_d = 5\,\text{kPa}$. For each case, you need to consider different values of the latitude.

Problem 12.14 Assuming $a_u = 0.01$ and $b_u = 0.025$, estimate the failure rate λ_{ij}, i.e. the probability of failure, under different combinations of hurricane categories and storm surge zones based on Eq. (12.4.6).

Problem 12.15 Assume that a transformer is maintained under an age-based policy. The following parameters are given $c_f = 5000$, $c_p = 1000$, $R(t) = \exp(-\,(0.0001\,t)^2)$. Determine the optimal maintenance interval such that the maintenance cost rate is minimized.

Problem 12.16 What wind turbine and solar PV units are treated as variable generation systems? What are the main challenges in deploying wind and solar generation? Which technologies eventually will assist the large-scale adoption of wind and solar power?

Problem 12.17 For a commercial 1-MW wind turbine, the cut-in speed $v_c = 3\,\text{m/s}$, the rate wind speed $v_r = 12\,\text{m/s}$, and the cut-off wind speed $v_s = 25\,\text{m/s}$. Also the wind speed follows the Weibull distribution and its probability density function is

$$f_W(y) = \frac{k}{c}\left(\frac{y}{c}\right)^{k-1} e^{-\left(\frac{y}{c}\right)^k}, \quad y \geq 0$$

where c is the scale parameter and k is the shape parameter. Estimate the mean and variance of the output power of the wind turbine assuming: (1) $c = 7\,\text{m/s}$ and $k = 3$ and (2) $c = 10\,\text{m/s}$ and $k = 4.5$.

Problem 12.18 Using the same turbine characteristic speed data estimate the mean and variance of the wind turbine output, assuming the wind speed follows the normal distribution. Case 1: $\mu = 7$ m/s and $\sigma = 4$ m/s. Case 2: $\mu = 9$ m/s and $\sigma = 5$ ms.

Problem 12.19 Eq. (12.5.6) shows that the PV generation decreases with the panel temperature T_o. Assuming $\eta_s = 20\%$, $A_s = 1$ m^2, and $s = 1000$ W m^{-2}, plot how $P_s(s)$ changes with the increase of T_o. Are you able to find some conclusion from the chart you obtained?

Problem 12.20 The repair process of damaged overhead lines depends on the failure rate of the lines and the availability of repair teams (see Section 12.6.4). We compare the recovery time where $\lambda = 2$ lines/hour during the hurricane attack period. Let the number of branches or wires $K = 20$. Compute the L and W assuming $R = 2$ teams and the results are compared to the case where $R = 3$ teams. In both cases, we assume the recovery rate is 0.5 lines/hour/team.

Index

Reliability Engineering and Services, First Edition. Tongdan Jin.
© 2019 John Wiley & Sons Ltd. Published 2019 by John Wiley & Sons Ltd.
Companion website: www.wiley.com/go/jin/serviceengineering